W9-CYY-679

AN INTRODUCTION TO LINEAR PROGRAMMING AND GAME THEORY

AN INTRODUCTION TO LINEAR PROGRAMMING AND GAME THEORY

Third Edition

Paul R. Thie

G. E. Keough

A JOHN WILEY & SONS, INC., PUBLICATION

Published by John Wiley & Sons, Inc., Hoboken, New Jersey.
Published simultaneously in Canada.

For general information on our other products and services or for technical support, please contact our Customer Care Department within the United States at (800) 762-2974, outside the United States at (317) 572-3993 or fax (317) 572-4002.

Wiley publishes in a variety of print and electronic formats and by print-on-demand. Some material included with standard print versions of this book may not be included in e-books or in print-on-demand. If this book refers to media such as a CD or DVD that is not included in the version you purchased, you may download this material at http://booksupport.wiley.com. For more information about Wiley products, visit www.wiley.com.

Library of Congress Cataloging-in-Publication Data:

Thie, Paul R., 1936–
An introduction to linear programming and game theory / Paul R. Thie, G. E. Keough. — 3rd ed.
p. cm.
Includes bibliographical references and index.
ISBN 978-0-470-23286-6 (cloth)
1. Linear programming. 2. Game theory. I. Keough, G. E. II. Title.
T57.74.T44 2008
519.7'2—dc22 2008004933

10 9 8 7 6 5 4 3

To Our Wives, Mary Lou and Dianne

and

In Memory of a Gentle Irishman
of Gifted Wit and Charm

Contents

Preface

PURPOSE

This textbook develops, at an introductory level, the theoretical concepts and computational techniques of linear programming and game theory, and also discusses applications of these topics in the social, life, and managerial sciences. Closely related to this development, it presents an introduction to the process of mathematical model building, which is discussed in two distinct settings. The chapters on linear programming contain various examples of real-world situations involving a single decision maker faced with some sort of deterministic (except in Sections 8.1 and 8.4) optimization problem. In the two chapters on game theory the emphasis is on the development of a different type of model, a model of a conflict situation involving two participants with opposing interests.

LEVEL AND PREREQUISITES

The text is written for students in mathematics, science, economics, and operations research. The presentation is, for the most part, mathematically complete, that is, in terms of definitions, theorems, and proofs. However, examples are used frequently, not only to motivate new ideas, but also to assist in the understanding of the theory and the associated proofs. The goal is to provide a book that the student will find rigorous and challenging, yet readable and helpful.

The prerequisites for reading the text are minimal. The material should be accessible to any student who has successfully completed one or two undergraduate mathematics courses. No use is made of the theoretical concepts from linear algebra such as the dimension and basis of a vector space or linear independence of vectors. Matrices and vectors are used only as notational tools, so any student familiar with these tools and their operations of addition and multiplication can read the text. Appendix A contains a brief list of the topics from linear algebra used in the book.

TECHNOLOGY

Two software tools for solving linear programming problems are introduced in the third edition of the text. The first tool is LP Assistant, a user-friendly program that performs the arithmetic of the pivot operation, the computational heavy step in each iteration of the simplex algorithm. To use the program, the user need only input the initial tableau, indicate the appropriate pivot point at each iteration, and be able to recognize and interpret a final tableau. It is an ideal teaching tool. It allows the student to master the steps of the algorithm without hindrance from minor errors in arithmetic, and it allows the instructor to ask students to solve larger and therefore more realistic linear programming problems without fear of student failure

An Introduction to Linear Programming and Game Theory, Third Edition. By P. R. Thie and G. E. Keough.

simply because of a computational error. The program, developed by coauthor G. E. Keough, is designed for use with the text. It emulates the presentation and use of the algorithm as it appears in the book. Its capabilities and operation are described briefly in Appendix D (full documentation is made available with the program). The software is platform-independent and available for download from the Internet.

The second software unit to be integrated into the book is the spreadsheet tool Solver, an add-in to Microsoft's Excel package. Solver can solve linear, nonlinear, and integer programming problems. It is used in the text to provide solutions, and sensitivity analysis where applicable, to linear and integer programming problems. Also, the data contained in Solver's sensitivity report is explained and verified, using the theory developed in Chapter 5. Appendix E, written for someone already familiar with spreadsheet operations, outlines the use of Excel and Solver to solve programming problems.

LENGTH AND ORGANIZATION

The book probably contains more material than can be taught in a one-semester course. However, once the central ideas of Chapters 3 and 4 have been developed, the instructor has considerable latitude in the selection of other topics to be discussed. Chapters 5, 6, 7, and 9 and the four sections of Chapter 8 are all independent of each other and can immediately follow upon Chapter 4, with the only provisos being that Sections 5.6 and 5.7 also be covered before Chapter 6 and Section 5.1 before Section 8.4. Chapter 10, on non-zero-sum games, has Chapter 9, on zero-sum games, as a prerequisite.

CONTENTS

Linear programming and game theory are introduced in Chapter 1 by means of examples. This chapter also contains some discussion on the application of mathematics and on the roles that linear programming and game theory can play in such applications. To introduce the reader to the broad scope of the theory, Chapter 2 (on model building) presents various real-world situations that lead to mathematical models involving linear optimization problems. Also, a two-variable problem is resolved geometrically, and with this example the ideas of sensitivity analysis are introduced. Several of the examples are revisited later in the text as tools are developed to resolve the questions raised here.

Chapters 3 and 4 are the core of the book. The simplex algorithm is presented in Chapter 3 and the concept of duality in Chapter 4. The development of the simplex algorithm is motivated algebraically, and all of Chapter 3 maintains an algebraic flavor. LP Assistant is introduced in the problem set following Section 3.5, where the reader is first asked to use the simplex algorithm. The convergence of the algorithm is proved inductively in Section 3.8. There are geometrical considerations throughout the chapter, however, to promote understanding of the development, and Section 3.9 is about convexity. The concept of convexity is used later in the text in Section 8.3 and Chapter 10. The use of Excel and Solver to solve linear programming models is demonstrated in the last section of Chapter 3.

The dual of any linear programming problem is defined in Section 4.2, and the Duality Theorem is proved in Section 4.4. Sections 4.1 and 4.3 develop examples demonstrating the relevance of the dual problem. The Complementary Slackness Theorem is discussed and proved in Section 4.5. The proof is an immediate consequence of a result preliminary to the proof of the Duality Theorem. No results in the text are contingent on the Complementary Slackness Theorem, but complementary slackness is referred to occasionally, especially in the problem sets.

Sensitivity analysis is presented at two levels in Chapter 5. In Section 5.1, three examples involving elementary sensitivity analysis are presented, and the problems raised are solved using the theory of duality. Also in this section Solver's sensitivity report is introduced, the constraints section explained, and some data corroborated. The more general study of sensitivity analysis begins in Section 5.2 with the development of the matrix representation of the simplex algorithm. Here it is assumed that the reader is familiar with matrix multiplication and the inverse of a matrix. Accompanying the development of the theory, the variables (Adjustable Cells) portion of Solver's sensitivity report is discussed and some results are corroborated in Section 5.3, and a similar correlation between the theory of the chapter and data of a sensitivity report occurs in Section 5.5. In Section 5.6 the Dual Simplex Algorithm is presented. Although the algorithm is motivated by problems raised in Section 5.5, Section 5.6 is independent of the theory of these preceding sections and could, in fact, be read directly after Chapter 4. The Dual Simplex Algorithm is used in Sections 5.7, 6.3, and 6.4.

Chapter 6 provides an introduction to integer programming. Two algorithms that can be used to solve integer programming problems are presented. Except for the fact that both of these algorithms use the Dual Simplex Algorithm as a tool, this chapter could be read after Chapter 3. The solution of integer programming models using Excel and Solver is presented in the last section of the chapter.

Chapter 7 deals with the transportation problem. A Ford-Fulkerson algorithm is developed for the solution of these problems, and in Section 7.3 various other models to which the algorithm can be applied are discussed. Variations on these models and sensitivity analysis questions are considered in Problem Set 7.3, along with several other models amenable to a solution using the algorithm.

Extensions of the general theory are introduced by means of examples in the first three sections of Chapter 8. The first example demonstrates one approach to a nondeterministic model. (The resulting optimization problem has many upper bound constraints, and so, as an auxiliary benefit, special solution techniques for such problems are illustrated.) In Section 8.2 a method of working with a problem with multiple goals is discussed, and in Section 8.3 the decomposition principle is illustrated. In Section 8.4, a different type of application of linear programming is presented. By means of an example, the problem of measuring the efficiencies of similar operating units is considered. The four sections are independent of each other. Sections 8.1 and 8.2 may be read after Chapter 3; Section 8.3 requires an understanding of duality (and convexity), and Section 8.4 an understanding of sensitivity analysis.

Two-person, zero-sum games are the subject of Chapter 9. First, the axioms that form the foundation of the theory are discussed at some length to help the reader understand not only the concept of a solution to a game, but also the limitations on the applicability of the theory. Then, using the Duality Theorem of linear programming, the existence of solutions to two-person, zero-sum games is demonstrated. Computational techniques and examples conclude the chapter.

Utility theory is introduced in the first section of Chapter 10. The remainder of the chapter is devoted to two-person, non-zero-sum games. These games provide excellent examples of some of the difficulties that can be encountered when attempting to formulate mathematical models of complicated situations that involve human behavior. In discussing these games, factors not relevant in the theory of two-person, zero-sum games, such as the possibility of cooperation between the participants, are considered, and various approaches and solution concepts are explored, primarily by means of examples. Added to the text in this third edition is J. Nash's proof in Section 10.3 of the existence of an equilibrium strategy for any noncooperative two-person, non-zero-sum matrix game.

Finally, in addition to Appendices A, D, and E mentioned above, Appendix B displays an example of simplex algorithm cycling, and Appendix C contains a brief discussion of the efficiency of the simplex algorithm and some theoretical advances in the field.

EXERCISES

Problem sets containing computational exercises, problems testing understanding, and examples motivating new material conclude each section of the text. There are over 450 problems in the text, almost half of which have multiple parts. The problems are placed in each section and not simply at the conclusion of each chapter, so the reader is constantly encouraged to test and develop his or her understanding of the material. Solutions to a selected set of the problems are given at the end of the book.

ACKNOWLEDGMENTS

We are grateful to the many people who offered valuable suggestions and constructive criticism of the text. They include Professors Joseph G. Ecker, James A. Murtha, and Edward J. Smerek, reviewers of the original manuscript; Professors Robert F. Brown, Gove Effinger, Bertram Mond, and Morris Weisfeld, reviewers of the second edition; and Professors Ed Keller, Maynard Thompson, and David Vella, reviewers of the prospectus for this edition. In addition, numerous users of the earlier editions of the text, along with colleagues Jenny Baglivo, Daniel Chambers, and John Smith of Boston College, have provided helpful comments and suggestions. Also, the many students at B.C. who have taken the mathematical programming course must be thanked, for they, with their questions, frowns, and comments, have contributed greatly to the development of the material. Lastly, we want to acknowledge the professional expertise provided by the staff of Wiley in the production of the text.

Paul R. Thie
G. E. Keough

CHAPTER 1

MATHEMATICAL MODELS

1.1 APPLYING MATHEMATICS

Recent history has shown us that many problems of our technically oriented society yield to mathematical descriptions and solutions. Problems as complex as sending people into space or maximizing the profit of a giant industrial conglomerate and problems as simple as balancing our own monthly budget or winning at the game of Nim are susceptible to mathematical formulations. This book is concerned with two specific fields of mathematics, linear programming and game theory, that offer insights into certain problems of the real world and techniques for solving some of these problems.

To understand best how one goes about applying a mathematical theory to the solution of some real-world problem, consider the stages that a problem passes through from organization to conclusion. We list four:

- recognition of the problem;
- formulation of a mathematical model;
- solution of the mathematical problem; and
- translation of the results back into the context of the original problem.

These four stages are by no means exclusive or well defined. Other authors have broken down the problem-solving operation in different ways, but the four steps listed indicate the framework in which the applied mathematician works.

The meaning of the first stage, recognition of the problem, is self-explanatory. The meaning of the second stage, formulation of a mathematical model, can be much more mysterious, conjuring visions of a precisely built representation of a small, snow-covered village at a scale of $\frac{1}{87}$. Actually, although the meaning of this step can be made quite clear, it is usually the most critical and difficult step to implement in the entire operation. The development of the mathematical model consists of translating the problem into mathematical terms, that is, into the language and concepts of mathematics. As an example of this process, consider what is called the "word problem" *word problem* of high school algebra. Here the mathematics is trivial and the problems are unrealistic, but many students stumble over the difficulties inherent in translating some concocted word problem into an algebraic equation, that is, in formulating the mathematical model. It was not always easy to determine how

An Introduction to Linear Programming and Game Theory, Third Edition. By P. R. Thie and G. E. Keough.

much 40% antifreeze solution to drain from the 20-qt cooling system to attain a 75% solution by adding a 90% antifreeze mixture.

In the development of a mathematical model of a complex situation, two basic and opposing elements are encountered. On the one hand, one seeks simplifying assumptions and overlooks minor details so that the resulting mathematical problem yields to a successful analysis. On the other hand, the model must adequately reflect reality so that the knowledge gained from the study of the model can be applied to the original problem. The ability to select those elements of a problem that are of major importance and disregard those of minor importance probably comes best from experience. Throughout the text and, in particular, in the next two sections, examples and problems requiring the development of a mathematical model are given. Although in many instances problems from a text may immediately single out the important elements and may seem somewhat artificial, much skill is to be gained by attempting them; practice model building and problem solving whenever possible.

Once the mathematical model has been formulated, one comes to the third stage in the process, the solution of the mathematical problem. It should be emphasized that this can entail much more than just computing the difference of a function at the end points of an interval or finding the solution to a system of equations. Even if the known theory does provide a complete theoretical solution to the problem, the specific answer to the problem at hand must still be calculated. It could very well be that further analysis does not provide any simplification of the problem, and only through involved computations can an estimate of the solution be made. Thus, finding a solution to a problem could mean determining a technique to approximate a solution that is financially feasible to implement within a given computer's capabilities and provides error estimates within given tolerance limits.

The meaning of the fourth step of the operation, the translation of the results back into the context of the original problem, is clear. Of course, more than a simple numerical answer is called for. The simplifying assumptions on which the solution is based must be understood, and the changes in the problem that would invalidate these assumptions should be considered.

We now give two examples of specific and well-known problems and begin the development of the associated mathematical models.

1.2 The Diet Problem

The diet problem is one of the classical illustrations of a problem that leads to a linear programming model. The problem is concerned with providing at minimal cost a diet adequate for a person to sustain himself or herself. Simply stated, what is the least expensive way of combining various amounts of available foods in a diet that meets a person's nutritional requirements?

To develop a mathematical model of this problem, first the various aspects of the problem must be considered. Here the two competing needs for simplification and realism come into play as one attempts to state in precise terms the different components of the problem. For example, just how does one determine the basic nutritional requirements? We must consider the age, sex, size, and activity of our subject. We must determine what nutrients, among the many known nutrients such as calories, proteins, and the multitude of vitamins and minerals, are essential. Can a need for one be met by a combination of others? Is it the case that too much of a certain nutrient is harmful and therefore forces an upper bound on the intake of that quantity? Should we provide for some variety in the diet, hopefully to meet nutritional requirements unknown to us at the present time?

Another component of the problem requiring study is consideration of the foods to be used in the diet. What foods can we assume are available? For example, can we assume that fresh fish, fruits, or vegetables or frozen foods are available? Once the foods that can be used in the problem are established, the nutrient values of these foods must be determined. Here again only approximations can be made, since the nutrient value of a certain type of food, say apples or hamburger, not only varies from sample to sample because of lack of uniformity, but is also contingent on the conditions and duration of storage and the method of preparation for consumption. The cost of a food can also fluctuate due to seasonal and geographical variances.

Once suitable approximations for the nutritional requirements of our subject and the nutrient values and cost of the available foods have been determined, a mathematical problem involving finding the minimum of a linear function can be formulated. To demonstrate this, we will consider a much simplified version of the diet problem.

Suppose we wish to minimize the cost of meeting our daily requirements of proteins, vitamin C, and iron with a diet restricted to apples, bananas, carrots, dates, and eggs. The nutrient values and cost of a unit of each of these five foods, along with the meaning of a unit of each, are given in the following table.

Food	*Measure of a Unit*	*Protein (g/unit)*	*Vitamin C (mg/unit)*	*Iron (mg/unit)*	*Cost (cents/unit)*
Apples	1 med.	0.4	6	0.4	8
Bananas	1 med.	1.2	10	0.6	10
Carrots	1 med.	0.6	3	0.4	3
Dates	$\frac{1}{2}$ cup	0.6	1	0.2	20
Eggs	2 med.	12.2	0	2.6	15

Our daily diet requires at least 70 g of protein, 50 mg of vitamin C, and 12 mg of iron. Since we are assuming that our supply of these foods is unlimited, it is obvious that we can find a diet that meets our needs; for example, a diet consisting of 6 units of eggs and 5 units of bananas would be more than adequate, as the reader can easily verify.

Our problem then is to determine the least expensive way of combining various amounts of the five foods to meet our three daily requirements. Hence the decision to be made involves the number of units of each of the five foods to consume daily. To translate this question into a mathematical problem, introduce five variables A, B, C, D, and E, where A is defined as the number of units of apples to be used in the daily diet, B the number of units of bananas, C the number of units of carrots, D the number of units of dates, and E the number of units of eggs. The cost in cents of such a diet is given by the function $f(A,B,C,D,E) = 8A + 10B + 3C + 20D + 15E$, found by using the cost column in the above table. It is this function that we wish to minimize.

However, there are clearly restrictions imposed by the problem on the possible values of the variables A, B, C, D, and E, that is, restrictions on the domain of the function f. First, all the variables must be nonnegative. And to guarantee that the daily nutritional requirements are fulfilled, the following three inequalities must be satisfied:

$$\begin{aligned} 0.4A + 1.2B + 0.6C + 0.6D + 12.2E &\geq 70 \\ 6A + 10B + 3C + 1D &\geq 50 \\ 0.4A + 0.6B + 0.4C + 0.2D + 2.6E &\geq 12 \end{aligned}$$

These inequalities are determined by considering the total input of the three required nutrients in a diet consisting of A units of apples, B units of bananas, and so on. For example, since 1 unit of apples contains 0.4 g of protein, A units contain $0.4A$ g. Similarly, B units of bananas contain $1.2B$ g of protein, C units of carrots $0.6C$ units, D units of dates $0.6D$ units, and E units of eggs $12.2E$ units. Adding these five terms gives the total intake of protein. Since our daily requirement of 70 g of protein is a minimal requirement and more is allowable, we have the first inequality. Similarly, the other two inequalities follow.

In sum, the resulting mathematical problem is to determine the minimum value of the function

$$f(A,B,C,D,E) = 8A + 10B + 3C + 20D + 15E$$

with the possible values of A, B, C, D, and E restricted by the inequalities

$$\begin{aligned} 0.4A + 1.2B + 0.6C + 0.6D + 12.2E &\geq 70 \\ 6A + 10B + 3C + 1D &\geq 50 \\ 0.4A + 0.6B + 0.4C + 0.2D + 2.6E &\geq 12 \\ A,B,C,D,E &\geq 0 \end{aligned}$$

In 1945 George Stigler [1] considered the general diet problem. Stigler discussed the questions we raised and others, and he justified modifications and simplifications. For human nutritional requirements, Stigler decided on nine common nutrients (calories, protein, calcium, iron, vitamins A, B_1, B_2, C, and niacin) and estimated their needs from data supplied by the National Research Council. Stigler initially considered 77 types of foods and determined their average nutrient values and costs. From this he was able to construct a diet that satisfied all the basic nutritional requirements and cost only \$39.93 a year (less than 11 cents/day) for the year 1939. The diet consisted solely of wheat flour, cabbage, and dried navy beans.

1.3 The Prisoner's Dilemma

In the context of game theory, the word game in general refers to a situation or contest involving two or more players with conflicting interests, with each player having partial but not total control over the outcome of the conflict. The following is an example of such a situation. However, at this stage we are not yet able to translate the conflicting interests represented in the example into a precise mathematical problem, in contrast to the example developed in the previous section. Indeed, one of the major contributions of game theory is the resulting study of the question of what it means to solve a game.

The situation we consider is as follows. A certain democratic republic has a unicameral legislature with a membership drawn primarily from two major political parties. Before the assembly is a bill sponsored by a citizens' group designed to restrict the power and influence of the senior members of each political party. On this issue the legislators can be divided into three approximately equal groups – two groups whose members will follow the directives of their respective party leaders and a third group of responsible representatives who consider passage of the bill more important than the maintenance of party loyalties and will support the bill regardless of circumstances.

Consider now this situation from the viewpoint of the leaders of the two parties. Due to the nature of things they would like to see the bill defeated, but their constituents overwhelmingly support the bill. However, an impending general election complicates matters. Because they are fairly adaptable people, the leaders know that they could, in fact, work moderately well within the limits set by the bill, so each group believes that the most beneficial outcome of the vote on the bill would be for their party to profess support for the bill while the opposition party opposes the bill. Of course, this would mean that the bill would pass, but the wave of public support generated for the one party voting for the bill would be a prevailing factor in the impending election. Thus the problem is, how should each group of leaders direct their respective faithful party members to vote on the bill?

To answer this question, the leaders of one of the parties gather to consider the various possible outcomes of the vote on the bill. The most favorable outcome, as far as they are concerned, is for their party to support the measure and the opposition to oppose it. They denote this outcome by the ordered pair (Y,N) (they vote "yea" and the opposition votes "nay"). The least favorable outcome is the reverse of this situation, with their party members opposing but the opposition favoring passage of the bill (the (N,Y) outcome). The two remaining possible outcomes are for both parties to support the bill (outcome (Y,Y)) and for both parties to oppose the bill (outcome (N,N)). Neither of these outcomes would be a factor in the election, since the public reaction, either good or bad, would be balanced evenly between the two parties. However, outcome (N,N) is preferred over outcome (Y,Y), on the grounds that if both parties oppose the bill, it would be defeated and so the power of the party leaders would remain unaffected. Thus the leaders of the party linearly order the four possible outcomes, from most to least favorable, as follows:

$$(Y,N) > (N,N) > (Y,Y) > (N,Y)$$

Wishing to make this analysis even more precise and, hopefully, instructive, some of the leaders propose to assign numerical weights to each of these outcomes. They claim that such an assignment not only could reflect the above linear ordering, but also could measure how much more one outcome is preferred over another. They point out, for example, that a consideration in some contest of the three outcomes win \$3, win \$2, and win \$1 would not be identical to a consideration of the three outcomes win \$100, win \$2, and win \$1. Seeing the merits of this proposition, the leaders continue their deliberations on the four possible outcomes of the vote on the bill. Since outcomes (Y,N), (Y,Y), and (N,Y) all result in passage of the bill, their relative merits can be measured only by their effects in the impending election. Moreover, because of the equivalent strengths across the country of the two parties, the leaders believe that the advantage of (Y,N) over (Y,Y) is equal to the advantage of (Y,Y) over (N,Y). In fact, they argue that public reaction to support of the bill by only one party could be the determining factor in the election contests in up to 12 representative districts. Accepting this as a general unit and arbitrarily assigning the value 0 to outcome (Y,Y), they set (Y,N) to be worth 12 units and (N,Y) to be worth -12 units. There remains to be considered outcome (N,N), which lies between (Y,N) and (Y,Y) in the linear ordering. The assigning of a weight to this outcome is not immediate but, after a subcommittee review, prolonged debate, and various trade-offs in other matters, the political leaders accept the value of 6 units for this outcome.

Suppose that the leaders of the other party conduct similar deliberations and, since the positions of the two parties are comparable, reach the same conclusions. Then, to each possible outcome is attached two numerical weights, the value of that outcome to each party. Let us denote this pair of weights by an ordered pair of numbers, with the first component being the value of that particular outcome to one fixed party, called Party D, and the second component being the value to the other party, Party R. Then this situation can be represented by the following tableau:

		Party R	
		Vote "yea"	*Vote "nay"*
Party D	*Vote "yea"*	$(0,0)$	$(12,-12)$
	Vote "nay"	$(-12,12)$	$(6,6)$

Thus, for example, the outcome of a "nay" vote by Party D and a "yea" vote by Party R is $(-12,12)$; that is, that outcome is worth -12 units for Party D and 12 units for Party R.

This completes our analysis of this situation for the time being. It will be resumed in Chapter 10. We have formulated a two-person, non-zero-sum game in which each player has two possible moves, but we do not yet have a precisely stated mathematical problem to be solved. A primary component of game theory is the analysis accompanying an attempt to define exactly what one would mean by a solution to the game or a resolution of the conflict. Such an analysis for a certain type of game is made in Chapter 9, where a complete mathematical model is formulated for finite, two-person, zero-sum games and the resulting mathematical problems are resolved (terms such as *zero-sum* are defined there).

The assigning of meaningful weights to the various possible outcomes is not properly a part of game theory but is the function of utility theory (see Section 10.1). In the example of this section the use of game theory actually begins with the above tableau. Moreover, it is assumed in the theory that the information contained in that tableau is known to both parties. However, the theory does distinguish various interpretations of the conflict situation, such as whether or not the players can communicate with each other before the event, whether or not they can cooperate with each other, and whether or not agreements made are actually binding.

A word of explanation as to the meaning of the title of this section is in order. The game that has been developed in the section is an example of a certain type of two-person game. The archetype of games in this category, and the game that lends its name to the category, is the following example of a *prisoner's dilemma.*

Two men are arrested on suspicion of armed robbery. The district attorney is convinced of their guilt but lacks sufficient evidence for conviction at a trial. He points out to each prisoner separately that he can either confess or not confess. If one prisoner confesses and the other does not, the district attorney promises immunity for the confessor and a 2-year jail sentence for the convicted partner. If both confess, he promises leniency and the probable result of a 1-year jail sentence for each prisoner. If neither confesses, he promises to throw the book at each of them on a concealed weapons charge, with a 6-month jail sentence resulting for each.

The possible actions and the corresponding outcomes for the two prisoners are given by the following tableau. The outcomes are stated in terms of ordered pairs, with the first component representing the length of a prison term in months for Prisoner 1 and the second component the length for Prisoner 2.

		Prisoner 2	
		Confess	*Not Confess*
Prisoner 1	*Confess*	$(-12,-12)$	$(0,-24)$
	Not Confess	$(-24,0)$	$(-6,-6)$

The negative signs indicate the undesirable nature of the outcomes (certainly a 12-month sentence is more favorable than a 24-month sentence, that is, $-12 > -24$). The similarity between this tableau and the previous one should be apparent, since the positions of the numbers in the linear ordering of the preferences and in the tableaux correspond. In fact, in this particular case, all the corresponding entries in the two tableaux differ by a fixed amount, 12.

1.4 The Roles of Linear Programming and Game Theory

Using as a base the four-step description of the operation of applying mathematics given in Section 1.1, an outline of how the fields of linear programming and game theory fit into this general scheme can be given.

In Section 1.2 an example of a linear programming problem was given. Many problems that occur in business, industry, warfare, economics, and so on can be reduced to problems of this type, problems of finding the optimal value of some given linear function while the domain of the function is restricted by a system of linear equations or inequalities. The major concern here is not to determine whether or not an optimal value exists, but to develop a technique to determine quickly and easily the optimal value and where it occurs. Thus, from a mathematical point of view, we wish to develop for linear programming problems a method to use in the third stage of the process, finding the solution of the mathematical problem; and in particular, because realistic problems arising from a complex situation may have many variables and many constraints, we need a computationally efficient method of solution. Moreover, since the users of an algorithm need to know if the algorithm will always work, the question of completeness of the solution technique must be addressed.

In Section 1.3 an example of a game theory problem was given. Our first concern with games will be with two-person, zero-sum games. Although the extent of our assumptions may seem to limit the applicability of the theory, this theory still serves as the foundation for the study of more complex games. Moreover, two-person, zero-sum games provide the opportunity to consider at a theoretical level the second stage in the process of applying mathematics, the formulation of the mathematical model. What one means by the solution to a game is not at all apparent, and axioms must be established that define this concept precisely and adequately reflect the economic or social situations to which game theory might be applied. This is in contrast to linear programming problems, where the desire to maximize profits or minimize costs translates immediately into a problem of optimizing a particular function.

From our discussion so far, the problems of game theory and linear programming may seem to be totally unrelated, but this is not the case. Once our mathematical model for two-person, zero-sum games is developed, the problems of existence and calculation of a solution to a game will be related to the theory of linear programming. Here the unifying concept will be the notion of duality. Duality will be introduced in Chapter 4, and the main theorem of that chapter, the Duality Theorem, will provide the answer to the principal question of our study of games, that is, the question of existence of a solution.

CHAPTER 2

THE LINEAR PROGRAMMING MODEL

2.1 HISTORY

The basic problem of linear programming, determining the optimal value of a linear function subject to linear constraints, arises in a wide variety of situations, but the theory that we will develop is of recent origin.

In 1939 the Russian mathematician L. V. Kantorovich published a monograph entitled Mathematical Methods in the Organization and Planning of Production [2]. Kantorovich recognized that a broad class of production problems led to the same mathematical problem and that this problem was susceptible to solution by numerical methods. However, Kantorovich's work went unrecognized.

In 1941 Frank Hitchcock [3] formulated the transportation problem, and in 1945 George Stigler [1] considered the problem referred to in Section 1.2 of determining an adequate diet for an individual at minimal cost. Through these problems and others, especially problems related to the World War II effort, it became clear that a feasible method for solving linear programming problems was needed. Then in 1951 George Dantzig [4] developed the simplex method. This technique is the basis of the next chapter. John von Neumann recognized the importance of the concept of duality, the mathematical thread uniting linear programming and game theory, and the first published proof of the Duality Theorem is that of Gale, Kuhn, and Tucker [5].

Since the late 1940s, many other computational techniques and variations have been devised, usually for specific types of problems or for use with certain types of computing hardware. The theory has been applied extensively in industry. On the one hand, management has been forced to define explicitly its desired objectives and given constraints. This has brought about a much greater understanding of the decision-making process. On the other hand, the actual techniques of linear programming have been successfully applied in the petroleum industry, the food processing industry, the iron and steel industry, and many more.

Theoretical developments in linear programming have attracted the attention of both theoreticians and the practitioners in the field (along with the readers of the *New York Times*). Some comments on these events are included in Appendix C on theory and efficiency in linear programming

An Introduction to Linear Programming and Game Theory, Third Edition. By P. R. Thie and G. E. Keough.

2.2 The Blending Model

The diet problem described in Section 1.2 is an example of a general type of linear programming problem that involves blending or combining various ingredients. The cost and composition or characteristics of the various ingredients are known, and the problem is to determine how much of each of the ingredients to blend together so that the total cost of the mixture is minimized while the composition of the mixture satisfies specified requirements. In the diet problem, foods were combined to form a diet minimizing costs and meeting basic nutritional requirements.

The construction of the mathematical model for problems of this type follows quickly once the usually more difficult task of defining the characteristics and cost of the ingredients and required composition of the blend has been accomplished. Assuming that all this information is at hand, the amounts of each of the ingredients to blend together must be decided. Thus, variables are assigned to represent these amounts. The cost function, the function to be optimized, can then be constructed by considering the cost of each of the ingredients and assuming that the total cost is the sum of the individual costs. The system of constraints, that is, the set of restrictions of the variables, follows by considering the requirements specified for the final blend.

Example 2.2.1. To feed her stock a farmer can purchase two kinds of feed. The farmer has determined that the herd requires 60, 84, and 72 units of the nutritional elements A, B, and C, respectively, per day. The contents and cost of a pound of each of the two feeds are given in the following table.

	Nutritional Elements (units/lb)			
	A	*B*	*C*	*Cost (cents/lb)*
Feed 1	3	7	3	10
Feed 2	2	2	6	4

Obviously, the farmer could use only one feed to meet the daily nutritional requirements. For example, it can easily be seen that 24 lb of the first feed would provide an adequate diet at a daily cost of \$2.40. However, the farmer wants to determine the least expensive way of providing an adequate diet by combining the two feeds. To do this, the farmer should consider all possible diets that satisfy the specified requirements and then select from this set the diet of minimal cost.

To translate this into a mathematical problem, let x be the number of pounds of Feed 1 and y the number of pounds of Feed 2 to be used in the daily diet. Then by definition, x and y must be nonnegative. Moreover, a diet consisting of x lb of Feed 1 and y lb of Feed 2 would contain $3x+2y$ units of nutritional element A. Since 60 units of element A are required daily, we must have $3x+2y \geq 60$. We are assuming that providing more than the minimal requirements of any of the nutritional elements will have no harmful effects, and so any diet providing at least 60 units of element A will satisfy this requirement. Thus the inequality and not an equality.

To provide insight into the nature of linear programming, this particular problem will be solved geometrically. The set of diets satisfying the above requirements can be illustrated graphically. All the points (x,y) in the first quadrant satisfying the inequality are shown in Figure 2.1.

The other two nutritional requirements demand that

$$7x + 2y \geq 84 \text{ and } 3x + 6y \geq 72$$

The corresponding regions in the first quadrant are sketched in Figure 2.2.

We must consider all feasible diets, that is, all diets that satisfy all three requirements. They are given graphically by the shaded region in Figure 2.3.

The cost in cents of a diet of x lb of Feed 1 and y lb of Feed 2 is $10x + 4y$. Thus we must determine the minimum of the function $f(x,y) = 10x + 4y$, while the x and y are restricted to the shaded region in Figure 2.3.

Consider the graphs of the family of lines determined by the equation $10x + 4y = c$, where c is constant. In Figure 2.4, some of these lines are graphed for various values of c. Note that all the lines have the same slope and that the lines move to the left as c decreases.

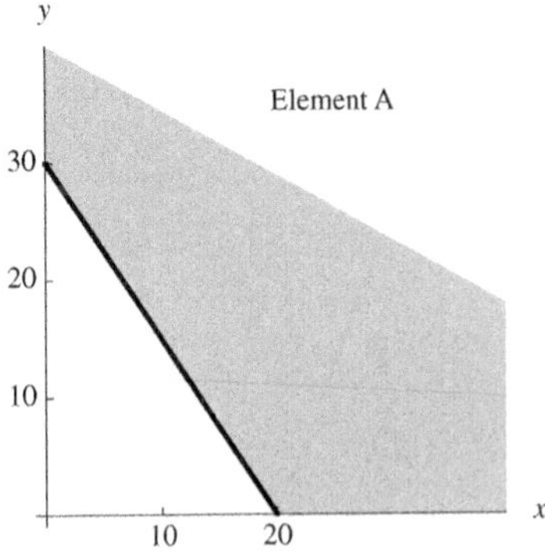

Figure 2.1

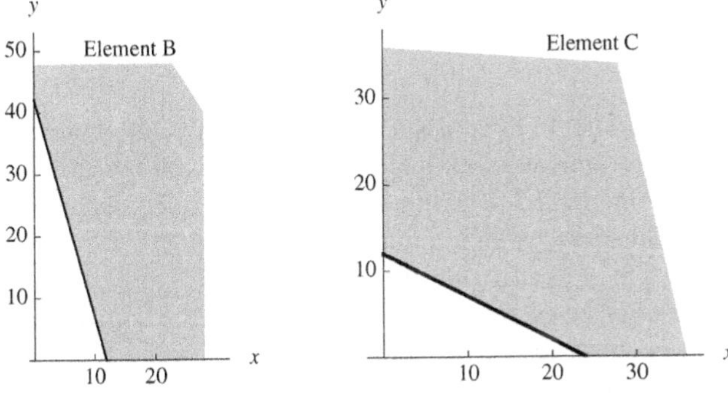

Figure 2.2

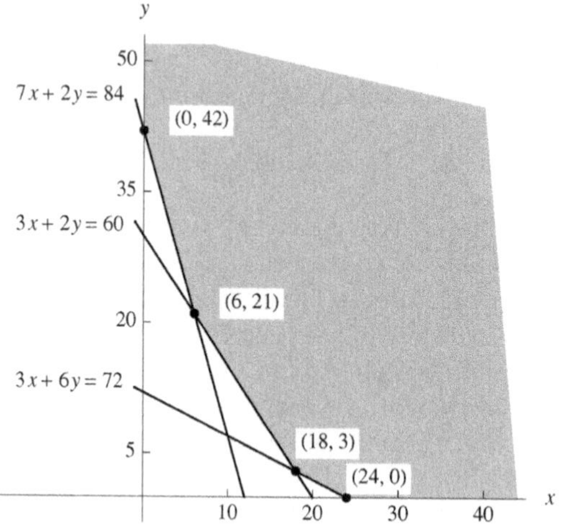

Figure 2.3

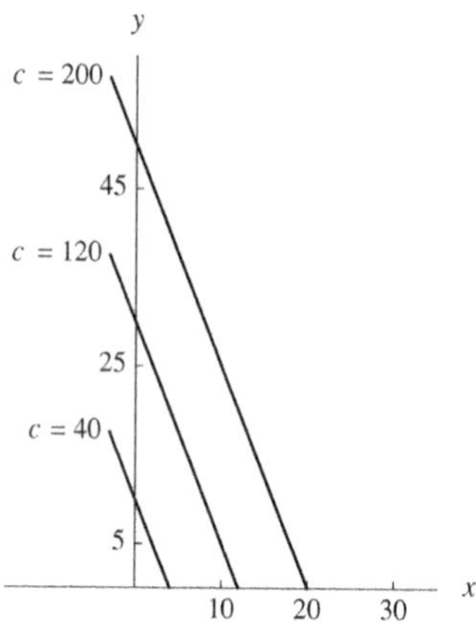

Figure 2.4

Each of the parallel lines consists of points that give the same value for the cost function $10x+4y$. Thus we seek that line farthest to the left that still intersects the shaded region of Figure 2.3. The line through point $(6,21)$ is that line, as illustrated in Figure 2.5. Thus the cost of a minimal diet is $10\cdot 6+4\cdot 21=144$ cents, and this diet consists of 6 lb of Feed 1 and 21 lb of Feed 2.

This analysis can be extended. As the value of c in the family of lines $10x+4y=c$ decreases and the lines slide down and to the left, from the geometry it follows that the line we seek will intersect the set of feasible solutions at a corner point (or vertex) of the set of feasible solutions. In this example we can therefore conclude that a minimal-cost diet, if it exists, must be attained at either point $(0,42)$, $(6,21)$, $(18,3)$,

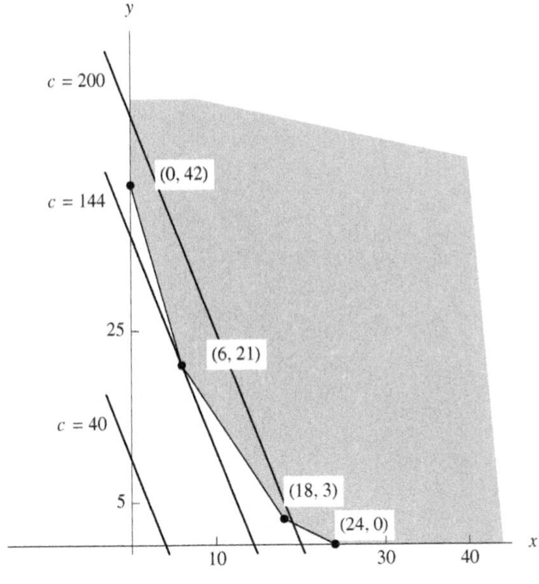

Figure 2.5

or $(24,0)$. Thus, if we have the corner points at hand, evaluating the cost function at each of these points and comparing values will yield the desired optimal diet:

corner points	$(0,42)$	$(6,21)$	$(18,3)$	$(24,0)$
$10x+4y$	168	144	192	240
		⇑		

Our above result is confirmed; the minimal-cost diet is to use daily 6 lb of Feed 1 and 21 lb of Feed 2 at a cost of 144 cents.

Suppose now that the price of Feed 1 increases from 10 cents/lb to 14 cents/lb, with all other data unchanged. Then the corner points of the set of feasible solutions is as above, and an evaluation of the new cost function at these points will yield the revised optimal solution.

corner points	$(0,42)$	$(6,21)$	$(18,3)$	$(24,0)$
$14x+4y$	168	168	264	336
	⇑	⇑		

Now the optimal diet is not unique. The minimal-cost line $14x+4y=168$ passes through the two corner points $(0,42)$ and $(6,21)$, and since any feasible point on this line delivers a diet of 168 cents/lb, the set of optimal feasible diets consists of the points on the line segment between the corner points $(0,42)$ and $(6,21)$, as displayed in Figure 2.6.

We have in the solution to the above problem a function with a unique minimum value (certainly there can be only one minimum value) but with multiple optimal solution points. And in the example, with only two variables, the geometry justifies

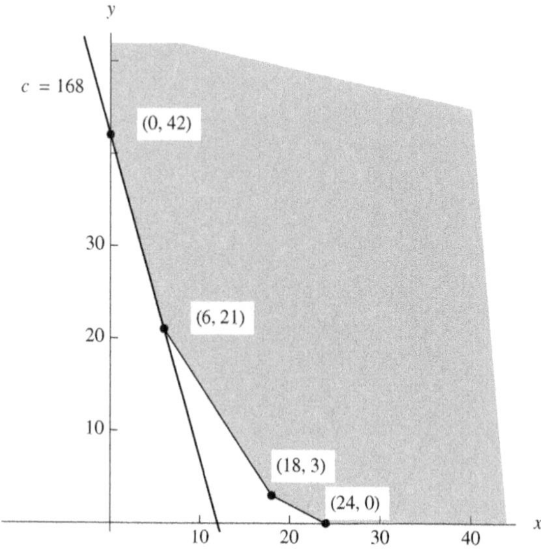

Figure 2.6

the result. The lines in the family $\{14x+4y=c : c \text{ a constant}\}$ and the boundary line $7x+2y=84$ are parallel, with common slope$-\frac{7}{2}$, and when c decreases, the line with a minimum value for c that intersects the set of feasible solutions will lie on the segment of the boundary corresponding to this constraining line.

The use of slopes can be extended. Consider the original cost function $10x+4y$. The slope of the associated family of lines $\{10x+4y=c : c \text{ a constant}\}$ is $-\frac{5}{2}$, and the optimal solution point to the problem, $(6,21)$, is at the intersection of the boundary lines $7x+2y=84$ (with slope$-\frac{7}{2}$) and $3x+2y=60$ (with slope$-\frac{3}{2}$). Thus from the geometry, the slope $-\frac{5}{2}$ of the function to be minimized must be between these two slopes. Indeed, $-\frac{7}{2}<-\frac{5}{2}<-\frac{3}{2}$.

In fact, we can say that if the cost function is c_1x+c_2y, where c_1 and c_2 are positive numbers, the minimum cost would be attained at the point $(6,21)$ if $-\frac{7}{2}\leq-\frac{c_1}{c_2}\leq-\frac{3}{2}$, that is, $\frac{3}{2}\leq\frac{c_1}{c_2}\leq\frac{7}{2}$, and the solution point would be unique if the inequalities are strict.

Thus, for example, if the cost c_2 of Feed 2 is fixed at 4 cents/lb but the cost c_1 of Feed 1 is variable, the farmer should continue to use the $(6,21)$ diet as long as $\frac{3}{2}\leq\frac{c_1}{4}\leq\frac{7}{2}$, that is, as long as $6\leq c_1\leq 14$, with a minimum daily cost of $6c_1+21\cdot 4=6c_1+84$ cents.

Example 2.2.2. A landscaper has on hand two grass seed blends. Blend I contains 60% bluegrass seed and 10% fescue and costs 80 cents/lb; Blend II contains 20% bluegrass seed and 50% fescue and costs 60 cents/lb. (Each also contains other types of seeds and inert materials.) The field about to be sowed requires a composition seed consisting of at least 30% bluegrass and 26% fescue. What is the least expensive combination of the two blends that meets these requirements?

To formulate a mathematical model for a problem involving percentages, ambiguities can arise. To avoid these, we can determine the optimal way to produce a fixed amount of the final product.

For example, let us determine the combination that minimizes costs and produces 100 lb of the required composition seed. Defining x as the number of pounds of Blend I used in this composition and y as the number of pounds of Blend II, the 30% bluegrass requirement translates into the inequality

$$0.60x + 0.20y \geq 30$$

as the 100 lb of the final composition must contain at least 30 lb of bluegrass. The fescue requirement yields the inequality

$$0.10x + 0.50y \geq 26$$

These inequalities simplify to $3x + y \geq 150$ and $x + 5y \geq 260$. The region in the first quadrant satisfying the inequalities is graphed in Figure 2.7.

Since 100 lb of the composition is to be produced, x and y must also satisfy the equation $x + y = 100$ (see Figure 2.8).

The cost in dollars of x lb of Blend I and y lb of Blend II is $c(x,y) = 0.8x + 0.6y$, and we seek the minimum of this linear function on the set of points represented by the heavy line in Figure 2.8. From the geometric argument of the previous example, it follows that the line in the family of parallel lines $\{(x,y) : 0.8x + 0.6y = c\}$, where c is a constant, with minimal c and intersecting this set must intersect the set at either $(25,75)$ or $(60,40)$. Evaluating,

$$c(25,75) = \$65 \text{ and } c(60,40) = \$72$$

Thus, to produce 100 lb of the composition at minimum cost, 25 lb of Blend I and 75 lb of Blend II should be used, and so the minimal-cost prescription for making any amount of the composition seed is to use 25% Blend I and 75% Blend II.

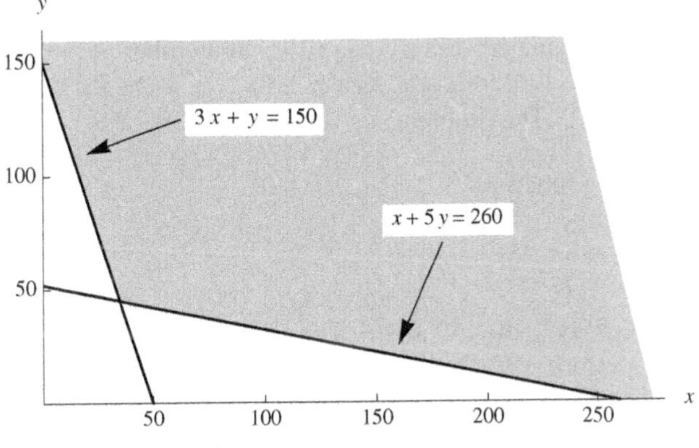

Figure 2.7

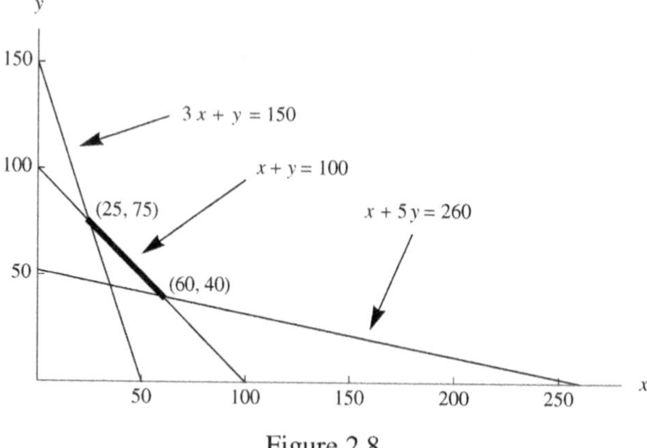

Figure 2.8

Example 2.2.3 (Continuation of Example 2.2.2). The operation of the landscaper of the above example has expanded. Now there are two fields to be maintained, Field X (the original field) and Field Y, with Field Y requiring a seed mixture that is at least 15% bluegrass and 35% fescue; and there is an additional grass seed blend to work with, Blend III, with a composition of 25% bluegrass and 15% fescue and a cost of 35 cents/lb. The relevant data are summarized in the following table.

		Bluegrass	*Fescue*	*Cost (cents/lb)*
	Blend I	60%	10%	80
Composition	*Blend II*	20%	50%	60
	Blend III	25%	15%	35
Requirements	*Field X*	$\geq 30\%$	$\geq 26\%$	
	Field Y	$\geq 15\%$	$\geq 35\%$	

Suppose the landscaper has an order for 100 lbs of seed for Field X and 160 lbs of seed for Field Y. To determine the minimum cost to meet these demands, the following model is formulated. Let x_1, x_2, x_3 be the number of pounds of Blends I, II, and III, respectively, used for Field X, and let y_1, y_2, y_3 be the number of pounds of each used for Field Y. The problem:

To minimize the function

$$(80x_1 + 60x_2 + 35x_3) + (80y_1 + 60y_2 + 35y_3)$$

subject to

$$\begin{array}{ll} x_1 + x_2 + x_3 = 100 & y_1 + y_2 + y_3 = 160 \\ .6x_1 + .2x_2 + .25x_3 \geq 30 & .6y_1 + .2y_2 + .25y_3 \geq .15(160) = 24 \\ .1x_1 + .5x_2 + .15x_3 \geq 26 & .1y_1 + .5y_2 + .15y_3 \geq .35(160) = 56 \\ x_1, x_2, x_3 \geq 0 & y_1, y_2, y_3 \geq 0 \end{array} \tag{2.2.1}$$

Unlike the optimization problems of Examples 2.2.1 and 2.2.2, each with only two variables, this problem, with six variables, cannot be solved graphically. The problems are essentially the same, with linear functions to be optimized subject to linear constraints. But any such problem with more than two variables is intractable to a graphical approach. The goal of Chapter 3 is to develop an efficient method of solving the general problem, regardless of size.

While we cannot complete problem (2.2.1) at this time, some further comments on the problem are in order. The reader may have already noted that (2.2.1) can be simplified. Meeting the demands for Field X and meeting the demands for Field Y are independent problems; the x's and the y's in (2.2.1) are not related in the family of constraints. We could solve each of these problems separately and then combine the solutions to resolve the two-field problem. (Of course, graphical solution techniques would remain out of reach for the two three-variable problems.)

On the other hand, further restrictions could easily eliminate this simplification. Suppose, for example, that only a limited amount of one of the blends is available — perhaps only 125 lbs of the new Blend III is on hand and can be used at this time. Then the constraint $x_3 + y_3 \leq 125$ would need to be added to (2.2.1), and the optimization problems for the two fields are no longer independent.

Another variation could be that, because of shipping restrictions, the producer of the seed can deliver Blends I and II only in a single drum containing a premixed combination of the two blends, with the customers specifying the ratio of Blend I to Blend II to be used in preparing their orders. In the landscaper model, this means that the ratios of Blend I to Blend II used in each of the fields are the same, that is, $\frac{x_1}{x_2} = \frac{y_1}{y_2}$ or $x_1 y_2 = x_2 y_1$. However, adding the simple equality $x_1 y_2 = x_2 y_1$ to (2.2.1) changes the optimization problem dramatically. The problem is no longer a linear programming problem, as $x_1 y_2 = x_2 y_1$ is not a linear constraint. The problem is in the domain of nonlinear programming, a topic not considered in this linear programming text.

Problem Set 2.2

Problems 1–5 refer to Example 2.2.1.

1. A salesperson offers the farmer a new feed for her stock. One pound of this feed contains 2, 4, and 4 units of the nutritional elements A, B, and C, respectively, and costs 7 cents. By considering a blend that consists of equal parts of Feeds 1 and 2, show that the use of this new feed cannot reduce the minimal cost of an adequate diet.

2. The farmer has determined that as long as the ratio of the cost of Feed 1 to the cost of Feed 2 is between $\frac{1}{2}$ and $\frac{3}{2}$, an adequate diet of minimal cost can be achieved by using 18 lb of Feed 1 and 3 lb of Feed 2. Explain.

3. What should the ratio of the costs of the feeds be to warrant the use of a diet consisting solely of Feed 1? When should the farmer use only Feed 2 for her stock?

4. After reviewing his mother's mathematical formulation of the feed problem, the farmer's son claims that in general the constraining inequalities should be equal-

ities. He reasons that money must be wasted if some of the nutritional elements are fed to the stock at a level above the minimal requirements. Is this true?

5. After some study, the farmer has decided that 40 units of nutritional element D are also critical for the daily feeding of his stock. One pound of Feeds 1 and 2 contains 4 and 2 units of element D, respectively. How does this change the analysis of the original problem?

6. Products X and Y are to be blended to produce a mixture that is at least 30% A and 30% B. Product X is 50% A and 40% B and costs \$10/gal; Product Y is 20% A and 10% B and costs \$2/gal. To formulate a model to be used to determine a minimal-cost blend, we let x and y equal the number of gallons of X and Y used, respectively, and write the following mathematical problems:

   (a) Our first attempt.

   $$\begin{aligned}&\text{Minimize } 10x+2y\\&\text{subject to}\\&.5x + .2y \geq .3\\&.4x + .1y \geq .3\\&x,y \geq 0\end{aligned}$$

   Note that $x = 0$, $y = 3$ satisfies the constraints. So should we use only Product Y? Explain.

   (b) We try again. Our final product is to be at least 30% A and 30% B and contain $x+y$ gal, so we want to

   $$\begin{aligned}&\text{Minimize } 10x+2y\\&\text{subject to}\\&.5x + .2y \geq .3(x+y)\\&.4x + .1y \geq .3(x+y)\\&x,y \geq 0\end{aligned}$$

   But does $x = 0$, $y = 0$ satisfy the constraints? Explain.

   (c) Formulate a correct model.

For Problems 7–10, formulate mathematical models and then solve the problems.

7. (a) A poultry producer's stock requires at least 124 units of nutritional element A and 60 units of nutritional element B daily. Two feeds are available for use. One pound of Feed 1 costs 16 cents and contains 10 units of A and 3 units of B. One pound of Feed 2 costs 14 cents and contains 4 units of A and 5 units of B. Determine for the producer the least expensive adequate feeding diet.

   (b) For what range on the ratio of the costs of Feed 1 to Feed 2 would the optimal diet be the above diet?

   (c) For what values of the ratio of the costs of Feed 1 to Feed 2 would the optimal diet for the problem of part (a) not be unique?

8. Premium loam is 60% soil and 40% domestic manure and costs $5/50 lb. Generic loam is 20% soil and 10% domestic manure (and 70% sand, stone, etc.) and costs $1/50 lb. We need loam for our backyard that is at least 36% soil and at least 20% domestic manure. What combination of the two loams should we use to minimize costs?

9. A crude insecticide used commercially is 40% Toxin A and 35% Toxin B. New federal regulations set upper limits on toxin levels for commercial insecticides: 36% for Toxin A and 28% for Toxin B. A compatible insecticide can be produced using a more refined process, but at an increased cost of $4 more than the crude insecticide for every 10 lb. This product would be only 15% Toxin A and 10% Toxin B. The two insecticides can be blended. What combination of the two minimizes production costs and meets federal standards?

10. (a) A cheese producer must feed her stock of Jersey cattle daily at least 550 units of nutritional element A, 500 units of nutritional element B, and 820 units of nutritional element C. She has available two feeds. One pound of Feed X costs 80 cents and contains 2 units of A, 5 units of B, and 7 units of C. One pound of Feed Y costs 30 cents and contains 3 units of A, 1 unit of B, and 2 units of C. The cheese producer wants to determine what combination of the two feeds will meet the dietary requirements of her Jerseys and keep costs at a minimum. Determine the least expensive adequate feeding diet.

(b) Generalize. Suppose Feed X costs c_1 cents/lb and Feed 2 costs c_2 cents/lb. For what range on the ratio of c_1 to c_2 would the optimal diet of part (a) remain optimal?

(c) In particular, assume that the cost of Feed Y is fixed at 30 cents/lb but that the cost of Feed X is increasing. By how much can this cost increase before the diet of part (a) is no longer optimal? If the cost of Feed X increases by more than this bound, what would be the new optimal diet?

(d) Determine the resolution of the original problem with the added restriction that no more than 215 lbs of Feed X may be used in the daily diet.

Formulate mathematical models for the following problems. (Do not attempt to solve the problems.)

11. A paint manufacturer must produce a base for its line of indoor domestic paints. Four chemicals, A, B, C, and D, are critical in its manufacture. The final composition of the base by weight must be at least 5% of Chemical A, 3% of Chemical B, 26% of Chemical C, and no more than 15% of Chemical D. The manufacturer can produce this base by combining three crude minerals. The compositions by weight and the costs of these minerals are given in the following table:

	% of Chemical				
	A	*B*	*C*	*D*	*Cost ($/lb)*
Mineral 1	0	5	30	20	4.00
Mineral 2	6	8	30	10	7.50
Mineral 3	7	0	25	16	3.00

The manufacturer could use just Mineral 2. However, he asks, "Is there some combination of the three minerals that will provide a base with the desired characteristics at a lower cost?"

12. A firm wants to market bags of lawn fertilizer that contain 23% nitrogen, 7% phosphoric acid, and 7% soluble potash. Chemicals A, B, C, D, and E are available and can be combined for the product. The contents in pounds and cost in dollars of 100 lb of each are:

	A	*B*	*C*	*D*	*E*
Nitrogen	18	28	0	30	16
Phosphoric Acid	12	5	6	7	3
Potash	0	5	18	8	2
Cost	10	23	10	30	15

How much of each chemical should be used to minimize costs?

13. A coin is to be minted containing at least 40% silver and at least 50% copper. The mint has available Alloys A, B, C, and D, with the following compositions and costs:

	A	*B*	*C*	*D*
% Silver	30	35	50	40
% Copper	60	35	50	45
Cost/lb ($)	11	12	16	14

What blend of these alloys provides the required composition at minimal cost?

14. The manager of a fleet of trucks needs an antifreeze solution containing at least 50% pure antifreeze and at least 5% anticorrosion additives. He has available three commercial products, A, B, and C, with characteristics and costs given in the following table. What blend will provide a suitable solution at minimal costs?

	A	*B*	*C*
% Antifreeze	60	18	75
% Additives	10	3	0
Cost (dollars/gal)	1.6	0.5	1.4

15. A firm produces a rare blend of scotch whiskey. The blend must contain exactly 43% alcohol, at least 25% Highland blend, and no more than 8% malt. Four distillery products can be combined for the blend. The contents are given below. Determine the combination that minimizes the cost.

	A	B	C	D
% Alcohol	46	40	45	40
% Highland	33	20	28	18
% Malt	10	5	12	2
Cost ($/gal)	12	8	11	7

16. The highway department requires a sand/salt mixture for spreading on its roads in the winter. The mixture must be no more than 70% sand and no less than 10% salt. (It can also contain gravel, dirt, etc.) Company A provides a mixture that is 75% sand and 2% salt and costs $5/ton; Company B provides a mixture that is 60% sand and 6% salt and costs $12/ton. Pure road salt costs $100/ton. What combination of the two mixtures and salt meets the requirements at minimal cost?

17. (a) A fuel additive must be at least 32% Chemical A, at least 15% Chemical B, and no more than 40% inert element C. Four products, W, X, Y, and Z, can be combined to produce the additive, composition, and cost ($/gallon) as listed. Determine what percentage of each of these products is contained in the minimal-cost blend.

	W	X	Y	Z
% A	45	25	28	26
% B	22	10	0	16
% C	20	42	44	27
Cost ($/gal)	35	5	0	15

(b) As in part (a), but with the additional restriction that the amount of X in the final blend cannot exceed the combined amounts of W and Z by more than 5% of the combined amounts of W and Z.

2.3 THE PRODUCTION MODEL

Production models and their variations occur frequently in linear programming applications. Central to these problems is an operation or production system, say a factory or a refinery. Commodities such as raw materials, capital, and labor are input into the system and are acted on by various productive processes. The results are the output or goods produced, and the basic problem is to operate the system in a way that maximizes profit using limited resources, or minimizes costs while meeting specified production requirements, or some combination of these goals.

Example 2.3.1. Suppose a boat manufacturer produces two types of boats for the sports and camping trade, a family rowboat and a sports canoe. The boats are molded from aluminum by means of a large pressing machine and are finished by hand labor. A rowboat requires 50 lb of aluminum, 6 min of machine time, and 3 hr of finishing labor; a canoe requires 30 lb of aluminum, 5 min of machine time, and 5 hr of finishing labor. For the next 3 months the company can commit up to 1 ton of aluminum,

5 hr of machine time, and 200 hr of labor for the manufacture of the small boats. The company realizes a \$50 profit on the sale of a rowboat and a \$60 profit on the sale of a canoe. Assuming that all boats made can be sold, how many of each type should be manufactured in the next 3 months in order to maximize profits?

Here the decision to be made involves the number of rowboats and the number of canoes to be produced in the next 3 months. Thus, let R and C denote these numbers, with R the number of rowboats and C the number of canoes. Then the profit for the company, measured in dollars, from its small boat line will be $50R+60C$, and this is the function to be maximized.

The quantities R and C cannot be negative. Moreover, they are limited by the amount of resources available for the production of the boats. Specifically, at most 1 ton of aluminum can be used, and so we must have $50R+30C \leq 2000$. Similarly, consideration of available machine time and finishing labor leads to the inequalities

$$6R+5C \leq 300 \text{ and } 3R+5C \leq 200$$

Thus the mathematical problem is to determine R and C that maximize the function $50R+60C$ and satisfy the constraints $R \geq 0$, $C \geq 0$,

$$\begin{aligned} 50R + 30C &\leq 2000 \\ 6R + \ \ 5C &\leq \ \ 300 \\ 3R + \ \ 5C &\leq \ \ 200 \end{aligned}$$

Example 2.3.2. In the above example, the \$50 and \$60 profit estimates would be determined by subtracting production and delivery costs from the selling price of each of the two boats. Suppose now that the cost to the manufacturer of the 1 ton of aluminum is not fixed. In particular, assume that the price per pound of the last 500 lb of aluminum is 20 cents/lb more than the price of the first 1500 lb, and that the price of the first 1500 lb is the cost used in determining the \$50 and \$60 profit estimates. With this increase in cost of the last 500 lb of aluminum, what is the optimal production schedule?

To account for this potential additional cost, the amount of aluminum used over 1500 lb must be measured. Define X to be this amount, in pounds, and, as above, define R and C to be the number of rowboats and canoes to be produced. The problem now is to determine R, C, and X that maximize the function

$$50R+60C-0.2X$$

and satisfy the constraints

$$\begin{aligned} 50R + 30C &\leq 1500+X \\ 6R + \ \ 5C &\leq \ \ 300 \\ 3R + \ \ 5C &\leq \ \ 200 \\ X &\leq \ \ 500 \\ R,C,X &\geq 0 \end{aligned}$$

Notice that the constraint involved with the amount of aluminum used is stated in terms of a less than or equal to inequality as opposed to an equality. The inequality allows for the possibility of using less than 1500 lb of aluminum in an optimal production schedule. If more than 1500 lb of aluminum is to be used, the $-0.2X$ term in the function to be maximized guarantees that at any optimal solution point in the problem, the value of $X \geq 0$ will be as small as possible, and so the 20 cents/lb additional cost will be assessed on the exact amount over 1500 lb required.

(In this example, the profit function needed to be altered once the amount of aluminum used exceeded 1500 lb. At this point, profits decreased, and we were able to model this unfavorable shift using the one additional variable X. However if the cost of aluminum were less when purchased in quantity, then the objective function could experience a favorable shift depending on the amount of aluminum used, and the formulation of a correct mathematical model would not have been as straightforward. In Chapter 6 we will present a technique for modeling favorable shifts in the function to be optimized.)

Example 2.3.3. A cabinet shop makes and sells two types of cabinets, type 1, for the kitchen, and type 2, for the bathroom. Manufacture of the cabinets consists of two steps, making the frames and drawers and then assembling and finishing the units. Labor requirements, in hr/unit, are as follows:

Cabinet	*Frame/Drawers (hr)*	*Assembly/Finishing (hr)*
Type 1 (kitchen)	2.6	2.1
Type 2 (bathroom)	1.5	1.8

Each week the shop has 480 hr of labor available for the manufacture of the cabinets. However, to conserve labor, frames and drawers completed and ready for assembly and finishing can also be bought from a local dealer at a cost of \$200 for a kitchen frame/drawer set and \$110 for a bathroom frame/drawer set.

The kitchen cabinets sell for \$350 each; the first 70 bathroom cabinets sell for \$250 per unit, but any more produced sell for only \$225 per unit. We assume that all units produced will be sold.

In order to determine a production schedule that maximizes net income (sales revenue less the cost of any frames and drawers bought), the shop manager first notes the decisions to be made, namely, how many of each type of cabinet to produce and how to generate the associated frames and drawers. Considering also the shift in selling price of the bathroom cabinets, the following variables are defined:

$$
\begin{aligned}
t_i &= \text{the } \textit{total} \text{ number of cabinets of type } i \text{ produced}, i = 1,2 \\
m_i &= \text{the number of frames/drawers } \textit{made} \text{ of type } i, i = 1,2 \\
b_i &= \text{the number of frames/drawers } \textit{bought} \text{ of type } i, i = 1,2 \\
u &= \text{the number of bathroom cabinets sold } \textit{up to } 70 \\
v &= \text{the number of bathroom cabinets sold } \textit{over } 70
\end{aligned}
$$

The mathematical model then is to maximize $350t_1 + 250u + 225v - 200b_1 - 110b_2$ subject to

$$
\begin{aligned}
&t_1 = m_1 + b_1 \\
&t_2 = m_2 + b_2 \\
&2.6m_1 + 1.5m_2 + 2.1t_1 + 1.8t_2 \leq 480 \\
&t_2 = u + v, u \leq 70 \\
&t_1, t_2, m_1, m_2, b_1, b_2, u, v \geq 0
\end{aligned}
$$

Note the roles of the variables u and v; u measures the number of bathroom cabinets produced up to and including 70, and v measures the number produced over 70. (For example, if 85 bathroom units are produced and sold, we would have $u = 70$ and $v = 15$.) The unfavorable shift in the function to be maximized reflected in the sum $250u + 225v$ guarantees that u will reach 70, the variable's bound, before v moves off 0.

Example 2.3.4. Consider the operation of one division in a large plant. The division is responsible for manufacturing two parts of the plant's final product. The division manager has available four different processes to produce these two parts; each process uses various amounts of labor and two raw materials. The inputs and outputs for 1 hr of each of the four processes are given in the following table.

	Input			***Output***	
Process	*Labor (worker-hours)*	*Raw Material A (lb)*	*Raw Material B (lb)*	*Units of Part 1*	*Units of Part 2*
1	20	160	30	35	55
2	30	100	35	45	42
3	10	200	60	70	0
4	25	75	80	0	90

The division is responsible for producing each week 2100 units of Part 1 and 1800 units of Part 2. The division manager has at her disposal each week 4 tons of Raw Material A and 2 tons of Raw Material B and 1000 hr of labor. One pound of Raw Material A costs the firm \$3, and one pound of Raw Material B costs \$7. Because of labor contracts, the plant must pay its employees a full week's salary, regardless of whether or not the employees are used that week, so the cost of the 1000 hr of labor is fixed. However, the division manager can request her workers to work up to an extra 200 hr per week in overtime at a cost of \$30/hr to the firm. The plant vice-president in charge of production wants to know if the division can meet its weekly production requirements with the material on hand without using overtime and, if so, the minimal cost of this operation. And, because the decision to allow overtime must be made at the plant level, the vice-president also wants some estimate on how much money, if any, the division can save by using overtime.

To respond to her supervisor's questions, the division manager must consider the problem in two stages and at each stage must determine the optimal use of her facilities. In the first stage overtime is not available, and thus the manager must decide only on how best to utilize the four available processes. Let x_i denote the number of hours a week that Process i is used, for $i = 1,2,3,4$. The constraints imposed by the limited amounts of labor and raw materials are the following.

$$\begin{aligned}
20x_1 + 30x_2 + 10x_3 + 25x_4 &\le 1000\\
160x_1 + 100x_2 + 200x_3 + 75x_4 &\le 8000\\
30x_1 + 35x_2 + 60x_3 + 80x_4 &\le 4000
\end{aligned}$$

The output requirements give

$$\begin{aligned}
35x_1 + 45x_2 + 70x_3 \qquad &\ge 2100\\
55x_1 + 42x_2 \qquad + 90x_4 &\ge 1800
\end{aligned}$$

Thus the initial question of determining whether or not the weekly production requirements can be met with the available materials is translated into the mathematical problem of determining if there exist four nonnegative numbers x_1, x_2, x_3, x_4 that satisfy these five inequalities. A solution to this problem would provide a suitable or feasible way of operating the division, and this suggests that any nonnegative solution to the system of constraints in a linear programming problem be called a *feasible solution.* In this particular problem, the existence of a feasible solution is easy to verify. The weekly use of 30 hr of Process 3 and 20 hr of Process 4 will produce the exact number of needed parts and will not even exhaust any of the supplies of labor and raw materials.

The cost of the operation when overtime is not employed depends only on the amounts of the raw materials used and is given in dollars by the function

$$\begin{aligned}
f(x_1,x_2,x_3,x_4) &= 3(160x_1 + 100x_2 + 200x_3 + 75x_4)\\
&\quad + 7(30x_1 + 35x_2 + 60x_3 + 80x_4)\\
&= 690x_1 + 545x_2 + 1020x_3 + 785x_4
\end{aligned}$$

Thus the first optimization problem is to determine x_1,x_2,x_3,x_4 that minimize

$$f(x_1,x_2,x_3,x_4) = 690x_1 + 545x_2 + 1020x_3 + 785x_4$$

subject to

$$\begin{aligned}
20x_1 + 30x_2 + 10x_3 + 25x_4 &\le 1000\\
160x_1 + 100x_2 + 200x_3 + 75x_4 &\le 8000\\
30x_1 + 35x_2 + 60x_3 + 80x_4 &\le 4000\\
35x_1 + 45x_2 + 70x_3 \qquad &\ge 2100\\
55x_1 + 42x_2 \qquad + 90x_4 &\ge 1800\\
x_1,x_2,x_3,x_4 &\ge 0
\end{aligned} \tag{2.3.1}$$

The possibility of using overtime introduces one more decision the division manager can make: how much if any overtime to use to reduce the total production cost. Let x_5 denote the number of hours of overtime employed. Then $0 \le x_5 \le 200$. As before, x_1, x_2, x_3, x_4 represent the hours of use of the four processes. Note, however, that this is a different problem, so the optimal schedule here may employ amounts of the processes different from those in the previous optimal schedule. The first constraint, the restriction on available hours, is the only inequality that needs to be changed. Since the total hours used cannot exceed $1000 + x_5$, this inequality becomes

$$20x_1 + 30x_2 + 10x_3 + 25x_4 \leq 1000 + x_5$$

The total cost function must also reflect the cost of the overtime, but the cost of the raw materials is measured as before. Thus the new total cost function, say g, can be defined by simply adding the cost of overtime to the original cost function f. Thus we must minimize

$$g(x_1, x_2, x_3, x_4, x_5) = 30x_5 + f(x_1, x_2, x_3, x_4)$$

Hence the second problem is to minimize the function

$$g(x_1, x_2, x_3, x_4, x_5) = 30x_5 + 690x_1 + 545x_2 + 1020x_3 + 785x_4$$

subject to

$$\begin{aligned} 20x_1 + 30x_2 + 10x_3 + 25x_4 &\leq 1000 + x_5 \\ 160x_1 + 100x_2 + 200x_3 + 75x_4 &\leq 8000 \\ 30x_1 + 35x_2 + 60x_3 + 80x_4 &\leq 4000 \\ 35x_1 + 45x_2 + 70x_3 &\geq 2100 \\ 55x_1 + 42x_2 + 90x_4 &\geq 1800 \\ x_1, x_2, x_3, x_4, x_5 \geq 0, x_5 &\leq 200 \end{aligned} \tag{2.3.2}$$

One final question. Suppose the vice-president in charge of production wants to make some estimate on the production costs of the firm's products and, to do this, requests the division manager to estimate the costs of manufacturing 1 unit of Parts 1 and 2. It would be easy to determine the cost of 1 unit of Part 1, for example, if the division produced only this type of part and Process 3 was used in its production. Then the total cost of 1 hr of operation of this process — and here the cost of the hours involved would need to be included — divided by the number of units of Part 1 produced would give a unit cost. However, this is not the situation. Not only can several processes be involved in the production of Part 1, but also the output of the processes can be mixed. Moreover, how can we measure the true costs of the labor and raw materials? It could be, for example, that a minimal-cost production schedule leaves a surplus of Raw Material A but exhausts the available supply of Raw Material B, and thus Raw Material B is more precious. Should this fact also be included in the costs of the raw materials? We will consider such questions later, in Chapters 4 and 5, after the concept of duality has been introduced. (See in particular Problem 14 of Section 5.1.)

Problem Set 2.3

1. Solve the problem of Example 2.3.1 graphically.

2. Extremum problems for functions with several variables are discussed in multivariable calculus. The standard technique involves taking first partials and setting them equal to zero. What information would this method provide for the problem of Example 2.3.1?

3. The following are other suggested mathematical models for the problem of Example 2.3.2 (with the variables R, C, and X defined as in the example). Determine why each model is not a proper representation of the problem.

(a) Maximize $50R + 60C - 0.2(500)$
subject to

$$\begin{aligned} 50R + 30C &\leq 2000 \\ 6R + 5C &\leq 300 \\ 3R + 5C &\leq 200 \\ R, C &\geq 0 \end{aligned}$$

(b) Maximize $50R + 60C - 0.2(50R + 30C - 1500)$
subject to

$$\begin{aligned} 50R + 30C &\leq 2000 \\ 6R + 5C &\leq 300 \\ 3R + 5C &\leq 200 \\ R, C &\geq 0 \end{aligned}$$

(c) Maximize $50R + 60C - 0.2X$
subject to

$$\begin{aligned} 50R + 30C &= 1500{+}X \\ 6R + 5C &\leq 300 \\ 3R + 5C &\leq 200 \\ X &\leq 500 \\ R, C, X &\geq 0 \end{aligned}$$

Problems 4–6 refer to Example 2.3.4.

4. As long as all workers in the division are interchangeable, there arises a restriction on the use of overtime: the total number of regular hours available must be exhausted before overtime is used. In this problem that would mean that, for an optimal schedule considering the use of overtime, if the quantity $20x_1 + 30x_2 + 10x_3 + 25x_4$ is less than or equal to 1000, x_5 must be 0. Prove that any optimal solution to (2.3.2) must have this property.

5. Suppose the firm can sell excess units of Part 1 on the market for \$3 and excess units of Part 2 for \$2. Modify (2.3.1), the mathematical problem not involving overtime, to incorporate this possibility.

6. Suppose the firm can supplement its supply of Raw Material A by purchasing the material from an outside source. Assume that an unlimited amount can be purchased and that the cost would be \$4/lb. Construct the associated mathematical problem, assuming that the use of overtime is also possible.

For Problems 7–10, formulate mathematical models and then solve the problems.

7. A plant has two processes it can use to produce Products A and B. An hour's operation of Process 1 produces 3 units of A and 6 of B and costs \$25; an hour's operation of Process 2 produces 5 units of A and 5 of B and costs \$20. The plant must produce at least 90 units of A and 120 of B over the next week. How many hours should each process be used so that demands are met and costs minimized?

8. A small plastics company makes novelty figures for sale at political conventions. This spring the company has available 450 hr of labor and 825 spare units of plastic for use in the production of donkeys and elephants. Each elephant requires 2 hr of labor and 7 units of plastic; each donkey requires 5 hr of labor and 5 units of plastic. Elephants sell for \$10, and up to 100 can be sold this summer; donkeys sell for \$7, with a market for 80. How many of each should the company produce to maximize income over the coming summer?

9. A bakery is preparing its weekend specials, this week to be bran muffins and/or brownies. The bakery can commit up to 81 lb of flour and up to 51 lb of sugar for the preparation of these specials. A dozen bran muffins require 12 oz of flour and 5 oz of sugar and sell for a profit of \$0.50. A box of brownies uses 20 oz of flour and 16 oz of sugar and sells for a profit of \$0.80. Assuming all muffins and brownies made can be sold, how many dozen muffins and how many boxes of brownies should be made to maximize profits?

10. A cabinet shop produces and installs cabinets. Business is good, and the shop has an unlimited number of customers willing to pay \$100 for each cabinet installed. However, for the next month, the shop has only 1750 hr of labor and 1032 units of wood that it can commit for cabinet production. Each installed cabinet requires 5 hr of labor, 3 units of wood, and one frame. The frames can be prepared in the shop before installation, with each frame requiring 2 hr of the shop's labor and 1 unit of its wood, or they can be bought ready for installation from the local mill for \$27 each. The shop pays \$6/hr for labor, \$5/unit for wood, and only pays for the labor and wood used. For the next month, how many cabinets should the shop install, and how should the necessary frames be generated so that net income is maximized?

Formulate mathematical models for Problems 11–22. (Do not try to solve the problems.)

11. An electronics firm manufactures integrated circuits for radios, televisions, and stereos. For the next month it has available 1500 units of materials and 920 units of labor. The requirements and selling price of one of each of the above products are given in the following table.

	Units of Material	*Units of Labor*	*Selling Price ($)*
Radio	2	1	8
TV	12	8	60
Stereo	15	6	45

Determine a production schedule that maximizes income.

12. An oil refinery has available three different processes to produce gasoline. Each process produces varying amounts of three grades of gasoline: Regular, Special, and Super. These amounts, in hundreds of gallons per hour of operation, are given in the following table, along with the cost in dollars of an hour's operation of each of the processes.

	Regular	*Special*	*Super*	*Cost ($)*
Process 1	3	4	2	160
Process 2	6	6	8	400
Process 3	6	3	4	300

Each week the refinery must produce at least 3600 gal of Regular, 2000 gal of Special, and 3000 gal of Super. Determine the operation of the refinery that satisfies these demands and minimizes costs.

13. A fruit grower has two systems for picking crops. In the first system, the pickers work individually and, because of their selectivity and care, this method yields more choice produce than regular produce. In the second system, four pickers work with a machine; while this method has a greater harvest, it yields less choice produce than regular produce and costs proportionately more due to the operating expense of the machine. The amounts of input and output for 1 hr of operation of each system are given in the following table:

	Input	*Output (bu)*		*Costs*
	(worker-hours)	*Choice*	*Regular*	*($)*
System 1	1	4	2	2
System 2	4	20	40	11

Weekly the grower must supply the retail outlet with 480 bu of choice produce and deliver 800 bu of regular produce to the local cannery. The grower has available 100 hr of labor per week. Determine an operating schedule that meets these demands and minimizes costs.

14. A small steel plant uses three processes for the production of steel. The processes require varying amounts of labor, ore, and coal and produce not only steel, but also one side product with limited salability. The relevant data for 1 hr of operation of each process is as follows.

	Input			Output	
	Labor (worker-hours)	Ore (lb)	Coal (lb)	Steel (lb)	Side Product (lb)
Process 1	8	200	145	550	35
Process 2	11	140	120	735	15
Process 3	7	300	225	600	75

For a week's operation the plant has available up to 5 tons of ore at \$43/ton, 350 hr of labor at \$15.75/hr, and an unlimited amount of coal at \$12/ton. All the steel produced can be sold for \$650/ton, and up to 1 ton of the side product can be sold for \$37/ton (any amount above 1 ton has no value). Because of operational restrictions, no one process can be employed for more than 40 hr in any week. Determine an operating schedule that maximizes net income. (*Suggestion.* To incorporate the value of the production of the side product into the function to be optimized, divide its total amount produced into two increments, say S_1 and S_2, where S_1, $0 \leq S_1 \leq 2000$ lb represents the first ton produced.)

15. Using carnations and roses, a florist can make up to three different floral arrangements for the Mother's Day trade. The composition (number of flowers of each type) and selling price (\$) of a single arrangement of each type are as follows:

	Carnations	Roses	Price (\$)
Type A	5	2	2.75
Type B	12	4	6.50
Type C	3	6	5.25

The florist can purchase from a local wholesaler up to 85 doz carnations at \$1.80/doz and up to 75 doz roses at \$4.80/doz. The florist can also purchase up to an additional 65 doz carnations at \$3/doz from a distant dealer. Assuming that all arrangements made can be sold, how many of each type should the florist make to maximize net income?

16. A company has three machines to make units of A. Input and output data for 1 hr of operation of each machine are as follows:

	Input		
	Raw Material (lb)	Labor (worker-hours)	Output (units of A)
Machine 1	80	16	37
Machine 2	50	35	43
Machine 3	76	33	52

The company must produce 2000 units of A weekly. The company can purchase up to 1 ton of the raw material for \$4/lb from one source and an unlimited amount from another source for \$5.50/lb. The firm has 900 hr of labor available at \$8/hr, and an additional 200 hr of overtime available at \$12/hr. The company pays only for the labor and raw material it uses. How many hours should each machine be used to meet demands at minimum cost?

17. A farmer has 100 acres of tillable land on which corn, tomatoes, beans, peas, and carrots can be planted. The labor requirements, plant costs, and gross income for 1 acre of each of these crops are as follows:

	Labor (worker-hours)	*Costs (\$)*	*Gross Income (\$)*
Corn	5	20	95
Tomatoes	120	200	1300
Beans	25	55	275
Peas	35	40	345
Carrots	40	75	435

The farmer has available up to 3600 hr of labor at \$7.25/hr. However, the labor and plant costs must be paid before any income from the sale of the produce is realized. The farmer has \$3000 in capital to invest in this year's planting and can borrow up to another \$12,000 if desired. Any such loan would have a 9% annual interest rate but would be repaid within 4 months. Land unused for any of these vegetables must be maintained by planting ground cover. One acre of cover crops requires 2 hr of labor and costs \$9. Determine a planting schedule that maximizes net income.

18. A subsidiary is contracted to deliver 300 units of Part A, 450 units of Part B, and 380 units of Part C to the parent enterprise. The subsidiary can either make the parts or purchase them from a distant wholesaler. The requirements if made and cost if bought of a unit of each part are as follows:

	Labor (hr)	*Q (units)*	*Purchase Price (\$)*
Part A	2	20	200
Part B	6	15	265
Part C	3	22	235

The subsidiary has available 3500 hr of labor at \$20/hr, another 550 hr of overtime at \$30/hr, and 5000 units of Q at \$5/unit. The shop only pays for the labor and Q's used. Management would like to know how many units of each part the subsidiary should make and how many to buy in order to minimize net costs in meeting their contractual obligation.

19. Using labor and Raw Material M, a shop can make and sell up to three different products, with the requirements and selling price per unit of each of the products as follows:

	Labor (hr)	M (lb)	Selling Price ($)
Product 1	3	6	325
Product 2	1	10	300
Product 3	5	8	415

(a) For the next week, the shop has available 2000 hr of labor at $25/hr, another 400 hr of overtime at $35/hr, and 3000 lb of M at $10/lb. The shop only pays for the labor and raw material used. How many units of each product should the shop make and sell in order to optimize net profit?

(b) How would you modify your answer above if in fact only the first 150 units of product 1 would sell for $325 and any others sold would sell for $280?

20. A company produces three types of tires for the SUV market. In their manufacture, the tires are processed on two machines, a molder and a capper. The time (in hours) required on each machine and the income (wholesale selling price less costs, including labor at the *regular* pay rate) per unit made of each type of tire are:

	Machine Time (hr)		
Type	Molder	Capper	Income ($)
1	8	4	45
2	10	7	53
3	5	6	37

Contractual demands for the next month call for the delivery of at least 75 units of each type of tire. To meet these demands while maximizing net income, the company has set aside 3400 hr of machine time at regular pay for the molder and 2700 hr at regular pay for the capper. There is also available up to a total of 1000 hr of overtime that can be divided in any manner and used among the two machines. Overtime pay is $12/hr more than regular-time pay, and this cost must be considered in determining net income. Determine an operating schedule that maximizes net income.

21. Blackstone Woodworkers has signed a contract with Lowe's Depot. They are committed to delivering 50 gazebos and 100 sheds next month for sales this spring. The manufacturing requirements and cost for a unit of each, along with the amount of each resource available next month, are as follows:

	Manufacturing Requirements			
	Wood (units)	Construction Time (hr)	Finishing time (hr)	Cost ($)
Gazebo	13	5	8	440
Shed	20	3	4	275
Supply	2400 units	500 hours	900 hours	

Blackstone can also purchase completely finished gazebos ready for delivery from a wholesale shop for \$600/unit and can purchase sheds from a local lumber yard for \$325/unit. However, the sheds come unfinished; before being ready for delivery, each shed purchased from the lumber yard requires the same 4 hr of finishing time as each of Blackstone's manufactured sheds. Formulate a model which could be used to determine how many gazebos and sheds Blackstone should make, and how many of each to purchase, in order to minimize the total cost.

22. Shakers Inc., a furniture manufacturer, markets two types of tables, a Country-style and a Mission-style. Next month the manufacturer is committed to meeting an order for 115 Country-style tables and 145 Mission-style tables. Using a sequence of three operations, Shakers can produce the tables themselves. Each table requires processing time on each operation. The time requirements and costs for a manufactured table are as follows:

Manufactured Tables

Operation (mins/table)	*Country*	*Mission*
Cutting & Routing	10	20
Sanding & Joining	25	45
Staining & Finishing	20	30
Cost (\$/table)	120	160

The company can also purchase unfinished tables from a commune of local woodcrafters. However, these tables need to be processed through the staining and finishing operation before they can be marketed. The relevant data, with costs including the necessary labor costs, are as follows:

Purchased Tables

Operation (mins/table)	*Country*	*Mission*
Staining & Finishing	25	40
Cost (\$/table)	145	200

Next month, Shakers has committed for this project 60 hr of labor for the cutting and routing operation, 120 hr for the sanding and joining operation, and 125 hr for the staining and finishing operation. Shakers would like to know how to meet next month's demand at minimum cost.

23. A shop with three furniture makers produces uniquely designed chairs and sofas from fabric and wood. The requirements for each piece are as follows:

	Fabric (yd)	*Wood (units)*	*Labor (hr)*
Chair	3	6	9
Sofa	8	5	4

For a week's operation, the shop has available 96 yd of fabric, 90 units of wood, and 120 hr of labor. A profit of \$70 is realized from the sale of a chair and a profit of \$60 from a sofa. Determine a weekly production schedule that maximizes profit.

(a) Formulate a mathematical model for this example.

(b) Show graphically that the solution to the problem calls for the manufacture of $7\frac{3}{11}$ chairs and $9\frac{3}{11}$ sofas.

(c) How would one implement such a schedule?

(d) Possible answer to part (c): Show that the weekly production of 8 chairs and 10 sofas, or of 8 chairs and 9 sofas, or of 7 chairs and 10 sofas is impossible within the given restrictions, but that 7 chairs and 9 sofas can be produced. Thus this feasible schedule, with integral components, appears to be the desired schedule.

(e) Show that it is also possible for the shop to produce weekly, within the given limitations, 10 chairs and 6 sofas. Compare the profit associated with this production schedule with the profit of the $(7,9)$ production schedule of part (d).

(f) *Conclusion*: Problems requiring integral answers may require special techniques.

2.4 The Transportation Model

Transportation problems were one of the first types of problems analyzed in the early history of linear programming. The general problem arises when goods available at several sources, such as warehouses or plants, must be shipped to various destinations, such as retail outlets or distribution centers. With fixed amounts available at the sources and fixed demands to be met at the destinations, the problem is to determine a shipping schedule that minimizes transportation costs. It is assumed that the costs of shipping goods from a source to a destination are directly proportional to the amount of goods shipped.

Example 2.4.1. A paper manufacturer having two mills must supply weekly three printing plants with newsprint. Mill 1 produces 350 tons of newsprint a week and Mill 2 550 tons. Plant 1 requires 275 tons/week, Plant 2 325 tons, and Plant 3 300 tons. The shipping costs, in dollars per ton, are as follows:

	Plant 1	*Plant 2*	*Plant 3*
Mill 1	17	22	15
Mill 2	18	16	12

The problem is to determine how many tons each mill should ship to each plant so that the total transportation cost is minimal.

To formulate the mathematical model, let x_{ij} denote the amount in tons to be shipped weekly from Mill i to Plant j, for $i = 1,2$ and $j = 1,2,3$. Then each x_{ij} must be nonnegative. Moreover, the amount shipped from each mill cannot exceed the supply. Thus we must have

$$x_{11} + x_{12} + x_{13} = 350$$
$$x_{21} + x_{22} + x_{23} = 550$$

We have equalities here, since the total weekly supply is equal to the total demand, and thus all the available newsprint at both mills must be shipped, leaving no surplus. So that the demands at each printing plant are met, the following equalities must be satisfied:

$$x_{11} + x_{21} = 275$$
$$x_{12} + x_{22} = 325$$
$$x_{13} + x_{23} = 300$$

The total shipping cost is $17x_{11} + 22x_{12} + 15x_{13} + 18x_{21} + 16x_{22} + 12x_{23}$, and it is this function of the six variables x_{ij}, $i = 1,2$, $j = 1,2,3$ that we wish to minimize.

One complication in problems of this type can occur when the commodity to be shipped is not divisible. For example, problems involving the shipment of automobiles, lawn tractors or refrigerators would require integral solutions. Moreover, integrally restricted variables have applications extending beyond such obvious situations. The following is an example.

Example 2.4.2. In the above transportation problem, suppose that the truck assigned to the Mill 2 to Plant 2 route is temporarily out of service, and that if this shipping link is to be utilized, a replacement vehicle must be rented, at a weekly rate of \$700. This is an example of a *fixed charge*. Now our model must consider both viable shipping schedules which do not use the Mill 2 to Plant 2 link and include only shipping costs in the cost function and schedules which use the link and therefore include the rental fee in the cost function. Note that if the link is used, the rental fee of \$700 is independent of the number of tons shipped; the product $700x_{22}$ cannot be simply added to the above cost function. What is needed is an "on/off" variable, a variable that is 1 if the link is used ($x_{22} > 0$) and 0 if the link is not used ($x_{22} = 0$). We can establish such a variable using integral restrictions. Using y to denote the variable, we add to the above constraints the restrictions $0 \le y \le 1, y \ge \frac{1}{325}x_{22}$, and y integral, and to the cost function the term $700y$.

The fixed charge is now properly accessed. The integrally restricted variable y can only equal 0 or 1, and the presence of the cost term $600y$ in the function to be minimized guarantees that y will be 0 if the constraints so allow ($x_{22} = 0$). But if $x_{22} > 0$, the inequality $y \ge \frac{1}{325}x_{22}$ forces y to be positive and therefore equal to 1. (Note that 325, the demand in tons of newsprint at Plant 2, is an upper bound for the possible values of x_{22}, so the quotient $\frac{1}{325}x_{22}$ will never exceed 1.)

We have an *integer programming problem.* The constraints and the function to be optimized are linear, and some (here only one) variables are restricted to be integral. Integer programming solution techniques are not as straightforward as one might hope. See, for example, Problem 23 of the previous section. Considerable study has been done in this field, and the theory has been found to have many applications. We will consider the topic in Chapter 6.

Problem Set 2.4

1. (a) Even though the optimization problem of Example 2.4.1 has six variables, the following elementary analysis does lead to the solution of the problem. It costs \$1/ton less to supply Plant 1 from Mill 1 instead of Mill 2. However, supplying Plant 2 from Mill 2 saves \$6/ton, and supplying Plant 3 from Mill 2 saves \$3/ton. The greatest relative savings comes from supplying Plant 2 as much as possible from Mill 2. Continuing this argument, show that the optimal shipping schedule has $x_{11} = 275$, $x_{12} = 0$, $x_{13} = 75$, $x_{21} = 0$, $x_{22} = 325$, $x_{23} = 225$.
   (b) Extend the above approach to determine a minimal-cost shipping schedule if the link from Mill 2 to Plant 2 is not used.
   (c) Use these results to determine the resolution of the fixed-charge problem of Example 2.4.2. Is the rental truck used?

Formulate mathematical models for the following problems. (Do not attempt to solve the problems.)

2. (a) A canned goods supplier has two warehouses serving four outlets. The East Coast Warehouse has 600 cases on hand and the West Coast Warehouse has 1000 cases on hand. The shipping costs, in cents per case, and the requirements for the four outlets, all located east of the Mississippi, are given in the following table.

	Outlet 1	*Outlet 2*	*Outlet 3*	*Outlet 4*
Shipping Costs				
East Coast Warehouse	20	16	30	20
West Coast Warehouse	45	39	50	44
Requirements (cases)	300	350	400	450

   Determine a shipping schedule that minimizes transportation costs.
   (b) As in part (a), but assume also that there are truck rental fees of \$50 if units are shipped from the East Coast Warehouse to Outlet 1 and \$60 if units are shipped to Outlet 2.

3. Three beverage plants supply five wholesale outlets with cases of soft drinks. The weekly production of each plant is as follows:

Plant 1	*Plant 2*	*Plant 3*
4000	2000	3000

The weekly demands and transportation costs (in cents per case) are as follows (dashed entries in the table indicate the impossibility of shipping cases between the corresponding plants and outlets):

	Outlets				
	1	2	*3*	4	5
Plant 1	6.2	—	5.1	10.1	8.0
Plant 2	6.5	10.5	4.3	11.3	6.5
Plant 3	6.3	9.0	—	10.8	—
Requirements (cases)	1000	1200	3000	400	2200

Suppose in addition that the weekly surplus at each plant can be sold locally for \$1.20/case at Plant 1, \$1.10/case at Plant 2, and \$1.14/ case at Plant 3. Determine a shipping schedule that minimizes transportation costs and takes into account the amount accrued from the sale of the surplus.

4. A commodity is to be shipped from three warehouses to four outlets, each outlet receiving 120 units. The shipping costs in dollars per unit are:

		Outlets			
		1	2	*3*	4
	1	12	15	10	25
Warehouses	2	10	19	11	30
	3	21	30	18	40

Warehouse 1 has available 100 units, Warehouse 2 150 units, and Warehouse 3 300 units. For any unit not shipped, there is a storage charge of \$6/unit at Warehouses 1 and 2 and \$12/unit at Warehouse 3. Moreover, because of labor contracts, Outlet 2 cannot receive more units from Warehouse 1 than from Warehouse 2, and Outlet 4 must receive at least half of its supply from Warehouse 3. Determine a minimal-cost shipping and storing schedule.

5. Three distribution centers supply four retail stores with a commodity. Each center has 150 units of the commodity on hand, and each store requires 100 units. Shipping costs in dollars per unit are:

		Retail Stores			
		1	2	*3*	4
	1	23	16	56	31
Distribution Centers	2	20	14	64	24
	3	21	19	58	28

No storage facilities exist at Center 1, so all of its units must be delivered. Undelivered units can be stored at Centers 2 and 3, but there is a \$3/unit storage charge at Center 3 (and no storage fee at Center 2). Determine a minimal-cost shipping and storing schedule.

6. Two sources supply three destinations with a commodity. Each source has a supply of 80 units, and each destination has a demand for 50 units. Shipping costs in dollars per unit are:

		Destinations		
		1	2	*3*
Sources	*1*	8	17	19
	2	–	21	22

The transportation costs from Source 2 to Destination 1 vary. The first 20 units shipped on this route cost \$10/unit, and each unit over 20 cost \$13/unit. Determine a minimal-cost shipping schedule.

7. Three distribution centers supply four retail stores with a commodity. The supplies at the centers, the demands at the stores, and the shipping costs (\$/unit) are as follows:

		Retail Stores				
		1	2	*3*	*4*	*Supplies*
	1	40	50	65	85	225
Distribution Centers	2	38	42	60	80	300
	3	35	54	55	76	375
	Demands	200	200	200	200	

All 225 units at Center 1 must be shipped. However, surplus units remaining at Center 2 may be sold at Center 2 for a profit of \$25/unit, and surplus units at Center 3 may be sold for a profit of \$27/unit for the first 30 sold and \$23/unit for any sold over 30. Determine a minimal-cost shipping schedule that takes into account the gain from the sale of the surplus units.

2.5 THE DYNAMIC PLANNING MODEL

The operation of many systems or processes can be divided into distinct time periods that allow for the flexibility of activities during each period and are such that decisions for one period affect not only that period but also subsequent periods. For example, the yearly operation of a giant steel plant can be divided into 12 monthly time periods. In each period labor, capital, and raw materials are combined to produce steel, with varying monthly demands for the product. At the beginning of each month, the amount of steel to be produced must be decided on. If demands are low, should employees be laid off or left idle, or should surplus steel be produced and stored? Should future high demands be met by immediately increased production and storage or by hiring and training personnel? If the cost and availability of raw materials vary due to, say, weather factors and mining conditions, then the decision on the amount of raw materials to purchase in any one month is influenced by present and future needs, storage capacity, and available capital.

Example 2.5.1. Consider the operation of a dealer of home heating oil. Suppose the dealer owns a storage tank with a capacity for 10,000 gal of oil that initially contains 3000 gal. Each month for the next 3 months the dealer can sell up to 8000 gal of oil per month, charging \$2.40/gal the first month, \$2.55/gal the second, and \$2.78/gal the third. Furthermore, the dealer can purchase up to 5000 gal of oil each month either for distribution during the month or for storage for later use. The cost to the dealer of this oil is \$2.17/gal the first month, \$2.29/gal the second month, and \$2.45/gal the third. The storage cost is 15 cents/gal for fuel stored at the end of any given month. How much oil should the dealer purchase, sell, and store during each month to maximize profits? Assume that any oil left in the storage tank after the third month has a value of \$2.05/gal.

To formulate a mathematical model, we must first, as before, assign variables to represent the amounts of each activity that the dealer performs. Since, at the beginning of each month, the dealer must decide on how much oil to buy, distribute, and store during that month, three variables will be needed for each period. In particular, let P_i denote the number of gallons of oil purchased by the dealer during Month i, where $i = 1,2,3$. Similarly, let D_i represent the number of gallons of oil distributed during Month i and S_i the number of gallons in storage at the end of the month.

There are many restrictions on these nine variables. Obviously, they must be nonnegative, and they all have fixed upper bounds, with $S_i \leq 10{,}000$, $P_i \leq 5{,}000$, and $D_i \leq 8{,}000$, for $i = 1,2,3$. However, the quantities are also interrelated. For each month, the oil purchased during that month plus the oil stored from the previous month must equal the total amount of oil delivered and stored during that month. Thus, for the first month, $3000 + P_1 = S_1 + D_1$. Similarly, for the next 2 months we must have

$$S_1 + P_2 = S_2 + D_2 \quad \text{and} \quad S_2 + P_3 = S_3 + D_3$$

The total profit for the operation is equal to the income from the oil sold plus the value of the oil left in the storage tank less the cost of the oil purchased and oil stored. This quantity, in dollars, is given by the following function, the function to be maximized:

$$\begin{aligned}&(2.40D_1 + 2.55D_2 + 2.78D_3) + 2.05S_3\\&\quad -(2.17P_1 + 2.29P_2 + 2.45P_3) - 0.15(S_1 + S_2 + S_3)\end{aligned}$$

Example 2.5.2. A shop must deliver 500 units of Q in Period 1, 650 in Period 2, and 625 in Period 3. The shop has two different processes that can be used to produce the Q's, each process using raw material M and labor. Input and output for 1 hr of operation of each are:

	Input		**Output**
	M (units)	*Labor (hr)*	*Units of Q*
Process X	8	3	25
Process Y	4	5	20

Each period the shop has available up to 175 units of M, but the material deteriorates quickly. Any units of M unused in one period cannot be saved for later use. The shop also has available each period up to 120 hr of regular-time labor at \$30/hr and another 40 hr of overtime labor at \$45/hr. Surplus units of Q made in one period may be stored for later delivery at a cost of \$25/unit-period, but space limitations restrict the number stored to be no more than 50/period, and no units are to remain in stock after the three periods. The shop pays only for the labor and storage space used. (The cost of the raw material is fixed by other constraints.)

To construct a model to be used to determine a production and storage schedule that minimizes labor plus storage costs, the shop manager first considers the decisions to be made. Each period she must determine the number of hours to run each process, the number of units of Q to be made and how many of them to be stored, and the allocation of the labor force. This suggests that we define the following set of variables:

x_i = number of hours that process X is used in Period i, $i = 1,2,3$
y_i = number of hours that process Y is used in Period i, $i = 1,2,3$
q_i = number of units of Q made in Period i, $i = 1,2,3$
s_i = number of units of Q stored at the end of Period i, $i = 1,2$
u_i = number of hours of regular-time labor used in Period i, $i = 1,2,3$
v_i = number of hours of overtime labor used in Period i, $i = 1,2,3$

With these variables at hand, the constraints for the model are:

for Period $i, i = 1,2,3$ $\quad 8x_i + 4y_i \le 175$ (units of M)
$3x_i + 5y_i = u_i + v_i, u_i \le 120, v_i \le 40$ (hr of labor)
$x_i, y_i, q_i, u_i, v_i \ge 0$ (nonnegativity)

for Period 1 $\quad 25x_1 + 20y_1 = 500 + s_1, 0 \le s_1 \le 50$ (units of Q)

for Period 2 $\quad 25x_2 + 20y_2 + s_1 = 650 + s_2, 0 \le s_2 \le 50$

for Period 3 $\quad 25x_3 + 20y_3 + s_2 = 625$

The cost function (in \$), the function to be minimized, is the sum of the costs accrued over the three periods: $30(u_1 + u_2 + u_3) + 45(v_1 + v_2 + v_3) + 25(s_1 + s_2)$

Example 2.5.3. A firm is given a short-term government contract to produce a total of 3700 units of some commodity over a period of 4 weeks. Producing the commodity can involve three new cost factors to the firm: the hiring of new workers, the purchasing of material for the commodity from an outside source, and the imposing of a penalty for late deliveries. The operation of the plant that keeps the total cost of these factors at a minimum is to be determined.

Specifically, the delivery schedule for the 3700 units is as follows:

End of Week	*1*	*2*	*3*	*4*
Number of Units	700	1200	1000	800

The plant has a stable workforce of 35. New workers can be hired but require a week of training, with one experienced worker capable of training five new workers each week. A worker can produce 25 units/week, but the trainees and trainers are not involved in production. All new workers hired for work on this contract cannot be retained by the firm, and so must be laid off by the end of the fourth week, if not sooner, at a cost of $125/worker. All workers receive $350/week.

One raw material is required in the production of this commodity, with each unit of the commodity needing 2 lb of the material. A subsidiary of the firm produces 1 ton of this material each week, but the material is perishable and can be used only during the week it is produced. However, the firm can also purchase an unlimited amount of the material on the market at a cost of $3/lb above its own production costs.

To ensure considerations for future contracts, the plant must deliver the 3700 units by the end of the fourth week. There is, however, a $5/unit/week penalty for all units not delivered on schedule. On the other hand, we will assume that there is no penalty or storage charge for any units delivered early.

To construct a mathematical model of this situation, the contingencies involving labor, raw materials, and the delivery schedule must all be considered. Moreover, these elements can and do vary from week to week, and so decisions on the operation of the plant must be made each week. Thus we have in essence four time intervals: the beginning of the first, second, third, and fourth weeks.

Let us first consider labor. Each week the firm must decide how to employ its labor force. The activities to be established at the beginning of each week are as follows:

Activity	*Denoted by, for Week i,* $i = 1,2,3,4$
1. New workers to be hired	H_i
2. Workers to be laid off	F_i
3. Workers to train and be trained	T_i
4. Workers to be idle	I_i
5. Workers to produce	P_i

Let M_i denote the number of pounds of raw material the firm should purchase during Week i from the outside source. Let D_i denote the number of units of the commodity produced and delivered during Week i. To measure the penalty costs for late deliveries, let U_i denote the accumulated number of units required but not delivered during Week i. Thus, if for some week the number of units delivered is less than the scheduled number required for that week plus any deficit accrued from previous weeks, the associated U_i would measure that difference. If the number delivered is greater than or equal to that sum, the U_i would be 0.

The constraints imposed by labor during the first week come from considering the employment of the total labor force and the training of the new workers. Note that F_i must be 0 because, at the beginning of the first week, there are no new workers who can be laid off. We have

$$35 + H_1 = T_1 + I_1 + P_1 \quad \text{and} \quad H_1 + \frac{H_1}{5} = T_1$$

Constraints imposed on the number of units produced are as follows:

$$\begin{aligned} 25P_1 &= D1 \\ 2000 + M_1 &\geq 2D_1 \\ D_1 + U_1 &\geq 700 \end{aligned}$$

The last inequality may need some clarification. If the optimal production schedule calls for D_1 to be less than 700, then U_1 will be the difference between 700 and D_1 and the quantity $D_1 + V_1$ will equal 700. On the other hand, if the optimal schedule calls for D_1 to be greater than 700, then U_1 will be zero and $D_1 + V_1$ will be greater than 700.

The production costs in dollars for the first week's operation are

$$350(35 + H_1) + 3M_1 + 5U_1$$

The constraints and costs for the next 3 weeks follow.
Second-week constraints:

$$\begin{aligned} 35 + H_1 + H_2 - F_2 &= T_2 + I_2 + P_2 \\ H_2 + \frac{H_2}{5} &= T_2 \\ F_2 &\leq H_1 \\ 25P_2 &= D_2 \\ 2000 + M_2 &\geq 2D_2 \\ D_2 + U_2 &\geq 1200 + (700 - D_1) \end{aligned}$$

Second-week costs:

$$350(35 + H_1 + H_2 - F_2) + 125F_2 + 3M_2 + 5U_2$$

Third-week constraints:

$$\begin{aligned} 35 + H_1 + H_2 - F_2 + H_3 - F_3 &= T_3 + I_3 + P_3 \\ H_3 + \frac{H_3}{5} &= T_3 \\ F_3 &\leq H_1 + H_2 - F_2 \\ 25P_3 &= D_3 \\ 2000 + M_3 &\geq 2D_3 \\ D_3 + U_3 &\geq 1000 + (1200 + 700 - D_1 - D_2) \end{aligned}$$

Third-week costs:

$$350(35 + H_1 + H_2 - F_2 + H_3 - F_3) + 125F_3 + 3M_3 + 5U_3$$

Fourth-week constraints:

$$35 + H_1 + H_2 - F_2 + H_3 - F_3 + H_4 - F_4 = T_4 + I_4 + P_4$$

$$H_4 + \frac{H_4}{5} = T_4$$

$$F_4 \leq H_1 + H_2 - F_2 + H_3 - F_3$$

$$25P_4 = D_4$$

$$2000 + M_4 \geq 2D_4$$

$$D_1 + D_2 + D_3 + D_4 = 3700$$

(all units must be delivered by the end of the fourth week)

Fourth-week costs:

$$350(35 + H_1 + H_2 - F_2 + H_3 - F_3 + H_4 - F_4) + 125F_3 + 3M_4$$

At the end of the fourth week, any worker hired for this project and still employed must be laid off. Let F_5 denote this number. Then $125F_5$ is the cost of this activity, while F_5 is given by the equation

$$F_5 = H_1 + H_2 + H_3 + H_4 - F_2 - F_3 - F_4$$

The function to be minimized (total cost) is given by the sum of these five costs. The nonnegative variables are restricted by all the above equalities and inequalities.

Problem Set 2.5

Problems 1 and 2 refer to Example 2.5.3.

1. The constraints relating the amount of raw material used and the number of units produced, $2000 + M_i \geq 2D_i$, $i = 1, 2, 3, 4$, are all inequalities. The use of an equality here would prevent the consideration of what flexibility in the firm's operation?
2. Intuitively, F_2 should be 0. Prove that any optimal solution to the resulting mathematical problem must have $F_2 = 0$. (*Hint*. Show that if $F_2 = k > 0$, then another solution can be found by letting $F_2 = 0$ and reducing H_1 by k. How does this change affect the value of the objective function?)

Formulate mathematical models for the following problems. (Do not attempt to solve the problems.)

3. An appliance dealer sells small refrigerators in the college market. This July, 25 units are on hand. For the next 3 months, the dealer can buy from the manufacturer up to 65 refrigerators each month, and can sell to the student population up to 100 units each month at the following prices:

Refrigerators	*Buy ($)*	*Sell ($)*
August	60	90
September	65	110
October	68	105

The dealer has storage facilities for 45 units but must pay a $7/unit/month storage charge for each refrigerator stored for sale in a subsequent month. Determine an optimal buying, selling, and storing plan.

4. Suppose the dealer in Problem 3 can also buy and sell microwave ovens. Each month up to 35 can be purchased, and up to 55 sold, at the following prices:

Ovens	*Buy ($)*	*Sell ($)*
August	150	200
September	175	250
October	200	240

The dealer presently has no ovens on hand. The ovens can be stored but, as above, the storage facility has space for at most 45 total units (either refrigerators or ovens or some combination) and storage costs remain $7/unit/month. Determine an optimal buying, selling, and storing program utilizing both the refrigerators and ovens.

5. For the next 3 months a dealer in Commodity A can buy from the producer and sell to the retailer units of A at the following prices per unit:

A	*Buy ($)*	*Sell ($)*
Month 1	31	40
Month 2	33	44
Month 3	36	48

During any particular month, the dealer can buy at most 450 units and sell at most 600 units and, moreover, can rent storage space from a local warehouse for up to 300 units at any one time at $2/unit/month. Determine an optimal buying, selling, and storing program, assuming the dealer has no units of A on hand initially and wants none on hand at the termination of the 3-month period.

6. (a) Suppose the agent in Problem 5 can also buy and sell Commodity B at the following prices per unit:

B	*Buy ($)*	*Sell ($)*
Month 1	80	95
Month 2	85	110
Month 3	95	125

The dealer can buy at most 200 units of B and sell at most 250 units during any one month and can also store B at the local warehouse, but space is limited. Assume that the warehouse has 300 yd^3 of space available, at $2/$yd^3$, and that a unit of A requires 1 yd^3 and a unit of B requires 2 yd^3. Again, the dealer has no stock on hand and wants none at the end of the 3 months. Determine an optimal buying, selling, and storing program utilizing both commodities.

(b) In the above problem, any units stored represent an investment of capital. Reconsider the problem, assuming that a maximum of $10,000 can be borrowed each month for this purpose, with an accompanying 2% per month interest rate.

7. A shop must deliver 300 units of A each month for the next 3 months. There are two different processes, each using labor and Raw Material M, which can be used to make the A's. Input and output for 1 hr of operation of each are:

	Input		***Output***
	M (lb)	*Labor (hr)*	*Units of A*
Process 1	6	3	12
Process 2	2	4	9

However, Process 2 is unavailable for use in the third and final month; it can be utilized only during the first 2 months. Each month the shop has available up to 110 lb of M at $36/lb and up to 95 hr of labor at $26/hr, but unused pounds of M and hours of labor in one month cannot be saved for use in a later month. However, extra units of A made in one month can be stored for later use at a cost of $20/unit-month. The shop pays only for storage, labor, and the raw material labor used. Determine a production schedule which minimizes total costs.

8. A subsidiary division of an automobile plant produces automobile engines. For the next four quarters, the demands of the plant are:

Quarter	*1*	2	3	4
Number of Engines	400	450	800	550

There is an initial inventory of 100 engines. The division can produce 475 engines in a quarter using its normal facilities. By the use of overtime, up to an additional 100 engines can be produced in any quarter, at a cost of $26/engine above the normal costs. Any engines on hand at the end of a quarter can be stored at a cost of $14/engine each quarter. Any quarterly demand not met by the division costs the main plant in underutilization $33/engine for each quarter of the deficiency. By the end of the fourth quarter, all the demands must be met. Determine an operating schedule that minimizes costs.

9. Using Material M, a firm produces Commodities A, B, and C, with the manufacture of a unit of A requiring 3 lb of M, a unit of B 7 lb, and a unit of C 12 lb.

For each of the next four quarters, the firm can sell up to 50 units of A, 40 units of B, and 30 units of C at the following prices:

Quarter	*1*	*2*	*3*	*4*
Units of A	\$20	\$24	\$ 26	\$ 32
Units of B	50	58	62	65
Units of C	90	95	105	120

Units of A, B, and C can be made and stored for later sale, with no storage costs, but a combined total of only 35 units can be stored over any one quarter. Each quarter, the firm can commit 700 lb of Material M to the production of A, B, and C. Unused M can also be stored for later use, but because of the volatile nature of the material, special storage facilities are required. At most 150 lb can be stored in the facility, and storage costs \$0.75/lb/quarter. Determine an operating schedule that optimizes profit for the coming year.

10. Using Material C, a firm produces Commodities A and B. The requirements for the manufacture of a unit of each are as follows:

Commodity	*Labor (hr)*	*C (units)*
A	3	5
B	2	8

The firm has available each month 400 hr of labor at a cost of \$12/hr and up to an additional 100 hr of overtime at \$18/hr. The firm has 3500 units of C in stock but can obtain no more. For the next 3 months, up to 100 units each of A and B can be sold each month at the following prices per unit:

	Price of A (\$)	*Price of B (\$)*
Month 1	60	50
Month 2	62	58
Month 3	64	65

The firm's warehouse, with a 3000-ft^3 capacity, can store units of A and B produced during one month for sale during a later month, but it must also store all the unused units of C. One unit of A requires 8 ft^3 of space, one unit of B 9 ft^3 of space, and one unit of C 0.8 ft^3 of space. At the end of the 3-month period, any units of C left in stock must be disposed of at a cost of \$1.50/unit. Determine an operating schedule that maximizes profit.

11. During the next 3 months, FMC must meet the following demands for vans and light-duty trucks: Month 1, 400 vans and 200 trucks; Month 2, 150 vans and 150 trucks; Month 3, 300 vans and 225 trucks.

- Because of plant capacities, each month at most a total of 520 vehicles can be produced.

- Each van uses 0.8 ton of steel, and each truck uses 1.3 tons. During Month 1, steel costs \$575/ton; during Month 2, \$625/ton; and during Month 3, \$650/ton. At most 500 tons of steel may be purchased each month and because of limited storage space, steel may be used only during the month it is purchased.
- At the beginning of the 3-month period, 100 vans and 50 trucks are in inventory.
- At the end of Month 1 and at the end of Month 2, vans and/or trucks on hand but not delivered may be stored for future delivery. Storage costs are \$150/vehicle-month for the first 40 trucks stored in a given month and \$175/vehicle-month for any number over 40.
- At the end of the 3-month period, the company wants no vehicles in inventory.

Determine a production and storage schedule that minimizes the steel plus storage cost.

2.6 SUMMARY

Now that we have seen some examples of problems leading to linear programming models, it should be emphasized that our list of categories and examples is by no means comprehensive, nor are our categories mutually exclusive. Many types of problems occurring in the world and amenable to linear programming techniques are not confined to the categories we have described. Moreover, our examples were somewhat straightforward. Real-life problems, usually with a multitude of interrelated components to be considered, tend to be much more complicated and may lend themselves to several different approaches.

However, all the examples we have considered led to the same mathematical problem: that of finding the maximum or minimum of a linear function on a set determined by a family of linear equations or inequalities. This is the basic problem of linear programming: the optimization of a linear function subject to a system of linear constraints.

If one is to establish a linear programming model for some real-life optimization problem, the system or operation under study must be amenable to some basic assumptions. First, the system must be decomposable into a number of elementary operations called *activities*. An activity is usually a transformation process that converts inputs such as labor or raw materials into the product of the operation, such as the manufactured goods. For example, in the feed problem of Example 2.2.1, an activity is the process of converting feed into three nutritional elements. For the paper manufacturer problem of Example 2.4.1, an activity is the transportation of newsprint from a mill to a printing plant. It is these activities that are combined in varying amounts to attain the stated objective. The amount or rate at which an activity operates or functions is called its *activity level*.

Second, the objective of the entire operation, when measured in terms of the activity levels, must be a linear function. This means simply that if x_j measures the level of activity j, there are constants c_j such that the product c_jx_j measures the attainment of the objective from the operation of activity j and such that the total output of the operation, if there are n activities, is given by the sum $c_1x_1 + c_2x_2 + \cdots + c_nx_n$. Thus the objective function is a linear function of the variables x_j.

Third, the restrictions on the various input and output items and the requirements on the relationships between these items must be linear in form, that is, given by either linear equations or inequalities in the variables x_j. Thus the component processes or activities of the problem must be linear, and so, for example, doubling the quantities of all the items input into an activity must have the effect of doubling the output of the activity. In Example 2.3.1, dealing with the production of small boats, the activity of producing a boat depends linearly on the input of raw material, machine time, and labor.

Certainly real-life situations can fail to satisfy the requirements listed above, but can still be open to a linear programming model that provides accurate and useful information. One simplifying principle is to neglect incidental details. For example, the paper manufacturer of Example 2.4.1 can probably realistically ignore the quantity of newsprint lost in shipment from mill to plant. Also, the linearity restrictions may only be an approximation of the situation at hand, but at least it would still lead to a good first estimate of the desired solution. For example, in transportation problems, the cost of shipping a unit of goods may decrease if the volume of goods shipped is increased, due to a more economical use of available equipment or due to the accessibility of other means of transportation suitable for large shipments. The problems that we have considered belong to what is called the *deterministic class*; that is, they have involved no uncertainty. For example, the output of newsprint at the two paper mills and the requirements for newsprint at the three printing plants are stated precisely. However, few real-life situations can be predicted with such certainty. One approach here is to work with average values for the quantities under study. Another approach is to develop a probabilistic or *stochastic* model. An elementary example of such a model is given in Section 8.1.

The actual construction of a mathematical model of a real-life problem involves several steps, steps that the reader may have already recognized in doing the examples and exercises of the preceding sections. First, the entire operation under study must be decomposed into its component activities. Then the items and units used to measure the activity levels must be determined. It is the rate of these activity levels that is subject to our control and that is represented by the variables in the problem. Finally, the objective function, the function to be optimized, and the linear constraints must be identified. The constraints on the system are usually identified by consideration of the input and output items in the system and the restrictions and relationships between them. And although it may seem that the execution of these steps may become somewhat straightforward, that is not the case. Experience with programming techniques and understanding of the real-life problem under study are necessary in order to be able to distinguish the significant elements of the problem from the inconsequential ones and to interpret and employ any solution found for the problem properly.

The goal of the next chapter will be to develop a mechanical technique for solving the general mathematic problem resulting from the execution of the above steps. This general problem is called the *basic problem* of linear programming.

Problem Set 2.6

Formulate mathematical models for the following problems. (Do not attempt to solve the problems.)

1. A soap manufacturer uses 1200 gal of Mineral Oil A and 2000 gal of Mineral Oil B weekly. These oils can be obtained from three products. The yield and cost of a drum of each are as follows:

Product	*Oil A (gal)*	*Oil B (gal)*	*Cost ($)*
1	10	15	13
2	9	16	7
3	12	25	8

   Supplies of these products are unlimited. However, Products 2 and 3 require special processing to separate out the desired mineral oils, with each drum of Product 2 requiring 1 hr of processing and each drum of Product 3 requiring 2 hr. Sixty hours of processing time are available weekly at $12/hr, and another 12 hr are available at $16/hr. Determine what combination of these products should be used to meet the weekly demands and minimize purchase plus processing time costs.

2. A farmer must determine a plan to feed his stock during the coming winter. He has two types of stock, each with distinct nutritional requirements. To feed the stock the farmer has available 1000 lb of grain harvested over the summer. However, this supply of grain is not adequate to meet the nutritional demands of the entire stock over the winter, and so the farmer must supplement this supply with feeds purchased from the local coop. Determine a feeding plan that utilizes all the available grain, satisfies the nutritional demands of the stock, and minimizes the amount spent on the supplementary feeds. The data follow.

   Nutritional demands (minimal number of units required/winter):

	Element A	*Element B*	*Element C*
Stock 1	150	360	650
Stock 2	90	700	450

   Nutritional contents (units/lb) and costs (cents/lb) of the grain and the two available feeds:

	Element A	*Element B*	*Element C*	*Cost*
Grain	0.2	0.9	0.8	0
Feed 1	1	5	10	15
Feed 2	3	7	13	23

3. A pet food manufacturer produces weekly 600 lb of dog food and 250 lb of cat food. The foods must contain minimal percentages of four nutritional elements, A, B, C, and D, as follows:

	A	*B*	*C*	*D*
Dog Food	10	5	8	8
Cat Food	8	16	5	12

The manufacturer has available six different meat by-products to combine to meet these demands, but only limited quantities of some are available. The available weekly supply, costs, and contents of the by-products are as follows:

	Supply	*Cost*	***Contents % Nutritional Element***			
By-product	*(lb/week)*	*(cents/lb)*	*A*	*B*	*C*	*D*
1	*Unlimited*	33	10	0	5	8
2	300	29	8	20	6	10
3	500	30	15	14	8	10
4	400	28	6	12	10	15
5	*Unlimited*	37	0	18	13	20
6	200	23	12	10	4	6

Determine how much of each by-product should be combined in the production of each food so that total costs are minimized.

4. A firm combines Raw Materials A, B, and C in the production of two products. The requirements (in pounds) for the manufacture of a unit of each product are as follows:

	Raw Material		
	A	*B*	*C*
Product 1	4	12	8
Product 2	7	9	10

The firm has available 1 ton of A, 2 tons of B, and $1\frac{1}{2}$ tons of C. All units of the products made can be sold. The firm realizes a profit of \$1.20/unit on the first 200 units of Product 1 sold and \$1/unit on the remainder sold; and a profit of \$1.40/unit on the first 150 units of Product 2 sold and \$1.05/unit on the remainder. Determine a production schedule that maximizes profit.

5. The firm of Problem 4 has changed owners. Because of a subsequent expanded market, the firm can now sell all units produced at the fixed prices of \$5.55/unit for Product 1 and \$6.30/unit for Product 2. However, now the firm must purchase the necessary raw materials from outside sources, and the costs, which must be considered in determining profits, vary. The firm can purchase up to 1 ton of Raw Material A at 20 cents/lb and any amount over 1 ton at 35 cents/lb; up to 2 tons of B at 10 cents/lb and any amount over at 20 cents/lb; and up to $1\frac{1}{2}$ tons of C at 15 cents/lb and any amount over at 25 cents/lb. Determine a production schedule that maximizes net profit, assuming that there is an additional overhead cost to the firm of \$1/unit for each unit of Product 1 and of Product 2 produced.

6. A machine shop assembles transuniversals for sale to the local automobile plant. Because of high demand, all units assembled can be sold for \$425/unit. Three major components, C_1, C_2, and C_3, are required in the assembly of each transuniversal, and the shop can either purchase these components from outside sources or manufacture the components themselves. In the internal manufacture of the components, and also in the final assembly of the transuniversals, labor and machine time on two machines, M_1 and M_2, are required. The requirements are as follows:

	Labor (hr/unit)	*Machine Time M_1 (min/unit)*	*Machine Time M_2 (min/unit)*
Transuniversal Assembly	7	35	25
Manufacture of C_1	0.3	10	8
Manufacture of C_2	0.5	15	20
Manufacture of C_3	1.0	13	12

The shop has available each week 1600 hr of labor at \$20/hr and 400 hr of overtime at \$30/hr; and 180 hr of machine time for M_1 and 200 hr for M_2. The costs per unit of the components, if purchased externally or produced internally (internal costs exclude labor costs), are as follows:

	C_1 (\$)	*C_2 (\$)*	*C_3 (\$)*
Purchase Price	65	81	73
Production Cost	49	60	50

Determine a production schedule that maximizes net income.

7. A poultry producer has available 112 rods^2 of land on which to raise during the next 12-week period chickens, ducks, and turkeys. The space and labor requirements and the profit — excluding labor costs — from the sale after the 12-week breeding period are as follows:

	Space *(rods^2/unit)*	*Labor* *(hr/week/unit)*	*Profit ($/unit)* *(excluding labor costs)*
Chickens	1.2	3	260
Ducks	1.0	2	172
Turkeys	0.8	1	88

The producer has available each week 200 hr of labor at $13/hr and up to 45 hr of overtime at $18/hr. What stock should the producer raise over the 12-week period in order to maximize net income (profits less labor costs)?

8. Labor, Material M, and units of Q are used to produce Products A and B, with requirements for a unit of each as follows:

	Labor (hr)	*Material M (lb)*	*Units of Q*
Unit of A	2	5	2
Unit of B	4	4	1

For the next month, up to 1000 hr of labor at $25/hr and 1 ton of M at $12/lb can be used in the production of A and B. Units of Q can be purchased externally for $50/unit and can be assembled internally, with each unit assembled requiring 1 hr of labor and parts which cost $15. A and B each sell for $350/unit. How many of each should be made, and how should the necessary units of Q be generated so that net profits are maximized?

9. A shop is responsible for making and delivering 225 differentials each month for the next 4 months. Manufacture of a differential requires 2 hr of labor and 3 units of A. Each month the shop has available 400 hr of labor at $18/hr, 150 hr of overtime at $26/hr, and an unlimited supply of A. However, the cost of a unit of A increases from month to month and is as follows:

Month	*1*	*2*	*3*	*4*
Cost ($)	12	17	25	26

Any differentials above 225 made in one month can be stored for later delivery at a cost of $10/month/unit. Determine a minimal cost production plus storage schedule.

10. A dealer supplies daily 500 units of Commodity C to an outlet in City A and 800 units to an outlet in City B. To meet these demands, the dealer can buy units of C from a distant manufacturer (at $8/unit) and then ship the units to the outlets. Delivery of a single unit requires the use of a long-distance vehicle (1000 are available daily with a use cost of $20/vehicle); to City A, 12 hr of labor/unit delivered, and to City B, 14 hr of labor/unit delivered. The dealer can also purchase the commodity from wholesalers in each city. The wholesaler in City A charges $86/unit, and 3 hr of labor/unit are required for the distribution of C from this wholesaler to the outlet in A. The wholesaler in City B charges \$94/unit, with 4 hr of labor/unit required for distribution to the outlet in B. The dealer has available daily 14,140 hr of labor at \$12/hr. Determine a purchase schedule that minimizes the cost of meeting the demands.

11. A firm must meet the demands of seven markets for a commodity. The firm has three plants at which limited amounts of the commodity can be produced, and the firm can also buy unlimited amounts of the finished commodity from an outside source. The production capabilities of the plants, the demands of the markets, and the total costs (production or purchase costs plus transportation costs) of supplying the markets from the varying sources are given the following tables. Determine a supply schedule that minimizes overall costs.

Supply

Plant	*1*	*2*	*3*
Number of Units	700	600	400

Demand

Market	*1*	*2*	*3*	*4*	*5*	*6*	*7*
Number of Units	150	300	425	325	200	250	250

Costs (\$/unit)

	Markets						
	1	*2*	*3*	*4*	*5*	*6*	*7*
Plant 1	5	7	8	7	9	12	3
Plant 2	13	16	10	12	14	18	9
Plant 3	10	12	11	9	13	14	7
Outside Source	21	25	35	26	27	38	20

12. A refinery produces two grades of gasoline, regular and premium, by blending three different crude oils, *A*, *B*, and *C*. Two ingredients, α and β, are critical in each grade. Regular grade must contain at least 20% of ingredient α and no more than 60% of ingredient β; premium grade must contain at least 30% of ingredient α and no more than 55% of ingredient β. Next week the refinery needs exactly 2000 gal of regular and 1000 gal of premium. These data are summarized in the following table.

	α	β	*Gallons Required*
Regular	$\geq$ 20%	$\leq$ 60%	2000
Premium	$\geq$ 30%	$\leq$ 55%	1000

The composition, available supply, and cost of each of the crude oils A, B, and C which can be blended to produce these two products are as follows:

	Composition of α	*Composition of β*	*Supply (gal)*	*Cost ($/gal)*
Oil A	30%	65%	1800	3.50
Oil B	45%	50%	800	5.50
Oil C	15%	55%	—	—

Oil C is available from two different suppliers; the first supplier can provide up to 500 gal of oil C at $3/gal, and the second supplier can provide an unlimited amount of the crude oil, but at $4/gal. The company would like to know how to meet next week's demands at minimum cost.

13. Using Raw Material A and units of Part B, firm LP Inc. makes units of three different products, with the requirements and selling price per unit of each product as follows:

	A (lb)	*B (units)*	*Selling Price ($)*
Product 1	5	10	95
Product 2	3	12	85
Product 3	6	9	100

For the next week, LP Inc. must produce at least 450 total units of Products 1, 2, and 3. The firm has available 2000 lb of A at $5/lb and another 700 lb at $7/lb and only pays for the pounds of A used. Part B is a by-product of another operation of the firm, and at this time 5000 units of B are available at no cost. However, any units of B not used in this production can be sold for a gain of $3/unit. Management would like to know how many units of each of the three products the firm should make and sell in order to optimize net profit.

14. A wholesale clock company produces two models, a floor model (the grandfather) and a wall model (the cuckoo). Each clock consists of two main components, the power mechanism and the case. Labor is required in the manufacture of each of these components and in the final assembly of the components to make the clock. The time requirements, along with the cost to the firm of the completed unit, are given in the following table. These cost estimates include the cost of the parts involved and the cost, computed at the regular pay rate, of labor.

	Mechanism (hr)	*Case (hr)*	*Assembly (hr)*	*Cost/unit ($)*
Floor Model	3.00	1.75	5.0	260
Wall Model	2.00	1.25	3.5	190

The firm can also subcontract out the manufacture of any number of cases of either or both types to a local shop. This saves labor (the time required to manufacture the cases) but increases the overall cost of the units. Floor model cases used from this outside source in the final assembly of the grandfather add $25 to the total cost of production of the model ($260 → $285), and the use of the shop's wall model case adds $20 ($190 → $210) to final costs. Also, the firm can buy from a competitor any number of the complete wall model clocks at a price of $255 each.

Next month, the firm must deliver 180 floor model units and 300 wall model units. To meet these demands, they have available 3000 hr of regular-time labor and up to an additional 650 hr of overtime, at a cost of $10/hr above regular-time pay. How can the firm meet these demands at minimum cost?

15. A company has plants P_1, P_2, and P_3 which produce units of Z needed at assembly centers C_1, C_2, C_3, and C_4. The annual output capacities of the plants and demands at the assembly centers are:

*Annual **Output** of Units of Z*		
P_1	P_2	P_3
10,500	18,800	13,200

*Annual **Demand** for Units of Z*			
C_1	C_2	C_3	C_4
7,700	9,900	12,200	11,100

The units can be delivered from the plants to the centers either by truck or by rail. However, for each route for which units are delivered by rail there is an annual route lease fee, independent of the number of units shipped through the route. The data follow.

			To Center			
	From Plant	*By*	C_1	C_2	C_3	C_4
Delivery Cost *($/unit)*	P_1	*Truck*	60	80	50	30
		Rail	35	62	22	25
	P_2	*Truck*	95	120	65	75
		Rail	75	89	45	55
	P_3	*Truck*	110	98	77	88
		Rail	53	35	32	38

		To Center			
	From Plant	C_1	C_2	C_3	C_4
Route Lease Fee *(in $1000)*	P_1	165	220	175	150
	P_2	250	200	220	210
	P_3	180	190	200	170

Determine a minimum cost delivery schedule for the next year; that is, for each plant and each center, determine how many units are to be shipped from the plant to the center by truck, and how many by rail, so that supplies are not exceeded, demands are met, and total cost is minimized.

16. A firm supplies six outlets with two commodities, A and B, produced at three plants. The transportation costs vary from plant to market and are also dependent on the commodity being shipped. Moreover, there are upper-bound capacities on the combined total number of units that can be shipped from each plant to each market. Determine if it is possible to meet the demands with the supplies given the restrictions on the shipping capacities and, if so, determine a minimal cost shipping schedule. The data are:

Supplies

	Plant 1	Plant 2	Plant 3
Units of A	1000	2500	1500
Units of B	1400	1500	1100

Demands

	Outlet 1	Outlet 2	Outlet 3	Outlet 4	Outlet 5	Outlet 6
Units of A	1200	600	1100	1000	500	600
Units of B	800	800	1000	500	300	600

Shipping Costs (cents/unit)

From Plant	*Commodity*	*To Outlet* 1	2	3	4	5	6
1	A	26	35	27	32	23	40
	B	17	19	13	20	12	25
2	A	48	56	70	45	55	60
	B	20	32	45	25	30	32
3	A	30	39	–	40	35	32
	B	21	32	–	33	29	25

Shipping Capacities

From Plant	*To Outlet* 1	2	3	4	5	6
1	600	500	1000	500	300	650
2	600	500	2000	2000	350	450
3	800	500	0	2000	250	400

CHAPTER 3

THE SIMPLEX METHOD

3.1 THE GENERAL PROBLEM

In the previous chapter, all examples led to one basic mathematical problem: the optimization of a linear function subject to a system of linear constraints. In this chapter we will develop a technique for solving this basic problem.

One minor complication in studying the problem is that the optimization problem can take various forms. For example, we have seen both maximization and minimization problems and constraint sets that have consisted of equalities and inequalities in both directions. However, this difficulty is easily resolved because all linear programming problems can be transformed into equivalent problems that are in what we call *standard form*.

Definition 3.1.1. The *standard form* of the linear programming problem is to determine a solution of a set of equations

$$\begin{aligned} a_{11}x_1 + a_{12}x_2 + \ldots + a_{1n}x_n &= b_1 \\ a_{21}x_1 + a_{22}x_2 + \ldots + a_{2n}x_n &= b_2 \\ &\vdots \\ a_{m1}x_1 + a_{m2}x_2 + \ldots + a_{mn}x_n &= b_m \end{aligned} \tag{3.1.1}$$

with

$$x_j \geq 0, j = 1, \ldots, n$$

that minimizes the function

$$z = c_1x_1 + c_2x_2 + \cdots + c_nx_n - z_0$$

(The $-z_0$ term allows for the inclusion of a constant in the expression for the function to be optimized. In an application such a constant could represent, for example, fixed costs or guaranteed benefits. We precede the constant with a negative sign for future convenience; z_0 can be positive, negative, or zero.)

It is this standard form of the linear programming problem, a minimization problem involving only equalities, that we will solve. Thus our first task is to show that any linear programming problem can be formulated as a problem in standard form, where the number of equalities, m, and the number of variables, n, are determined by the problem.

Consider first a linear programming problem with a system of constraints that contains inequalities. For example, suppose a particular diet problem reduces to the mathematical problem of minimizing $3x_1 + 2x_2 + 4x_3$ subject to the constraints

$$\begin{aligned} 30x_1 + 100x_2 + 85x_3 &\leq 2500 \\ 6x_1 + \quad 2x_2 + \quad 3x_3 &\geq \quad 90 \\ x_1, x_2, x_3 &\geq 0 \end{aligned}$$

Such a problem could result from seeking minimal-cost diet that places an upper bound on calorie intake and a lower bound on protein intake. We will show that this problem is equivalent to the following problem derived from the original problem by the addition of two new nonnegative variables, x_4 and x_5.

$$\begin{aligned} &\text{Minimize } 3x_1 + 2x_2 + 4x_3 \\ &\text{subject to} \\ &30x_1 + 100x_2 + 85x_3 + x_4 \qquad = 2500 \\ &\;\;6x_1 + \quad 2x_2 + \quad 3x_3 \qquad - x_5 = \quad 90 \\ &x_1, x_2, x_3, x_4, x_5 \geq 0 \end{aligned}$$

Notice that if $(x_1^*, x_2^*, x_3^*, x_4^*, x_5^*)$ is a solution to the second constraint set, then, since x_4^* and x_5^* are restricted to nonnegative values, $30x_1^* + 100x_2^* + 85x_3^* = 2500 - x_4^* \leq 2500$ and $6x_1^* + 2x_2^* + 3x_3^* = 90 + x_5* \geq 90$. Therefore (x_1^*, x_2^*, x_3^*) is a solution to the first constraint set. Similarly, if (x_1^*, x_2^*, x_3^*) is a solution to the first constraint set, there exist x_4^* and x_5^* [let $x_4^* = 2500 - (30x_1^* + 100x_2^* + 85x_3^*)$ and $x_5^* = 6x_1^* + 2x_2^* + 3x_3^* - 90$] that are nonnegative and such that $(x_1^*, x_2^*, x_3^*, x_4^*, x_5^*)$ is a solution to the second constraint set. Thus solutions of the two constraint sets correspond, with corresponding solutions having the same first three coordinates. At the same time, the form to be minimized, $3x_1 + 2x_2 + 4x_3$, depends only on the first three coordinates. Hence the minimal value of the linear function for both problems will be the same, and points where this minimum is achieved for one problem will correspond to points with this same property for the other problem.

Clearly, this technique generalizes. Given any problem with a system of constraints containing inequalities, by adding additional nonnegative variables, an equivalent problem can be formulated with a constraint system consisting only of equalities. The number of variables added would equal the number of inequalities in the system of constraints. The variables added are called *slack variables*. In fact, they usually can be interpreted as measuring the slack or surplus of the items or requirements of the problem. For example, in the preceding diet problem, suppose the first restriction comes from consideration of the calorie intake and the second from the protein intake. Then, for a fixed diet, the slack variable x_4 measures the number of calories below the maximum calorie requirement, and x_5 measures the number of units of protein above the minimum protein requirement for that diet.

Second, suppose a linear programming problem seeks to maximize the linear function $c_1x_1 + c_2x_2 + \cdots + c_nx_n$. But the problem of maximizing this function is equivalent to the problem of minimizing its negative: $-c_1x_1 - c_2x_2 - \cdots - c_nx_n$.

Thus a maximization problem can be easily formulated as a minimization problem by multiplying the function to be optimized by (-1).

The last restriction on the standard form of the linear programming problem is that all the variables be nonnegative. For most problems this restriction comes naturally from the physical interpretation of the variables. In all the examples we have considered, the variables could assume only nonnegative values. However, for some complicated production systems involving various processes and options, it could be that some commodity that is input for some process is output for another, and it is not clear whether this commodity will be input or output in the optimal operation of the system. Thus we may wish to formulate the problem with a variable not restricted in sign. (Problems with unrestricted variables also appear when discussing duality, as we will see in Chapter 4.)

Suppose that x_1 is a variable unrestricted in sign for a linear optimization problem. However, any number can be written as the difference of two (not unique) nonnegative numbers. (For example, $7 = 7-0 = 8-1, -7 = 0-7 = 1-8$.) Hence we can introduce into the problem two nonnegative variables, say x_1' and x_1'', and replace x_1 everywhere in the problem with the difference $x_1' - x_1''$. This will give an equivalent problem with the unrestricted variable replaced by two nonnegative variables.

As a result of these methods, for any linear programming problem, an equivalent problem can be constructed that is in standard form.

Example 3.1.1. The problem of maximizing $3x_1 - 2x_2 - x_3 + x_4 - 87$ subject to

$$\begin{aligned}
4x_1 - x_2 + x_4 &\le 6 \\
-7x_1 + 8x_2 + x_3 &\ge 7 \\
x_1 + x_2 + 4x_4 &= 12 \\
x_1, x_2, x_3 \ge 0,\ x_4 &\text{ unrestricted}
\end{aligned}$$

is equivalent to

$$\begin{aligned}
&\text{Minimize } -3x_1 + 2x_2 + x_3 - (x_4' - x_4'') + 87 \\
&\text{subject to} \\
&4x_1 - x_2 + x_4' - x_4'' + x_5 = 6 \\
&-7x_1 + 8x_2 + x_3 - x_6 = 7 \\
&x_1 + x_2 + 4x_4' - 4x_4'' = 12 \\
&x_1, x_2, x_3, x_4', x_4'', x_5, x_6 \ge 0
\end{aligned}$$

In a linear programming problem, the function to be optimized is called the *objective function*. Any point $(x_1, x_2, \ldots, x_n)$ with nonnegative coordinates that satisfies the system of constraints is called a *feasible solution* to the problem. For a particular problem, a feasible solution can be interpreted as a way of operating the system under study so that all of the requirements are fulfilled, that is, as a feasible way of operation.

Thus our basic problem is to determine, from among the set of all feasible solutions, a point that minimizes the objective function. Moreover, to be able to handle involved real-life problems, we need a solution algorithm easily programmed for computer use. Existence theorems derived from, say, the theory of continuous functions on compact sets or the theory of linear functions on convex sets, although mathematically quite attractive, do not provide an efficient means for actually finding a desired solution.

The method that will be developed in this chapter for solving the basic linear programming problem is called the *simplex method.* It is credited to George Dantzig [4], and this method and its various modifications remain among the primary means used today to solve linear optimization problems. One additional feature of this method that is useful for practical application and also very attractive mathematically is that the method can handle exceptional cases. For example, the method can determine if a problem has, in fact, any feasible solutions and, if so, whether the objective function actually assumes a minimum value.

The basic step in the simplex method is derived from the pivot operation used to solve linear equations. In the next section we pause briefly from our consideration of the standard linear programming problem to consider linear equations.

Problem Set 3.1

1. (a) In Example 3.1.1, $x_1 = 4, x_2 = 12, x_3 = 0, x_4' = 21, x_4'' = 22, x_5 = 3, x_6 = 61$ is a solution to the second constraint set. Find the corresponding solution to the first constraint set.

   (b) Conversely, $x_1 = 1, x_2 = 3, x_3 = 5, x_4 = 2$ is a solution to the first constraint set. Find a corresponding solution to the second. In this case, is your answer unique?

2. Explain why the following constraint sets are not equivalent.

Set A	Set B
$x_1 + x_2 \leq 6$	$x_1 + x_2 + x_3 = 6$
$x_1 + 2x_2 \leq 10$	$x_1 + 2x_2 + x_3 = 10$
$x_1, x_2 \geq 0$	$x_1, x_2, x_3 \geq 0$

   Hint. $x_1 = 3$ and $x_2 = 3$ satisfy the inequalities of Set A. Can you find an x_3 such that $(3, 3, x_3)$ satisfies the equalities of Set B?

   This shows that when introducing slack variables, the same variable cannot be used for different inequalities.

3. Put the following problems into standard form.

   (a) Maximize $3x_1 - 2x_2$
   subject to

$$
\begin{aligned}
5x_1 + 2x_2 - 3x_3 + x_4 &\leq 7 \\
3x_2 - 4x_3 &\leq 6 \\
x_1 + x_3 - x_4 &\geq 11 \\
x_1, x_2, x_3, x_4 &\geq 0
\end{aligned}
$$

(b) Minimize $x_2 + x_3 + x_4$
subject to

$$\begin{aligned} x_1 + x_2 \qquad\qquad\quad & \geq 6 \\ x_2 + x_3 - x_4 & \leq 1 \\ 5x_1 - 6x_2 + 7x_3 - 8x_4 & \geq 2 \end{aligned}$$

$x_1 \geq 0, x_2 \leq 0, x_3, x_4$ unrestricted

(c) Minimize $x_1 + x_3 - x_4 + 48$
subject to

$$\begin{aligned} -3x_1 + x_2 - x_3 + 2x_4 & = -50 \\ x_1 - x_2 \qquad + x_4 & \leq 100 \\ 2x_2 - x_3 - x_4 & \geq -150 \end{aligned}$$

$x_1, x_2, x_3, x_4 \geq 0$

(d) Maximize $6x_1 - 2x_2 + 9x_3 + 300$
subject to

$$\begin{aligned} 2x_1 - 6x_2 - x_3 & \leq 100 \\ x_1 + x_2 + 9x_3 & \leq 200 \end{aligned}$$

$0 \leq x_1 \leq 50, x_2 \geq -60, x_3 \geq 5$

(e) Minimize $6x_1 + x_2$
subject to

$$\begin{aligned} -5x_1 + 8x_2 & \leq 80 \\ x_1 + 2x_2 & \geq 4 \end{aligned}$$

$x_1 \leq 10, x_2 \geq 0$

(f) Maximize $x_1 + 2x_2 + 4x_3$
subject to

$|4x_1 + 3x_2 - 7x_3| \leq x_1 + x_2 + x_3$

$x_1, x_2, x_3 \geq 0$

(g) Maximize $x_1 + 6x_2 + 12x_3$
subject to

$-x_1 - x_2 + x_4 \geq$ maximum of $7x_1 + 2x_2$ and $5x_2 + x_3 + x_4$

$x_1, x_2, x_3, x_4 \geq 0$

(h) $-x_1 - x_2 + 2x_3 + x_5$
subject to

$$\begin{aligned} x_1 + 7x_2 + 16x_3 & \leq 4x_4 + x_5 \\ x_3 + 12x_4 & \geq x_1 + 6x_2 \\ 9x_5 & \leq x_2 + 3x_4 \end{aligned}$$

$x_1, x_2, x_3, x_4, x_5 \geq 0$

4. Determine all feasible solutions to the linear programming problem of Problem 3(a) for which

(a) $x_1 = x_2 = x_4 = 0$
(b) $x_2 = 0, x_3 = 6$
(c) $x_3 = 0$

5. Many times the amount of slack or surplus of a commodity enters into the initial formulation of the problem; it is a factor in the function to be optimized. For example, in a production problem, there could be a cost associated with the storage of the surplus production of a commodity. For another example, formulate the mathematical model for the following.

Two warehouses supply two retail outlets with 100-lb bags of lime. Warehouse A has 1000 bags, and Warehouse B has 2000 bags. Both outlets need 1200 bags. The transportation costs in cents per bag are given in the following table.

From	*Outlet 1*	*Outlet 2*
Warehouse A	5	4
Warehouse B	12	9

However, there is a storage charge of 2 cents/bag for all bags left at Warehouse A and 8 cents/bag for those left at Warehouse B. Determine a shipping schedule that minimizes the total cost.

6. In the text it was suggested that when putting a linear programming problem with unrestricted variables into standard form, each unrestricted variable is to be replaced by a pair of nonnegative variables. Actually, this method is inefficient if the problem has more than one unrestricted variable; we need introduce only one additional variable to handle all the unrestricted variables. For example, if a problem has unrestricted variables x_1 and x_2, show that replacing x_1 with $x_1' - x_0$ and x_2 with $x_2' - x_0$ where x_1', x_2' and x_0 are new nonnegative variables leads to an equivalent problem.

7. Show that the following problems are equivalent.

Problem A: Minimize $x_1 + 2x_2 - 3x_3 + 4x_4$
subject to

$$\begin{aligned} 3x_1 - 2x_2 + 5x_3 - 6x_4 &= 20 \\ x_1 + 7x_2 - 6x_3 + 9x_4 &= 30 \end{aligned}$$

$x_1 \geq 0, x_2, x_3, x_4$ unrestricted

Problem B: Minimize $x_1 + 2x_2' - 3x_3' + 4x_4' - 3x_0$
subject to

$$\begin{aligned} 3x_1 - 2x_2' + 5x_3' - 6x_4' + 3x_0 &= 20 \\ x_1 + 7x_2' - 6x_3' + 9x_4' - 10x_0 &= 30 \end{aligned}$$

$x_1, x_2', x_3', x_4', x_0 \geq 0$

8. Using the technique suggested in Problem 6, determine a linear programming problem in standard form with only eight variables and equivalent to the linear programming problem of Problem 3(b).

3.2 Linear Equations and Basic Feasible Solutions

The pivot operation used in solving linear equations consists of replacing a system of equations with an equivalent system in which a selected variable is eliminated from all but one of the equations. The operation revolves around what is called the *pivot term*. The pivot term can be the term in any one of the equations that contains the selected variable with a nonzero coefficient. In the first step of the pivot operation, the equation containing the pivot term is divided by the coefficient in that term, thus producing an equation in which the selected variable has coefficient 1. Multiples of this equation are added to the remaining equations in such a way that the selected variable is eliminated from these remaining equations.

It is easy to show that the solution set of the system of equations resulting from the pivot operation is identical to the solution set of the original system, that is, that the systems are equivalent (Problem 9). In general, repeated use of this pivot operation can lead to a system of equations whose solution set is obvious.

Example 3.2.1. Solve

$$\begin{aligned} x_1 + 4x_2 + 2x_3 &= 6 \\ 3x_1 + 14x_2 + 8x_3 &= 16 \\ 4x_1 + 21x_2 + 10x_3 &= 28 \end{aligned}$$

We arbitrarily select x_1 as the first variable to be eliminated from two of the equations and the $1x_1$ term of the first equation as the pivot term. Notice that we could have also selected the $3x_1$ term of the second equation or the $4x_1$ term of the third equation for the pivot term. However, the arithmetic associated with the selection of the $1x_1$ term is less involved because of the unit coefficient. The pivot operation at this term consists of dividing the first equation by 1, subtracting three times the first equation from the second, and subtracting four times the first equation from the third. The resulting equivalent system is

$$\begin{aligned} x_1 + 4x_2 + 2x_3 &= 6 \\ 2x_2 + 2x_3 &= -2 \\ 5x_2 + 2x_3 &= 4 \end{aligned}$$

Continuing, we arbitrarily select x_2 as the next variable to be eliminated from two of the equations. Since we are striving to simplify the system, the next pivot term should not be the $4x_2$ term of the first equation; pivoting here would reinstate in the last two equations the x_1 variable. Pivoting at the x_2 term of either of the other two equations, however, will isolate the x_2 variable to that pivoting equation without destroying the isolated status of the x_1 variable. Using the $2x_2$ term of the second equation as the pivot term (i.e., we divide the second equation by 2, then subtract four times the result from the first equation and five times the result from the third equation), we obtain

$$\begin{aligned} x_1 \qquad - 2x_3 &= 10 \\ x_2 + x_3 &= -1 \\ - 3x_3 &= 9 \end{aligned}$$

At this stage, one might solve the third equation for x_3 and use this value and the first two equations to compute the associated values for x_1 and x_2. Actually, that operation is essentially equivalent to the pivot operation with the $-3x_3$ term of the third equation as pivot term. Pivoting at this term gives

$$\begin{aligned} x_1 &= 4 \\ x_2 &= 2 \\ x_3 &= -3 \end{aligned}$$

and this system of equations is equivalent to the original system. However, the solution set for the system obviously consists only of the point $(4, 2, -3)$, so we have proven that this point is the unique solution to the original problem.

As we have seen in this example, repeated use of the pivot operation led to a system of three equations with three unknowns in a special form, where each variable appeared in one and only one equation and in that equation had coefficient 1. This form, called the *canonical form*, is crucial to the simplex method. We now define it, along with the associated term *basic variable*.

Definition 3.2.1. A system of m equations and n unknowns, with $m \leq n$, is in *canonical form* with a distinguished set of m *basic variables* if each basic variable has coefficient 1 in one equation and 0 in the others, and each equation has exactly one basic variable with coefficient 1.

Given a linear programming problem in standard form, one way of simplifying the problem would be to replace the set of constraints with an equivalent system of equations in canonical form. Indeed, this step is necessary before the simplex algorithm can be initiated on the linear programming problem. To apply the algorithm, the system of constraints must be in canonical form and the associated basic solution must be feasible. We define the terms *basic solution* and *basic feasible solution* in the following example.

Example 3.2.2. Consider the linear programming problem in standard form of

$$\text{Minimizing } x_1 - x_2 + 2x_3 - 5x_4 = f(x_1, x_2, x_3, x_4) \tag{3.2.1}$$

subject to

$$\begin{aligned} x_1 + x_2 + 2x_3 + x_4 &= 6 \\ 3x_2 + x_3 + 8x_4 &= 3 \\ x_1, x_2, x_3, x_4 &\geq 0 \end{aligned}$$

The system of constraints consists of two equations in four unknowns. Pivoting at the $3x_2$ term of the second equation gives the equivalent system

$$\begin{array}{rcrcrcl} x_1 & & + \tfrac{5}{3}x_3 & - & \tfrac{5}{3}x_4 & = & 5 \\ & x_2 & + \tfrac{1}{3}x_3 & + & \tfrac{8}{3}x_4 & = & 1 \end{array} \tag{3.2.2}$$

This system is in canonical form with basic variables x_1 and x_2. One particular solution to this system of equations is obvious: set the nonbasic variables x_3 and x_4 equal to 0, and set x_1 equal to the constant term 5 and x_2 equal to the constant term 1. This solution point is called a *basic feasible solution.*

Given a system of equations in canonical form with a specified set of basic variables, the associated *basic solution* is that solution to the system with the values of the basic variables given by the constant terms in the equations and the values of the nonbasic variables equal to zero.

In a linear programming problem we are interested in solutions to the system of constraints with nonnegative coordinates. Those basic solutions with this property we call *basic feasible solutions*. These will prove to be the critical points when using the simplex method to determine the optimal value of the objective function.

The point $(5,1,0,0)$ is not the only basic feasible solution for the problem in our example. Returning to the constraints of (3.2.1), if we pivot at the $8x_4$ term of the second equation instead of the $3x_2$ term (or if we pivot in (3.2.2) at the $\frac{8}{3}x_4$ term of the second equation), we get

$$\begin{array}{rcrcrcl} x_1 & + & \tfrac{5}{8}x_2 & + & \tfrac{15}{8}x_3 & & = \tfrac{45}{8} \\ & & \tfrac{3}{8}x_2 & + & \tfrac{1}{8}x_3 & + x_4 & = \tfrac{3}{8} \end{array} \tag{3.2.3}$$

Here the constraint set is represented by a system of equations in canonical form with basic variables x_1 and x_4, and the associated basic solution $(\frac{45}{8},0,0,\frac{3}{8})$ is another basic feasible solution.

Pivoting at the $\frac{5}{8}x_2$ term of the first equation in (3.2.3) yields the equivalent system

$$\begin{array}{rcrcrcl} \tfrac{8}{5}x_1 & + & x_2 & + & 3x_3 & & = 9 \\ -\tfrac{3}{5}x_1 & & & - & x_3 & + x_4 & = -3 \end{array}$$

This system is in canonical form with basic variables x_2 and x_4, but the associated basic solution $(0,9,0,-3)$ is not feasible. The value of x_4 is negative. Obviously, randomly selecting the variables to serve as basic variables can lead to a system of equations with some negative constant terms and thus an associated basic solution that is not feasible. As we will see, the simplex method provides a systematic way to resolve the problem of starting with and maintaining feasibility.

We return now to the original linear programming problem of (3.2.1), but with the system of constraints replaced by the equivalent system of (3.2.2), a system in canonical form with a basic feasible solution. In order to apply the simplex method to the problem, one final step involving the objective function is necessary. The expression for the objective function needs to be coordinated with the canonical form of the system of the constraints. In particular, the expression for the objective function must be in terms of only the nonbasic variables. This step can be considered an extension of the pivot operation used to put the system of constraints into canonical form, and is easily accomplished here using the system of constraints. We demonstrate.

The objective function of the example is

$$f(x_1,x_2,x_3,x_4) = x_1 - x_2 + 2x_3 - 5x_4$$

and the system of constraints in canonical form with basic variables x_1 and x_2, from (3.2.2), is

$$\begin{aligned} x_1 \quad + \tfrac{5}{3}x_3 - \tfrac{5}{3}x_4 &= 5 \\ x_2 + \tfrac{1}{3}x_3 + \tfrac{8}{3}x_4 &= 1 \end{aligned}$$

From these equations, it is obvious that the value of the objective function f at any point (x_1,x_2,x_3,x_4) satisfying the constraints can be given by

$$\begin{aligned} x_1 - x_2 + 2x_3 - 5x_4 &= \left[5 - \tfrac{5}{3}x_3 + \tfrac{5}{3}x_4\right] - \left[1 - \tfrac{1}{3}x_3 - \tfrac{8}{3}x_4\right] + 2x_3 - 5x_4 \\ &= \tfrac{2}{3}x_3 - \tfrac{2}{3}x_4 + 4 \end{aligned}$$

Thus on this system of constraints, the problem of minimizing f is equivalent to the problem of minimizing the function $\frac{2}{3}x_3 - \frac{2}{3}x_4 + 4$. With this new function our goal of expressing the function to be optimized in terms of only the nonbasic variables is attained.

Through these operations we have replaced the linear programming problem of (3.2.1) with the following equivalent linear programming problem.

$$\begin{aligned} &\text{Minimize } \tfrac{2}{3}x_3 - \tfrac{2}{3}x_4 + 4 \\ &\text{subject to} \\ &x_1 \quad + \tfrac{5}{3}x_3 - \tfrac{5}{3}x_4 = 5 \\ &\quad x_2 + \tfrac{1}{3}x_3 + \tfrac{8}{3}x_4 = 1 \\ &x_1,x_2,x_3,x_4 \geq 0 \end{aligned}$$

This problem is said to be in *canonical form* with basic variables x_1 and x_2.

Definition 3.2.2. The standard linear programming problem is in *canonical form* with a distinguished set of basic variables if:

(a) The system of constraints is in canonical form with this distinguished set of basic variables.

(b) The associated basic solution is feasible.

(c) The objective function is expressed in terms of only the nonbasic variables.

If the first two conditions of this definition are satisfied for a linear programming problem, the system of constraints can be used, as in the above example, to eliminate the basic variables from the objective function. While organizing and maintaining a problem in canonical form, we will abuse the language somewhat and always speak of one fixed objective function. Certainly in the above example the function x_1 −

$x_2 + 2x_3 - 5x_4$ does not equal the function $\frac{2}{3}x_3 - \frac{2}{3}x_4 + 4$. However, the problems of optimizing these functions on the given constraint set are equivalent, that is, the functions have the same minimum value, and the sets of feasible solutions on which this common optimal value is attained are the same. It is this equivalency that we have in mind when we say, for example, that the objective function is now given by $\frac{2}{3}x_3 - \frac{2}{3}x_4 + 4$.

The question of feasibility of a basic solution can be stated geometrically using the column vectors associated with the coefficient matrix of the system of equations. We demonstrate.

Example 3.2.3. The system of constraints for the linear programming problem of (3.2.1) can be expressed in vector form as follows:

$$x_1 \begin{bmatrix} 1 \\ 0 \end{bmatrix} + x_2 \begin{bmatrix} 1 \\ 3 \end{bmatrix} + x_3 \begin{bmatrix} 2 \\ 1 \end{bmatrix} + x_4 \begin{bmatrix} 1 \\ 8 \end{bmatrix} = \begin{bmatrix} 6 \\ 3 \end{bmatrix}$$

Thus the system of two equations and four variables is equivalent to the problem of expressing the vector $\begin{bmatrix} 6 \\ 3 \end{bmatrix}$ as a linear combination of the vectors $\begin{bmatrix} 1 \\ 0 \end{bmatrix}$, $\begin{bmatrix} 1 \\ 3 \end{bmatrix}$, $\begin{bmatrix} 2 \\ 1 \end{bmatrix}$, and $\begin{bmatrix} 1 \\ 8 \end{bmatrix}$. Moreover, for our purposes, we are restricted to solutions with nonnegative coordinates.

Suppose now we wish to determine geometrically if x_1 and x_2 can serve as basic variables for a basic feasible solution. If so, the nonbasic variables x_3 and x_4 will equal zero, and the resulting vector equation reduces to

$$x_1 \begin{bmatrix} 1 \\ 0 \end{bmatrix} + x_2 \begin{bmatrix} 1 \\ 3 \end{bmatrix} = \begin{bmatrix} 6 \\ 3 \end{bmatrix}, \quad x_1, x_2 \geq 0$$

Using the notation

$$A^{(1)} = \begin{bmatrix} 1 \\ 0 \end{bmatrix}, A^{(2)} = \begin{bmatrix} 1 \\ 3 \end{bmatrix}, \text{ and } b = \begin{bmatrix} 6 \\ 3 \end{bmatrix}$$

these vectors in $\mathbb{R}^2$ are sketched in Figure 3.1.

Now the set of points of the form $x_1A^{(1)}$ for $x_1 \geq 0$ is the line ray emanating from the origin in $\mathbb{R}^2$ in the direction of $A^{(1)}$, and similarly for the points $x_2A^{(2)}$ with $x_2 \geq 0$. The set of points of the form $x_1A^{(1)} + x_2A^{(2)}$, x_1 and $x_2 \geq 0$, can be determined using the usual rule for addition of vectors. This region (the *convex cone* of $A^{(1)}$ and $A^{(2)}$) is illustrated in Figure 3.2. Since b lies in this region, a solution to the system of equations with x_1 and x_2 nonnegative and x_3 and x_4 equal to 0 must exist. This solution is the point $(5, 1, 0, 0)$ found previously.

To extend these ideas, let $A^{(3)} = \begin{bmatrix} 2 \\ 1 \end{bmatrix}$ and $A^{(4)} = \begin{bmatrix} 1 \\ 8 \end{bmatrix}$. From the graph in Figure 3.3 we see that b cannot be expressed as a sum of the form $x_2A^{(2)} + x_4A^{(4)}$ with x_2 and $x_4 \geq 0$. Thus x_2 and x_4 cannot serve as basic variables for a basic feasible solution.

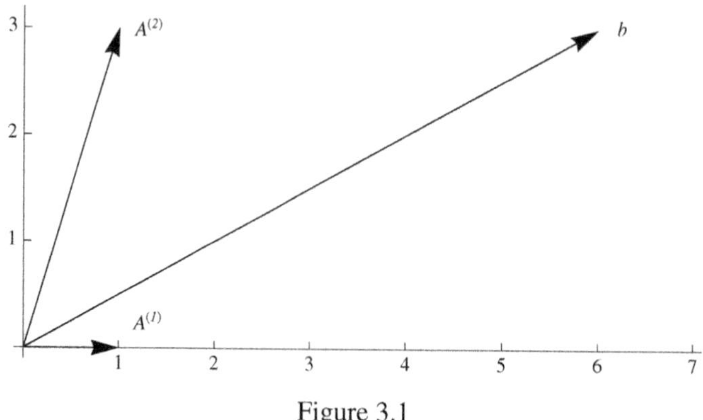

Figure 3.1

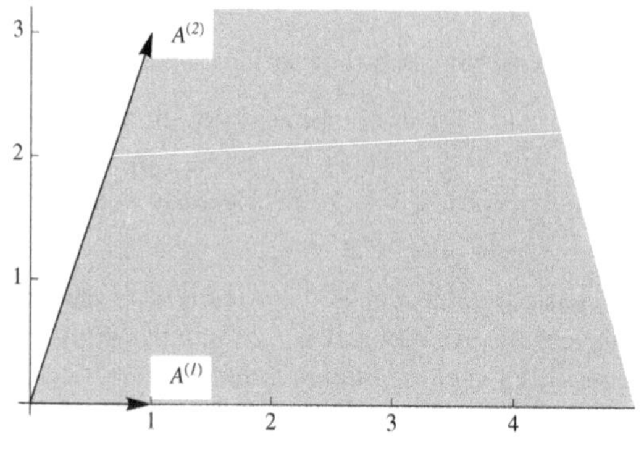

Figure 3.2

(Recall, the associated basic solution is $(0,9,0,-3)$.) Furthermore, it can be seen that any other pair of variables can serve as basic variables for a basic feasible solution. Note also that b is a multiple of $A^{(3)}$ alone. Thus in any basic feasible solution with x_3 as a basic variable, only the x_3 coordinate will be nonzero. Indeed, pivoting at the $1x_3$ term in the second equation in the constraints of (3.2.1) yields the equivalent system

$$\begin{aligned} x_1 - 5x_2 \qquad\quad - 15x_4 &= 0 \\ 3x_2 + x_3 + \ \ 8x_4 &= 3 \end{aligned}$$

This system is in canonical form with basic variables x_1 and x_3, and the associated basic (feasible) solution is $(0,0,3,0)$, with the basic variable x_1 equal to zero. A basic solution with some basic variables equal to zero is said to be *degenerate*. As we will see later when developing the simplex method, theoretical complications arise from the possibility of degeneracy.

The reader may be somewhat puzzled by our earlier remark that, when determining the minimum of the objective function of a linear programming problem, the

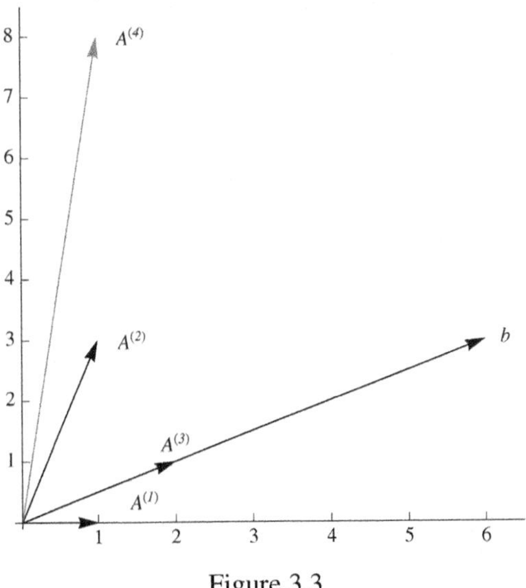

Figure 3.3

basic feasible solutions are the critical points to be considered. Why, when trying to minimize a function, should we wish to restrict our attention to only those feasible solutions of the constraint set that are basic and therefore have at least $n-m$ zero coordinates? For example, in a diet problem with five nutritional requirements and 15 foods from which to choose, is it possible to find a minimal-cost diet that uses at most only 5 of the foods? As we will show in this chapter, the answer to this question is "yes." In fact, we will show by an algebraic argument that if the objective function does have a minimum value, that value is assumed by at least one basic feasible solution.

Actually, the role played by the basic feasible solutions in the resolution of a two-variable problem is apparent from the geometry of such a problem. Consider, for example, the solution procedure used to solve the blending problem developed in Example 2.2.1 on page 10. The problem there was to determine a blend of two feeds that minimized costs and met three nutritional requirements. Letting x_1 denote the amount of Feed 1 and x_2 the amount of Feed 2 in a diet, the associated mathematical problem was to

$$\begin{aligned}
&\text{Minimize } 10x_1 + 4x_2 \\
&\text{subject to} \\
&3x_1 + 2x_2 \geq 60 \\
&7x_1 + 2x_2 \geq 84 \\
&3x_1 + 6x_2 \geq 72 \\
&x_1, x_2 \geq 0
\end{aligned}$$

Putting this into standard form gives the following:

$$\begin{array}{l} \text{Minimize } 10x_1 + 4x_2 \\ \text{subject to} \\ 3x_1 + 2x_2 - x_3 \qquad\qquad = 60 \\ 7x_1 + 2x_2 \qquad - x_4 \qquad = 84 \\ 3x_1 + 6x_2 \qquad\qquad - x_5 = 72 \\ x_1, x_2, x_3, x_4, x_5 \geq 0 \end{array} \tag{3.2.4}$$

The slack variables x_3, x_4, and x_5 measure the surplus amounts of the nutritional elements A, B, and C in a given diet. Now the geometric argument based on Figure 2.5 on page 13 showed that if the linear function had a minimal value, the function would assume that value at a corner or vertex of the region shaded in Figure 2.3. The four vertices of the shaded region in Figure 2.3 are the points $(0,42)$, $(6,21)$, $(18,3)$, and $(24,0)$. They occur on the boundaries of the regions defined by the original three inequalities, that is, when some of the inequalities are actually equalities and the corresponding slack variables therefore equal zero. In fact, the solutions to the constraint set in standard form corresponding to these four points are:

$$\begin{array}{l} (0,42) \leftrightarrow (0,42,24,0,180) \\ (6,21) \leftrightarrow (6,21,0,0,72) \\ (18,3) \leftrightarrow (18,3,0,48,0) \\ (24,0) \leftrightarrow (24,0,12,84,0) \end{array}$$

Note that each of the four points in the right column has two coordinates at zero level. These four points are basic feasible solutions to the constraint set in standard form. Therefore, if the objective function is bounded below, the minimal value must occur at a basic feasible solution.

This geometrical analysis extends to the general problem, yielding another proof that for a linear programming problem, if the set of optimal solution points is not empty, the set of basic feasible solutions provides the foundation for this set. However, we do not use these ideas in the algebraic development which follows, and so we will postpone discussion of the geometry of the general problem until Section 3.9.

Problem Set 3.2

1. Solve the following using the pivot operation.

(a)
$$\begin{aligned} 3x_2 - 3x_3 &= 15 \\ x_1 + x_2 + x_3 &= 0 \\ 3x_1 + 5x_2 + 3x_3 &= 4 \end{aligned}$$

(b)
$$\begin{aligned} 3x_1 + 2x_2 - 7x_3 &= 1 \\ x_1 - 5x_2 - 6x_3 &= -4 \end{aligned}$$

(c) $$\begin{aligned} x_1 + 2x_2 \qquad\quad - 2x_4 &= 5 \\ - 3x_2 + x_3 + 4x_4 &= 2 \end{aligned}$$

2. A system of equations is said to be *redundant* if one of the equations in the system is a linear combination of the other equations. Show by using the pivot operation that the following system is redundant. Is this system equivalent to a system of equations in canonical form?

$$\begin{aligned} x_1 + x_2 - 3x_3 &= 7 \\ -2x_1 + x_2 + 5x_3 &= 2 \\ 3x_2 - x_3 &= 16 \end{aligned}$$

3. A system of equations is said to be *inconsistent* if the system has no solution. Show by using the pivot operation that the following systems are inconsistent. Is either of these systems equivalent to a system in canonical form?

(a) $$\begin{aligned} x_1 + 2x_2 &= 3 \\ x_1 + 2x_2 &= 4 \end{aligned}$$

(b) $$\begin{aligned} x_1 + x_2 - 3x_3 &= 7 \\ -2x_1 + x_2 + 5x_3 &= 2 \\ 3x_2 - x_3 &= 15 \end{aligned}$$

4. (a) Solve the following system of equations by finding an equivalent system in canonical form with basic variables x_1 and x_2.

$$\begin{aligned} 2x_1 + x_2 - 2x_3 &= 17 \\ x_1 \qquad - x_3 &= 4 \end{aligned}$$

(b) Is this system equivalent to a system in canonical form with basic variables x_1 and x_3?

(c) Interpret these results geometrically.

5. Suppose a system of equations contains the following terms:

$$\begin{aligned} &ax_1 + bx_2 \\ &cx_1 + dx_2 \end{aligned}$$

where a, b, c, and d are constants, $a \neq 0$.

The system is then replaced with an equivalent system by pivoting at the ax_1 term. Show that these four terms become

$$x_1 + \frac{b}{a}x_2$$

$$0x_1 + \left(d - \frac{bc}{a}\right)x_2$$

The expression $d - bc/a$ provides a way of remembering the effect of the pivot operation on any term not in the row or column of the pivot term.

6. For the linear programming problem of

$$\begin{aligned}&\text{Minimizing } 5x_1+2x_2+3x_3+x_4\\&\text{subject to}\\&\quad x_1+x_2-2x_3+3x_4=2\\&-2x_1\qquad\;\; +\; x_3\qquad\;\; =2\\&x_1,x_2,x_3,x_4\geq 0\end{aligned}$$

(a) Show geometrically that there can be only two basic feasible solutions to the problem.
(b) Compute these two basic feasible solutions.
(c) Show that the objective function is bounded below.
(d) Assume that the minimal value of the objective function is attained at a basic feasible solution and determine this minimal value.

7. Following the outline in Problem 6, complete the problem of Example 3.2.3.

8. (a) Put the constraint set from the standard form of the blending problem considered in this section (the problem of (3.2.4)) into canonical form with basic variables x_1, x_2, and x_5. The associated basic feasible solution is $(6,21,0,0,72)$.
(b) The objective function for this problem is $10x_1+4x_2$. By eliminating the x_1 and x_2 variables by using the equations found in part (a), this function can be expressed in terms of only x_3 and x_4. Verify that the form reduces to $144+x_3+x_4$.
(c) Since we are considering only feasible solutions to the constraint set, using part(b), give another proof that the minimal value of the objective function is 144.

9. Prove that the system of equations resulting from a given system by applying the pivot operation is equivalent to (has the same solution set as) the original system.

10. Prove that although there may be different ways of driving a system of equations into canonical form with a specified set of basic variables, there is a unique basic solution associated with this specified set of basic variables.

11. True or false: A system of equations is equivalent to a system of equations in canonical form if and only if the original system has at least one solution.

12. Construct a linear programming problem with four variables and three equations for which there exist degenerate feasible solutions with exactly two nonzero coordinates.

3.3 Introduction to the Simplex Method

In this section the simplex method for solving linear programming problems will be introduced. The basic ideas behind the technique will be demonstrated by means of a specific example. The goal of this section is to develop motivation and understanding; the theorems related to the simplex method will be proven in subsequent sections of this chapter.

Let us consider the following problem in standard form:

$$\begin{aligned}
&\text{Minimize } -4x_1 + x_2 + x_3 + 7x_4 + 3x_5 = z \\
&\text{subject to} \\
&-6x_1 \qquad\quad + x_3 - 2x_4 + 2x_5 = 6 \\
&\ \ 3x_1 + x_2 - x_3 + 8x_4 + \ \ x_5 = 9 \\
&x_1, x_2, x_3, x_4, x_5 \geq 0
\end{aligned} \tag{3.3.1}$$

The simplex method can begin only with the problem in canonical form. To put the problem into canonical form, we could first arbitrarily select two variables to be basic variables and then, by pivoting, attempt to put the system of constraints into canonical form with these variables as basic variables, with the hope that the associated basic solution would be feasible. Or, because here we have a problem with only two constraints, we could determine, using elementary vector geometry, a pair of variables that would serve as basic variables for a feasible solution.

In general, however, finding an initial basic feasible solution to a problem can be a major difficulty. This problem will be solved in Section 3.6. For now, assume that we know that for the problem at hand, the variables x_2 and x_3 can serve as basic variables for a feasible solution. Pivoting at the $1x_3$ term of the first equation will put the system of constraints into canonical form. This gives

$$\begin{aligned}
-6x_1 \qquad\quad + x_3 - 2x_4 + 2x_5 &= \ \ 6 \\
-3x_1 + x_2 \qquad\quad + 6x_4 + 3x_5 &= 15
\end{aligned} \tag{3.3.2}$$

The associated basic solution, $(0, 15, 6, 0, 0)$, is feasible, as promised. Now these two equations can be used to eliminate the basic variables x_2 and x_3 from the expression for the objective function z, given by

$$-4x_1 + x_2 + x_3 + 7x_4 + 3x_5 = z \tag{3.3.3}$$

In fact, simply subtracting the two equations in (3.3.2) from the equation in (3.3.3) gives

$$5x_1 + 0x_2 + 0x_3 + 3x_4 - 2x_5 = z - 21$$

Hence the objective function can be given by the form

$$5x_1 + 3x_4 - 2x_5 + 21 = z$$

Thus the problem in canonical form with basic variables x_2 and x_3 is to

$$\begin{aligned}
&\text{Minimize } z \text{ with} \\
&-6x_1 \qquad\quad + x_3 - 2x_4 + 2x_5 = \quad 6 \\
&-3x_1 + x_2 \qquad\quad + 6x_4 + 3x_5 = \quad 15 \\
&\ \ 5x_1 \qquad\qquad\quad + 3x_4 - 2x_5 = -21 + z
\end{aligned} \tag{3.3.4}$$

The objective function has the value 21 at the associated basic feasible solution $(0,15,6,0,0)$. Now the key idea behind the simplex method is to move to another basic feasible solution that gives a smaller value for z by replacing exactly one basic variable from the present set. As we will see, the mechanics for this replacement will be provided by the pivot operation. However, what variable from the set of nonbasic variables x_1, x_4, and x_5 to insert into the basis, and what basic variable, x_2 or x_3, to replace in order to reduce the value of z are not obvious.

These questions are answered first by considering the objective function $z = 5x_1 + 3x_4 - 2x_5 + 21$. In this expression for z, the x_5 variable has a negative coefficient. Thus a feasible solution to the constraint set with x_1 and x_4 still equal to zero, but with x_5 greater than zero, will give a smaller value for z. This suggests that we move x_5 into the set of basic variables and attempt to make x_5 as large as possible.

But what basic variable, x_2 or x_3, should we replace? To answer this question, consider the constraint set with the conditions imposed by this situation, that the nonbasic variables x_1 and x_4 equal zero. From (3.3.4) we have

$$\begin{aligned} x_3 + 2x_5 &= 6 \\ x_2 + 3x_5 &= 15 \end{aligned}$$

Solving for x_3 and x_2 gives

$$\begin{aligned} x_3 &= 6 - 2x_5 \\ x_2 &= 15 - 3x_5 \end{aligned} \tag{3.3.5}$$

Clearly, x_5 cannot be arbitrarily large. To have a solution to the constraint set with $x_1 = x_4 = 0$, x_2 and x_3 must satisfy these equations and would possibly become negative. In fact, since x_2 and x_3 must be nonnegative, x_5 is restricted by the inequalities

$$0 \le 6 - 2x_5 \quad \text{and} \quad 0 \le 15 - 3x_5$$

that is, $x_5 \le 3 = \frac{6}{2}$ and $x_5 \le 5 = \frac{15}{3}$. Since x_5 must satisfy both these inequalities, the maximum possible value for x_5 is 3. Letting $x_5 = 3$ and using (3.3.5) to calculate x_3 and x_2, we have the feasible solution $x_1 = x_4 = 0$, $x_5 = 3$, $x_3 = 0$, and $x_2 = 6$. The value of z at this point is 15, six less than the value at the first basic feasible solution. At the point $(0,6,0,0,3)$, $x_2 = 6$ and $x_3 = 0$. Thus x_3, being at zero level, is the variable that should be replaced in the basis, giving x_2 and x_5 as the basic variables for this second solution point. (Note also that at $(0,6,0,0,3)$, x_2 and x_5 are the two variables assuming positive values.)

In fact, by letting x_5 equal the minimum of 3 and 5, we are guaranteed that x_3 will assume the value 0, because the minimum value 3 is the bound coming from the x_3 equation in (3.3.5). To determine the variable to extract from the basis, then, we need only determine the basic variable of that equation in the modified constraint set (3.3.5) that leads to the minimal bound. And each of these bounds of $3 = \frac{6}{2}$ and $5 = \frac{15}{3}$ is the ratio of the constant term in the equation to the coefficient of the x_5 variable. This suggests a simple procedure for determining the variable to extract from the basis, a procedure that will be spelled out in detail in the next section.

The simplex method is the continuation of this process. To proceed, however, the problem must be in canonical form with basic variables x_2 and x_5. To do this, we use the pivot operation. With the system of constraints expressed as in (3.3.4), the first equation contains the basic variable x_3, which is to be replaced with the variable x_5. Hence pivoting at the $2x_5$ term of this equation will put the system of constraints into canonical form with basic variables x_2 and x_5. Moreover, the effect of this pivot operation on the third equation in (3.3.4) would be to eliminate the variable x_5 from that equation also. Then the objective function z would be expressed in terms of only the variables x_1, x_3, and x_4. Thus the effect of the pivot operation at the $2x_5$ term of the first equation in (3.3.4) applied to all three equations would be to transform the entire problem into the desired canonical form. Pivoting here gives

$$\begin{array}{rcrcrcrcl} -3x_1 & & & + & \frac{1}{2}x_3 & - & x_4 & + & x_5 = 3 \\ 6x_1 & + & x_2 & - & \frac{3}{2}x_3 & + & 9x_4 & & = 6 \\ -x_1 & & & + & x_3 & + & x_4 & & = -15+z \end{array} \tag{3.3.6}$$

Now we proceed exactly as before. The variable x_1 has a negative coefficient in the expression for the objective function and so should be inserted into the basis. Letting $x_3 = x_4 = 0$, the constraint set of (3.3.6) becomes

$$\begin{array}{c} -3x_1 + x_5 = 3 \\ 6x_1 + x_2 = 6 \end{array} \quad \text{or} \quad \begin{array}{c} x_5 = 3 + 3x_1 \\ x_2 = 6 - 6x_1 \end{array} \tag{3.3.7}$$

Since x_2 and x_5 must be nonnegative, we have

$$\begin{array}{c} 0 \le 3 + 3x_1 \\ 0 \le 6 - 6x_1 \end{array} \quad \text{or} \quad \begin{array}{c} -1 \le x_1 \\ x_1 \le 1 \end{array}$$

The first inequality places no upper bound on x_1, so the upper limit for x_1 is determined solely by the second inequality, the inequality resulting from the x_2 equation in (3.3.7). Thus x_1 should replace x_2 in the basis. Letting $x_1 = 1$ gives the basic feasible solution $(1,0,0,0,6)$, and the value of the objective function at this point is 14.

One lingering question that we have so far avoided is the following: When do we know that the minimal value of the objective function has been achieved and the process can terminate? Our example will now provide the answer to this question.

We have seen that a reduced value for z can be determined by using x_1 and x_5 as basic variables instead of x_2 and x_5. Accordingly, we put the system into canonical form with these as basic variables by pivoting at the $6x_1$ term of the second equation in (3.3.6). This gives

$$\begin{array}{rcrcrcrcl} & & \frac{1}{2}x_2 & - & \frac{1}{4}x_3 & + & \frac{7}{2}x_4 & + & x_5 = 6 \\ x_1 & + & \frac{1}{6}x_2 & - & \frac{1}{4}x_3 & + & \frac{3}{2}x_4 & & = 1 \\ & & \frac{1}{6}x_2 & + & \frac{3}{4}x_3 & + & \frac{5}{2}x_4 & & = -14+z \end{array}$$

The objective function is given by $z = \frac{1}{6}x_2 + \frac{3}{4}x_3 + \frac{5}{2}x_4 + 14$. In contrast to the two previous situations, here the coefficients of the nonbasic variables are all positive. This means in fact that the value of the objective function at any feasible solution to the constraint set must be at least 14, since all the coordinates of a feasible solution are nonnegative. Thus our process is terminated. The minimal value of the objective function can be no less than 14, and this value is attained at the point $(1,0,0,0,6)$.

To summarize, the simplex method begins with the problem in canonical form. We move from one basic feasible solution to another by replacing exactly one basic variable at each step, with the new basic feasible solution providing a reduced value of the objective function (except possibly when there is degeneracy, a complication to be discussed later). Consideration of the coefficients of the objective function tells us if the minimal value has been achieved and, if not, what variable to insert into the basis. Consideration of the modified constraint set tells us what variable to extract from the basis. And one simple pivot operation at each step keeps the entire system in proper form.

In the next section, we will make precise the simplex method for the general problem and will consider the case where the objective function is not bounded below. (See also Problem 3.) In Section 3.6 a method based on the simplex method for determining an initial basic feasible solution will be discussed.

Problem Set 3.3

1. Consider the system of equations

$$\begin{aligned} x_1 \qquad\qquad + 2x_4 &= 8 \\ x_2 \qquad + 3x_4 &= 6 \\ x_3 + 6x_4 &= 18 \end{aligned} \tag{3.3.8}$$

The system is in canonical form with basic variables x_1, x_2, and x_3, and the associated basic solution is feasible.

(a) Express the set of solutions to the system in terms of x_4, that is, solve for x_1, x_2, and x_3 in terms of x_4.

(b) Determine the set of values for the parameter x_4 for which the corresponding solutions to the system are feasible.

(c) Let x_4 be the largest value in this set. What variable assumes the value zero?

(d) Suppose we wish to express the system in canonical form with x_4 in the basis, and such that the associated basic solution is feasible. From (c), what variable should be extracted from the basis and become the nonbasic variable? Thus, at what term in (3.3.8) should we pivot?

(e) Show that pivoting here has the desired effect.

(f) For each equation in (3.3.8), compute the ratio of the constant term to the coefficient of x_4. Relate these values to the choice of pivoting term in (d).

2. Consider the problem of

$$\begin{aligned} &\text{Minimizing } x_1 + x_2 + 4x_3 + 7x_4 \\ &\text{subject to} \\ &x_1 + x_2 + 5x_3 + 2x_4 = 8 \\ &2x_1 + x_2 + 8x_3 \qquad = 14 \\ &x_1, x_2, x_3, x_4 \geq 0 \end{aligned}$$

(a) The variables x_1 and x_2 can serve as basic variables for a basic feasible solution. Show that the problem expressed with these as basic variables is

$$\begin{aligned} x_1 \quad + 3x_3 - 2x_4 &= 6 \\ x_2 + 2x_3 + 4x_4 &= 2 \\ - x_3 + 5x_4 &= -8+z \end{aligned}$$

(b) Entering x_3 into the basis will reduce the value of z. Why? Show that the variable to be replaced is x_2.

(c) Perform the pivot operation. Show that the minimal value of the objective function is 7 and is achieved at $(3,0,1,0)$.

3. Use the simplex method to do the following problem. The problem is stated in canonical form with basic variables x_2 and x_3. Notice that at the first step in the simplex method, either x_1 or x_4 can enter the basis.

$$\begin{aligned} &\text{Minimize } -x_1 - 2x_4 + x_5 \\ &\text{subject to} \\ &x_1 \quad + x_3 + 6x_4 + 3x_5 = 2 \\ &-3x_1 + x_2 \quad + 3x_4 + x_5 = 3 \\ &x_1, x_2, x_3, x_4, x_5 \geq 0 \end{aligned}$$

4. In the following problem, the objective function does not have a minimum. However, the problem is stated in canonical form with basic variables x_1 and x_2, and the simplex method can be initiated.

$$\begin{aligned} &\text{Minimize } 4x_3 - 6x_4 \\ &\text{subject to} \\ &x_2 - 6x_3 + 2x_4 = 6 \\ &x_1 \quad + 2x_3 - x_4 = 5 \\ &x_1, x_2, x_3, x_4 \geq 0 \end{aligned}$$

(a) What occurs after the first pivot operation that makes this problem different from our other examples?

(b) Can you prove, using the resulting equations, that the objective function is in fact not bounded below?

3.4 Theory of the Simplex Method

In this section we develop the simplex method for a general linear programming problem. To initiate the algorithm, the problem must be in canonical form. In Section 3.1 we showed that any linear programming problem is equivalent to a problem in standard form, and in Section 3.6 we will show how to drive a problem in standard form into canonical form. In fact, the technique developed in Section 3.6 will make use of the ideas developed in this section. Thus, for the time being, we assume that our general problem is in canonical form.

Suppose the problem has m constraints and n variables, with the first m variables as basic variables. The problem is then:

$$
\begin{array}{l}
\text{Minimize } z \text{ where} \qquad\qquad\qquad\qquad (3.4.1)\\
\begin{array}{rcl}
x_1 + \quad \ldots + a_{1,m+1}x_{m+1} + \cdots + a_{1n}x_n &=& b_1 \\
+ x_2 + \ldots + a_{2,m+1}x_{m+1} + \cdots + a_{2n}x_n &=& b_2 \\
\ddots \qquad\qquad & & \\
x_m + a_{m,m+1}x_{m+1} + \cdots + a_{mn}x_n &=& b_m \\
c_{m+1}x_{m+1} + \cdots + c_nx_n &=& z_0 + z
\end{array}\\
x_1, x_2, \ldots, x_n \geq 0
\end{array}
$$

a_{ij}, b_i, c_j, and z_0 are constants and, since the associated basic solution is feasible, $b_i \geq 0$, $i = 1, ..., m$.

Example 3.4.1. We wish to minimize z with

$$
\begin{array}{rcl}
x_1 \qquad + 2x_3 - x_4 &=& 10 \\
x_2 - x_3 - 5x_4 &=& 20 \\
2x_3 + 3x_4 &=& 60 + z \\
x_1, x_2, x_3, x_4 \geq 0 & &
\end{array}
$$

Here we have a problem with $m = 2$ constraints, $n = 4$ variables, and in canonical form. The associated basic feasible solution is $(10, 20, 0, 0)$, and the value of the objective function z at this point is -60. Note that in this particular problem the coefficients $c_3 = 2$ and $c_4 = 3$ are nonnegative. Since x_3 and x_4 are restricted to be nonnegative, the smallest value $z = 2x_3 + 3x_4 - 60$ can possibly attain is -60, the value of the objective function at the $(10, 20, 0, 0)$ solution point. This suggests our first theorem.

Theorem 3.4.1 (optimality criterion). *For the linear programming problem of (3.4.1), if* $c_j \geq 0$, $j = m+1, \ldots, n$, *then the minimal value of the objective function is* $-z_0$ *and is attained at the point* $(b_1, b_2, \ldots, b_m, 0, \ldots, 0)$.

Proof. For any point satisfying the set of constraints, the value of the objective function is given by $z = c_{m+1}x_{m+1} + \cdots + c_nx_n - z_0$. Since any feasible solution to the constraints has nonnegative coordinates, the smallest possible value for the sum $c_{m+1}x_{m+1} + \cdots + c_nx_n$ is zero. Thus the minimal possible value for z is $-z_0$, and this value is assumed at the point $(b_1, b_2, \ldots, b_m, 0, \ldots, 0)$. □

Hence the problem is resolved if all the c_j's are nonnegative. Assume now that at least one c_j, say c_s, is negative. Then we attempt to enter the variable x_s into the basis. In order to determine what basic variable to replace, we consider the constraint set with all the nonbasic variables except x_s equal to zero. This gives

$$\begin{aligned} x_1 + a_{1s}x_s &= b_1 \\ x_2 + a_{2s}x_s &= b_2 \\ &\vdots \\ x_m + a_{ms}x_s &= b_m \end{aligned} \qquad \text{or} \qquad \begin{aligned} x_1 &= b_1 - a_{1s}x_s \\ x_2 &= b_2 - a_{2s}x_s \\ &\vdots \\ x_m &= b_m - a_{ms}x_s \end{aligned} \tag{3.4.2}$$

Example 3.4.2. Here we wish to minimize z with

$$\begin{aligned} x_1 \quad + 2x_3 - x_4 &= 10 \\ x_2 - x_3 - 5x_4 &= 20 \\ 2x_3 - 3x_4 &= 60 + z \\ x_1, x_2, x_3, x_4 &\geq 0 \end{aligned}$$

Except for a change in sign in c_4, this is exactly the problem of Example 3.4.1. As before, $(10, 20, 0, 0)$ is a feasible solution, and the value of the objective function $z = 2x_3 - 3x_4 - 60$ at this point is -60. However, here we could reduce the value of z if we could find feasible solutions to the constraint set with x_4 positive and x_3 equal to zero, since $c_4 = -3$ is negative. Setting $x_3 = 0$, the constraints reduce to

$$\begin{aligned} x_1 - x_4 &= 10 \\ x_2 - 5x_4 &= 20 \end{aligned} \qquad \text{or} \qquad \begin{aligned} x_1 &= 10 + x_4 \\ x_2 &= 20 + 5x_4 \end{aligned}$$

Note that if we fix x_4 at any positive number and then use these two equations to solve for x_1 and x_2, the resulting values will be positive. Thus all points in the set

$$\{(x_1, x_2, 0, x_4) : x_4 \geq 0, x_1 = 10 + x_4, x_2 = 20 + 5x_4\}$$

are feasible solutions to the system of constraints. But the function $z = 2x_3 - 3x_4 - 60$ is unbounded below on this set. This suggests our next theorem.

Theorem 3.4.2. *For the linear programming problem of (3.4.1), if there is an index s, $m+1 \leq s \leq n$, such that $c_s < 0$ and $a_{is} \leq 0$ for all $i = 1, 2, \ldots, m$, then the objective function is not bounded below.*

Proof. Assume there is an index s satisfying the conditions of the theorem. Since the coefficients a_{is} are all nonpositive, the m equations of (3.4.2) can be used to find a set S of feasible solutions to the constraints with x_s assuming arbitrarily large values, the original basic variables x_1 to x_m positive values, and the remaining variables value zero. But the objective function is given by the form

$$z = c_m x_{m+1} + \cdots + c_s x_s + \cdots + c_n x_n - z_0,$$

and on S, this reduces to $z = c_s x_s - z_0$. Since $c_s < 0$, z is unbounded below on S. □

Assume now that $c_s < 0$ and that at least one $a_{is} > 0$. Then the argument above breaks down, because if $a_{is} > 0$, the equation $x_i = b_i - a_{is}x_x$ places a limit on how large x_s can become. In fact, for x_i to remain nonnegative, we must have $0 \leq b_i - a_{is}x_s$, that is, $x_s \leq b_i / a_{is}$ for $a_{is} > 0$. Thus our goal now is simply to replace in the basis one of the basic variables $x_1,\ldots,x_m$ with the variable x_s. Because of the term $c_s x_s$ in the expression for the objective function, the value of z at this new basic feasible solution hopefully will be reduced. Our one demand on this new basis is that the associated basic solution be feasible. Hence the equations of (3.4.2) for which $a_{is} > 0$ restrict our choice of the variable to extract from the basis. Since we must have $x_s \leq \frac{b_i}{a_{is}}$ for all i with $a_{is} > 0$, the largest possible value for x_s is

$$\text{Min}\left\{\frac{b_i}{a_{is}} : a_{is} > 0\right\}.$$

Suppose this minimum value is attained when $i = r$. Then letting $x_s = \frac{b_r}{a_{rs}}$ will give $x_i \geq 0$ for $i = 1,\ldots,m$ and, in particular, $x_r = b_r - \frac{b_r}{a_{rs}} = 0$. Since x_r takes on the value zero here, it appears that x_r is the variable to be replaced in the basis. And since in (3.4.1) the rth equation of the constraints isolates x_r, the problem can be put into canonical form with basic variables $x_1, \ldots, x_{r-1}, x_{r+1}, \ldots, x_m, x_s$ by a single pivot operation at the $a_{rs}x_s$ term of the rth equation. Before formally stating and proving these results, we give an example.

Example 3.4.3. Minimize z with

$$\begin{array}{rrrrrrl}
x_1 & & & +\ 2x_4 & -\ \ x_5 & = & 10 \\
& x_2 & & -\ \ x_4 & -\ 5x_5 & = & 20 \\
& & x_3 & +\ 6x_4 & -\ 12x_5 & = & 18 \\
& & & -\ 2x_4 & +\ 3x_5 & = & 60+z \\
\multicolumn{7}{l}{x_1,x_2,x_3,x_4,x_5 \geq 0}
\end{array}$$

The problem is in canonical form with basic variables x_1, x_2, and x_3. The associated basic feasible solution is $(10,20,18,0,0)$, and the value of the objective function at this point is -60. However, $c_4 = -2$ is negative, and so we attempt to reduce the value of z by inserting x_4 into the basis. Letting $x_5 = 0$, the constraints reduce to

$$\begin{array}{rl}
x_1 + 2x_4 = 10 & \\
x_2 - \ \ x_4 = 20 & \quad\text{or} \\
x_3 + 6x_4 = 18 &
\end{array}
\qquad
\begin{array}{l}
x_1 = 10 - 2x_4 \\
x_2 = 20 + \ \ x_4 \\
x_3 = 18 - 6x_4
\end{array}$$

The second equation places no restriction on x_4. However, the first requires that $x_4 \leq \frac{10}{2} = 5$ and the third that $x_4 \leq \frac{18}{6} = 3$. The largest possible value for x_4 with $x_5 = 0$ is the minimum of 3 and 5, that is, 3. Letting $x_4 = 3$ gives $x_3 = 0$. Thus x_4 should replace x_3 in the basis and, since the third equation of the constraints isolates x_3, pivoting at the $6x_4$ term of this equation should keep the problem in canonical form, but with basic variables x_1, x_2, and x_4. In fact, pivoting here yields the following equivalent problem:

Minimize z with

$$\begin{array}{rrrrrl} x_1 & & -\frac{1}{3}x_3 & & +3x_5 & = 4 \\ & x_2 & +\frac{1}{6}x_3 & & -7x_5 & = 23 \\ & & \frac{1}{6}x_3 & +x_4 & -2x_5 & = 3 \\ & & \frac{1}{3}x_3 & & -x_5 & = 66+z \end{array}$$

$$x_1, x_2, x_3, x_4, x_5 \geq 0$$

The problem remains in canonical form, but with basic variables x_1, x_2, and x_4. The associated basic solution $(4,23,0,3,0)$ is feasible, and the value of the objective function at this point is -66. Although the optimal value of z has not yet been attained, we have, as promised, moved to a basic feasible solution yielding a reduced value for z while maintaining the problem in canonical form.

Theorem 3.4.3. *In the problem of (3.4.1), assume that there is an index s such that $c_s < 0$ and that at least one $a_{is} > 0$, $i = 1, \ldots, m$. Suppose*

$$\frac{b_r}{a_{rs}} = \text{Min}\left\{\frac{b_i}{a_{is}} : 1 \leq i \leq m \text{ and } a_{is} > 0\right\}.$$

Then the problem can be put into canonical form with basic variables

$$x_1, x_2, \ldots, x_{r-1}, x_{r+1}, \ldots, x_m, x_s.$$

The value of the objective function at the associated basic feasible solution is

$$-z_0 + \frac{c_s b_r}{a_{rs}}$$

Proof. Consider the problem of (3.4.1) under the assumptions of the theorem. The coefficient $a_{rs} \neq 0$ (it is, in fact, positive), and so the term $a_{rs}x_s$ of the rth equation can be used as the pivot term in the pivot operation applied to the $m+1$ equations. By pivoting here, the system of constraints will be expressed in canonical form with basic variables $x_1, \ldots, x_{r-1}, x_{r+1}, \ldots, x_m, x_s$. The constant terms, b_i^* say, $i = 1, \ldots, m$, on the right side of the equations, become

$$b_i^* = b_i - \frac{a_{is}b_r}{a_{rs}}, \text{ for } i = 1, \ldots, m \text{ and } i \neq r \quad \text{and} \quad b_r^* = \frac{b_r}{a_{rs}} \tag{3.4.3}$$

Clearly $b_r^* \geq 0$. If $a_{is} \leq 0$ then, since $b_r \geq 0$ and $a_{rs} > 0$, $b_i^* \geq b_i \geq 0$. If $a_{is} > 0$ and $i \neq r$, by the choice of r, $b_i/a_{is} \geq b_r/a_{rs}$, and so $b_i \geq a_{is}b_r/a_{rs}$. Hence $b_i^* \geq 0$. Therefore the basic solution associated with these basic variables is feasible.

Now the objective function is given in (3.4.1) by the form $c_{m+1}x_{m+1} + \cdots + c_s x_s + \cdots + c_n x_n = z_0 + z$. The effect of the pivot operation on this equation will be to eliminate the x_s term from the equation, producing the equation

$$c_r^* x_r + c_{m+1}^* x_{m+1} + \cdots + c_{s-1}^* x_{s-1} + c_{s+1}^* x_{s+1} + \cdots + c_n^* x_n = z_0^* + z \tag{3.4.4}$$

with $z_0^* = z_0 - c_s b_r / a_{rs}$.

Thus the objective function is expressed in terms of only the new nonbasic variables and the value of this function at the new basic feasible solution is $-z_0 + c_s b_r / a_{rs}$. □

Notice the result of this pivot operation applied to the system of constraints and the objective function. The problem remains in canonical form with the original basic variable x_r replaced with the variable x_s. The value of the objective function at this new basic feasible solution is equal to the value $-z_0$ at the original basic feasible solution plus the quantity $c_s b_r / a_{rs}$. Since we have assumed that $c_s < 0$ and $a_{rs} > 0$, $c_s b_r / a_{rs}$ is less than or equal to zero, and is strictly less than zero if b_r is strictly positive. Thus, if $b_r > 0$, the pivot operation has left the system in canonical form at a basic feasible solution with a smaller value for the objective function. Let us assume for the time being that this is always the case, that any basic feasible solution to the system of constraints has no basic variable equal to zero. A basic solution with some basic variables equal to zero is called a *degenerate solution*, so we are assuming that all basic feasible solutions are nondegenerate.

Under this nondegeneracy hypothesis, Theorem 3.4.3 states that if at least one of the coefficients c_j, $m+1 \leq j \leq n$, is negative, say c_s, and if at least one of the coefficients a_{is}, $1 \leq i \leq m$, is positive, then a specific pivot operation leaves the problem in canonical form at a basic feasible solution that gives a reduced value for the objective function. Now we can continue. If the new coefficients of the objective function are all nonnegative, we are at the minimal value for the objective function, as Theorem 3.4.1 applies. If one of these coefficients is negative and if all of the coefficients of the associated variable are nonpositive in the constraint set, the objective function is unbounded below, as Theorem 3.4.2 applies. Otherwise, we can apply Theorem 3.4.3 again, driving to another basic feasible solution with an even smaller value for the objective function. Since at each step the value of the objective function is reduced (due to the nondegeneracy assumption), there can be no repetition of basic feasible solutions. The different values for the objective function guarantee that a particular basic feasible solution can appear at most once in the process (see Problem 10 of Section 3.2). Now there are at most a finite number of basic solutions, as there are only $\binom{n}{m} = n!/[m!(n-m)!]$ ways of selecting m basic variables from a set of n variables. Thus this process must eventually terminate. Either the minimum value of the objective function will be reached or the function will be proven to be unbounded.

This is the simplex method, with a proof, using the nondegeneracy hypothesis, that the process must terminate after a finite number of steps with either Theorems 3.4.1 or 3.4.2 applying. The nondegeneracy assumption is quite critical. If some basic feasible solutions were degenerate, the pivot operation of Theorem 3.4.3 applied in a row with $b_i = 0$ would leave the value of the objective function unchanged. After several steps of this, we would have no assurance that basic feasible solutions would not reappear, possibly causing the process to cycle indefinitely. In fact, examples of cycling have been constructed (see Appendix B). Thus, from a mathematical point of view, our proof of convergence of the process is inadequate. In Section 3.8 we will provide a complete proof that, for any linear programming problem, there exists a sequence of pivot operations that will drive the problem to completion.

From a practical point of view, however, a pleasant phenomenon occurs. The cliché "whatever can go wrong will go wrong" does not seem to apply. Although degeneracy occurs quite frequently in linear programming applications, very rarely will cycling occur. Simple rules such as those described below usually are sufficient to prevent cycling. The rules are certainly adequate to prevent cycling in the examples of this text (except, of course, for the example of Appendix B). Moreover, more precise rules for the selection of the pivoting term can be given that will guarantee that cycling does not occur (see Section 3.8).

We now summarize the steps of the simplex method, starting with the problem in canonical form.

1. If all $c_j \geq 0$, the minimum value of the objective function has been achieved (Theorem 3.4.1).
2. If there exists an s such that $c_s < 0$ and $a_{is} \leq 0$ for all i, the objective function is not bounded below (Theorem 3.4.2).
3. Otherwise, pivot (Theorem 3.4.3). To determine the pivot term:
   (a) Pivot in any column with a negative c_j term. If there are several negative c_j's, pivoting in the column with the smallest c_j may reduce the total number of steps necessary to complete the problem. Assume that we pivot in column s.
   (b) To determine the row of the pivot term, find that row, say row r, such that

$$\frac{b_r}{a_{rs}} = \text{Min}\left\{\frac{b_i}{a_{is}} : a_{is} > 0\right\}$$

   Notice that here only those ratios b_i/a_{is} with $a_{is} > 0$ are considered. If the minimum of these ratios is attained in several rows, a simple rule such as choosing the row with the smallest index can be used to determine the pivoting row.
4. After pivoting, the problem remains in canonical form at a different basic feasible solution. Now return to step 1.

If the problem contains degenerate basic feasible solutions, proceed as above. These steps should still be adequate to drive the problem to completion.

Problem Set 3.4

1. Complete the problem of Example 3.4.3.
2. Solve the following using the ideas developed in this section.
   (a) Minimize $x_3 + x_4$ subject to

$$\begin{aligned} x_1 \qquad\qquad\quad - \; x_4 &= 5 \\ x_2 + 2x_3 - 3x_4 &= 10 \\ x_1, x_2, x_3, x_4 &\geq 0 \end{aligned}$$

(b) Minimize x_3 subject to the constraints of part (a).
(c) Minimize $x_3 - x_4$ subject to the constraints of part (a).
(d) Minimize $x_3 - x_4$ subject to

$$\begin{aligned} x_1 \qquad\qquad - x_4 &= 5 \\ x_2 + 2x_3 \qquad &= 10 \\ x_1, x_2, x_3, x_4 &\geq 0 \end{aligned}$$

(e) Minimize $-x_3 + x_4$ subject to the constraints of part (d).
(f) Minimize $-x_3 + x_4$ subject to

$$\begin{aligned} x_1 \quad + x_3 - x_4 &= 0 \\ x_2 + 2x_3 \qquad &= 10 \\ x_1, x_2, x_3, x_4 &\geq 0 \end{aligned}$$

(g) Minimize $-x_3 - x_4$ subject to the constraints of part (f).

3. Calculate the coefficient c_r^* in (3.4.4) on page 81. Can the variable removed from the basis at one step of the pivot operation return to the basis on the next step?

4. Using the form for the objective function given in (3.4.1) on page 78 and the coordinates of the new basic feasible solution given in (3.4.3) on page 81, by direct calculation show that the value of the objective function at the new basic feasible solution is as stated in Theorem 3.4.3.

5. Using (3.4.3) on page 81, determine when the pivot operation will go from a nondegenerate basic feasible solution to a degenerate basic feasible solution.

6. Suppose a problem is in canonical form and the associated basic feasible solution is degenerate, and x_1 is a basic variable with the value zero. The pivot operation is performed with the x_1 variable extracted from the basis. Describe the new basic feasible solution.

7. In Chapter 2 we saw linear programming problems with multiple optimal solution points. We do, however, have a uniqueness condition for problems in canonical form. Show that if a problem is driven to the canonical form in (3.4.1) and $c_j > 0$ for $m+1 \leq j \leq n$, then the minimal value $-z_0$ of the objective function is attained only at the point $(b_1, \ldots, b_m, 0, \ldots, 0)$.

8. Extend the formulas in the proof of Theorem 3.4.3 expressing the results of the pivot operation at the a_{rs} term. Show that for any $j \neq s$,

$$\begin{aligned} a_{ij}^* &= a_{ij} - \frac{a_{is}a_{rj}}{a_{rs}}, \quad i \neq r \\ a_{rj}^* &= \frac{a_{rj}}{a_{rs}} \\ c_j^* &= c_j - \frac{c_s a_{rj}}{a_{rs}} \end{aligned}$$

9. Consider the linear programming problem of (3.4.1). Suppose that the value of the function

$$z' = c'_{m+1}x_{m+1} + \cdots + c'_n x_n - z'_0$$

equals the value of the objective function

$$z = c_{m+1}x_{m+1} + \cdots + c_n x_n - z_0$$

in all solutions to the system of constraints of (3.4.1). Prove that

$$z_0' = z_0 \text{ and } c_j' = c_j \text{ for all } j, m+1 \leq j \leq n$$

Conclusion. Given a linear programming problem in canonical form with a specified set of basic variables, the coefficients in the expression for the objective function are unique.

3.5 THE SIMPLEX TABLEAU AND EXAMPLES

At each step of the simplex method, it is crucial to know only the basic variables and the values of the coefficients in the system of equations. To facilitate computation of a solution, at each step all we need do is record this information. This suggests a notation similar to the *detached coefficient* notation used for solving linear equations. We illustrate with the example of Section 3.3 [see equation (3.3.1)]. The problem, expressed in canonical form with basic variables x_2 and x_3, was, as in (3.3.4), to minimize z with

$$\begin{aligned}
-6x_1 \phantom{{}+x_2} + x_3 - 2x_4 + 2x_5 &= 6 \\
-3x_1 + x_2 \phantom{{}+x_3} + 6x_4 + 3x_5 &= 15 \\
5x_1 \phantom{{}+x_2+x_3} + 3x_4 - 2x_5 &= -21 + z \\
x_1, x_2, x_3, x_4, x_5 &\geq 0
\end{aligned}$$

This information is recorded in tableau form in Table 3.1.

The initial line of x's in the array simply labels the columns of the tableau with the variables of the problem. The first column identifies the basic variables. The first two rows correspond to the system of constraints, with the constant terms given in the last column. The last row corresponds to the equation defining the objective function, with the constant term on the right side of that equation in the last column and the z term suppressed from the tableau because it remains fixed throughout the simplex method.

We now apply the simplex method. As noted in Section 3.3, the -2 in the x_5 column of the last row indicates that we should pivot in that column. To determine the pivoting row, we compare the ratios b_i/a_{is} for $a_{is} > 0$, as in Theorem 3.4.3, and

Table 3.1

	x_1	x_2	x_3	x_4	x_5	
x_3	-6	0	1	-2	2	6
x_2	-3	1	0	6	3	15
	5	0	0	3	-2	-21
					↑	

Table 3.2

	x_1	x_2	x_3	x_4	x_5	
x_3	-6	0	1	-2	2	6
x_2	-3	1	0	6	3	15
	5	0	0	3	-2	-21
x_5	-3	0	$\frac{1}{2}$	-1	1	3
x_2	6	1	$-\frac{3}{2}$	9	0	6
	-1	0	1	1	0	-15

Table 3.3

x_5	0	$\frac{1}{2}$	$-\frac{1}{4}$	$\frac{7}{2}$	1	6
x_1	1	$\frac{1}{6}$	$-\frac{1}{4}$	$\frac{3}{2}$	0	1
	0	$\frac{1}{6}$	$\frac{3}{4}$	$\frac{5}{2}$	0	-14

find the row in which the minimum is attained. In this case $\frac{6}{2}$ is less than $\frac{15}{3}$ and, therefore, we should pivot at the 2 in the first row, replacing the basic variable x_3 with the variable x_5. The tableau representing the result of this pivot operation can be constructed from the present tableau by dividing the first row by 2 and then adding multiples of this row to the remaining rows in such a way as to generate zeros in the x_5 column. We illustrate in Table 3.2, placing this new tableau directly below the original tableau.

The second tableau represents the problem as stated in (3.3.6) on page 75. The associated basic feasible solution is $(0,6,0,0,3)$, and the value of the objective function at this point is the negative of the constant in the lower right-hand corner of the tableau, $-(-15) = 15$.

Pivoting now at the 6 in the x_1 column of the second row gives the tableau of Table 3.3. Since all the constants in the last row, excluding the -14, are nonnegative, the minimum value of the objective function has been attained. This value, $-(-14) = 14$, is attained at the basic feasible solution $(1,0,0,0,6)$, as can be read from the final tableau.

Hereafter the steps of the simplex method for any example will be recorded using this tableau notation. We emphasize that if at any time you are confused or bewildered by a statement based on the tableau presentation of a problem, simply translate the information in the tableau back into a clearly stated problem with the system of constraints and the objective function defined as usual, that is, "attach back" the variables. The tableau remains just a notation for a linear programming problem and the associated equations.

Table 3.4

	x_1	x_2	x_3	x_4	x_5	x_6	
x_4	3	2	0	1	0	0	60
x_5	−1	(1)	4	0	1	0	10
x_6	2	−2	5	0	0	1	50
	−2	−3	−3	0	0	0	0
x_4	(5)	0	−8	1	−2	0	40
x_2	−1	1	4	0	1	0	10
x_6	0	0	13	0	2	1	70
	−5	0	9	0	3	0	30
x_1	1	0	$-\frac{8}{5}$	$\frac{1}{5}$	$-\frac{2}{5}$	0	8
x_2	0	1	$\frac{12}{5}$	$\frac{1}{5}$	$\frac{3}{5}$	0	18
x_6	0	0	13	0	2	1	70
	0	0	1	1	1	0	70

Example 3.5.1. Maximize $2x_1 + 3x_2 + 3x_3$ subject to

$$\begin{aligned} 3x_1 + 2x_2 &\leq 60 \\ -x_1 + x_2 + 4x_3 &\leq 10 \\ 2x_1 - 2x_2 + 5x_3 &\leq 50 \\ x_1, x_2, x_3 &\geq 0 \end{aligned}$$

Introducing three slack variables and putting the problem into standard form gives the following:

$$\begin{aligned} &\text{Minimize } -2x_1 - 3x_2 - 3x_3 \\ &\text{subject to} \\ &3x_1 + 2x_2 + x_4 = 60 \\ &-x_1 + x_2 + 4x_3 + x_5 = 10 \\ &2x_1 - 2x_2 + 5x_3 + x_6 = 50 \\ &x_1, x_2, x_3, x_4, x_5, x_6 \geq 0 \end{aligned}$$

The system of constraints for this problem is in canonical form with basic variables x_4, x_5, and x_6, the associated basic solution, $(0,0,0,60,10,50)$, is feasible, and the objective function is written in terms of the nonbasic variables. Thus the simplex method can be initiated. Table 3.4 gives the resulting tableaux.

Note that the first pivot could have been made in either the first, second, or third column. From the last tableau we see that, for the problem as stated in standard form, the minimal value of the objective function is -70, and this value is attained at the point $(8,18,0,0,0,70)$. Since the original problem was a maximization problem with no slack variables, the optimal value for the original objective function is 70 and is attained at the point $(8,18,0)$.

Table 3.5

	x_1	x_2	x_3	x_4	x_5	
x_4	1	1	-2	1	0	7
x_5	-3	(1)	2	0	1	3
	0	-2	-1	0	0	0
x_4	(4)	0	-4	1	-1	4
x_2	-3	1	2	0	1	3
	-6	0	3	0	2	6
x_1	1	0	-1	$\frac{1}{4}$	$-\frac{1}{4}$	1
x_2	0	1	-1	$\frac{3}{4}$	$\frac{1}{4}$	6
	0	0	-3	$\frac{3}{2}$	$\frac{1}{2}$	12

Example 3.5.2. Maximize $2x_2 + x_3$ subject to

$$\begin{aligned} x_1 + x_2 - 2x_3 &\leq 7 \\ -3x_1 + x_2 + 2x_3 &\leq 3 \\ x_1, x_2, x_3 &\geq 0 \end{aligned}$$

The standard form of the problem is

$$\begin{aligned} &\text{Minimize } -2x_2 - x_3 \\ &\text{subject to} \\ &\quad x_1 + x_2 - 2x_3 + x_4 = 7 \\ &-3x_1 + x_2 + 2x_3 + x_5 = 3 \\ &x_1, x_2, x_3, x_4, x_5 \geq 0 \end{aligned}$$

This problem is in canonical form with basic variables x_4 and x_5, and the steps of the simplex algorithm are displayed in Table 3.5. The three negative entries in the third column of the previous tableau indicate that the objective function is unbounded below.

Example 3.5.3. Finally, we consider the problem of

$$\begin{aligned} &\text{Minimizing } -4x_1 + x_2 + 30x_3 - 11x_4 - 2x_5 + 3x_6 \\ &\text{subject to} \\ &-2x_1 + 6x_3 + 2x_4 - 3x_6 + x_7 = 20 \\ &-4x_1 + x_2 + 7x_3 + x_4 - x_6 = 10 \\ &\quad - 5x_3 + 3x_4 + x_5 - x_6 = 60 \\ &x_1, x_2, x_3, x_4, x_5, x_6, x_7 \geq 0 \end{aligned}$$

The system of constraints, as given, is in canonical form with basic variables x_7, x_2, and x_5, and the associated basic solution, $(0, 10, 0, 0, 60, 0, 20)$, is feasible. However,

Table 3.6

	x_1	x_2	x_3	x_4	x_5	x_6	x_7	
x_7	-2	0	6	2	0	-3	1	20
x_2	-4	1	7	(1)	0	-1	0	10
x_5	0	0	-5	3	1	-1	0	60
	0	0	13	-6	0	2	0	110
x_7	(6)	-2	-8	0	0	-1	1	0
x_4	-4	1	7	1	0	-1	0	10
x_5	12	-3	-26	0	1	2	0	30
	-24	6	55	0	0	-4	0	170
x_1	1	$-\frac{1}{3}$	$-\frac{4}{3}$	0	0	$-\frac{1}{6}$	$\frac{1}{6}$	0
x_4	0	$-\frac{1}{3}$	$\frac{5}{3}$	1	0	$-\frac{5}{3}$	$\frac{2}{3}$	10
x_5	0	1	-10	0	1	(4)	-2	30
	0	-2	23	0	0	-8	4	170
x_1	1	$-\frac{7}{24}$	$-\frac{7}{4}$	0	$\frac{1}{24}$	0	$\frac{1}{12}$	$\frac{5}{4}$
x_4	0	$\frac{1}{12}$	$-\frac{5}{2}$	1	$\frac{5}{12}$	0	$-\frac{1}{6}$	$\frac{45}{2}$
x_6	0	$\frac{1}{4}$	$-\frac{5}{2}$	0	$\frac{1}{4}$	1	$-\frac{1}{2}$	$\frac{15}{2}$
	0	0	3	0	2	0	0	230

the expression for the objective function contains the basic variables x_2 and x_5. By subtracting the second equation and adding twice the third equation to the equation

$$-4x_1 + x_2 + 30x_3 - 11x_4 - 2x_5 + 3x_6 = z$$

we have

$$13x_3 - 6x_4 + 2x_6 = 110 + z$$

Using this expression to define the objective function, the problem is in canonical form with basic variables x_7, x_2, and x_5, and the simplex method can be initiated. The corresponding tableaux are given in Table 3.6. As can be seen, the minimal value of the objective function is -230 and is attained at the point $(\frac{5}{4}, 0, 0, \frac{45}{2}, 0, \frac{15}{2}, 0)$. Note the presence of degeneracy in the second and third steps.

Problem Set 3.5

1. Each of the following tableaux corresponds to a linear programming problem in canonical form with three equality constraints, an objective function to be minimized, seven nonnegative variables $x_1, \ldots, x_7$, and with variables x_5, x_3, x_1 serving as basic variables. For each, either (i) the solution of the problem can be determined from the given tableau or (ii) one or more iterations of the simplex algorithm are necessary to complete the problem. If (i), state the complete resolution of the problem; if (ii), determine all valid pivot points for the tableau.

(a)

	x_1	x_2	x_3	x_4	x_5	x_6	x_7	
x_5	0	5	0	3	1	−1	8	39
x_3	0	6	1	−1	0	0	−6	10
x_1	1	9	0	8	0	−3	4	88
	0	6	0	−4	0	2	3	$-75+z$

(b)

	x_1	x_2	x_3	x_4	x_5	x_6	x_7	
x_5	0	5	0	−3	1	−1	8	39
x_3	0	6	1	1	0	−1	−6	10
x_1	1	9	0	−8	0	−3	4	88
	0	6	0	4	0	2	0	$-75+z$

(c)

	x_1	x_2	x_3	x_4	x_5	x_6	x_7	
x_5	0	5	0	−3	1	−1	8	3
x_3	0	6	1	1	0	0	−6	2
x_1	1	9	0	−8	0	−3	4	0
	0	−6	0	0	0	0	3	$-75+z$

(d)

	x_1	x_2	x_3	x_4	x_5	x_6	x_7	
x_5	0	5	0	−3	1	−1	8	3
x_3	0	6	1	1	0	0	−6	2
x_1	1	9	0	−8	0	−3	4	1
	0	−6	0	0	0	−2	3	$-75+z$

(e)

	x_1	x_2	x_3	x_4	x_5	x_6	x_7	
x_5	0	5	0	−3	1	1	8	60
x_3	0	6	1	−1	0	0	−6	30
x_1	1	9	0	−8	0	−3	7	50
	0	−6	0	0	0	−2	−3	$-75+z$

(f)

	x_1	x_2	x_3	x_4	x_5	x_6	x_7	
x_5	0	−5	0	−3	1	−1	8	39
x_3	0	−6	1	−1	0	0	−6	0
x_1	1	9	0	−8	0	−3	4	88
	0	6	0	0	0	2	3	$-75+z$

2. For each of the following, put the problem into canonical form, set up the initial tableau, and solve by hand using the simplex method. At most, two pivots should be required for each. Along the way, objective functions requiring some initial adjustments and unbounded objective functions should be encountered.

(a) Minimize $2x_1 + 4x_2 - 4x_3 + 7x_4$
subject to
$$\begin{aligned} 8x_1 - 2x_2 + x_3 - x_4 &\le 50 \\ 3x_1 + 5x_2 \quad\quad + 2x_4 &\le 150 \\ x_1 - x_2 + 2x_3 - 4x_4 &\le 100 \\ x_1, x_2, x_3, x_4 &\ge 0 \end{aligned}$$

(b) Maximize $x_1 + 2x_2 - x_3$
subject to
$$\begin{aligned} x_2 + 4x_3 &\le 36 \\ 5x_1 - 4x_2 + 2x_3 &\le 60 \\ 3x_1 - 2x_2 + x_3 &\le 24 \\ x_1, x_2, x_3 &\ge 0 \end{aligned}$$

(c) Minimize $-5x_1 + 4x_2 + x_3$
subject to
$$\begin{aligned} x_1 + x_2 - 3x_3 &\le 8 \\ 2x_2 - 2x_3 &\le 7 \\ -x_1 - 2x_2 + 4x_3 &\le 6 \\ x_1, x_2, x_3 &\ge 0 \end{aligned}$$

(d) Maximize $9x_2 + 2x_3 - x_5$
subject to
$$\begin{aligned} x_1 - 3x_2 \quad - 4x_4 \quad + 2x_6 &= 60 \\ 2x_2 \quad - x_4 - x_5 + 4x_6 &= -20 \\ x_2 + x_3 \quad\quad + 3x_6 &= 10 \\ x_1, x_2, x_3, x_4, x_5, x_6 &\ge 0 \end{aligned}$$

(e) Maximize $x_1 + 12x_2 + 9x_3$
subject to
$$\begin{aligned} 3x_1 + 2x_2 - 6x_3 &\le 20 \\ 2x_1 + 6x_2 + 3x_3 &\le 30 \\ 6x_1 \quad + 2x_3 &\le 16 \\ x_1, x_2, x_3 &\ge 0 \end{aligned}$$

(f) Minimize $x_3 - x_4$
subject to
$$\begin{aligned} x_1 \quad - 3x_4 + x_5 &= 1 \\ x_2 \quad + 6x_4 - 5x_5 &= 6 \\ x_3 - 3x_4 + 2x_5 &= 5 \\ x_1, x_2, x_3, x_4, x_5 &\ge 0 \end{aligned}$$

For the remaining problems, the use of the software LP Assistant, as described in Appendix D, is strongly encouraged. The program facilitates considerably the implementation of the simplex method. The user needs to enter a valid initial tableau and appropriate pivots points, and needs to recognize a final tableau and interpret the results, but the machine completes the arithmetic of each pivot step.

3. Solve. Maximize $x_4 - x_5$
subject to

$$\begin{aligned} x_1 \quad\quad + x_4 - 2x_5 &= 1 \\ x_2 \quad + x_4 \quad\quad &= 6 \\ x_3 + 2x_4 - 3x_5 &= 4 \\ x_1, x_2, x_3, x_4, x_5 &\geq 0 \end{aligned}$$

Note that in this example a variable removed from the basis in one step of the pivot operation eventually returns to the basis. Compare with Problem 3 of Section 3.4.

4. Solve. Maximize $10x_3 + 3x_4$
subject to

$$\begin{aligned} x_1 \quad + 10x_3 + 2x_4 &= 20 \\ x_2 - x_3 + x_4 &= 12 \\ x_1, x_2, x_3, x_4 &\geq 0 \end{aligned}$$

(If in your first iteration you put x_3 into the basis, you will have an example of a variable inserted into the basis in one step of the simplex algorithm being removed from the basis in the very next step.)

5. Consider the problem of Example 3.5.3. The minimum value of the objective function is -230 and is attained at $(\frac{5}{4}, 0, 0, \frac{45}{2}, 0, \frac{15}{2}, 0)$. However, this optimal value is attained at other solution points to the system of constraints.

(a) The previous tableau for the solution to this problem suggests that optimal basic feasible solutions exist with either x_2 or x_7 in the basis. Why?

(b) Use the previous tableau to determine an optimal basic feasible solution with x_7 in the basis.

(c) Find an optimal solution with x_2 in the basis.

6. For each of the following, determine two distinct basic feasible solutions at which the optimal value of the objective function is attained.

(a) Maximize $4x_1 + 12x_2 + 8x_3$
subject to

$$\begin{aligned} 3x_1 + 2x_2 - 6x_3 &\leq 20 \\ 3x_1 + 6x_2 + 4x_3 &\leq 30 \\ x_1, x_2, x_3 &\geq 0 \end{aligned}$$

(b) Minimize $x_1 - 3x_2 - 6x_3$
subject to

$$\begin{aligned} 2x_1 - x_2 + x_3 + x_4 &\le 60 \\ 3x_1 + 4x_2 + 2x_3 - 2x_4 &\le 150 \\ x_1, x_2, x_3, x_4 &\ge 0 \end{aligned}$$

7. Consider the problem of Example 3.5.2.

(a) Show that any point of the form $(t, 0, t)$, for $t \ge 0$, is a feasible solution.
(b) Using this, show that the objective function is unbounded.

8. Compute the solution to Problem 11 of Section 2.3.

9. Compute the solution to Problem 7 of Section 2.6.

10. Compute the solution to Problem 5 of Section 2.6

3.6 Artificial Variables

As we have seen, many linear programming problems can be put into canonical form with little or no effort. For example, the addition of slack variables with positive coefficients can provide the basic variables necessary for the initial basic feasible solution. On the other hand, the system of constraints for many other problems contains no obvious basic feasible solutions. Problems of this type occur, for example, in production models involving output requirements and therefore ($\ge$) inequalities in the constraint set, such as we saw in Example 2.3.4 on page 24, or in transportation problems involving fixed demands and therefore equalities in the constraint set, such as in Example 2.4.1 on page 34. In fact, in any application of linear programming to a real-world problem, it would be rare to find the original formulation of the problem in canonical form.

What must be developed is a technique for determining an initial basic feasible solution for an arbitrary system of equations. This technique must also be capable of handling problems having no feasible solution. Such a problem could arise, for example, in a model containing an error in formulation or in a complicated production model where it is not obvious that the various output requirements can be met with the limited resources available. In this section we will introduce such a technique; in the next section we will discuss some of the complications that can occur.

The basic idea behind the method used to find an initial basic feasible solution is simple. We introduce into the problem a sufficient number of variables, called *artificial variables*, to put the system of constraints into canonical form with these variables as the basic variables. Then we apply the simplex method, not to the objective function of the original problem, but to a new function defined in such a way that its minimal value is attained at a feasible solution to the original problem. Thus the method of the previous three sections applied to this new function drives the original problem to a basic feasible solution.

Consider the standard linear programming problem of (3.1.1) of finding a nonnegative solution to the system

$$\begin{aligned} a_{11}x_1 + a_{12}x_2 + \ldots + a_{1n}x_n &= b_1 \\ a_{21}x_1 + a_{22}x_2 + \ldots + a_{2n}x_n &= b_2 \\ &\vdots \\ a_{m1}x_1 + a_{m2}x_2 + \ldots + a_{mn}x_n &= b_m \end{aligned} \tag{3.6.1}$$

that minimizes the function $z = c_1x_1 + c_2x_2 + \ldots + c_nx_n$. By multiplication of an equation by (-1) if necessary, we may assume that all the constant terms b_i, $i = 1, \ldots, m$, are nonnegative. Now introduce into the system of constraints m new variables, x_{n+1}, $\ldots$, x_{n+m}, called *artificial variables*, one to each equation. The resulting system is

$$\begin{aligned} a_{11}x_1 + a_{12}x_2 + \ldots + a_{1n}x_n + x_{n+1} \qquad\qquad &= b_1 \\ a_{21}x_1 + a_{22}x_2 + \ldots + a_{2n}x_n \qquad + x_{n+2} \qquad &= b_2 \\ &\vdots \\ a_{m1}x_1 + a_{m2}x_2 + \ldots + a_{mn}x_n \qquad\qquad + x_{m+n} &= b_m \end{aligned} \tag{3.6.2}$$

Note that this system is in canonical form with basic variables x_{n+1}, $\ldots$, x_{m+n}, and that the associated basic solution is feasible, since we have assumed that the b_i's are nonnegative.

Now consider the problem of determining the minimal value of the function $w = x_{n+1} + x_{n+2} + \cdots + x_{n+m}$ on the set of all nonnegative solutions to the system of equations in (3.6.2). Since all variables are nonnegative, w can never be negative. The function w would assume the value zero at any feasible solution to (3.6.2) in which all the artificial variables are at zero level. Thus the simplex method applied to this function should replace the artificial variables as basic variables with the variables from the original problem and will hopefully drive the system in (3.6.2) into canonical form with basic variables from the original set x_j, $j = 1, \ldots, n$. The value of w at the associated basic feasible solution would be zero, its minimal value, and the simplex method could then be initiated on the original problem as stated in (3.6.1). Furthermore, if the system of constraints in (3.6.1) does have at least one feasible solution, the system in (3.6.2) must have feasible solutions in which all the artificial variables equal zero. In this case the minimal value of w would be, in fact, zero. Thus, when applying the simplex method to the function w, if we reach a step at which we can pivot no more but the associated value of w is greater than zero, we can conclude that the original problem has no feasible solutions.

Before we present examples, some remarks of a technical nature are in order. First, before the simplex method can be applied to the function $w = x_{n+1} + x_{n+2} + \cdots + x_{n+m}$, the problem must be in canonical form. The system of constraints in (3.6.2) is in canonical form with the artificial variables as basic variables and the associated basic solution is feasible, but the function w is not expressed in terms of only the nonbasic variables. To rectify this, we subtract from the equation defining w each constraining equation containing an artificial variable. (In the general problem above, artificial variables have been introduced into every constraint. However, this need not always be the case. In some instances, some of the original problem variables may be used in the initial basic variable set. An example will be seen in Example 3.6.2 shortly.)

Second, if the pivot operations dictated by the problem of minimizing w are also simultaneously performed on the equation $c_1x_1 + c_2x_2 + \cdots + c_nx_n = z$ which defines the original objective function, this function will be expressed in terms of nonbasic variables at each step. Thus, if an initial basic feasible solution is found for the original problem, the simplex method can be initiated immediately on z. Therefore we incorporate this z equation into the notation and operations of the problem of minimizing w.

In the sum, the first step in solving the general problem of (3.6.1) is to consider the problem of minimizing w with

$$\begin{array}{lllllllll} a_{11}x_1 & + & a_{12}x_2 & + \dots + & a_{1n}x_n & + x_{n+1} & & & = b_1 \\ a_{21}x_1 & + & a_{22}x_2 & + \dots + & a_{2n}x_n & & + x_{n+2} & & = b_2 \\ \vdots \\ a_{m1}x_1 & + & a_{m2}x_2 & + \dots + & a_{mn}x_n & & & + x_{m+n} & = b_m \\ c_1x_1 & + & c_2x_2 & + \dots + & c_nx_n & & & & = z \\ d_1x_1 & + & d_2x_2 & + \dots + & d_nx_n & & & & = w_0 + w \end{array} \tag{3.6.3}$$

where $d_j = -(a_{1j} + a_{2j} + \cdots + a_{mj})$ and $w_0 = -(b_1 + b_2 + \cdots + b_m)$.

Example 3.6.1. Consider the problem to

$$\begin{array}{l} \text{Minimize } 2x_1 - 3x_2 + x_3 + x_4 \\ \text{subject to} \\ x_1 - 2x_2 - 3x_3 - 2x_4 = 3 \\ x_1 - x_2 + 2x_3 + x_4 = 11 \\ x_1, x_2, x_3, x_4 \geq 0 \end{array} \tag{3.6.4}$$

Introducing artificial variables x_5 and x_6, we now instead consider the problem of minimizing w where

$$\begin{array}{l} x_1 - 2x_2 - 3x_3 - 2x_4 + x_5 \qquad = 3 \\ x_1 - x_2 + 2x_3 + x_4 \qquad + x_6 = 11 \\ 2x_1 - 3x_2 + x_3 + x_4 \qquad = z \\ \qquad\qquad x_5 + x_6 = w \\ x_1, x_2, x_3, x_4, x_5, x_6 \geq 0 \end{array} \tag{3.6.5}$$

Subtracting the first two equations from the w equation gives the system

$$\begin{array}{l} x_1 - 2x_2 - 3x_3 - 2x_4 + x_5 \qquad = 3 \\ x_1 - x_2 + 2x_3 + x_4 \qquad + x_6 = 11 \\ 2x_1 - 3x_2 + x_3 + x_4 \qquad = z \\ -2x_1 + 3x_2 + x_3 + x_4 \qquad = -14 + w \end{array}$$

Table 3.7

	x_1	x_2	x_3	x_4	x_5	x_6	
x_5	1	−2	−3	−2	1	0	3
x_6	1	−1	2	1	0	1	11
	2	−3	1	1	0	0	0
	−2	3	1	1	0	0	−14

Table 3.8

	x_1	x_2	x_3	x_4	x_5	x_6	
x_5	(1)	−2	−3	−2	1	0	3
x_6	1	−1	2	1	0	1	11
	2	−3	1	1	0	0	0
	−2	3	1	1	0	0	−14
x_1	1	−2	−3	−2	1	0	3
x_6	0	1	(5)	3	−1	1	8
	0	1	7	5	−2	0	−6
	0	−1	−5	−3	2	0	−8
x_1	1	$-\frac{7}{5}$	0	$-\frac{1}{5}$	$\frac{2}{5}$	$\frac{3}{5}$	$\frac{39}{5}$
x_3	0	$\frac{1}{5}$	1	$\frac{3}{5}$	$-\frac{1}{5}$	$\frac{1}{5}$	$\frac{8}{5}$
	0	$-\frac{2}{5}$	0	$\frac{4}{5}$	$-\frac{3}{5}$	$-\frac{7}{5}$	$-\frac{86}{5}$
	0	0	0	0	1	1	0

This information can be recorded in tableau form by simply augmenting the notation of the previous section (see Table 3.7). The last row corresponds to the w equation, with the w suppressed from the notation. Now the simplex method is initiated, with the entries in the last row determining the pivoting column at each step. The second to last row, the z row, is operated on at each pivot operation but is otherwise ignored for the time being. Table 3.8 gives the resulting tableaux.

Thus the minimal value of w is 0, and one point at which this value is attained is $(\frac{39}{5}, 0, \frac{8}{5}, 0, 0, 0)$. Since this point is a solution to the system of constraints in (3.6.5) and has as its last two coordinates zero, $(\frac{39}{5}, 0, \frac{8}{5}, 0)$ is a basic feasible solution to the system in (3.6.4), and the data for the tableau corresponding to the original problem expressed in canonical form with basic variables x_1 and x_3 are contained in the last tableau. In fact, translating these data back into equation form gives the following system, equivalent to (3.6.4).

$$\begin{aligned} x_1 - \tfrac{7}{5}x_2 \qquad\;\; - \tfrac{1}{5}x_4 &= \tfrac{39}{5} \\ \tfrac{1}{5}x_2 + x_3 + \tfrac{3}{5}x_4 &= \tfrac{8}{5} \\ - \tfrac{2}{5}x_2 \qquad\;\; + \tfrac{4}{5}x_4 &= -\tfrac{86}{5} + z \end{aligned}$$

Table 3.9

	x_1	x_2	x_3	x_4	
x_1	1	$-\frac{7}{5}$	0	$-\frac{1}{5}$	$\frac{39}{5}$
x_3	0	$\textcircled{\frac{1}{5}}$	1	$\frac{3}{5}$	$\frac{8}{5}$
	0	$-\frac{2}{5}$	0	$\frac{4}{5}$	$-\frac{86}{5}$
x_1	1	0	7	4	19
x_2	0	1	5	3	8
	0	0	2	2	-14

The second stage of the problem, the application of the simplex process to the problem of minimizing z, can be initiated immediately (Table 3.9). The minimal value of z is 14 and is attained at the point $(19, 8, 0, 0)$.

The above computational procedure can be streamlined somewhat. First, there is no need to make a formal break in the tableau notation when passing from the first stage of a linear programming problem, the minimization of the w function, to the second stage, the minimization of the z function. Once a basic feasible solution to the original problem has been found, the w row of the augmented tableau notation can be dropped and the problem continued directly using the z row.

Second, once an artificial variable is extracted from the basis, there is no need to reenter it in any future step. To see this, consider the above example after the first pivot operation. The data of the first two rows of the second tableau of Table 3.8 correspond to the following two equations:

$$\begin{aligned} x_1 - 2x_2 - 3x_3 - 2x_4 + x_5 \quad &= 3 \\ x_2 + 5x_3 + 3x_4 - x_5 + x_6 &= 8 \end{aligned} \tag{3.6.6}$$

Setting x_5, the artificial variable removed from the basis in the first iteration, equal to zero yields the system of equations

$$\begin{aligned} x_1 - 2x_2 - 3x_3 - 2x_4 \quad &= 3 \\ x_2 + 5x_3 + 3x_4 + x_6 &= 8 \end{aligned} \tag{3.6.7}$$

a system equivalent to the constraints of (3.6.5) with $x_5 = 0$, that is, the system of equations

$$\begin{aligned} x_1 - 2x_2 - 3x_3 - 2x_4 \quad &= 3 \\ x_1 - x_2 + 2x_3 + x_4 + x_6 &= 11 \end{aligned} \tag{3.6.8}$$

Now the constraints of the original problem (3.6.4) have feasible solutions if and only if (3.6.8) has feasible solutions with $x_6 = 0$ if and only if (3.6.7) has feasible solutions with $x_6 = 0$. Thus, if (3.6.4) has feasible solutions, the simplex algorithm applied the problem of minimizing the function "w" $= x_6$ subject to the constraints of (3.6.7) would drive this modified w function to zero using only the variables of (3.6.7). (Notice that to apply the algorithm to the function "w" $= x_6$ subject to the constraints of (3.6.7), the basic variables of (3.6.7), x_1 and x_6, would first need to be extracted from the expression for the objective function. Thus the second equation of (3.6.7) would be subtracted from this expression; the resulting form is exactly that of the bottom row of the second tableau of Table 3.8, with x_5 set equal to zero.) Hence the artificial variable x_5 need never return to the basis after the first iteration. As a result, in applying the simplex algorithm, it is never necessary to use the information in the artificial variable columns of the tableau, and so these data need not be calculated at each pivot step.

Example 3.6.2. Minimize $x_1 + x_2 + x_3 = z$ subject to

$$\begin{aligned} -x_1 + 2x_2 + x_3 &\leq 1 \\ -x_1 + 2x_3 &\geq 4 \\ x_1 - x_2 + 2x_3 &= 4 \\ x_1, x_2, x_3 &\geq 0 \end{aligned}$$

Adding two slack variables, the problem in standard form becomes

$$\begin{aligned} &\text{Minimize } x_1 + x_2 + x_3 = z \\ &\text{subject to} \\ &-x_1 + 2x_2 + x_3 + x_4 = 1 \\ &-x_1 + 2x_3 - x_5 = 4 \\ &x_1 - x_2 + 2x_3 = 4 \\ &x_1, x_2, x_3, x_4, x_5 \geq 0 \end{aligned}$$

Note that the x_4 variable can serve as a basic variable. Thus it is sufficient to add only two artificial variables, say x_6 and x_7, to the problem and at the first stage minimize the function $w = x_6 + x_7$. The problem is then

$$\begin{aligned} -x_1 + 2x_2 + x_3 + x_4 &= 1 \\ -x_1 + 2x_3 - x_5 + x_6 &= 4 \\ x_1 - x_2 + 2x_3 + x_7 &= 4 \\ x_1 + x_2 + x_3 &= z \\ x_6 + x_7 &= w \end{aligned}$$

Subtracting the second and third equations from the w equation gives the equation $x_2 - 4x_3 + x_5 = -8 + w$. Now the expression for w does not contain the initial basic variables x_4, x_6, and x_7, and the simplex method can be initiated. The resulting tableaux are given in Table 3.10. The minimal value for the function $w = x_6 + x_7$ is $\frac{4}{3}$, and this value is attained at the point $(\frac{2}{3}, 0, \frac{5}{3}, 0, 0, \frac{4}{3}, 0)$. Therefore we can conclude that the original problem has no feasible solution.

Problem Set 3.6

Note: Again the use of the LP Assistant software is strongly recommended. The program provides easy designation of artificial variables and automatically computes the relevant w-function data into the working tableau.

Table 3.10

	x_1	x_2	x_3	x_4	x_5	x_6	x_7	
x_4	-1	2	(1)	1	0	0	0	1
x_6	-1	0	2	0	-1	1	0	4
x_7	1	-1	2	0	0	0	1	4
	1	1	1	0	0			0
	0	1	-4	0	1			-8
x_3	-1	2	1	1	0			1
x_6	1	-4	0	-2	-1			2
x_7	(3)	-5	0	-2	0			2
	2	-1	0	-1	0			-1
	-4	9	0	4	1			-4
x_3	0	$\frac{1}{3}$	1	$\frac{1}{3}$	0			$\frac{5}{3}$
x_6	0	$-\frac{7}{3}$	0	$-\frac{4}{3}$	-1			$\frac{4}{3}$
x_1	1	$-\frac{5}{3}$	0	$-\frac{2}{3}$	0			$\frac{2}{3}$
	0	$\frac{7}{3}$	0	$\frac{1}{3}$	0			$-\frac{7}{3}$
	0	$\frac{7}{3}$	0	$\frac{4}{3}$	1			$-\frac{4}{3}$

1. Using the technique described in this section, find solutions with nonnegative coordinates to the following systems of equations.

(a)
$$\begin{aligned} x_1 - x_2 \qquad &= 1 \\ 2x_1 + x_2 - x_3 &= 3 \end{aligned}$$

(b)
$$\begin{aligned} x_1 + x_2 \qquad &= 1 \\ 2x_1 + x_2 - x_3 &= 3 \end{aligned}$$

2. Solve the following.

(a) Minimize $2x_1 + 2x_2 - 5x_3$
subject to
$$\begin{aligned} 3x_1 + 2x_2 - 4x_3 &= 7 \\ x_1 - x_2 + 3x_3 &= 2 \end{aligned}$$
$x_1, x_2, x_3 \geq 0$

(b) Minimize $x_1 - 3x_3$
subject to
$$\begin{aligned} x_1 + 2x_2 - x_3 &\leq 6 \\ x_1 - x_2 + 3x_3 &= 3 \end{aligned}$$
$x_1, x_2, x_3 \geq 0$

(c) Minimize $x_1 + x_2 - x_4$
subject to

$$\begin{aligned} 4x_1 + x_2 + x_3 + 4x_4 &= 8 \\ x_1 - 3x_2 + x_3 + 2x_4 &= 16 \\ x_1, x_2, x_3, x_4 &\geq 0 \end{aligned}$$

(d) Maximize $3x_1 - x_2$
subject to

$$\begin{aligned} x_1 - x_2 &\leq 3 \\ 2x_1 &\leq x_2 \\ x_1 + x_2 &\geq 12 \\ x_1, x_2 &\geq 0 \end{aligned}$$

(e) Maximize $x_1 + 2x_2 + 3x_3 + 4x_4$
subject to

$$\begin{aligned} x_1 + x_3 - 4x_4 &= 2 \\ x_2 - x_3 + 3x_4 &= 9 \\ x_1 + x_2 - 2x_3 - 3x_4 &= 21 \\ x_1, x_2, x_3, x_4 &\geq 0 \end{aligned}$$

(f) Minimize $8x_1 - 2x_2 - x_3 - 6x_4$
subject to

$$\begin{aligned} x_1 + x_2 - x_3 + x_4 &= 12 \\ -2x_1 + 3x_2 + 2x_4 &= 42 \\ x_1, x_2, x_3, x_4 &\geq 0 \end{aligned}$$

(g) Minimize $3x_1 - x_2 + 2x_3 + 5x_4 + 6x_5$
subject to

$$\begin{aligned} 12x_1 - 3x_2 + 5x_3 - 2x_4 + 4x_5 &= 100 \\ 8x_1 - 2x_2 - 4x_3 + 5x_5 &= 150 \\ x_1, x_2, x_3, x_4, x_5 &\geq 0 \end{aligned}$$

3. Using a combination of birdseed mixtures A, B, and C, a blend of minimum cost which is at least 20% thistle and 30% corn is desired. Given the data which follow, determine the percentages of each of the mixtures in the final blend.

	% Thistle	*% Corn*	*Cost (cents/lb)*
A	25	40	57
B	0	30	13
C	10	15	20

4. Consider the tableaux for the first stage of the problem discussed in Example 3.6.1. The very last row, the w row after the second pivot step in Table 3.8,

contains four 0's, two 1's, and one 0. This row corresponds to what function? Why was this result expected?

5. Suppose that the objective function z for a linear programming problem is unbounded. Show that this can be learned from the simplex method only after the first stage of the method is completed.

6. Show that at each step of the first stage of the simplex method, the coefficients d_j^* of the w function are equal to $-\sum a_{ij}^*$, where the sum is over those rows i that isolate the remaining artificial variables.

7. Prove that Theorem 3.4.2 can never apply to the w function (i.e., if any $d_s^* < 0$, there must exist an $a_{is}^* > 0$).

8. Compute the solution to Problem 12 of Section 2.3.

9. In the first tableau for the problem of Example 3.6.1, the second column contains three negative entries: the a_{12} and a_{22} entries and the c_2 entry. Evidently the objective function of the problem is bounded below and Theorem 3.4.2 does not apply. Why not?

3.7 REDUNDANT SYSTEMS

In the previous section it was seen that by introducing artificial variables, a linear programming problem could be put into canonical form by means of the simplex method applied to the function w, defined to be the sum of the artificial variables. If the original problem has no feasible solutions, this method would also make evident that fact. In this section we will discuss one minor complication, the problem of redundancy, that could occur with the original system of constraints.

It could very well be that an equation or some equations in the original system of constraints are linear combinations of the remaining equations in the system. This often occurs when, for ease of formulation, more than the minimal number of necessary equations are introduced into a problem. For example, the five constraints of the transportation problem formulated in Section 2.4 (Example 2.4.1) contain one redundant equation. Now the simplex method described in Sections 3.3–3.5 could begin only with the original problem in canonical form. Clearly, however, if a system of equations is in canonical form, there can be no redundant equations because of the isolated nature of the basic variables. It would seem at first glance that it would be necessary to ferret out redundant equations from the system of constraints before the machinery of the simplex method could be applied to a linear programming problem. Fortunately, this is not the case. In this section we will show that from the first stage of the simplex method using artificial variables, redundancies in the original system of constraints can be discovered and deficiencies caused by the lack of a complete set of basic variables from the original set of variables can be compensated for by the presence of artificial variables.

Suppose we now apply the two-stage simplex method to an arbitrary linear programming problem. We have already shown that if the simplex process applied to the w function stops without driving w to zero (i.e., we reach a step in which Theorem 3.4.1 applies to the w function but $w_0 > 0$), the original problem has no feasible solution. Thus suppose that after several pivot operations, the minimal value of w is determined to be zero. If at this time no artificial variables remain in the basis, the original system of constraints must be in canonical form and so contains no redundant equations. Stage two of the simplex method can now begin.

Consider the remaining case: the value of w is driven to zero, but artificial variables remain as basic variables. Note that since w is the sum of the artificial variables, and since we have reached a point in the simplex process at which the value of w at the associated basic feasible solution is zero, those artificial variables remaining in the basis must be at zero level; that is, the constant terms b_i^* in those constraining equations containing the artificial variables must be zero. We now attempt to replace these artificial variables in the basis with variables from the original set.

Suppose the ith equation in the set of constraints defined by this tableau we have reached contains one of the remaining artificial variables. Consider the coefficients, say a_{ij}^*, $1 \leq j \leq n$, in this ith row. If any $a_{ij}^* \neq 0$, pivot at this term, replacing the artificial variable associated with that row in the basis with a variable from the original set. Since the artificial variable was at zero level, this pivot operation leaves the constant-term column, the right-hand column, unchanged (see Problem 6 of Section 3.4). Continue this process wherever possible. If, however, a point is reached at which the ith row contains a remaining artificial variable but $a_{ij}^* = 0$ for all j, $1 \leq j \leq n$, we can conclude that because of redundancies, it is impossible to find a complete set of m basic variables from the original set. In fact, the number of redundant equations would equal the number of artificial variables remaining with coefficient row zero. However, stage two of the simplex method can still be initiated on the tableau at hand, and the rows of zeros corresponding to the remaining artificial variables can be ignored. In essence, this tableau, ignoring the zero rows, corresponds to an independent system of equations in canonical form and equivalent to the original system of constraints.

We now summarize the procedure for driving a linear programming problem in standard form to a problem in canonical form.

1. Add artificial variables to each constraint where necessary.
2. Define the auxiliary objective function w equal to the sum of the artificial variables.
3. Using the constraints of the problem, express w in terms of the nonartificial variables.
4. Apply the simplex algorithm to find the minimum value of the w function.
5. If $\text{Min}\, w > 0$, stop; the original problem has no feasible solution.
6. (a) If $\text{Min}\, w = 0$ and no artificial variables remain in the basis, the original problem is in canonical form; apply the simplex algorithm to the problem.

   (b) If $\text{Min}\, w = 0$ but artificial variables remain in the basis, use the pivot operation to replace with variables from the original set all those artificial variables which have suitable nonzero a_{ij} coefficients. After replacing all that can be replaced, the original problem is in canonical form and the simplex algorithm can be applied.

Example 3.7.1. Minimize $z = 2x_1 - x_2 + x_3$ subject to

$$\begin{aligned} x_1 - 2x_2 + 3x_3 + x_4 &= 6 \\ -x_1 + x_2 + 2x_3 + \tfrac{2}{3}x_4 &= 4 \\ x_1, x_2, x_3, x_4 &\geq 0 \end{aligned}$$

Adding two artificial variables, x_5 and x_6, for initial basic variables and expressing $w = x_5 + x_6$ in terms of the original variables, we have the system

$$\begin{aligned} x_1 - 2x_2 + 3x_3 + x_4 + x_5 &= 6 \\ -x_1 + x_2 + 2x_3 + \tfrac{2}{3}x_4 + x_6 &= 4 \\ 2x_1 - x_2 + x_3 &= z \\ x_2 - 5x_3 - \tfrac{5}{3}x_4 &= -10 + w \end{aligned}$$

From the tableaux of Table 3.11, we see that the minimal value of the objective function z is zero, and is attained at the point $(0,0,0,6)$. Notice that the first pivot term could have been either the 3 or 2 of the x_3 column (or either term in the x_4 column, for that matter). The purpose of the second pivot step is to eliminate the artificial variable x_6 from the basis, and this pivot could have been made at either nonzero entry in the x_6 row of the second tableau. Since both artificial variables were extracted from the basis, the original system of constraints contained no redundancies.

Table 3.11

	x_1	x_2	x_3	x_4	x_5	x_6	
x_5	1	−2	(3)	1	1	0	6
x_6	−1	1	2	$\frac{2}{3}$	0	1	4
	2	−1	1	0			0
	0	1	−5	$-\frac{5}{3}$			−10
x_3	$\frac{1}{3}$	$-\frac{2}{3}$	1	$\frac{1}{3}$			2
x_6	$-\frac{5}{3}$	($\frac{7}{3}$)	0	0			0
	$\frac{5}{3}$	$-\frac{1}{3}$	0	$-\frac{1}{3}$			−2
	$\frac{5}{3}$	$-\frac{7}{3}$	0	0			0
x_3	$-\frac{1}{7}$	0	1	($\frac{1}{3}$)			2
x_2	$-\frac{5}{7}$	1	0	0			0
	$\frac{10}{7}$	0	0	$-\frac{1}{3}$			−2
	0	0	0	0			0
x_4	$-\frac{3}{7}$	0	3	1			6
x_2	$-\frac{5}{7}$	1	0	0			0
	$\frac{9}{7}$	0	1	0			0

Example 3.7.2. Minimize $z = x_1 + 4x_2 + 3x_3 + 2x_4$ subject to

$$\begin{aligned} x_1 + 2x_2 \quad\quad\quad + x_4 &= 20 \\ 2x_1 + x_2 + x_3 \quad\quad &= 10 \\ -x_1 + 4x_2 - 2x_3 + 3x_4 &= 40 \\ x_1, x_2, x_3, x_4 \geq 0 \end{aligned}$$

Adding artificial variables x_5, x_6, and x_7 and expressing $w = x_5 + x_6 + x_7$ in terms of x_1, x_2, x_3, and x_4, the system becomes

$$\begin{aligned} x_1 + 2x_2 \quad\quad + x_4 + x_5 \quad\quad\quad &= 20 \\ 2x_1 + x_2 + x_3 \quad\quad\quad + x_6 \quad &= 10 \\ -x_1 + 4x_2 - 2x_3 + 3x_4 \quad\quad\quad + x_7 &= 40 \\ x_1 + 4x_2 + 3x_3 + 2x_4 \quad\quad\quad &= z \\ -2x_1 - 7x_2 + x_3 - 4x_4 \quad\quad\quad &= -70 + w \end{aligned}$$

The steps of the simplex method are displayed in the tableaux of Table 3.12. The minimal value of the objective function z is 35, and this value is attained at the

Table 3.12

	x_1	x_2	x_3	x_4	x_5	x_6	x_7	
x_5	1	2	0	1	1	0	0	20
x_6	2	(1)	1	0	0	1	0	10
x_7	−1	4	−2	3	0	0	1	40
	1	4	3	2				0
	−2	−7	1	−4				−70
x_5	−3	0	−2	(1)				0
x_2	2	1	1	0				10
x_7	−9	0	−6	3				0
	−7	0	−1	2				−40
	12	0	8	−4				0
x_4	−3	0	−2	1				0
x_2	(2)	1	1	0				10
x_7	0	0	0	0				0
	−1	0	3	0				−40
	0	0	0	0				0
x_4	0	$\frac{3}{2}$	$-\frac{1}{2}$	1				15
x_1	1	$\frac{1}{2}$	$\frac{1}{2}$	0				5
x_7	0	0	0	0				0
	0	$\frac{1}{2}$	$\frac{7}{2}$	0				−35

point $(5,0,0,15)$. Since an artificial variable cannot be removed from the basis, the original system of constraints contains one redundant equation. Notice that any nonzero term in the x_5 or x_7 row of the second tableau could have been used as the pivot term of the second step.

Problem Set 3.7

1. Show that the third equation in the set of constraints for the problem of Example 3.7.2 is a linear combination of the other two equations.

2. Show that the system of constraints for the transportation problem formulated in Example 2.4.1 on page 34 is redundant by:

   (a) Exhibiting a relationship between the equations.
   (b) Solving the problem using the simplex method.

3. For each of the following, determine the optimal value of the objective function, an optimal solution point, and whether or not the system of constraints contains any redundancies.

   (a) Minimize $x_1 + x_2 + x_3 + 3x_4$
   subject to
   $$\begin{aligned} 3x_1 - x_2 + 3x_3 + 6x_4 &= 150 \\ 2x_1 + 2x_2 - x_3 + 4x_4 &= 100 \\ x_1, x_2, x_3, x_4 &\geq 0 \end{aligned}$$

   (b) Maximize $5x_1 + 3x_2 + 3x_3$
   subject to
   $$\begin{aligned} 2x_1 + x_2 + x_3 &= 12 \\ 3x_1 + x_2 + 2x_3 &= 18 \\ x_1, x_2, x_3 &\geq 0 \end{aligned}$$

   (c) Minimize $x_1 + 2x_2 - x_4$
   subject to
   $$\begin{aligned} x_1 - x_2 + 3x_3 &= 1 \\ x_2 - 2x_3 + x_4 &= 1 \\ 3x_1 + x_2 + x_3 + x_4 &= 7 \\ x_1, x_2, x_3, x_4 &\geq 0 \end{aligned}$$

   (d) Maximize $3x_1 - 2x_2 - 2x_3 + 2x_4$
   subject to
   $$\begin{aligned} -x_1 + 3x_2 - x_3 + 2x_4 &= 1 \\ -2x_1 + 4x_3 - x_4 &= 8 \\ 2x_1 - 2x_2 + 2x_3 + x_4 &= 2 \\ x_1, x_2, x_3, x_4 &\geq 0 \end{aligned}$$

(e) Minimize $x_1 + 2x_2 + 3x_3 + 4x_4$
subject to

$$\begin{aligned} x_1 - x_2 + 2x_3 + x_4 &= 6 \\ x_1 + 2x_2 + 2x_3 &= 12 \\ 3x_1 + 3x_2 + 6x_3 + x_4 &= 30 \\ x_1, x_2, x_3, x_4 &\geq 0 \end{aligned}$$

(f) Minimize $x_1 + 3x_2 + 2x_3 + x_4$
subject to

$$\begin{aligned} -10x_1 + 5x_2 + 5x_3 + x_4 &= 30 \\ x_1 + x_3 &= 1 \\ 2x_2 + x_4 &= 16 \\ x_1, x_2, x_3, x_4 &\geq 0 \end{aligned}$$

(g) Minimize $x_1 + 6x_2 + x_3 + x_4$
subject to

$$\begin{aligned} x_1 + 3x_2 - x_3 &= 15 \\ x_1 + x_2 + 2x_3 - x_4 &= 5 \\ x_1 + 7x_2 - 7x_3 + 2x_4 &= 35 \\ x_1, x_2, x_3, x_4 &\geq 0 \end{aligned}$$

(h) Maximize $x_1 + x_2 + x_3$
subject to

$$\begin{aligned} x_1 + 2x_2 + 3x_3 &= 42 \\ 3x_1 + x_3 &= 6 \\ x_1 + 4x_2 + 7x_3 &= 90 \\ x_1, x_2, x_3 &\geq 0 \end{aligned}$$

4. True or false: Suppose that the system of constraints for a linear programming problem in standard form is not redundant and has no degenerate solutions. Then at that step in the first stage of the simplex method when the w function first attains a value of zero, no artificial variables can remain in the basis.

5. (Requires linear algebra.) Show that when applying the simplex method to a linear programming problem with m constraining equations and n unknowns, if, after driving w to zero, it is impossible to drive r artificial variables from the basis, then the rank of the coefficient matrix of the original system of constraints is $m - r$.

3.8 A Convergence Proof

Consider a linear programming problem presented in canonical form, as in (3.4.1) on page 78. We proved in Section 3.4 that, under the assumption that no basic feasible solution to the problem was degenerate, the simplex process must terminate after a finite number of steps, with the process driving to either the minimal value of the objective function (Theorem 3.4.1 applying) or a set of feasible solutions on which the objective function is unbounded below (Theorem 3.4.2 applying). However, the proof breaks down if degeneracy is present because there could exist a sequence of pivot steps for which the associated value of the objective function remains fixed and the basic feasible solutions repeat, that is, the process cycles. An example of such a problem is given in Appendix B.

However, as already shown in our examples, at any step in the simplex process there may be more than one term qualified to serve as the pivot term. For example, pivoting can occur in any column with a negative c_j (and a positive a_{ij}), and if c_s is negative and the minimum of $\{b_i/a_{is} : a_{is} > 0\}$ is not attained in a unique row, pivoting can occur in the s column at any row attaining this minimum. We will show in this section that for any problem, degenerate or not, it is always possible to select a finite sequence of pivot steps that leads either to the minimal value of the objective function or to a set of feasible solutions on which the objective function is unbounded. Thus we will prove that although cycling is possible, by a proper choice of pivot terms it can be avoided and a step reached in the simplex process where either Theorem 3.4.1 or 3.4.2 applies. The proof that we give is by induction on m, the number of equations in the system of constraints, and is due to Dantzig [6] (or see also [7]).

In the following, we will continue our present use of the notation. Thus the constants a_{ij}, b_i, and c_j, $1 \leq i \leq m$, $1 \leq j \leq n$, refer to the coefficients of the constraining equations, the constant terms for these equations, and the coefficients for the objective function, as in (3.4.1). After the system has been modified by the application of the simplex method, the corresponding constants will be denoted by a_{ij}^*, b_i^*, and c_j^*, respectively.

Theorem 3.8.1. *For any linear programming problem presented in canonical form, there exists a finite sequence of pivot steps that leads to either of the following:*

(a) The minimal value of the objective function with Theorem 3.4.1 applying.

(b) A set of feasible solutions on which the objective function is unbounded below, with Theorem 3.4.2 applying applying.

That is, the sequence leads to a presentation of the problem in canonical form with either of the following:

(a) All the coefficients c_j^* of the nonbasic variables in the expression for the objective function are nonnegative.

(b) For some column index s, $c_s^* < 0$ and $a_{is}^* \leq 0$ for all i.

Before proving the theorem, we state two lemmas. The proofs are left to the reader.

Lemma 3.8.1. *If all the constant terms b_i, $1 \leq i \leq m$, of the system of constraints for a linear programming problem equal zero, then after a pivot operation, all the $b_i^* = 0$. If initially at least one $b_i \neq 0$, then after a pivot operation at least one $b_i^* \neq 0$.*

Lemma 3.8.2. *Given a linear programming problem with at least one nonzero term b_i, if there exists a sequence of pivot steps leading to the completion of the problem* *(reaching either (a) or (b) of the above theorem), this same sequence of pivot steps leads to the completion of that problem derived from the given problem by replacing all the nonzero b_i's with zero.*

Proof of Theorem 3.8.1. The proof proceeds by induction on m, the number of equations in the system of constraints. For the case $m = 1$, we have a problem with only one constraining equation. If the constant term b_1 is initially nonzero, Lemma 3.8.1 guarantees that this constant term will remain nonzero after a pivot step. Thus the former argument, valid in the nondegenerate case (see the discussion which follows Theorem 3.4.3 and its proof on page 81), applies here and shows that the simplex process must terminate after a finite number of steps. Now if $b_1 = 0$, we can replace it with any positive constant and then apply Lemma 3.8.2.

Now we prove the induction step. Thus assume that the theorem is valid for any problem with a system of constraints containing $m-1$ or fewer equations. With this assumption, the theorem will be proven for a system with m constraining equations.

Consider first any linear programming problem with m constraints and at least one $b_i \neq 0$. We apply the simplex process until, due to degeneracy, it is not possible to find a pivot operation that reduces the value of the objective function at that step. Rearrange the constraints so that the constant terms for exactly the first r equations are zero, that is, $b_i = 0$, $1 \leq i \leq r$, and $b_i > 0$, $r+1 \leq i \leq m$. Note that from Lemma 3.8.1, $r < m$. Let us call this canonical form of the problem Form I.

Consider the linear programming problem derived from the problem of Form I by deleting the last $m-r$ equations in the system of constraints. Notice that the last $m-r$ basic variables from Form I do not appear anywhere in this problem, and the problem is in canonical form with basic variables consisting of the first r basic variables of Form I. To this problem we can apply the induction assumption. Using it, we find a sequence of pivot steps that leads to a canonical form of the problem with either of the following:

(a) All $c_j^* \geq 0$, $1 \leq j \leq n$.

(b) At least one $c_s^* < 0$ and all $a_{is}^* \leq 0$, $1 \leq i \leq r$.

Now apply these same pivot steps to the full problem in Form I. The resulting problem will be in canonical form because, first, the last $m-r$ basic variables from Form I combine with the r basic variables from the first r equations to give m distinct basic variables and, second, since each pivot term is in a row with a zero constant term, the constant-term column remains unchanged and the associated basic solution feasible. Notice also that the effect of these pivot operations on the c_j^* row is completely independent of the addition of the $m-r$ constraints. There are now three possibilities.

1. If the sequence of pivot steps on the r equations reached condition (a) above, then all $c_j^* \geq 0$ implies that the minimal value of the objective function on the full constraint set has been attained.
2. If condition (b) was reached and $a_{is}^* \leq 0$ also for $r+1 \leq i \leq m$, the objective function is unbounded below on the full system of constraints.

3. If condition (b) was reached but $a_{is}^* > 0$ for some i, $r+1 \leq i \leq m$, then a new pivot term can be found in the s column at that row where

$$\text{Min}\left\{\frac{b_i^*}{a_{is}^*} : a_{is} > 0\right\}$$

is attained. Since $b_i^* > 0$, pivoting here reduces the value of the objective function at the associated basic feasible solution.

In sum we have shown that, for any problem with at least one $b_i \neq 0$, at a given step of the simplex process, either:

1. The minimum value of the objective function or a set on which the objective function is unbounded is attained, or
2. A single pivot term or a sequence of pivot steps can be found that leads to a reduced value for the objective function.

Since there are only a finite number of basic feasible solutions and the reduced value for the objective function guarantees that they cannot repeat, the simplex process must eventually terminate.

Finally, to complete the proof, we must prove the theorem for a problem with m constraints and all constant terms equal to zero. But in this case we can simply replace a constant term with any positive constant and use what we have already proven and Lemma 3.8.2. □

Corollary 3.8.1. *Given a linear programming problem with a system of constraints that has feasible solutions and an objective function to be minimized that is bounded below, there exists at least one feasible solution (in fact, a basic feasible solution) at which the objective function attains its minimal value.*

Thus we have shown that a bounded objective function of a linear programming problem with feasible solutions must attain its optimal value, a property shared by continuous functions on closed and bounded sets (compact sets) and in contrast to the problem of optimizing $f(x) = x$ on the set $0 < x < 1$. Moreover, we know that this optimal value can be attained at a point with, at most, m nonzero coordinates, where m is the number of equations in the constraints.

One final note in passing. The proof of convergence of the simplex algorithm given in this section is an existence proof but not a constructive proof. The proof, by induction, only demonstrates the existence of a sequence of pivot steps leading to the termination of the simplex algorithm applied to any linear programming problem and does not prescribe a constructive method to use to actually determine this sequence. However, constructive proofs certainly do exist. In fact, a constructive procedure that is related to the inductive proof of this section and involves the modification (perturbation) of the constant terms of the problem has been developed by Philip Wolfe [8]. Another constructive technique to prevent cycling, and one of the most curious, is a simple pivoting rule stated in terms of only the indices of the involved variables. It was developed by R. Bland [9]. His algorithm: to avoid cycling, at each pivot step to determine both the exiting and entering variables, when there is more than one eligible variable, use the variable with the smallest index.

Although these procedures solve the cycling problem in theory, cycling in practice is another question. Various factors influencing cycling can be involved in a computer implementation of the simplex algorithm, such as roundoff errors, special pivoting rules, data scaling, and built-in perturbation techniques; and in fact, some linear programming problems have caused cycling in some programmed versions of the algorithm (see, e.g., [10]). However, the issue of cycling in practice is just part of the broader question of the efficiency of a given solution algorithm being implemented on a particular computer system to resolve the specific class of problems under consideration.

Problem Set 3.8

1. Prove Lemma 3.8.1.
2. Prove Lemma 3.8.2. *Hint.* Consider the effect or noneffect of these pivot operations on the b_i column and the c_j row.
3. Prove Corollary 3.8.1. Note that Theorem 3.8.1 applies only to a problem presented in canonical form.
4. True or false: Suppose the simplex method is applied to a linear programming problem presented in canonical form and that, at each step, there is at most one term that could serve as a pivot term. Then for this problem, cycling is impossible.
5. True or false: Given a linear programming problem with $n = m + 1$ and presented in canonical form, at most one step in the simplex method is necessary to drive the process to termination.
6. Using Lemma 3.8.2, solve the linear programming problem of:
   (a) Example 3.5.1, but with the constant terms 60, 10, and 50 replaced with zeros.
   (b) Example 3.5.2, but with the constant terms 7 and 3 replaced with zeros.
7. True or false: Given a linear programming problem with all the constant terms of the system of constraints equal to zero, either the objective function is unbounded or it attains its optimal value at the point zero.

3.9 Linear Programming and Convexity

In Section 2.2 we considered a linear programming problem involving only two variables. We were able to graph the set of feasible solutions to the set of constraints (Figure 2.3) and, by a geometric argument, show that the optimal value of the linear objective function must be attained at a corner or vertex to this solution set. This result generalizes, as suggested at the end of Section 3.2. In this section we will first define the concept of convexity and show that the solution set to a system of

Figure 3.4

equations and inequalities is convex. Then we will define the concept of a vertex of a convex set and relate the basic feasible solutions of a system of constraints to the vertices of the solution set to this system. The corollary of the previous section will then give directly the generalization of the above result.

Only the concept of convexity will be used later in the book, and then not until Section 8.3 and Chapter 10. We present these ideas here primarily to initiate an appreciation of some of the geometry underlying the linear programming problem.

For two points P and Q in $\mathbb{R}^n$, the *line segment* between P and Q is that set of points in $\mathbb{R}^n$ of the form $tP+(1-t)Q$ for $0 \leq t \leq 1$ (see Problem 1). A subset S of $\mathbb{R}^n$ is said to be *convex* if, for any two points of S, the line segment between these two points is also in S.

Example 3.9.1. Of the six subsets of $\mathbb{R}^2$ shown in Figure 3.4, each of the three on the left is convex, while none of the three on the right is convex.

Example 3.9.2. Let $S = \{(x_1,x_2) \in \mathbb{R}^2 : x_1+x_2 \geq 2\}$. Then S is convex, a fact obvious from a graph of S. To prove this algebraically using only our definitions, take any two points $P=(p_1,p_2)$ and $Q=(q_1,q_2)$ in S. Then $p_1+p_2 \geq 2$ and $q_1+q_2 \geq 2$. Take any point

$$tP+(1-t)Q = (tp_1+(1-t)q_1, tp_2+(1-t)q_2), \text{ with } 0 \leq t \leq 1$$

on the line segment between P and Q. We have

$$\begin{aligned} tp_1+(1-t)q_1+tp_2+(1-t)q_2 &= t(p_1+p_2)+(1-t)(q_1+q_2) \\ &\geq 2t+2(1-t) \\ &= 2 \end{aligned}$$

using the fact that t and $1-t$ are nonnegative. Thus $tP+(1-t)Q$ is in S, and we have an algebraic proof that S is convex.

The set of feasible solutions to a linear programming problem is convex, since it is the intersection of a collection of hyperplanes and half-spaces. We state these results in the following, leaving the proofs of the theorems for the reader.

Definition 3.9.1. A subset of $\mathbb{R}^n$ of the form

$$X = \{(x_1,\ldots,x_n) : a_1x_1+a_2x_2+\cdots+a_nx_n = b\}$$

for constants $a_1,a_2,\ldots,a_n$ and b is called a *hyperplane*.

A subset of the form

$$X = \{(x_1, \ldots, x_n) : a_1x_1 + a_2x_2 + \cdots + a_nx_n \leq b\}$$

for constants $a_1, a_2, \ldots, a_n$ and b is called a *half-space*.

Theorem 3.9.1. *A half-space is convex.*

Theorem 3.9.2. *The intersection of two convex sets is convex.*

Corollary 3.9.1. *The set of feasible solutions to a linear programming problem is convex.*

Intuitively, the corners or vertices of a convex set are those points of the set that do not lie on the interior of a line segment contained in the set. This suggests the following.

Definition 3.9.2. A point P of a convex set S is a *vertex* of S if P is not the midpoint of a line segment connecting two other points of S.

Example 3.9.3. For the three convex figures of Example 3.9.1, the line segment has two vertices (the two end points), the triangle has three (the three corners), and the home plate has five.

Theorem 3.9.3. *Let S be the set of feasible solutions to the system of constraints of a linear programming problem in a standard form. Then any basic feasible solution to the problem is a vertex of S.*

Proof. Let X be a basic feasible solution, and suppose the first m variables are the basic variables, with n the total number of variables. Assume $X = (P+Q)/2$, where $P = (p_1, \ldots, p_n)$ and $Q = (q_1, \ldots, q_n)$ are in S. Then

$$\begin{aligned} X &= (x_1, \ldots, x_n, 0, \ldots, 0) \\ &= \frac{1}{2}(p_1 + q_1, \ldots, p_m + q_m, p_{m+1} + q_{m+1}, \ldots, p_n + q_n). \end{aligned}$$

Since all the coordinates of P and Q are nonnegative,

$$p_j = q_j = 0 \text{ for } j = m+1, \ldots, n$$

But there is only one basic feasible solution, X, with all these coordinates equal to zero (see Problem 10 of Section 3.2). Thus $P = Q = X$. Hence X is a vertex of S. □

Corollary 3.9.2. *If the objective function of a linear programming problem has a finite optimal value, this value is assumed by at least one vertex of the set of feasible solutions to the system of constraints.*

Proof. This follows directly from Theorem 3.9.3 and Corollary 3.8.1. □

In the simplex algorithm we move from basic feasible solution to basic feasible solution by replacing at each step one variable in the basis. From Theorem 3.9.3, we see that we are simply moving from vertex to vertex in the convex set of feasible solutions to the system of constraints. In fact, since at each step exactly one basic variable is replaced, we are actually moving between adjacent vertices. See Problem 10 for a development of these ideas.

By using the corollary of the previous section in the proof of the above corollary, we have made use of the central theorems of this chapter, theorems that have been proved algebraically. In fact, the above result can also be proved independently using only the theory of convex sets. (See, for example, Problem 11.) This suggests an alternative, theoretically sound approach to the linear optimization problem. First, compute all the basic feasible solutions to the problem; second, compare the value of the objective function at each of these points. As long as we know that the function has an optimal value, it must be the optimal value in this set. However, this technique is far from practical; if the constraint system has m equations and n unknowns, there could be up to $\binom{n}{m}$ basic feasible solutions, where

$$\binom{n}{m} = \frac{n!}{m!(n-m)!}$$

is the binomial coefficient. For example,

$$\binom{15}{5} = 3003 \text{ and } \binom{20}{10} = 184,756$$

Problem Set 3.9

1. Suppose P and Q are points in $\mathbb{R}^n$. Show geometrically that the set $tP+(1-t)Q = Q+t(P-Q)$, $0 \le t \le 1$, is the line segment connecting P and Q.

2. Prove Theorem 3.9.1. (*Hint.* Use Example 3.9.2 as a model.)

3. Prove Theorem 3.9.2.

4. Prove Corollary 3.9.1.

5. Theorems 3.9.1 and 3.9.2 imply immediately that a hyperplane is convex. Why?

6. True or false:

   (a) The union of two convex sets is convex.
   (b) The complement of a convex set is convex.

7. True or false: A point P is a vertex of a convex set S if and only if P is not the interior point of any line segment in S. (An interior point of a line segment L is any point of L other than the two end points.)

8. Prove that if P and Q are vertices of a convex set S and $X = P+t(Q-P)$ is a point of S, then $0 \le t \le 1$.

9. Consider the general linear programming problem (3.4.1) on page 78. Suppose $P = (b_1, \ldots, b_m, 0, \ldots, 0)$ and $Q = (0, b_2^*, \ldots, b_m^*, b_{m+1}^*, \ldots, 0)$ are distinct basic

feasible solutions, and $X = (x_1, x_2, \ldots, x_m, x_{m+1}, 0, \ldots, 0)$ is a feasible solution. Show that $X = P + t(Q - P)$ for some t, $0 \leq t \leq 1$. (*Hints*. For each solution Q and X, use the equations of (3.4.1) to express the first m coordinates of the solution in terms of the $(m+1)$ coordinate. Problem 8 is also of use.)

10. Let S be the set of feasible solutions to a linear programming problem. A line segment L joining two vertices of S is an *edge* of S if no point of L is the midpoint of a line segment between two points in S but not on L. Two such vertices of S joined by an edge are said to be *adjacent* vertices. Show that in each step of the simplex algorithm, we move from a vertex of S to an adjacent vertex. (*Hint*. Use Problem 9.)

11. Define a function $f(X) = c \cdot X = c_1x_1 + c_2x_2 + \cdots + c_nx_n$ for $X = (x_1, \ldots, x_n) \in \mathbb{R}^n$, where $c = (c_1, \ldots, c_n) \in \mathbb{R}^n$ is a constant

(a) Show that for any P and Q in $\mathbb{R}^n$ and $t \in \mathbb{R}$, $0 \leq t \leq 1$, $f(tP + (1-t)Q) = tf(P) + (1-t)f(Q)$. Note that $tf(P) + (1-t)f(Q)$ is a real number between $f(P)$ and $f(Q)$.

(b) What does this suggest about the optimal value of f on a convex set S in $\mathbb{R}^n$?

12. Suppose T is the set of those feasible solutions to a linear programming problem at which the objective function of the problem attains its optimal value. What does Problem 11(a) say about T?

13. Let S be the set of solutions to

$$\begin{array}{llcrcrcl}
x_1 & + a_{1,k+1}x_{k+1} & + & a_{1,k+2}x_{k+2} + \ldots & + & a_{1n}x_n & = & b_1 \\
\quad \ddots & & & & & & & \\
\qquad x_k & + a_{k,k+1}x_{k+1} & + & a_{k,k+2}x_{k+2} + \ldots & + & a_{kn}x_n & = & b_k \\
 & & & a_{k+1,k+2}x_{k+2} + \ldots & + & a_{k+1,n}x_n & = & b_{k+1} \\
 & & & \vdots & & & & \\
 & & & a_{m,k+2}x_{k+2} + \ldots & + & a_{m,n}x_n & = & b_n \\
x_1, \ldots, x_n \geq 0 & & & & & & &
\end{array}$$

where $k \leq m < n$ and at least one $a_{i,k+1} \neq 0$, $1 \leq i \leq k$. Now suppose that $P = (p_1, \ldots, p_n) \in S$ and $p_i > 0$ for $1 \leq i \leq k+1$. Show that P is not a vertex of S. (*Hint*. Use the system of equations to generate points P^+ and P^- of S with the $(k+1)$ coordinate equal to $p_{k+1} \pm \varepsilon$, $\varepsilon > 0$ and such that $P = (P^+ + P^-)/2$.)

14. *Converse to Theorem 3.9.3*. Let S be the set of feasible solutions to the system of constraints of a linear programming problem in standard form. Show that any vertex of S is a basic feasible solution to the problem. (*Hint*. Given any vertex P of S, by rearranging variables if necessary, we can assume that $P = (p_1, \ldots, p_\ell, 0, \ldots, 0)$ with $p_i > 0$ for $1 \leq i \leq \ell$. Now try to make $x_1, \ldots, x_\ell$ basic variables using Problem 13.)

3.10 Spreadsheet Solution of a Linear Programming Problem

While the simplex method can be used to solve linear programming problems of any size, if we are restricted to working by hand or with LP Assistant, large problems can easily become unmanageable. However, there are many commercial products that can solve large and realistic problems. In this section we demonstrate via examples the use of one such product, Microsoft Excel's spreadsheet tool Solver. A description of this application and an outline of how to use it are presented in Appendix E. Here we present only the final spreadsheet resolution using Solver for three examples. In subsequent chapters we will discuss Solver's associated sensitivity report.

Example 3.10.1. Using units of the component materials A, B, C, and D, Company Zeta produces Products 1, 2, and 3. The input (units of each component material) and profit per unit produced of the products, and the available supplies for the next month of the component materials, are as follows.

	Product			
Component	1	2	3	*Supply (units)*
A	16	30	28	1550
B	24	40	36	2044
C	30	50	32	2438
D	10	20	15	975
Profit/unit	\$78	\$136	\$104	

To determine the optimal production schedule and profit for the next month, the company analyst defines variables x_1, x_2, x_3 to be the number of units of product i to be produced, $i = 1,2,3$, and formulates the following model:

$$
\begin{aligned}
&\text{Maximize profit } z \text{ (in \$)}, \ z = 78x_1 + 136x_2 + 104x_3 \\
&\text{subject to} \\
&16x_1 + 30x_2 + 28x_3 \le 1550 \\
&24x_1 + 40x_2 + 36x_3 \le 2044 \\
&30x_1 + 50x_2 + 32x_3 \le 2438 \\
&10x_1 + 20x_2 + 15x_3 \le \ \ 975 \\
&x_1, x_2, x_3 \ge 0
\end{aligned}
$$

The spreadsheet resolution appears in Figure 3.5. Company Zeta's optimal profit for next month is \$6,748, attained by making 10 units of Product 1, 37 units of Product 2, and 9 units of Product 3. The component materials constraints show that with this production schedule, surplus units remain only for material A. However, with this, as with any spreadsheet resolution of a problem, much of the action is behind the scenes. For example, besides what is seen on the spreadsheet, formulas define the values of the objective function cell and the cells for the left-hand and right-hand sides of the constraints. Furthermore, beyond the spreadsheet, the actual mathematical problem

	A	B	C	D	E	F
1	**Company Zeta**					
2				**Product**		
3		**Component**	1	2	3	**Supply**
4		A	16	30	28	1550
5		B	24	40	36	2044
6		C	30	50	32	2438
7		D	10	20	15	975
8		**Profit/unit**	$78	$136	$104	
9						
10				**Variables**		
11		Product #	1	2	3	
12		Units made	10	37	9	
13						
14		**Maximize Profit**	$6,748			
15						
16		Comp. Materials	LHS		RHS	
17		A	1522	≤	1550	
18		B	2044	≤	2044	
19		C	2438	≤	2438	
20		D	975	≤	975	

Figure 3.5

is established on the tool Solver, and then Solver is invoked to resolve the problem. This is all explained in Appendix E.

Example 3.10.2 (Similar to Example 2.2.3). A landscaper has two fields to maintain, Field X and Field Y, with each field requiring grass seed mixtures of specified percentages of bluegrass and fescue. To meet these needs, the landscaper has three grass seed blends with which to work. The relevant data are summarized in the following table.

		Bluegrass	*Fescue*	*Cost (cents/lb)*
	Blend I	60%	10%	80
Composition	*Blend II*	20%	50%	95
	Blend III	25%	15%	35
Requirements	*Field X*	$\geq 30\%$	$\geq 10\%$	
	Field Y	$\geq 25\%$	$\geq 45\%$	

The landscaper has an order for 200 lb of seed for Field X and 180 lb of seed for Field Y; and on hand to fill the order there are unlimited amounts of Blends I and II but only 125 lb of Blend III.

To determine the minimum cost to meet these demands, the following model is formulated. Let x_1, x_2, x_3 be the number of pounds of Blends I, II, and III, respectively, used for Field X, and let y_1, y_2, y_3 be the number of pounds of each used for Field Y. The problem:

	A	B	C	D	E	F
1	Landscaper					
2			Composition Data for Blends			
3			Bluegrass	Fescue	Cost (cents/lb)	
4		Blend I	60%	10%	$0.80	
5		Blend II	20%	50%	$0.95	
6		Blend III	25%	15%	$0.35	
7						
8			Min Requirements Data for Fields			
9			Bluegrass	Fescue	Pounds	
10		Field X	30%	10%	200	
11		Field Y	25%	45%	180	
12						
13			Variables (lb by Blend and Field)			
14			Field X	Field Y		
15		Blend I	75	22.5		
16		Blend II	0	157.5		
17		Blend III	125	0.00		
18						
19		Minimize Cost	$271.38			
20						
21		Constraints	LHS		RHS	
22		Field X Bluegrass	76.25	≥	60	
23		Field X Fescue	26.25	≥	20	
24		Field X Total	200	=	200	
25		Field Y Bluegrass	45	≥	45	
26		Field Y Fescue	81	≥	81	
27		Field Y Total	180	=	180	
28		Blend III Maximum	125	≤	125	

Figure 3.6

To minimize the function $(80x_1 + 95x_2 + 35x_3) + (80y_1 + 95y_2 + 35y_3)$ subject to

$$x_1 + x_2 + x_3 = 200 \qquad y_1 + y_2 + y_3 = 180$$

$$.6x_1 + .2x_2 + .25x_3 \geq 0.3(200) = 60 \qquad .6y_1 + .2y_2 + .25y_3 \geq .25(180) = 45$$

$$.1x_1 + .5x_2 + .15x_3 \geq 0.1(200) = 20 \qquad .1y_1 + .5y_2 + .15y_3 \geq .45(180) = 81$$

$$x_3 + y_3 \leq 125$$

$$x_1, x_2, x_3 \geq 0 \qquad y_1, y_2, y_3 \geq 0$$

The spreadsheet resolution is shown in Figure 3.6. The minimum cost for the landscaper is $271.38, attained by using 75 lb of Blend I and 125 lb of Blend III in preparing the 200-lb mix for Field X and using 22.5 lb of Blend I and 157.5 lb of Blend II in preparing the 180-lb mix for Field Y. All of the available 125 pounds of Blend III are utilized.

This suggests an obvious question. How much money might be saved if more of Blend III were available? The answer to this question, and similar ones, is available from the final tableau of the simplex algorithm resolution of the problem and on

Solver's associated sensitivity report accompanying the spreadsheet of Figure 3.6. We shall see all this in the next two chapters.

Example 3.10.3 (A second look at Example 2.3.3)**.** A cabinet shop makes and sells two types of cabinets: type 1, for the kitchen, and type 2, for the bathroom. Manufacture of the cabinets consists of two steps, making the frames and drawers, and then assembling and finishing the units. Labor requirements, in hours/unit, are as follows:

Cabinet	*Frame/Drawers (hr)*	*Assembly/Finishing (hr)*
Type 1 (kitchen)	2.6	2.1
Type 2 (bathroom)	1.5	1.8

Each week the shop has 480 hr of labor available for the manufacture of the cabinets. Frames and drawers, completed and ready for assembly and finishing, can also be bought from a local dealer at a cost of \$200 for a kitchen frame/drawer set and \$110 for a bathroom frame/drawer set. The kitchen cabinets sell for \$350 each; the first 70 bathroom cabinets sell for \$250 per unit, but any more produced sell for only \$225 per unit. All units produced can be sold.

To determine a production schedule that maximizes net income, and noting that the total number of each type of cabinet produced is the sum of the corresponding number of frames made and the number bought, we define the primary decision variables

$$\begin{aligned} m_i &= \text{number of frame/drawers } \textit{made} \text{ of type } i, i = 1,2 \\ b_i &= \text{number of frames/drawers } \textit{bought} \text{ of type } i, i = 1,2 \end{aligned}$$

and the auxiliary quantities

$$\begin{aligned} u &= \text{number of bathroom cabinets sold } \textit{up} \text{ to } 70 \\ v &= \text{number of bathroom cabinets sold } \textit{over } 70 \end{aligned}$$

and formulate the following model:

$$\begin{aligned} &\text{Maximize } 350(m_1+b_1)+250u+225v-200b_1-110b_2 \\ &\text{subject to} \\ &\qquad 2.6m_1+1.5m_2+2.1(m_1+b_1)+1.8(m_2+b_2) \leq 480 \\ &\qquad m_2+b_2 = u+v \\ &\qquad u \leq 70 \\ &\qquad m_1,m_2,b_1,b_2,u,v \geq 0 \end{aligned}$$

The spreadsheet resolution is shown in Figure 3.7. The optimal production schedule calls for the shop to produce 75.32 kitchen cabinets, making all the component frames and doors in the shop, while producing 70 bathroom cabinets, buying the component frames and doors from the local dealer. The associated net revenue is \$36,161.69.

	A	B	C	D	E
1	**Cabinet Shop**				
2			**Time Required (hr)**		
3			Kitchen	Bathroom	
4		Frame/Drawers	2.6	1.5	
5		Assembly/Finishing	2.1	1.8	
6					
7		Labor Available (hr):	480		
8					
9			All	First 70	All Other
10		Selling Price	Kitchen	Bathroom	Bathroom
11		of Cabinets	$350	$250	$225
12					
13		Cost of Frames	Kitchen	Bathroom	
14		at the Local Dealer	$200	$110	
15					
16			**Variables**		
17			Kitchen	Bathroom	
18		Frames Made	75.32	0.00	
19		Frames Bought	0.00	70.00	
20		Bath Cabinets Sold Up to 70		70	<- u
21		Bath Cabinets Sold After 70		0	<- v
22					
23			Kitchen	Bathroom	
24		(Total Frames Sold)	75.32	70.00	
25					
26		Maximize Net Revenue	$36,161.69		
27					
28		**Constraints**	LHS		RHS
29		Labor	480.00	≤	480
30		Bath Frames Made/Bought = u+v	70.00	=	70
31		u ≤ 70	70.00	≤	70

Figure 3.7

To implement this program, the shop manager could round down and produce 75 kitchen units, frames and all, and 70 bathroom units, buying the associated components. This would generate a profit of \$36,050, \$111.69 less than the absolute maximum (and use 1.5 fewer hours of labor). This may be the optimal integral solution, or maybe not. Integer programming could be used to resolve the issue. (See also Problem 2.)

Problem Set 3.10

1. For readers with access to and facility with Microsoft Excel and Solver, enter into Microsoft Excel the problem of Example 3.10.1.
   (a) Confirm that your model generates the same solution as in the text.
   (b) Decrease the number of available units of component material A by 20 to 1530 and solve the problem. Note that nothing changes. But this was expected. Why?
   (c) Returning the number of available units of A to 1550, show that if the number of units of material B is increased by 20 to 2064, profit increases \$25,

and if the number is increased by 40 to 2084, profit increases by \$50; but if the number is increased by 60 to 2104, profit increases by less than \$75. (This too was expected, but only with the sensitivity report for the original problem in hand, as we will see in Chapter 5.)

2. For the model of Example 3.10.3, show that it is feasible to produce 75 kitchen cabinets and 70 bathroom cabinets if all the required kitchen frames and doors are made along with exactly one bathroom frame and door. What is the associated net revenue?

CHAPTER 4

DUALITY

4.1 INTRODUCTION TO DUALITY

Frequently in mathematics there exist relationships between concepts, systems, or problems that are not immediately apparent but, once understood, reap many dividends. For example, consider in calculus the relationship between the integral and the derivative expressed in the Fundamental Theorem of Calculus, or in linear algebra, the relationship between linear transformations and matrices. Relationships such as these not only can be used for practical or computational purposes, but also can provide a unified and coherent theory, so that insights and techniques from one system can contribute to the understanding and usefulness of another.

In this chapter we will develop one such unifying notion, the concept of *duality*. For any linear programming problem, the associated dual linear programming problem will be defined. In Section 4.3 it will be shown that in certain optimization situations, the dual problem arises quite naturally; and in Sections 4.4 and 4.5 important theoretical results relating the two problems will be developed. In particular, in Section 4.4 the fundamental Duality Theorem will be proved.

The concept of duality plays an important role in the remainder of the text. In Section 5.1, we will expand upon the ideas in Sections 4.3 and 4.4 to yield a sensitivity analysis procedure useful in a variety of applications. In Section 5.6 the Dual Simplex Algorithm will be developed, and in Section 7.2 the Transportation Problem Algorithm, a primal-dual algorithm, will be developed. Later, in Chapter 9, when we consider two-person zero-sum games, we will see that the problem of solving such a game is equivalent to solving a linear programming problem and its dual problem, and that the question of the existence of a solution to these games is answered using the Duality Theorem.

We conclude this section with an example that should provide some motivation for the definitions to follow in Section 4.2.

Example 4.1.1. To obtain favorable bulk rates, a soft ice cream producer negotiates 6-month contracts in early summer with distant wholesalers for the weekly purchase of fixed quantities of cream, skim milk, and chocolate syrup. However, in the fall, when the demand for soft ice cream decreases, the producer will be left with a surplus of these three quantities. In particular, suppose that in the fall there is weekly 100 gal of cream unused in the production of the ice cream, 300 gal of skim milk, and 60 lb of chocolate syrup.

An Introduction to Linear Programming and Game Theory, Third Edition. By P. R. Thie and G. E. Keough.

To utilize this surplus, the producer bottles and delivers cases of whole and chocolate milk to a local school. A case of whole milk uses 1 gal of cream and 2 gal of skim milk and yields a net gain of \$3 (selling price less bottling and delivery costs); a case of chocolate milk uses 0.4 gal of cream, 2.5 gal of skim milk, and 0.6 lb of chocolate syrup and yields a gain of \$4. Hoping to maximize the net gain attainable with this surplus, the producer formulates the following linear programming problem, with x_1 the number of cases of whole milk and x_2 the number of cases of chocolate milk to be produced each week.

$$\begin{aligned} &\text{Maximize } 3x_1 + 4x_2 \\ &\text{subject to} \\ &\quad x_1 + 0.4x_2 \leq 100 \\ &\quad 2x_1 + 2.5x_2 \leq 300 \\ &\quad 0.6x_2 \leq 60 \\ &\quad x_1, x_2 \geq 0 \end{aligned} \tag{4.1.1}$$

However, before this problem is solved and contracts are signed with the local school, the producer is contacted by the manager of the town dairy. The dairy also supplies milk to the local school system and, in fact, strives to be the sole such supplier. This would increase the dairy's presence in the town and would also allow the dairy some freedom in negotiating prices for the school contract. To accomplish this, the manager of the dairy offers to simply buy from the ice cream producer his surplus milk and syrup, which the dairy would then use in its own bottling plant.

The offer intrigues the ice cream producer. It would allow him to focus his company on the making and selling of ice cream and, if the dairy's offer is financially sound, to continue the economical bulk rate contracts with the distant wholesalers. But what prices for the surplus ingredients are financially sound to the producer?

To attempt to answer this question, the dairy manager notes that the only value to the producer that the surplus milk and syrup have is in bottling and selling cases of whole milk and chocolate milk. In particular, suppose the manager offers the producer y_1 dollars for each gallon of surplus cream, y_2 dollars for each gallon of skim, and y_3 dollars for each pound of chocolate syrup. Then, since the bottling and delivery of a case of whole milk requires 1 gal of cream and 2 gal of skim milk and yields a gain of \$3, the dairy manager realizes that to be competitive, y_1 and y_2 must be set so that $y_1 + 2y_2 \geq 3$. Similarly, consideration of the input and gain associated with a case of chocolate milk yields the inequality $0.4y_1 + 2.5y_2 + 0.6y_3 \geq 4$. Of course, the dairy manager also wants to keep her total costs down and so, in determining these prices, she is led to the following linear programming problem:

$$\begin{aligned} &\text{Minimize } 100y_1 + 300y_2 + 60y_3 \\ &\text{subject to} \\ &\quad y_1 + 2y_2 \geq 3 \\ &\quad 0.4y_1 + 2.5y_2 + 0.6y_3 \geq 4 \\ &\quad y_1, y_2, y_3 \geq 0 \end{aligned} \tag{4.1.2}$$

The linear programming problem (4.1.2) is the dual of the problem (4.1.1). We have been led to these problems by considering the disposal of surplus goods from two different but related perspectives. Other examples in which the dual arises quite naturally will be discussed in Section 4.3. For the time being, let us note some relationships between the two problems (4.1.1) and (4.1.2). (As we will see, these relationships constitute the definition of the dual linear programming problem.)

1. Problem (4.1.2) is a minimization problem with ($\geq$) constraints; (4.1.1) is a maximization problem with ($\leq$) constraints.
2. The number of nonnegative variables in (4.1.2) equals the number of constraints in (4.1.1). (A price was to be set using (4.1.2) for each limited resource in (4.1.1).)
3. The number of constraints in (4.1.2) equals the number of nonnegative variables in (4.1.1). (The y_1, y_2, y_3 had to compare favorably with each of the two processes of (4.1.1).)
4. (a) The coefficients of the objective function of (4.1.2) are the constant terms of the constraints of (4.1.1).
   (b) The constant terms of the constraints of (4.1.2) are the coefficients of the objective function of (4.1.1).
   (c) The coefficients of the constraints of (4.1.2) are the coefficients of the constraints of (4.1.1), with the rows and columns interchanged (transposed).

Problem Set 4.1

The following problems refer to the example of this section.

1. Solve (4.1.1) graphically. What is the maximum the ice cream producer can earn each week with his surplus?
2. (a) Solve (4.1.2) using the simplex algorithm.
   (b) How much should the dairy manager offer the producer for each gallon of cream? Each gallon of skim? Each pound of syrup?
   (c) What is the total amount the dairy manager would be paying the producer each week? Would he accept the offer?

4.2 DEFINITION OF THE DUAL PROBLEM

The definition of the dual problem will initially be given in terms of a linear programming problem expressed in a special form, called the *max form* of the problem. Problems in another special form, a *min form*, are equally useful. We first define these terms.

Definition 4.2.1. A linear programming problem stated in the following form is said to be in *max form*:

$$\begin{aligned}
&\text{Maximize } z = c_1x_1 + c_2x_2 + \cdots + c_nx_n \\
&\text{subject to} \\
&a_{11}x_1 + a_{12}x_2 + \ldots + a_{1n}x_n \leq b_1 \\
&a_{21}x_1 + a_{22}x_2 + \ldots + a_{2n}x_n \leq b_2 \\
&\vdots \\
&a_{m1}x_1 + a_{m2}x_2 + \ldots + a_{mn}x_n \leq b_m \\
&x_1, x_2, \ldots, x_n \geq 0
\end{aligned} \tag{4.2.1}$$

Thus the max form of a linear programming problem, called simply the *max problem*, is a maximization problem with nonnegative variables and a system of constraints consisting of only ($\leq$) inequalities. Note that there are no restrictions on the signs of the coefficients a_{ij}, constant terms b_i, and coefficients c_j.

Definition 4.2.2. A linear programming problem stated in the following form is said to be in *min form*:

$$\begin{aligned}
&\text{Minimize } z = c_1x_1 + c_2x_2 + \cdots + c_nx_n \\
&\text{subject to} \\
&a_{11}x_1 + a_{12}x_2 + \ldots + a_{1n}x_n \geq b_1 \\
&a_{21}x_1 + a_{22}x_2 + \ldots + a_{2n}x_n \geq b_2 \\
&\vdots \\
&a_{m1}x_1 + a_{m2}x_2 + \ldots + a_{mn}x_n \geq b_m \\
&x_1, x_2, \ldots, x_n \geq 0
\end{aligned} \tag{4.2.2}$$

The *min problem* is a minimization problem with nonnegative variables and a system of constraints consisting of only ($\geq$) inequalities. Again, no restrictions have been placed on the signs of the a_{ij}, b_i, and c_j.

We now define the dual to the max problem (4.2.1). Then we will build on this definition to extend the definition of duality to an arbitrary linear programming problem. As we will see, both the max problem and the min problem (4.2.2) will play equal roles in the summarizing definitions.

Definition 4.2.3. The *dual* of the max problem (4.2.1) is the following linear programming problem:

$$\begin{aligned}
&\text{Minimize } v = b_1y_1 + b_2y_2 + \cdots + b_my_m \\
&\text{subject to} \\
&a_{11}y_1 + a_{21}y_2 + \ldots + a_{m1}y_m \geq c_1 \\
&a_{12}y_1 + a_{22}y_2 + \ldots + a_{m2}y_m \geq c_n \\
&\vdots \\
&a_{1n}y_1 + a_{2n}y_2 + \ldots + a_{mn}y_m \geq c_n \\
&y_1, y_2, \ldots, y_n \geq 0
\end{aligned} \tag{4.2.3}$$

Thus the dual to the max problem (4.2.1) with m ($\leq$) constraints and n nonnegative variables is a minimization problem with m nonnegative variables and n ($\geq$) constraints. For each i, $1 \leq i \leq m$, variable y_i of the dual corresponds to the ith constraint of the max problem. The coefficients of y_i in the ith column of the constraints of (4.2.3) are the coefficients of the ith constraint in (4.2.1). Reciprocally, for each j, $1 \leq j \leq n$, the jth constraint in the dual corresponds to the jth variable x_j in (4.2.1); the coefficients of the variables in the jth constraint in the dual are the coefficients of x_j in the constraints of (4.2.1). Note also the interchange between the constant terms of the constraints and the coefficients of the objective functions for the two problems. (Compare the above with the list of relationships given at the end of the example of the previous section.)

Example 4.2.1. The linear programming problem of

$$\begin{aligned}
&\text{Maximizing } 6x_1 + x_2 + 4x_3 \\
&\text{subject to} \\
&3x_1 + 7x_2 + x_3 \leq 15 \\
&x_1 - 2x_2 + 3x_3 \leq 20 \\
&x_1, x_2, x_3 \geq 0
\end{aligned} \tag{4.2.4}$$

has as its dual the problem of

$$\begin{aligned}
&\text{Minimizing } 15y_1 + 20y_2 \\
&\text{subject to} \\
&3y_1 + y_2 \geq 6 \\
&7y_1 - 2y_2 \geq 1 \\
&y_1 + 3y_2 \geq 4 \\
&y_1, y_2 \geq 0
\end{aligned} \tag{4.2.5}$$

Matrix notation can be used to express any linear programming problem and, in particular, the max problem and its dual problem, succinctly. Using (4.2.1), we will define the coefficient matrix A and column vectors b, c, and X as follows:

$$A = \begin{bmatrix} a_{11} & a_{12} & \dots & a_{1n} \\ a_{21} & a_{22} & \dots & a_{2n} \\ \vdots & & & \\ a_{m1} & a_{m2} & \dots & a_{mn} \end{bmatrix}, \quad b = \begin{bmatrix} b_1 \\ b_2 \\ \vdots \\ b_m \end{bmatrix}, \quad c = \begin{bmatrix} c_1 \\ c_2 \\ \vdots \\ c_n \end{bmatrix}, \quad X = \begin{bmatrix} x_1 \\ x_2 \\ \vdots \\ x_n \end{bmatrix}$$

Let A^t denote the transpose of matrix A, and let $c \cdot X$ denote the dot or scalar product of the vectors c and X. Then

$$A^t = \begin{bmatrix} a_{11} & a_{21} & \dots & a_{m1} \\ a_{12} & a_{22} & \dots & a_{m2} \\ \vdots & & & \\ a_{1n} & a_{2n} & \dots & a_{mn} \end{bmatrix}$$

and

$$c \cdot X = c_1x_1 + c_2x_2 + \cdots + c_nx_n = c^tX = X^tc = X \cdot c$$

The max problem of (4.2.1) is simply to maximize $z = c \cdot X$ subject to $AX \leq b, X \geq 0$, where $AX \leq b$ means that each component of the column vector AX is less than or equal to the corresponding component of the vector b, and $X \geq 0$ is defined similarly, with 0 in this case being the n-dimensional zero vector. Let Y be the m-dimensional column vector $(y_1, y_2, \ldots, y_m)^t$. Then the problem of (4.2.3) is to minimize $v = b \cdot Y$ subject to $A^tY \geq c, Y \geq 0$.

In summary, we have the following:

$$\begin{array}{ll} \text{Max problem:} & \text{Maximize } z = c \cdot X \text{ subject to } AX \leq b, X \geq 0 \\ \text{Dual problem:} & \text{Minimize } v = b \cdot Y \text{ subject to } A^tY \geq c, Y \geq 0 \end{array} \tag{4.2.6}$$

To extend the definition of duality to an arbitrary problem, first note that any linear programming problem is equivalent to a problem stated in max form. For example, we have already seen how a minimization problem can be transformed into an equivalent maximization problem and unrestricted variables replaced by variables restricted in sign. A constraint involving an equality can be replaced by two inequalities in opposite directions. For example, the set of points $(x_1, x_2) \in \mathbb{R}^2$ such that $3x_1 + 2x_2 = 5$ equals the set of (x_1, x_2) such that $3x_1 + 2x_2 \geq 5$ and $3x_1 + 2x_2 \leq 5$. Finally, the direction of an inequality can be changed by multiplication by (-1).

With this, the dual to any linear programming problem can be constructed. To determine this dual, first express the given problem as an equivalent linear programming problem in max form and then use the above definition.

As an application, let us determine the dual to the min problem of (4.2.3), the dual of (4.2.1). The problem as stated is to minimize $b \cdot Y$ subject to $A^tY \geq c$, $Y \geq 0$. Letting $-M$ denote the matrix found by multiplying all the entries of a matrix M by (-1), the problem of (4.2.3) is equivalent to the problem of

$$\text{Maximizing } (-b) \cdot Y \text{ subject to } (-A^t)Y \leq -c,\ Y \geq 0$$

But this problem is in max form, and its dual is, using (4.2.6), to

$$\text{Minimize } (-c) \cdot X \text{ subject to } (-A^t)^tX \geq -b,\ X \geq 0$$

Using the fact that for any matrix M, $(M^t)^t = M$, this problem is equivalent to the problem of

$$\text{Maximizing } c \cdot X \text{ subject to } AX \leq b,\ X \geq 0$$

Note that this is precisely the problem of (4.2.1). We have proven that the dual of the min problem is a max problem and that for any linear programming problem, the dual of the dual is the original problem. Hence, repeated application of this operation of constructing the dual problem to a given problem does not lead to a chain of distinct problems but, instead, cycles after two steps, resulting in exactly two problems, each the dual of the other.

Example 4.2.2. The linear programming problem of

$$\begin{aligned}
&\text{Minimizing } 12x_1 + 9x_2 - 2x_3\\
&\text{subject to}\\
&8x_1 + 3x_2 + 5x_3 \geq 6\\
&x_1 \qquad\quad - 3x_3 \geq -4\\
&x_1, x_2, x_3 \geq 0
\end{aligned}$$

is in min form, and thus, from the above, we can write immediately that its dual is to

$$\begin{aligned}
&\text{Maximize } 6y_1 - 4y_2\\
&\text{subject to}\\
&8y_1 + y_2 \leq 12\\
&3y_1 \qquad \leq 9\\
&5y_1 - 3y_2 \leq -2\\
&y_1, y_2 \geq 0
\end{aligned}$$

We consider now the steps involved in the construction of the dual of a problem first, having an equality constraint, and second, having an unrestricted variable.

Example 4.2.3. To determine the dual of the problem of

$$\begin{aligned}
&\text{Maximizing } 6x_1 + x_2 + 4x_3\\
&\text{subject to}\\
&3x_1 + 7x_2 + x_3 \leq 15\\
&x_1 - 2x_2 + 3x_3 = 20\\
&x_1, x_2, x_3 \geq 0
\end{aligned} \tag{4.2.7}$$

notice that this is the problem of Example 4.2.1 with the second constraint changed to an equality. We replace the equality constraint by two inequalities and multiply the resulting ($\geq$) inequality by (-1) to find the equivalent problem in max form of

$$\begin{aligned}
&\text{Maximizing } 6x_1 + x_2 + 4x_3\\
&\text{subject to}\\
&3x_1 + 7x_2 + x_3 \leq 15\\
&x_1 - 2x_2 + 3x_3 \leq 20\\
&-x_1 + 2x_2 - 3x_3 \leq -20\\
&x_1, x_2, x_3 \geq 0
\end{aligned} \tag{4.2.8}$$

Using variables y_1, y_2, y_3, the dual to (4.2.8) is to

$$\begin{aligned}
&\text{Minimize } 15y_1 + 20y_2 - 20y_3\\
&\text{subject to}\\
&3y_1 + y_2 - y_3 \geq 6\\
&7y_1 - 2y_2 + 2y_3 \geq 1\\
&y_1 + 3y_2 - 3y_3 \geq 4\\
&y_1, y_2, y_3 \geq 0
\end{aligned} \tag{4.2.9}$$

which can be rewritten as

$$\begin{aligned}&\text{Minimize } 15y_1 + 20(y_2 - y_3) \\ &\text{subject to} \\ &3y_1 + \ \ (y_2 - y_3) \geq 6 \\ &7y_1 - 2(y_2 - y_3) \geq 1 \\ &\ \ y_1 + 3(y_2 - y_3) \geq 4 \\ &y_1, y_2, y_3 \geq 0\end{aligned} \tag{4.2.10}$$

which is equivalent to

$$\begin{aligned}&\text{Minimize } 15y_1 + 20y_4 \\ &\text{subject to} \\ &3y_1 + \ \ y_4 \geq 6 \\ &7y_1 - 2y_4 \geq 1 \\ &\ \ y_1 + 3y_4 \geq 4 \\ &y_1 \geq 0,\ y_4 \text{ unrestricted}\end{aligned} \tag{4.2.11}$$

Note that (4.2.11), the dual to (4.2.7), is almost (4.2.5), the dual to (4.2.4). The difference is that in (4.2.11), the variable y_4 corresponding to the equality constraint in (4.2.7) is unrestricted. Clearly, the algebra above generalizes. When defining a dual, any variable in the dual corresponding to an equality constraint in the original problem is unrestricted in sign.

Example 4.2.4. To determine the dual of (4.2.11), a problem in min form except for an unrestricted variable, we first replace the unrestricted variable with the difference of two nonnegative variables (4.2.10), simplify to a problem in min form (4.2.9), write the dual (4.2.8), and replace the last two inequalities with the equivalent equality. This yields (4.2.7), the dual to (4.2.11); and the constraint in the dual generated by the unrestricted variable y_4 in the original problem is an equality. Again, we can generalize. Constraints in a dual corresponding to unrestricted variables in the original problem are equality constraints.

Combining these observations, we summarize the construction of the dual to an arbitrary linear programming problem. First, express the problem, using nonnegative and unrestricted variables, as either a maximization problem with ($\leq$) and equality constraints or a minimization problem with ($\geq$) and equality constraints. The dual can then be immediately formulated.

The dual to a maximization problem is a minimization problem with ($\geq$) and equality constraints, and the dual to a minimization problem is a maximization problem with ($\leq$) and equality constraints. In both cases, unrestricted variables in the original problem generate equality constraints in the associated dual; and reciprocally, equality constraints in the original generate unrestricted variables in the dual problem. Table 4.1 summarizes the relationships.

Table 4.1

Max Problem	$\leftarrow$ *dual* $\rightarrow$	*Min Problem*
ith ($\leq$) inequality		ith nonnegative variable
ith ($=$) constraint		ith unrestricted variable
jth nonnegative variable		jth ($\geq$) inequality
jth unrestricted variable		jth ($=$) constraint
Objective function coefficients		Constant terms of constraints
Constant terms of constraints		Objective function coefficients
Coefficient matrix of constraints A		Coefficient matrix of constraints A^t

Example 4.2.5. The linear programming problem of

$$\begin{aligned}&\text{Minimizing } x_1 - 2x_2 + 3x_3\\&\text{subject to}\\&4x_1 + 5x_2 - 6x_3 = 7\\&8x_1 - 9x_2 + 10x_3 \leq 11\\&x_1, x_2 \geq 0,\ x_3 \text{ unrestricted}\end{aligned}$$

is equivalent to the problem of

$$\begin{aligned}&\text{Minimizing } x_1 - 2x_2 + 3x_3\\&\text{subject to}\\&\ \ 4x_1 + 5x_2 - 6x_3 = 7\\&-8x_1 + 9x_2 - 10x_3 \geq -11\\&x_1, x_2 \geq 0,\ x_3 \text{ unrestricted}\end{aligned}$$

and therefore has as its dual the problem of

$$\begin{aligned}&\text{Maximizing } 7y_1 - 11y_2\\&\text{subject to}\\&\ \ 4y_1 - 8y_2 \leq 1\\&\ \ 5y_1 + 9y_2 \leq -2\\&-6y_1 - 10y_2 = 3\\&y_1 \text{ unrestricted},\ y_2 \geq 0\end{aligned}$$

Example 4.2.6. The linear programming problem of

$$\begin{aligned}&\text{Maximizing } 12x_1 + 2x_2\\&\text{subject to}\\&8x_1 - x_2 \leq 21\\&\ \ x_1 - 6x_2 \geq 13\\&3x_1 - 4x_2 = -11\\&x_1 \text{ unrestricted},\ x_2 \geq 0\end{aligned}$$

is equivalent to the problem of

$$\begin{aligned}
&\text{Maximizing } 12x_1 + 2x_2 \\
&\text{subject to} \\
&\begin{array}{rcrcr}
8x_1 & - & x_2 & \leq & 21 \\
-x_1 & + & 6x_2 & \leq & -13 \\
3x_1 & - & 4x_2 & = & -11
\end{array} \\
&x_1 \text{ unrestricted, } x_2 \geq 0
\end{aligned}$$

and therefore has as its dual the problem of

$$\begin{aligned}
&\text{Minimizing } 21y_1 - 13y_2 - 11y_3 \\
&\text{subject to} \\
&\begin{array}{rcrcrcr}
8y_1 & - & y_2 & + & 3y_3 & = & 12 \\
-y_1 & + & 6y_2 & - & 4y_3 & \geq & 2
\end{array} \\
&y_1, y_2 \geq 0,\ y_3 \text{ unrestricted}
\end{aligned}$$

Problem Set 4.2

1. Determine the dual of each of the following linear programming problems.

(a) Maximize $20x_1 + 30x_2$
subject to

$$\begin{array}{rcrcr}
5x_1 & - & 4x_2 & \leq & 100 \\
-x_1 & + & 12x_2 & \leq & 90 \\
 & & x_2 & \leq & 500
\end{array}$$

$x_1, x_2 \geq 0$

(b) Minimize $4x_1 - 3x_2$
subject to

$$\begin{array}{rcrcr}
6x_1 & + & 11x_2 & \geq & -30 \\
2x_1 & - & 7x_2 & \leq & 50 \\
 & & x_2 & \leq & 80
\end{array}$$

$x_1, x_2 \geq 0$

(c) Maximize $-x_1 + 2x_2$
subject to

$$\begin{array}{rcrcr}
5x_1 & + & x_2 & \leq & 60 \\
3x_1 & - & 8x_2 & \geq & 10 \\
x_1 & + & 7x_2 & = & 20
\end{array}$$

$x_1, x_2 \geq 0$

(d) Minimize $6x_1 + 12x_2 - 18x_3$
subject to

$$\begin{aligned} x_1 - 3x_2 + 6x_3 &= 30 \\ 2x_1 + 8x_2 - 16x_3 &= 70 \end{aligned}$$

$x_1, x_2 \geq 0$, x_3 unrestricted

(e) Maximize $x_1 - 7x_2 + 3x_3$
subject to

$$\begin{aligned} 2x_2 + 5x_3 &= 20 \\ 8x_1 \quad - 3x_3 &= 40 \\ x_2 + 4x_3 &\geq 60 \end{aligned}$$

$x_1, x_3 \geq 0$, x_2 unrestricted

(f) Minimize $2y_1 - 3y_2 + 4y_3$
subject to

$$\begin{aligned} 8y_1 \quad - y_3 &= 50 \\ 6y_2 + y_3 &\leq 60 \end{aligned}$$

$y_1, y_2 \geq 0$, $-15 \leq y_3 \leq 0$

2. (a) Determine the dual to the problem of
Maximizing $x_1 - 2x_2$
subject to

$$\begin{aligned} x_2 &\geq 1 \\ x_1 \quad &\leq 2 \end{aligned}$$

$x_1, x_2 \geq 0$

(b) Rewrite your answer to part (a) as an equivalent maximization problem.

(c) Compare your response in part (b) to the original problem of part (a). Observation?

(d) Show that the following problem is also its own dual.
Maximizing $x_1 - 2x_2 - 3x_3$
subject to

$$\begin{aligned} x_2 + 2x_3 &\geq 1 \\ x_1 \quad + 3x_3 &\leq 2 \\ 2x_1 - 3x_2 \quad &= 3 \end{aligned}$$

$x_1, x_2 \geq 0$, x_3 unrestricted

3. Consider the linear programming problem of Example 4.2.1 of this section.

(a) Show that the objective function of the dual problem is bounded below.

(b) Solve the dual problem graphically.

(c) Solve the maximization problem using the simplex method. Note that the optimal values of the objective functions are equal.

(d) Compare the bottom two entries in the slack variable columns of the last simplex tableau of part (c) with the point in part (b) that yielded the minimal value.

4.3 Examples and Interpretations

In Section 4.1, the dual to a production problem involving profits to be maximized was developed. In this section the dual problems to other specific linear programming examples will be defined and discussed. The examples, using the categories of Chapter 2, are from the classes of blending problems, production problems (minimizing costs while meeting given demands), and transportation problems. Additional examples are contained in the problems at the end of this section.

Example 4.3.1 (A Blending Problem)**.** The diet problems that we have already seen lead to dual problems that have a standard but still extremely interesting interpretation. Consider, for example, the situation described in Example 2.2.1 on page 10 of the farmer wishing to feed her stock. The farmer's problem was to determine a diet using two feeds that minimized cost and satisfied three nutritional requirements. Here, letting x_1 and x_2 denote the amounts in pounds of Feeds 1 and 2 to use, respectively, the mathematical problem was to

$$\begin{aligned}
&\text{Minimize } 10x_1 + 4x_2 \\
&\text{subject to} \\
&3x_1 + 2x_2 \geq 60 \\
&7x_1 + 2x_2 \geq 84 \\
&3x_1 + 6x_2 \geq 72 \\
&x_1, x_2 \geq 0
\end{aligned} \tag{4.3.1}$$

The three inequalities in the system of constraints result from the requirement that the diet provide specified amounts of the nutritional elements A, B, and C. The dual to this problem is the problem of

$$\begin{aligned}
&\text{Maximizing } 60y_1 + 84y_2 + 72y_3 \\
&\text{subject to} \\
&3y_1 + 7y_2 + 3y_3 \leq 10 \\
&2y_1 + 2y_2 + 6y_3 \leq 4 \\
&y_1, y_2, y_3 \geq 0
\end{aligned} \tag{4.3.2}$$

To provide an interpretation of the dual, consider the problem of a traveling salesman dealing in nutrition tablets for cattle. Suppose the salesman has to offer the farmer three types of pure tablet: one type containing 1 unit of nutritional element A and nothing else, one containing 1 unit of B and nothing else, and the last containing 1 unit of C and nothing else. Now the salesman hopes to convince the farmer that it is to her advantage to nourish her cattle by using these tablets instead of any combination of Feeds 1 and 2. Although the farmer is probably somewhat set in her ways, the salesman believes that due to the problems of maintaining a small farm today, he can still appeal to her frugality. Thus the salesman attempts to set the prices for the three types of tablets in such a way that the tablets can compete favorably with the two feeds and he can realize the greatest income. To do this, he lets y_1, y_2, and y_3

Table 4.2

	y_1	y_2	y_3	y_4	y_5	
y_4	3	7	3	1	0	10
y_5	(2)	2	6	0	1	4
	−60	−84	−72	0	0	0
y_4	0	(4)	−6	1	$-\frac{3}{2}$	4
y_1	1	1	3	0	$\frac{1}{2}$	2
	0	−24	108	0	30	120
y_2	0	1	$-\frac{3}{2}$	$\frac{1}{4}$	$-\frac{3}{8}$	1
y_1	1	0	$\frac{9}{2}$	$-\frac{1}{4}$	$\frac{7}{8}$	1
	0	0	72	6	21	144

denote the cost in cents to the farmer of one tablet of nutritional elements A, B, and C, respectively.

Now 1 lb of Feed 1 provides 3, 7, and 3 units of A, B, and C, respectively, and costs 10 cents. To replace 1 lb of this feed with tablets, the farmer would need three tablets each of the first and third types and seven of the second type. This would cost $3y_1 + 7y_2 + 3y_3$ cents and so, to be competitive, the salesman must have

$$3y_1 + 7y_2 + 3y_3 \leq 10$$

Similarly, 1 lb of Feed 2 provides 2, 2, and 6 units of A, B, and C, respectively, and cost 4 cents. Thus we have the inequality

$$2y_1 + 2y_2 + 6y_3 \leq 4$$

Since the farmer has determined that the daily requirements of elements A, B, and C are 60, 84, and 72 units, respectively, the cost of meeting these requirements by using the tablets would be $60y_1 + 84y_2 + 72y_3$. Thus the salesman wishes to maximize this function subject to the above two inequalities. This problem is precisely the dual of the original problem.

Being a former mathematician, the salesman does not stop here but sets out to solve the linear programming problem (4.3.2). Adding two slack variables and using the simplex method, he generates the tableaux of Table 4.2. From the final tableau, in which we see that $y_1 = y_2 = 1$ and $y_3 = 0$, the salesman notes that he should charge the farmer 1 cent for each of the tablets of A and B and nothing for the tablets of C ("Place your order today and receive the C tablets at no extra charge"), and in doing this, he will realize his maximum income of \$1.44. Observe that this maximum income of \$1.44 equals the minimum cost to the farmer of an adequate diet using Feeds 1 and 2, as determined in Section 2.2.

In the next section, we will show that the above result is not just coincidental. The Duality Theorem, as we will see, states that the min problem of (4.3.1) and its dual, the max problem of (4.3.2), must have a common optimal value (as long as at least one of the problems has a finite optimal value). The alert reader may have already noticed another curious fact here. The minimal value of the objective function of (4.3.1), determined graphically in Section 2.2, was attained at the point $x_1 = 6, x_2 = 21$ (see Figure 2.5 on page 13). Note that these values are the last entries in the slack variable columns of the final tableau of the dual problem (Table 4.2). This is also not coincidental and, in fact, it is this relationship on which our proof of the Duality Theorem will rest.

Example 4.3.2. Consider the situation described in Problem 12 of Section 2.3 (see also Problem 8 of Section 3.6). An oil refinery, with three processes, produces three grades of gasoline. The problem is to determine the operation that minimizes cost and satisfies specified demands. Using the data from the problem of Section 2.3 and letting x_j denote the number of hours of operation of Process j, $j = 1, 2$, and 3, the resulting linear programming problem is to

$$\begin{aligned}&\text{Minimize } 160x_1 + 400x_2 + 300x_3\\&\text{subject to}\\&3x_1 + 6x_2 + 6x_3 \geq 36\\&4x_1 + 6x_2 + 3x_3 \geq 20\\&2x_1 + 8x_2 + 4x_3 \geq 30\\&x_1, x_2, x_3 \geq 0\end{aligned}$$

(As determined in Problem 8 of Section 3.6, the minimal cost is \$1950 and is attained by using Process 2 for $1\frac{1}{2}$ hr and Process 3 for $4\frac{1}{2}$ hr.)

The dual of this minimization problem is to

$$\begin{aligned}&\text{Maximize } 36y_1 + 20y_2 + 30y_3\\&\text{subject to}\\&3y_1 + 4y_2 + 2y_3 \leq 160\\&6y_1 + 6y_2 + 8y_3 \leq 400\\&6y_1 + 3y_2 + 4y_3 \leq 300\\&y_1, y_2, y_3 \geq 0\end{aligned}$$

To provide an interpretation of the dual, suppose that the management of the refinery wishes to determine a market price for each grade of gasoline. Although the market prices certainly should reflect the level of the grade, another factor to be considered is the actual cost of production. Thus, with the refinery operating to meet these specified demands, some estimate of the cost of each grade of gasoline must be made. Let y_1, y_2, and y_3 denote the cost in dollars of 100 gal of regular, special, and super gasoline, respectively. Now 1 hr of operation of Process 1 produces 300, 400, and 200 gal of regular, special, and super gasoline and costs \$160. Thus it is reasonable to demand that these estimated production costs satisfy the inequality

$$3y_1 + 4y_2 + 2y_3 \leq 160$$

(We have an inequality here instead of an equality because one or both of the other processes might produce comparable output at less cost.) Similarly, considering the output and costs of Processes 2 and 3, we see that y_1, y_2, and y_3 must satisfy

$$\begin{aligned} 6y_1 + 6y_2 + 8y_3 &\le 400 \\ 6y_1 + 3y_2 + 4y_3 &\le 300 \end{aligned}$$

To justify a price increase to the government, say, the management would desire that these operational costs be as high as possible. More precisely, since the refinery is delivering weekly 3600, 2000, and 3000 gal of regular, special, and super gasolines, it should choose y_1, y_2, and y_3 to maximize the function $36y_1 + 20y_2 + 30y_3$. Thus we have the dual problem.

Example 4.3.3. Consider the transportation problem in Problem 2 of Section 2.4. Letting x_{1j} denote the number of cases shipped from the Eastern Warehouse to Outlet j, $1 \le j \le 4$, and letting x_{2j} be the number shipped from the Western Warehouse, the resulting mathematical problem is to

$$\begin{aligned} \text{Minimize } z = {} & 20x_{11} + 16x_{12} + 30x_{13} + 20x_{14} \\ & +45x_{21} + 39x_{22} + 50x_{23} + 44x_{24} \end{aligned}$$

subject to

$$\begin{array}{lllllllllcr} x_{11} & + x_{12} & + x_{13} & + x_{14} & & & & & \le & 600 \\ & & & & x_{21} & + x_{22} & + x_{23} & + x_{24} & \le & 1000 \\ x_{11} & & & & + x_{21} & & & & = & 300 \\ & x_{12} & & & & + x_{22} & & & = & 350 \\ & & x_{13} & & & & + x_{23} & & = & 400 \\ & & & x_{14} & & & & + x_{24} & = & 450 \end{array}$$

$$x_{ij} \ge 0, 1 \le i \le 2, 1 \le j \le 4$$

To construct the dual, we first change the direction of the two inequalities by multiplying each by (-1). The dual then is to

$$\text{Maximize } -600y_1 - 1000y_2 + 300y_3 + 350y_4 + 400y_5 + 450y_6 \tag{4.3.3}$$

subject to

$$\begin{array}{lllllll} -y_1 & & + y_3 & & & & \le 20 \\ -y_1 & & & + y_4 & & & \le 16 \\ -y_1 & & & & + y_5 & & \le 30 \\ -y_1 & & & & & + y_6 & \le 20 \\ & - y_2 & + y_3 & & & & \le 45 \\ & - y_2 & & + y_4 & & & \le 39 \\ & - y_2 & & & + y_5 & & \le 50 \\ & - y_2 & & & & + y_6 & \le 44 \end{array}$$

$y_1, y_2 \ge 0$; y_3, y_4, y_5, y_6 unrestricted

Suppose now that a national shipping company, wanting to expand, offers to deliver the canned goods for the supplier. Instead of charging normal transportation costs, however, the shipper proposes to buy from the supplier all the available cases of canned goods, paying y_1 cents/case for those cases at the Eastern Warehouse and y_2 cents for the cases at the Western Warehouse. He guarantees delivery of the required number of cases at each of the four outlets, selling the cases back to the supplier at a cost of y_{i+2} cents/case at Outlet i, $1 \leq i \leq 4$.

Now the shipper must determine these six prices in such a way that they are competitive and realize the maximum income. To be competitive, for example, since it costs the supplier 20 cents to ship a case from the Eastern Warehouse to Outlet 1, y_1 and y_3 must be chosen so that $y_3 - y_1 \leq 20$. Consideration of the other seven shipping costs of the supplier leads to the other seven inequalities in the above dual problem. And as long as the y_i satisfy these inequalities, the shipper can assure the supplier that his offer certainly can cost the supplier no more than she is already paying for transportation and may save her money.

The income the shipper will realize from this venture is simply the difference between the total amount he pays at the two warehouses and the total amount he receives at the four outlets. But this difference is precisely the quantity measured by the objective function in the above problem. Thus, in determining the values of the variables y_i, the shipper encounters the dual of the original transportation problem.

Problem Set 4.3

1. Using the simplex method, show that the dual problem of Example 4.3.2 has as solution a maximum value of 1950 attained at the point $y_1 = 33\frac{1}{2}$, $y_2 = 0$, $y_3 = 25$. Interpret the fact that the production cost of special gasoline is zero. How would the vice president in charge of sales react?

2. (a) The shipper of Example 4.3.3 is concerned about the four unrestricted variables y_3, y_4, y_5, and y_6 of (4.3.3). If he determines an optimal solution point to (4.3.3) with negative values for any of these four components, it would mean that he would also pay the supplier for the supplier taking delivery at the corresponding outlets. He wonders if this is the best way to run a shipping business. Relieve his anxieties. Show that any optimal solution point $(y_1, \ldots, y_6)$ to (4.3.3) must have $y_3, y_4, y_5, y_6 \geq 0$.

   (b) In this example, if the supplier lets the shipper transport the canned goods as described, there will result 100 extra cases in the hands of the shipper. How might the two parties resolve this difficulty?

3. Consider the problem of Example 2.3.1 on page 21, a production problem of maximizing profits using limited resources.

   (a) Show that the dual problem is to

$$
\begin{aligned}
&\text{Minimize } 2000y_1 + 300y_2 + 200y_3 \\
&\text{subject to} \\
&50y_1 + 6y_2 + 3y_3 \geq 50 \\
&30y_1 + 5y_2 + 5y_3 \geq 60 \\
&y_1, y_2, y_3 \geq 0
\end{aligned}
$$

Now suppose a competitor approaches the boat manufacturer, offering to operate the small boat division under a lease agreement. The competitor claims that she will pay the manufacturer y_1 dollars for each pound of aluminum available, y_2 dollars for each minute of machine time, and y_3 dollars/hr for the finishing labor, and that these prices will guarantee that the manufacturer realize at least as much income as he could by operating the plant himself. Thus the competitor offers the manufacturer an income comparable to any profit the manufacturer could realize himself while freeing him from determining optimal schedules, running the plant, and selling the boats.

(b) Considering separately the profit the manufacturer realizes from the sale of a rowboat and a canoe, construct two inequalities that the prices y_1, y_2, and y_3 should satisfy. Considering the total amount the competitor pays the manufacturer, compare the problem of determining y_1, y_2, and y_3 with the problem of part (a).

(c) Solve the dual minimization problem of part (a) by the simplex method. Comparing the solution of this problem to the solution to the original problem (see Problem 1 of Section 2.3), show that the competitor's offer of a comparable, if not favorable, income will be realized.

(d) Compare the entries in the slack variable columns in the bottom row of the last tableau computed in part (c) with the values for R and C found in Problem 1 of Section 2.3.

4. Consider Problem 13 of Section 2.3. Letting x_1 and x_2 denote the number of hours of operation of Systems 1 and 2, the resulting mathematical problem is to

$$\begin{aligned}
&\text{Minimize } 2x_1 + 11x_2\\
&\text{subject to}\\
&x_1 + 4x_2 \leq 100\\
&4x_1 + 20x_2 \geq 480\\
&2x_1 + 40x_2 \geq 800\\
&x_1, x_2 \geq 0
\end{aligned}$$

(a) Show that the dual problem is to

$$\begin{aligned}
&\text{Maximize } -100y_1 + 480y_2 + 800y_3\\
&\text{subject to}\\
&-y_1 + 4y_2 + 2y_3 \leq 2\\
&-4y_1 + 20y_2 + 40y_3 \leq 11\\
&y_1, y_2, y_3 \geq 0
\end{aligned}$$

(b) A college student, working for the fruit grower for the summer, believes she can have the fruit picked more efficiently than the grower by using her own system and equipment. Fearing that she has nothing to gain financially by simply revealing her plan to the grower, she suggests to the grower that she will supervise the picking of the crop, paying the grower a set amount for each available hour of labor and then selling back to the grower the fruit, using two prices: one for a bushel of choice produce and the other for regular produce. Considering that the student must convince the grower that it is to his advantage to let her supervise the harvest, how should she set these three costs?

5. Consider Problem 11 of Section 2.3.
   (a) Formulate the associated linear programming problem.
   (b) Determine the dual problem.
   (c) Suppose the manager of the electronics firm wants to assess the value of a unit of material and a unit of labor in the production and sale of the circuits. To do this, she lets $\$y_1$ and $\$y_2$ denote these two values. The circuit for a radio requires 2 units of material and 1 unit of labor and sells for \$8. The manager reasons, therefore, that 2 units of material plus 1 unit of labor must be worth at least \$8, but could be worth more if these units can be used in the production of other types of circuits that are more profitable. Thus she sets $2y_1 + y_2 \geq 8$. The manager continues in this manner. Compare the resulting problem with the problem determined in part (b). (Note that the Duality Theorem guarantees that the optimal values for the problems of parts (a) and (b) are equal.)

4.4 THE DUALITY THEOREM

In this section we prove the celebrated Duality Theorem. It is generally accepted that John von Neumann was the first mathematician to recognize the significance of the duality principle in this setting and endeavor to develop a proof of the Duality Theorem.

We start with the max problem of (4.2.1), the problem of maximizing $z = c \cdot X$ subject to $AX \leq b$, $X \geq 0$. The dual to this problem is to minimize $v = b \cdot Y$ subject to $A^t Y \geq c$, $Y \geq 0$. We will show first that the set of possible values for the objective function z of the max problem lies to the left of the set of possible values for the function v. Then, with this result, we will prove the Duality Theorem using the simplex method and, in particular, Theorem 3.8.1.

Theorem 4.4.1. *Suppose X_0 is a feasible solution to the problem of maximizing $c \cdot X$ subject to $AX \leq b$, $X \geq 0$ and Y_0 is a feasible solution to the dual problem of minimizing $b \cdot Y$ subject to $A^t Y \geq c$, $Y \geq 0$. Then*

$$c \cdot X_0 \leq b \cdot Y_0$$

Proof. Since X_0 is a feasible solution to the max problem with constraints $AX \leq b$, where A is an $m \times n$ matrix, the $m \times 1$ vector $u = b - AX_0 \geq 0$. In fact, the m components of u are the slack in the m inequalities of $AX_0 \leq b$. Similarly, Y_0 a feasible solution to the dual implies that $A^t Y_0 \geq c$, and so the column vector $v = A^t Y_0 - c$ of slack in this set of n inequalities also has nonnegative components. Using these vectors, we can write

$$AX_0 = b - u \quad \text{and} \quad A^t Y_0 = c + v$$

Now since the product $\underbrace{Y_0^t}_{1\times m}\underbrace{A}_{m\times n}\underbrace{X_0}_{n\times 1}$ is a real number, we have $Y_0^tAX_0 = (Y_0^tAX_0)^t = X_0^tA^tY_0$, and so

$$Y_0^tAX_0 = Y_0^t(AX_0) = Y_0^t(b-u) \text{ equals } X_0^tA^tY_0 = X_0^t(A^tY_0) = X_0^t(c+v)$$

that is,

$$Y_0^tb - Y_0^tu = X_0^tc + X_0^tv$$

Thus, since $u, v, X_0, Y_0 \geq 0$,

$$b\cdot Y_0 - c\cdot X_0 = \underbrace{u\cdot Y_0}_{\geq 0} + \underbrace{v\cdot X_0}_{\geq 0} \geq 0$$ □

We state the first corollary below for future reference in Section 4.5. The two subsequent corollaries are for immediate use in this section.

Corollary 4.4.1. *If X_0 is a feasible solution to the problem of maximizing $c\cdot X$ subject to $AX \leq b$, $X \geq 0$ and Y_0 is a feasible solution to the problem of minimizing $A^tY \geq c$, $Y \geq 0$, then*

$$b\cdot Y_0 - c\cdot X_0 = (b - AX_0)\cdot Y_0 + (A^tY_0 - c)\cdot X_0$$

Proof. This is the equality statement of the last line of the above proof. □

Corollary 4.4.2. *If X_0 and Y_0 are feasible solutions to the max and min problems, respectively, and if $c\cdot X_0 = b\cdot Y_0$, then the optimal values of the objective functions z and v equal this common value; that is, maximum $z = c\cdot X_0 = b\cdot Y_0 =$ minimum v and X_0 and Y_0 are optimal solution points for their respective problems.*

Proof. Suppose X_1 is any feasible solution to the max problem. Then, from the theorem, $c\cdot X_1 \leq b\cdot Y_0$, so $c\cdot X_1 \leq c\cdot X_0$. Thus the maximum value of the function $z = c\cdot X$ is $c\cdot X_0$. Similarly for the dual problem. □

Corollary 4.4.3. *If the objective function z of the max problem is not bounded above, the min problem has no feasible solutions. Similarly, if the objective function v of the min problem is not bounded below, the max problem has no feasible solutions.*

The proof of Corollary 4.4.3 is left to the reader (Problem 1). The converse to this corollary is false. Examples can be constructed for which neither the max problem nor its dual, the min problem, have feasible solutions (see Problem 2).

Theorem 4.4.2 (Duality Theorem)**.** *Suppose either the problem of*

$$\textit{Maximizing } z = c\cdot X \textit{ subject to } AX \leq b,\ X \geq 0$$

or the problem of

$$\textit{Minimizing } v = b\cdot Y \textit{ subject to } A^tY \geq c,\ Y \geq 0$$

has a finite optimal solution. Then so does the other problem, and the optimal values of the objective functions are equal, that is,

$$\text{Max } z = \text{Min } v$$

Proof. Assume first that the max problem has a finite optimal solution. Thus we assume the existence of an X_0 such that $AX_0 \leq b$, $X_0 \geq 0$ and, for any other X with $AX \leq b, X \geq 0$, we have $c \cdot X \leq c \cdot X_0$.

Now the solution to the min problem will be found by applying the simplex method to the max problem. To do this, we first write the max problem in standard form by adding m slack variables x_j, $n+1 \leq j \leq n+m$, and multiply the objective function by -1. This gives the problem of

$$\begin{aligned}
&\text{Minimizing } -c_1x_1 - c_2x_2 - \cdots - c_nx_n = -z \qquad (4.4.1)\\
&\text{subject to}\\
&a_{11}x_1 + a_{12}x_2 + \ldots + a_{1n}x_n + x_{n+1} = b_1\\
&a_{21}x_1 + a_{22}x_2 + \ldots + a_{2n}x_n \quad + x_{n+2} = b_2\\
&\vdots\\
&a_{m1}x_1 + a_{m2}x_2 + \ldots + a_{mn}x_n \quad + x_{m+n} = b_m\\
&x_j \geq 0, 1 \leq j \leq n+m
\end{aligned}$$

We now assume in our proof that the constants b_i, $1 \leq i \leq m$, are nonnegative. If this is the case, the above problem is in canonical form with basic variables x_{n+1}, x_{n+2}, $\ldots$, x_{n+m}, since the associated basic solution is feasible, and the simplex method can be initiated directly commencing with the second stage.

(Recall that in Section 4.2 when the max and min problems were defined, no restrictions were placed on the constants. Thus, with this assumption, our proof loses some generality. The extension of the proof to the general case is developed in Problem 8.)

From Theorem 3.8.1, we know that there is a finite sequence of pivot operations driving the problem of (4.4.1) to the optimal value of the objective function. The initial tableaux for such a sequence would have a form such as

	x_1	x_2	$\ldots$	x_n	x_{n+1}	$\ldots$	x_{n+m}	
x_{n+1}	a_{11}	a_{12}	$\ldots$	a_{1n}	1	0	$\ldots$	b_1
x_{n+2}	a_{21}	a_{22}	$\ldots$	a_{2n}	0	1	$\ldots$	b_2
$\vdots$								
x_{n+m}	a_{m1}	a_{m2}	$\ldots$	a_{mn}	0	$\ldots$	1	b_m
	$-c_1$	$-c_2$	$\ldots$	$-c_n$	0	0	0	0

and the final tableau would assume the form

	x_1	x_2	$\ldots$	x_n	x_{n+1}	$\ldots$	x_{n+m}	
	r_1	r_2	$\ldots$	r_n	s_1	$\ldots$	s_m	$c \cdot X_0$

Since our concern will be with only the bottom row of this last tableau, we have used the symbols r_j, $1 \le j \le n$ and s_i, $1 \le i \le m$ to denote the numbers appearing in these positions and have left the other positions of the tableau blank. Since this tableau represents the final step of the simplex process in the problem of (4.4.1), we have $r_j \ge 0$ and $s_i \ge 0$ for $1 \le j \le n$, $1 \le i \le m$, and the minimum of $-z$ is $-c \cdot X_0$.

Let Y_0 be the column vector $(s_1, s_2, \dots, s_m)^t$. We will show that

(a) $Y_0 \ge 0$

(b) $A^t Y_0 \ge c$

(c) $b \cdot Y_0 = c \cdot X_0$

As has already been mentioned, $Y_0 \ge 0$. To show (b) and (c), consider the equation represented by the bottom row of the final tableau:

$$r_1x_1 + \cdots + r_nx_n + s_1x_{n+1} + \cdots + s_mx_{n+m} = c \cdot X_0 + (-z)$$

This equation represents the result of all the pivot operations on the initial equation for the objective function

$$-c_1x_1 - c_2x_2 - \cdots - c_nx_n = 0 + (-z)$$

And, at each pivot step, some linear combination of the original constraining equations was added to this equation for the objective function. Thus there exist m constants, t_i, $1 \le i \le m$, such that when the $(m+1)$ equations

$$\begin{array}{llllllll}
t_1(a_{11}x_1 & + \; a_{12}x_2 & + \; \dots & + \; a_{1n}x_n & + \; x_{n+1} & & & = b_1) \\
t_2(a_{21}x_1 & + \; a_{22}x_2 & + \; \dots & + \; a_{2n}x_n & & + \; x_{n+2} & & = b_2) \\
\quad\vdots & & & & & & & \\
t_m(a_{m1}x_1 & + \; a_{m2}x_2 & + \; \dots & + \; a_{mn}x_n & & & + \; x_{m+n} & = b_m) \\
\quad(-c_1x_1 & - \quad c_2x_2 & - \; \dots & - \quad c_nx_n & & & & = -z)
\end{array}$$

are added together, the result is the equation

$$r_1x_1 + \cdots + r_nx_n + s_1x_{n+1} + \cdots + s_mx_{n+m} = c \cdot X_0 + (-z)$$

Comparing the coefficients of the slack variables, we see that $s_i = t_i$ for $1 \le i \le m$. Using this result and comparing the coefficients of x_1, we have

$$s_1a_{11} + s_2a_{21} + \cdots + s_ma_{m1} - c_1 = r_1 \ge 0$$

and so

$$s_1a_{11} + s_2a_{21} + \cdots + s_ma_{m1} \ge c_1$$

Similarly, comparing the coefficients of x_j for any j, $1 \le j \le n$, we have

$$s_1a_{1j} + s_2a_{2j} + \cdots + s_ma_{mj} - c_j = r_j \ge 0$$

and so

$$s_1a_{1j} + s_2a_{2j} + \cdots + s_ma_{mj} \ge c_j$$

Thus

$$A^t Y_0 \geq c$$

To show (c), consider the constant terms in the above equations. We must have

$$s_1 b_1 + s_2 b_2 + \cdots + s_m b_m = c \cdot X_0$$

that is,

$$b \cdot Y_0 = c \cdot X_0$$

Since $Y_0 \geq 0$ and $A^t Y_0 \geq c$, the point Y_0 is a feasible solution to the min problem. The value of the objective function v at Y_0, $b \cdot Y_0$, is equal to the value of the objective function z at X_0. Thus, from Corollary 4.4.2, the minimal value of v is $b \cdot Y_0$, so the optimal values of both problems are equal.

Finally, suppose that we know initially that it is the min problem that has the finite optimal solution. But in Section 4.2 it was shown that this problem is equivalent to a problem expressed in max form. Thus we can apply what we have already proved to this equivalent problem and conclude that the dual to the min problem, the max problem, has the same optimal solution. □

Corollary 4.4.4. *If both the max and min problems have feasible solutions, then both objective functions have optimal solutions and* $\text{Max}\, z = \text{Min}\, v$.

Proof. Since both problems have feasible solutions, it follows from Theorem 4.4.1 that the objective function z is bounded above and the objective function v is bounded below. From Corollary 3.8.1, both objective functions attain their optimal values and, from the Duality Theorem, these optimal values must be equal. □

In summary, we have shown that there are exactly four different categories into which solutions to the max and min problems can fall.

1. Both problems have feasible solutions. Then the sets of possible values for the objective functions z and v relate on the real line as follows:

$$\begin{array}{c|c} z = c \cdot X & v = b \cdot Y \\ \hline \multicolumn{2}{c}{\downarrow} \\ \multicolumn{2}{c}{\text{optimal value for both}} \end{array}$$

2. The objective function z is unbounded above and the min problem has no feasible solutions.
3. The objective function v is unbounded below and the max problem has no feasible solutions.
4. Both problems have no feasible solutions.

The following example demonstrates an important application of the duality theorem.

Example 4.4.1. Suppose we apply the simplex algorithm to the problem of

$$\begin{aligned}
&\text{Maximizing } -5x_1 + 18x_2 + 6x_3 - x_4 \\
&\text{subject to} \\
&\begin{array}{rcrcrcrcr}
2x_1 & & & - & x_3 & + & 3x_4 & \leq & 20 \\
 & & x_2 & - & 2x_3 & - & x_4 & \leq & 30 \\
-3x_1 & + & 6x_2 & + & 3x_3 & + & 4x_4 & \leq & 24
\end{array} \\
&x_1, x_2, x_3, x_4 \geq 0
\end{aligned} \tag{4.4.2}$$

and the resulting final tableau suggests a maximum value of 112 for the objective function attained at the point $(10,9,0,0)$ (and an optimal solution point of $(2,0,3)$ for the dual). We can now easily check the accuracy of our calculations.

First, is the point $(10,9,0,0)$ a feasible solution to (4.4.2), and is the value of the objective function at this point 112? (It might be hoped that this part of the test procedure is already standard practice.)

Second, consider the dual to (4.4.2)

$$\begin{aligned}
&\text{Minimize } 20y_1 + 30y_2 + 24y_3 \\
&\text{subject to} \\
&\begin{array}{rcrcrcr}
2y_1 & & & - & 3y_3 & \geq & -5 \\
 & & y_2 & + & 6y_3 & \geq & 18 \\
-y_1 & - & 2y_2 & + & 3y_3 & \geq & 6 \\
3y_1 & - & y_2 & + & 4y_3 & \geq & -1
\end{array} \\
&y_1, y_2, y_3 \geq 0
\end{aligned}$$

Now we determine whether the point $(2,0,3)$ is a feasible solution to this problem and whether the value of the associated objective function at $(2,0,3)$ is also 112.

The answers to the above questions are all positive, as the reader may confirm. Corollary 4.4.2 guarantees then that we have calculated accurately and that our proposed optimal value and solution points are correct. The Duality Theorem guarantees that this test procedure is always available.

From the proof of the Duality Theorem, we know that when the simplex algorithm is applied to a maximization problem with ($\leq$) constraints, the entries in the bottom row of the final tableau in the slack variable columns give the optimal solution point to the corresponding dual minimization problem. (We had already seen an example of this in the tableau solution of the maximization problem of (4.3.2), the dual problem constructed in Example 4.3.1.) The following example exploits this result.

Example 4.4.2. Consider the linear programming problem of

$$\begin{array}{l} \text{Minimizing } 20x_1 + 15x_2 + 54x_3 \\ \text{subject to} \\ \begin{array}{rcrcrcr} x_1 & - & 2x_2 & + & 6x_3 & \geq & 30 \\ & & x_2 & + & 2x_3 & \geq & 6 \\ 2x_1 & & & - & 3x_3 & \geq & -5 \\ x_1 & - & x_2 & & & \geq & 18 \end{array} \\ x_1, x_2, x_3 \geq 0 \end{array}$$

To solve this problem using the simplex method, we would first add 4 slack variables, then 3 artificial variables (the slack variable in the third constraint could serve as a basic variable), and use the full two stages of the algorithm on the resulting problem of 4 constraints and 10 variables. However, the dual to this problem is to

$$\begin{array}{l} \text{Maximize } 30y_1 + 6y_2 - 5y_3 + 18y_4 \\ \text{subject to} \\ \begin{array}{rcrcrcrcr} y_1 & & & + & 2y_3 & + & y_4 & \leq & 20 \\ -2y_1 & + & y_2 & & & - & y_4 & \leq & 15 \\ 6y_1 & + & 2y_2 & - & 3y_3 & & & \leq & 54 \end{array} \\ y_1, y_2, y_3, y_4 \geq 0 \end{array}$$

Applying the simplex algorithm to this dual problem is somewhat easier. Adding three slack variables and solving, we have the tableaux of Table 4.3. The maximum value of the objective function $30y_1 + 6y_2 - 5y_3 + 18y_4$ is 522, and therefore the minimum value of the objective function of the original problem also is 522. Moreover, from the bottom row of the final tableau, we see that the point $(18, 0, 3)$ is an optimal solution point to the original problem. (Of course, the application of the simplex algorithm to the dual of the minimization problem is facilitated here by the fact that the coefficients in the original objective function, 20, 15, and 54, are all nonnegative. If this had not been the case, computing the solution to the dual with the simplex algorithm would also have required the use of artificial variables.)

These observations suggest a general question. If we solve any linear programming problem with a finite optimal solution using the simplex algorithm, can we always find in the final tableau an optimal solution point to the dual? We address this issue in the following examples, considering first the resolution of a minimization problem.

Example 4.4.3. Consider the problem of Example 4.3.1 of

$$\begin{array}{l} \text{Minimizing } 10x_1 + 4x_2 \\ \text{subject to} \\ \begin{array}{rcrcr} 3x_1 & + & 2x_2 & \geq & 60 \\ 7x_1 & + & 2x_2 & \geq & 84 \\ 3x_1 & + & 6x_2 & \geq & 72 \end{array} \\ x_1, x_2 \geq 0 \end{array}$$

Table 4.3

	y_1	y_2	y_3	y_4	y_5	y_6	y_7	
y_5	1	0	2	1	1	0	0	20
y_6	-2	1	0	-1	0	1	0	15
y_7	6	2	-3	0	0	0	1	54
	-30	-6	5	-18	0	0	0	0
y_5	0	$-\frac{1}{3}$	$\frac{5}{2}$	1	1	0	$-\frac{1}{6}$	11
y_6	0	$\frac{5}{3}$	-1	-1	0	1	$\frac{1}{3}$	33
y_1	1	$\frac{1}{3}$	$-\frac{1}{2}$	0	0	0	$\frac{1}{6}$	9
	0	4	-10	-18	0	0	5	270
y_4	0	$-\frac{1}{3}$	$\frac{5}{2}$	1	1	0	$-\frac{1}{6}$	11
y_6	0	$\frac{4}{3}$	$\frac{3}{2}$	0	1	1	$\frac{1}{6}$	44
y_1	1	$\frac{1}{3}$	$-\frac{1}{2}$	0	0	0	$\frac{1}{6}$	9
	0	-2	35	0	18	0	2	468
y_4	1	0	2	1	1	0	0	20
y_6	-4	0	$\frac{7}{2}$	0	1	1	$-\frac{1}{2}$	8
y_2	3	1	$-\frac{3}{2}$	0	0	0	$\frac{1}{2}$	27
	6	0	32	0	18	0	3	522

Table 4.4

	x_1	x_2	x_3	x_4	x_5	x_6	x_7	x_8	
x_6	3	2	-1	0	0	1	0	0	60
x_7	7	2	0	-1	0	0	1	0	84
x_8	3	6	0	0	-1	0	0	1	72
	10	4	0	0	0	0	0	0	0
	-13	-10	1	1	1	0	0	0	-216
x_5	0	0	$-\frac{9}{2}$	$\frac{3}{2}$	1	$\frac{9}{2}$	$-\frac{3}{2}$	-1	72
x_1	1	0	$\frac{1}{4}$	$-\frac{1}{4}$	0	$-\frac{1}{4}$	$\frac{1}{4}$	0	6
x_2	0	1	$-\frac{7}{8}$	$\frac{3}{8}$	0	$\frac{7}{8}$	$-\frac{3}{8}$	0	21
	0	0	1	1	0	-1	-1	0	-144

Subtracting three slack variables, adding three artificial variables, and then using the simplex algorithm yields the initial and final tableaux of Table 4.4.

We know from Example 4.3.1 that the optimal solution point for the corresponding dual maximization problem is $y_1 = 1$, $y_2 = 1$, $y_3 = 0$. Note that these values are precisely the numbers in the bottom row of the final tableau in the slack variable columns for the first, second, and third constraints, respectively.

This is always the case when starting with a minimization problem with ($\geq$) constraints: a solution point to the dual is given in the bottom row of the final tableau in the slack variable columns. A proof of this fact is called for in Problem 11. The proof essentially duplicates the proof in the Duality Theorem, with some minor adjustments (here, for example, $s_j = -t_j$, $1 \leq j \leq m$).

These results can be generalized. In a final tableau presenting the optimal value and an optimal solution point for a linear programming problem, the values of the variables in an optimal solution point to the dual for those variables that correspond to inequalities in the original problem are found in the bottom row of the final tableau in the associated slack variable columns.

Example 4.4.4. The dual to the problem of

$$\begin{array}{l} \text{Maximizing } 3x_1 + x_2 - x_3 \\ \text{subject to} \\ \begin{array}{rcrcrcrcr} x_1 & + & x_2 & + & 5x_3 & + & x_4 & \leq & 200 \\ -x_1 & & & + & 2x_3 & & & \geq & 20 \\ & & 2x_2 & - & x_3 & + & 5x_4 & \geq & 50 \end{array} \\ x_1, x_2, x_3, x_4 \geq 0 \end{array}$$

is the problem of

$$\begin{array}{l} \text{Minimizing } 200y_1 - 20y_2 - 50y_3 \\ \text{subject to} \\ \begin{array}{rcrcrcr} y_1 & + & y_2 & & & \geq & 3 \\ y_1 & & & - & 2y_3 & \geq & 1 \\ 5y_1 & - & 2y_2 & + & y_3 & \geq & -1 \\ y_1 & & & - & 5y_3 & \geq & 0 \end{array} \\ y_1, y_2, y_3 \geq 0 \end{array}$$

The initial and final tableau resolution of the maximization problem is in Table 4.5.

The dual variables y_1, y_2, and y_3 correspond to the first, second, and third inequalities, with slack variables x_5, x_6, and x_7, respectively, of the original problem. Thus an optimal solution point to the dual is $y_1 = 1$, $y_2 = 3$, $y_3 = 0$. This is easy to verify. Note that the point $(1,3,0)$ satisfies the dual constraints and has the required optimal value of 140 at the objective function.

The last two examples in this section contain equality constraints in the original problem and thus unrestricted variables in the dual.

Example 4.4.5. The problem of

$$\begin{array}{l} \text{Maximizing } 3x_1 + 5x_2 + 9x_3 \\ \text{subject to} \\ \begin{array}{rcrcrcr} 4x_1 & + & 12x_2 & + & 15x_3 & = & 900 \\ -x_1 & + & 2x_2 & + & 3x_3 & = & 120 \end{array} \\ x_1, x_2, x_3 \geq 0 \end{array}$$

Table 4.5

	x_1	x_2	x_3	x_4	x_5	x_6	x_7	x_8	x_9	
x_5	1	1	5	1	1	0	0	0	0	200
x_8	-1	0	2	0	0	-1	0	1	0	20
x_9	0	2	-1	5	0	0	-1	0	1	50
	-3	-1	1	0	0	0	0	0	0	0
	1	-2	-1	-5	0	1	1	0	0	-70
x_7	$\frac{15}{2}$	0	0	-3	2	$\frac{11}{2}$	1	$-\frac{11}{2}$	-1	240
x_3	$-\frac{1}{2}$	0	1	0	0	$-\frac{1}{2}$	0	$\frac{1}{2}$	0	10
x_2	$\frac{7}{2}$	1	0	1	1	$\frac{5}{2}$	0	$-\frac{5}{2}$	0	150
	1	0	0	1	1	3	0	-3	0	140

Table 4.6

	x_1	x_2	x_3	x_4	x_5	
x_4	4	12	15	1	0	900
x_5	-1	2	3	0	1	120
	-3	-5	-9	0	0	0
	-3	-14	-18	0	0	-1020
x_1	1	$\frac{2}{9}$	0	$\frac{1}{9}$	$-\frac{5}{9}$	$\frac{100}{3}$
x_3	0	$\frac{20}{27}$	1	$\frac{1}{27}$	$\frac{4}{27}$	$\frac{460}{9}$
	0	$\frac{7}{3}$	0	$\frac{2}{3}$	$-\frac{1}{3}$	560

with the dual problem of

$$\begin{aligned}
&\text{Minimizing } 900y_1 + 120y_2 \\
&\text{subject to} \\
&\ \ 4y_1 - \ \ y_2 \geq 3 \\
&12y_1 + 2y_2 \geq 5 \\
&15y_1 + 3y_2 \geq 9 \\
&y_1, y_2 \text{ unrestricted}
\end{aligned}$$

has a maximum value of 560 and an optimal solution point of $(\frac{100}{3}, 0, \frac{400}{9})$, as seen in what we'll refer to as the *reduced tableaux resolution* of the problem in Table 4.6, where only the first and last tableaux are displayed. The unrestricted variables y_1 and y_2 of the dual correspond to the two equalities in the constraints of the maximization problem, and to initiate the simplex algorithm for this problem, artificial variables needed to be introduced. As the reader may have guessed, these artificial variable columns provide the data for the optimal solution point of the dual. Indeed, the required value for the dual objective function of 560 is attained at the point $(\frac{2}{3}, -\frac{1}{3})$, a feasible solution point to the dual, as the reader may confirm.

In general, when solving a *maximization* problem containing equality constraints, the coordinates of the unrestricted dual variables at an optimal solution point to the dual are in the bottom row of the final tableau resolution of the maximization problem in the corresponding artificial variable columns. Problem 12 addresses the proof of this statement.

However, when solving a *minimization* problem containing equality constraints, a sign change adjustment is necessary when determining an optimal solution point to the dual. The value of each unrestricted variable in the optimal solution point to the dual is the *negative* of the entry in the bottom row of the associated artificial variable column. (Why this difference, one might ask? But note that the situations are not identical. For example, our algorithm has been designed for minimization problems; for such a problem, the coefficients of the objective function are entered directly into the initial tableau. To adapt the algorithm to a maximization problem, the corresponding minimization problem is considered, which necessitates an initial sign change in the objective function coefficients when entered into the initial tableau.)

Example 4.4.6. The reduced tableaux resolution for the problem of

$$
\begin{aligned}
&\text{Minimizing } z = 16x_1 + 32x_2 + 12x_3 \\
&\text{subject to} \\
&x_1 + 5x_2 + x_3 \geq 2 \\
&4x_1 + 4x_2 - 2x_3 = 1 \\
&x_1, x_2, x_3 \geq 0
\end{aligned}
$$

is in Table 4.7. We have Min $z = 14$ attained at $(0, \frac{5}{14}, \frac{3}{14})$. The dual problem is to

Table 4.7

	x_1	x_2	x_3	x_4	x_5	x_6	
x_5	1	5	1	-1	1	0	2
x_6	4	4	-2	0	0	1	1
	16	32	12	0	0	0	0
	-5	-9	1	1	0	0	-3
x_3	$-\frac{8}{7}$	0	1	$-\frac{2}{7}$	$\frac{2}{7}$	$-\frac{5}{14}$	$\frac{3}{14}$
x_2	$\frac{3}{7}$	1	0	$-\frac{1}{7}$	$\frac{1}{7}$	$\frac{1}{14}$	$\frac{5}{14}$
	16	0	0	8	-8	2	-14

$$\begin{aligned}
&\text{Maximize } v = 2y_1 + y_2\\
&\text{subject to}\\
&\quad y_1 + 4y_2 \leq 16\\
&5y_1 + 4y_2 \leq 32\\
&\quad y_1 - 2y_2 \leq 12\\
&y_1 \geq 0,\ y_2 \text{ unrestricted}
\end{aligned}$$

From the final tableau, the point $y_1 = 8$ (using the slack variable x_4 column) and $y_2 = -(2) = -2$ (using the artificial variable x_6 column) is an optimal solution point of the dual, as the reader may easily verify.

Problem Set 4.4

1. Prove Corollary 4.4.3.
2. Show that both the following linear programming problem and its dual do not have any feasible solutions.

$$\begin{aligned}
&\text{Maximize } x_1\\
&\text{subject to}\\
&\quad x_1 - x_2 \leq 1\\
&-x_1 + x_2 \leq -2\\
&x_1, x_2 \geq 0
\end{aligned}$$

3. Consider the linear programming problem of

$$\begin{aligned}
&\text{Maximizing } 4x_1 + 10x_2 - 3x_3 + 2x_4\\
&\text{subject to}\\
&\quad 3x_1 - 2x_2 + 7x_3 + x_4 \leq 26\\
&\quad x_1 + 6x_2 - x_3 + 5x_4 \leq 30\\
&-4x_1 + 8x_2 - 2x_3 - x_4 \leq 10\\
&x_1, x_2, x_3, x_4 \geq 0
\end{aligned}$$

   (a) Show that $(\frac{54}{5}, \frac{16}{5}, 0, 0)$ is a feasible solution to this problem. Compute the value of the objective function at this point.
   (b) Write out the dual problem. Show that $(\frac{7}{10}, \frac{19}{10}, 0)$ is a feasible solution to this problem. What is the value of the objective function of the dual at this point?
   (c) Using Corollary 4.4.2, what can you conclude?
4. Verify that $(0, 5\frac{2}{3}, 8\frac{1}{3}, \frac{1}{3})$ is an optimal solution point to the problem of

$$\begin{aligned}
&\text{Minimizing } 7x_1 + 11x_2 - 3x_3 - x_4\\
&\text{subject to}\\
&2x_1 + 2x_2 - x_3 - 3x_4 \geq 2\\
&-x_1 + 5x_2 - 2x_3 + x_4 \geq 12\\
&x_1 - 4x_2 + 3x_3 + 5x_4 \geq 4\\
&x_1, x_2, x_3, x_4 \geq 0
\end{aligned}$$

and $(3\frac{1}{2}, 2, 1\frac{1}{2})$ is an optimal solution point to the dual.

5. Verify that $(0, 3, 2\frac{1}{7}, 0, 1\frac{2}{7})$ is an optimal solution point to the problem of

$$\begin{aligned}
&\text{Maximizing } 3x_1 + 2x_2 + 5x_3 - 2x_4 + x_5\\
&\text{subject to}\\
&4x_1 + x_2 - x_3 + 2x_4 + 4x_5 \leq 6\\
&3x_1 + 3x_2 + 2x_3 - x_4 - x_5 \leq 12\\
&x_1 - 2x_2 + 5x_3 - x_4 + x_5 \leq 6\\
&x_1, x_2, x_3, x_4, x_5 \geq 0
\end{aligned}$$

and that $(\frac{1}{3}, 1, \frac{2}{3})$ is an optimal solution point to the dual.

6. Consider the problem of

$$\begin{aligned}
&\text{Minimizing } z = 13x_1 + 15x_2 + 12x_3 + 8x_4\\
&\text{subject to}\\
&4x_1 + 8x_2 - 5x_3 + 3x_4 = 32\\
&3x_1 - 2x_2 + 6x_3 - x_4 \geq 3\\
&x_1, x_2, x_3, x_4 \geq 0
\end{aligned}$$

(a) Determine which of the following points are feasible solutions to this min problem: $(9, 0, 2, 2)$, $(4, 1, -1, 1)$, and $(5, 1, 1, 3)$.
(b) Evaluate the function z at those points in part (a) that are feasible solutions to the problem.
(c) Write out the dual to the min problem.
(d) Determine which of the following points are feasible solutions to this dual problem: $(-1, 1)$, $(0, 2)$, and $(1, 3)$.
(e) Evaluate the dual objective function at those points in part (d) that are feasible solutions to the problem.
(f) Using the information above, and only this information, what can you say about the minimum value of z?

7. Solve the following by applying the simplex algorithm to the dual:

$$
\begin{array}{l}
\text{Minimize } 8x_1 + 13x_2 + 20x_3 \\
\text{subject to} \\
\begin{array}{rcrcrcr}
3x_1 & + & 2x_2 & + & x_3 & \geq & 2 \\
x_1 & - & x_2 & + & 2x_3 & \geq & 4 \\
 & & 2x_2 & + & 2x_3 & \geq & -1 \\
-2x_1 & + & 3x_2 & & & \geq & 0 \\
4x_1 & & & - & x_3 & \geq & -2
\end{array} \\
x_1, x_2, x_3 \geq 0
\end{array}
$$

8. *Generalization of the proof of the Duality Theorem.* Suppose some of the constant terms b_j in (4.4.1) are negative. By rearranging the constraining equations if necessary, assume that $b_i < 0$ for $1 \leq i \leq k$ and $b_i \geq 0$ for $k+1 \leq i \leq m$. Then, to apply the simplex method to (4.4.1), the first k equations must be multiplied by (-1), resulting in all nonnegative terms in the right column. However, now an initial basic feasible solution may not be apparent; if not, artificial variables must be introduced and the simplex method initiated at stage one. Thus the initial tableau would look something like the following:

x_1		x_n	x_{n+1}		x_{n+k}	x_{n+k+1}		x_{n+m}	*Art. Vars.*			
$-a_{11}$	$\cdots$	$-a_{1n}$	-1	$\cdots$	0	0	$\cdots$	0	1	$\cdots$	0	$-b_1$
$\vdots$												
$-a_{k1}$	$\cdots$	$-a_{kn}$	0	$\cdots$	-1	0	$\cdots$	0	0	$\cdots$	1	$-b_k$
$a_{k+1,1}$	$\cdots$	$a_{k+1,n}$	0	$\cdots$	0	1	$\cdots$	0	0	$\cdots$	0	b_{k+1}
$\vdots$												
$a_{m,1}$	$\cdots$	$a_{m,n}$	0	$\cdots$	0	0	$\cdots$	1	0	$\cdots$	0	b_m
$-c_1$	$\cdots$	$-c_n$	0	$\cdots$	0	0	$\cdots$	0				0

Since we have assumed that the problem of (4.4.1) has feasible solutions, the simplex method initiated on the above tableau will first drive the artificial variables from the basis and then drive to the optimal value of the objective function. Let r_j, s_i, and t_i be defined as in the proof of the Duality Theorem for $1 \leq j \leq n$ and $1 \leq i \leq m$. Show that the proof given there can be extended to this case, with the only difference being that here $s_i = -t_i$ for $1 \leq i \leq k$.

9. Show that the r_j's as defined in the proof of the Duality Theorem measure the slack in the constraints of the dual problem at the $Y_0 = (s_1, s_2, \ldots, s_n)^t$ solution point.

10. The simplex algorithm has been used to resolve the following problems, and the corresponding initial and final tableaux are given (with the w row omitted). For each, construct the dual, determine an optimal solution point to the dual using the data from the tableaux, and verify that your solution point is feasible and optimal.

(a) Minimize $100x_1 + 150x_2$
subject to
$2x_1 + x_2 \geq 13$
$6x_1 - 9x_2 \leq 2$
$7x_1 - 8x_2 \geq 5$
$x_1, x_2 \geq 0$

	x_1	x_2	x_3	x_4	x_5	x_6	x_7	
x_6	2	1	-1	0	0	1	0	13
x_4	6	-9	0	1	0	0	0	2
x_7	7	-8	0	0	-1	0	1	5
	100	150	0	0	0	0	0	0
x_5	0	0	$-\frac{5}{8}$	$\frac{23}{24}$	1	$\frac{5}{8}$	-1	$\frac{121}{24}$
x_1	1	0	$-\frac{3}{8}$	$\frac{1}{24}$	0	$\frac{3}{8}$	0	$\frac{119}{24}$
x_2	0	1	$-\frac{1}{4}$	$-\frac{1}{12}$	0	$\frac{1}{4}$	0	$\frac{37}{12}$
	0	0	75	$\frac{25}{3}$	0	-75	0	$-\frac{2875}{3}$

(b) Maximize $3x_1 - 4x_2 + 5x_3$
subject to
$4x_1 - x_2 + 6x_3 \leq 9$
$x_1 + 2x_2 - x_3 = 54$
$x_1, x_2, x_3 \geq 0$

	x_1	x_2	x_3	x_4	x_5	
x_4	4	-1	6	1	0	9
x_5	1	2	-1	0	1	54
	-3	4	-5	0	0	0
x_1	1	0	$\frac{11}{9}$	$\frac{2}{9}$	$\frac{1}{9}$	8
x_2	0	1	$-\frac{10}{9}$	$-\frac{1}{9}$	$\frac{4}{9}$	23
	0	0	$\frac{28}{9}$	$\frac{10}{9}$	$-\frac{13}{9}$	-68

(c) Minimize $-2x_1 + 5x_2 + 9x_3$
subject to
$2x_2 + 5x_3 \geq 1$
$3x_1 - x_2 - x_3 \leq 6$
$2x_1 - 4x_2 + x_3 = 3$
$x_1, x_2, x_3 \geq 0$

	x_1	x_2	x_3	x_4	x_5	x_6	x_7	
x_6	0	2	5	-1	0	1	0	1
x_5	3	-1	-1	0	1	0	0	6
x_7	2	-4	1	0	0	0	1	3
	-2	5	9	0	0	0	0	0
x_3	0	0	1	$-\frac{1}{6}$	$-\frac{1}{15}$	$\frac{1}{6}$	$\frac{1}{10}$	$\frac{1}{15}$
x_2	0	1	0	$-\frac{1}{12}$	$\frac{1}{6}$	$\frac{1}{12}$	$-\frac{1}{4}$	$\frac{1}{3}$
x_1	1	0	0	$-\frac{1}{12}$	$\frac{11}{30}$	$\frac{1}{12}$	$-\frac{1}{20}$	$\frac{32}{15}$
	0	0	0	$\frac{7}{4}$	$\frac{1}{2}$	$-\frac{7}{4}$	$\frac{1}{4}$	2

(d) Minimize $10x_1 + 20x_2 + 15x_3 + 21x_4 + 5x_5$
subject to
$7x_1 - 10x_2 + 8x_3 - 5x_4 + 3x_5 = 730$
$3x_1 + x_2 + 4x_3 - 2x_4 - x_5 = 350$
$x_1, x_2, x_3, x_4, x_5 \geq 0$

	x_1	x_2	x_3	x_4	x_5	x_6	x_7	
x_6	7	-10	8	-5	3	1	0	730
x_7	3	1	4	2	-1	0	1	350
	10	20	15	21	5	0	0	0
x_1	1	-12	0	-9	5	1	-2	30
x_3	0	$\frac{37}{4}$	1	$\frac{29}{4}$	-4	$-\frac{3}{4}$	$\frac{7}{4}$	65
	0	$\frac{5}{4}$	0	$\frac{9}{4}$	15	$\frac{5}{4}$	$-\frac{25}{4}$	-1275

(e) Maximize $10x_1 - 12x_2 + 11x_3$
subject to
$6x_1 - 7x_2 + 8x_3 = 90$
$-x_1 + 3x_3 \geq 42$
$x_1, x_2, x_3 \geq 0$

	x_1	x_2	x_3	x_4	x_5	x_6	
x_5	6	-7	8	0	1	0	90
x_6	-1	0	3	-1	0	1	42
	-10	12	-11	0	0	0	0
x_3	$-\frac{1}{3}$	0	1	$-\frac{1}{3}$	0	$\frac{1}{3}$	14
x_2	$-\frac{26}{21}$	1	0	$-\frac{8}{21}$	$-\frac{1}{7}$	$\frac{8}{21}$	$\frac{22}{7}$
	$\frac{25}{21}$	0	0	$\frac{19}{21}$	$\frac{12}{7}$	$-\frac{19}{21}$	$\frac{814}{7}$

11. (a) Consider the linear programming problem of

$$\begin{aligned}
&\text{Minimizing } c_1x_1 + c_2x_2 + \cdots + c_nx_n \\
&\text{subject to} \\
&a_{11}x_1 + a_{12}x_2 + \ldots + a_{1n}x_n \geq b_1 \\
&a_{21}x_1 + a_{22}x_2 + \ldots + a_{2n}x_n \geq b_2 \\
&\qquad\vdots \\
&a_{m1}x_1 + a_{m2}x_2 + \ldots + a_{mn}x_n \geq b_m \\
&x_1, x_2, \ldots, x_n \geq 0
\end{aligned}$$

Assume that $b_i \geq 0$, $1 \leq i \leq m$, and that the problem has a finite optimal solution. To find this solution, suppose the simplex method is used, first adding m slack variables to the problem (each with coefficient (-1)) and then m artificial variables. Let $s_1, s_2, \ldots, s_m$ denote the m entries in the bottom row of the final tableau in the m slack variable columns. Show that $(s_1, s_2, \ldots, s_m)$ is an optimal solution point to the dual, modeling your proof on the proof of the Duality Theorem.

(b) Show that the above result also follows from Problem 8.

12. Consider the linear programming problem of

$$\begin{aligned}
&\text{Maximizing } c_1x_1 + c_2x_2 + \cdots + c_nx_n \\
&\text{subject to} \\
&a_{11}x_1 + \ldots + a_{1n}x_n \leq b_1 \\
&\qquad\vdots \\
&a_{k1}x_1 + \ldots + a_{kn}x_n \leq b_k \\
&a_{k+1,1}x_1 + \ldots + a_{k+1n}x_n = b_{k+1} \\
&\qquad\vdots \\
&a_{m1}x_1 + \ldots + a_{mn}x_n = b_m \\
&x_1, x_2, \ldots, x_n \geq 0
\end{aligned}$$

Assume that $b_i \geq 0$ for $1 \leq i \leq m$. Suppose k slack variables and $m-k$ artificial variables are added to the problem and the simplex algorithm is applied, driving to a finite optimal solution. Denote by s_i, $1 \leq i \leq m$, the entries in the bottom row (the z row) of the final tableau in the slack variable $(1 \leq i \leq k)$ and artificial variable $(k+1 \leq i \leq m)$ columns. Show that $(s_1, s_2, \ldots, s_m)$ is an optimal solution point to the dual.

4.5 The Complementary Slackness Theorem

In this section we discuss the Complementary Slackness Theorem. The theorem relates optimal solution points of a linear programming problem and its dual. The theorem will not be needed in any further theoretical developments in the text. However, the relationships prescribed by the theorem are certainly interesting and useful, and will be referred to occasionally in the problem sets and in the development of the transportation problem algorithm in Chapter 7. Those readers who continue their studies in linear programming at a more advanced level may well encounter complementary slackness again.

In Example 4.4.1 of the previous section, it was verified that the point $(10,9,0,0)$ is an optimal solution point to the problem of

$$\begin{array}{l} \text{Maximizing } f(x_1,x_2,x_3,x_4) = -5x_1 + 18x_2 + 6x_3 - x_4 \\ \text{subject to} \\ \begin{array}{rcrcrcrcl} 2x_1 & & & - & x_3 & + & 3x_4 & \leq & 20 \\ & & x_2 & - & 2x_3 & - & x_4 & \leq & 30 \\ -3x_1 & + & 6x_2 & + & 3x_3 & + & 4x_4 & \leq & 24 \end{array} \\ x_1,x_2,x_3,x_4 \geq 0 \end{array} \tag{4.5.1}$$

and the point $(2,0,3)$ is an optimal solution point to the dual,

$$\begin{array}{l} \text{Minimize } g(y_1,y_2,y_3) = 20y_1 + 30y_2 + 24y_3 \\ \text{subject to} \\ \begin{array}{rcrcrcr} 2y_1 & & & - & 3y_3 & \geq & -5 \\ & & y_2 & + & 6y_3 & \geq & 18 \\ -y_1 & - & 2y_2 & + & 3y_3 & \geq & 6 \\ 3y_1 & - & y_2 & + & 4y_3 & \geq & -1 \end{array} \\ y_1,y_2,y_3 \geq 0 \end{array} \tag{4.5.2}$$

Since $(10,9,0,0)$ is an optimal solution to (4.5.1), it certainly satisfies the constraints of (4.5.1). In fact, evaluating the three constraints at this point, we find slack of 0, 21, and 0 at the first, second, and third inequalities, respectively. Now the three dual variables y_1,y_2,y_3 of (4.5.2) correspond to these three constraints; and note that where there is positive slack in the constraints of (4.5.1) at the point $(10,9,0,0)$, the value of the corresponding dual variable at the optimal solution point $(2,0,3)$ is 0.

Conversely, evaluating the four constraints of (4.5.2) at $(2,0,3)$ yields slack of 0, 0, 1, and 19. Again, for each inequality at which the slack is positive, the value of the corresponding dual variable at the optimal solution point $(10,9,0,0)$ is 0.

These results are guaranteed by the Complementary Slackness Theorem. Moreover, the converse is also true. In terms of (4.5.1) and (4.5.2), this means that if X^* and Y^* are feasible solutions to (4.5.1) and (4.5.2), respectively, and satisfy the complementary slackness conditions described, they are optimal solution points to their respective problems.

The statement and proof of the general theorem follow.

Theorem 4.5.1 (Complementary Slackness Theorem). *Suppose $X^* = (x_1^*,\ldots,x_n^*)$ is a feasible solution to the problem of*

$$\textit{Maximizing } c \cdot X \textit{ subject to } AX \leq b, X \geq 0 \tag{4.5.3}$$

and $Y^* = (y_1^*, \ldots, y_m^*)$ *is a feasible solution to the dual problem of*

$$\text{Minimizing } b \cdot Y \text{ subject to } A^t Y \geq c, Y \geq 0 \tag{4.5.4}$$

Then X^* *and* Y^* *are optimal solution points to their respective problems if and only if, for each i,* $1 \leq i \leq m$*, either*

$$(\text{slack in the ith constraint of (4.5.3) evaluated at } X^*) = b_i - \sum_j a_{ij} x_j^* = 0$$

or

$$y_i^* = 0$$

and, for each j, $1 \leq j \leq n$*, either*

$$(\text{slack in the jth constraint of (4.5.4) evaluated at } Y^*) = \sum_i a_{ij} y_i^* - c_j = 0$$

or

$$x_j^* = 0$$

Proof. Corollary 4.4.1 says it all, essentially. Since X^* is a feasible solution to the max problem (4.5.3) and Y^* is a feasible solution to the min problem (4.5.4), from the corollary we have

$$b \cdot Y^* - c \cdot X^* = (b - AX^*) \cdot Y^* + (A^t Y^* - c) \cdot X^*$$

If X^* and Y^* satisfy the complementary slackness hypothesis, then for each i, with $1 \leq i \leq n$, the product

$$y_i^* \left(b_i - \sum_j a_{ij} x_j^* \right) = 0$$

that is, each multiplication in the dot product $(b - AX^t) \cdot Y^*$ equals 0, and so $(b - AX^t) \cdot Y^* = 0$. Similarly, from complementary slackness, $(A^t Y^* - c) \cdot X^* = 0$. Thus $b \cdot Y^* = c \cdot X^*$, and so, from Corollary 4.4.2, X^* and Y^* are optimal solution points for their respective problems.

Conversely, if X^* and Y^* are optimal solution points for their respective problems, we have

$$0 = b \cdot Y^* - c \cdot X^* = (b - AX^*) \cdot Y^* + (A^t Y^* - c) \cdot X^*$$

But each dot product on the right side of the equation consists of a sum of products of nonnegative numbers, and so each dot product is nonnegative. Hence both $(b - AX^*) \cdot Y^* = 0$ and $(A^t Y^* - c) \cdot X^* = 0$, that is, the points X^* and Y^* satisfy the complementary slackness conditions. □

Example 4.5.1. The problem of

$$\begin{aligned}
&\text{Minimizing } 12x_1 + 5x_2 + 10x_3 \\
&\text{subject to} \\
&\quad x_1 - x_2 + 2x_3 \geq 10 \\
&-3x_1 + x_2 + 4x_3 \geq -9 \\
&-x_1 + 2x_2 + 3x_3 \geq 1 \\
&2x_1 - 3x_2 \geq -2 \\
&7x_1 - x_2 - 5x_3 \geq 34 \\
&x_1, x_2, x_3 \geq 0
\end{aligned} \tag{4.5.5}$$

has $(7,0,3)$ as an optimal solution point. To determine an optimal solution point to the dual,

$$\begin{aligned}
&\text{Maximize } 10y_1 - 9y_2 + y_3 - 2y_4 + 34y_5 \\
&\text{subject to} \\
&\quad y_1 - 3y_2 - y_3 + 2y_4 + 7y_5 \leq 12 \\
&-y_1 + y_2 + 2y_3 - 3y_4 - y_5 \leq 5 \\
&2y_1 + 4y_2 + 3y_3 - 5y_5 \leq 10 \\
&y_1, y_2, y_3, y_4, y_5 \geq 0
\end{aligned} \tag{4.5.6}$$

we can use complementary slackness. Evaluating the inequalities of (4.5.5) at the point $(7,0,3)$, we find positive slack in the first, third, and fourth constraints (and zero slack in the other two). Thus any optimal solution $Y^* = (y_1^*, y_2^*, y_3^*, y_4^*, y_5^*)$ to (4.5.6) must have $y_1^* = y_3^* = y_4^* = 0$. And the first and third components of $(7,0,3)$ positive implies that Y^* must yield zero slack in the first and third constraints of (4.5.6). Hence $Y^* = (0, y_2^*, 0, 0, y_5^*)$ and

$$\begin{aligned}
-3y_2^* + 7y_5^* &= 12 \\
4y_2^* - 5y_5^* &= 10
\end{aligned} \tag{4.5.7}$$

The (unique) solution to (4.5.7) is $y_2^* = 10$, $y_5^* = 6$, and so $Y^* = (0,10,0,0,6)$ is an (and the only) optimal solution point to (4.5.6). (In fact, the existence of this feasible solution to (4.5.6) satisfying complementary slackness now certifies the optimality of $(7,0,3)$.)

Example 4.5.2. Suppose it is claimed that the point $(3,0,1,0)$ is an optimal solution to the problem of

$$\begin{aligned}
&\text{Maximizing } 9x_1 + 3x_2 + 5x_3 + 22x_4 \\
&\text{subject to} \\
&2x_1 - x_2 + 2x_3 + 6x_4 \leq 8 \\
&5x_1 + 3x_2 + x_3 + 2x_4 \leq 16 \\
&4x_1 + x_2 - x_3 + 3x_4 \leq 12 \\
&x_1, x_2, x_3, x_4 \geq 0
\end{aligned} \tag{4.5.8}$$

We can use complementary slackness to attempt to ratify the claim. First, we verify that $(3,0,1,0)$ is a feasible solution to (4.5.8), noting that the point yields zero slack

in the first two constraints of (4.5.8) and positive slack in the third. Now consider the dual.

$$\begin{aligned} &\text{Minimize } 8y_1 + 16y_2 + 12y_3 \\ &\text{subject to} \\ &2y_1 + 5y_2 + 4y_3 \geq 9 \\ &-y_1 + 3y_2 + y_3 \geq 3 \\ &2y_1 + y_2 - y_3 \geq 5 \\ &6y_1 + 2y_2 + 3y_3 \geq 22 \\ &y_1, y_2, y_3 \geq 0 \end{aligned} \tag{4.5.9}$$

If (4.5.8) has a finite optimal solution, so does (4.5.9), and any optimal solution point $Y^* = (y_1^*, y_2^*, y_3^*)$ must satisfy the complementary slackness conditions with $(3,0,1,0)$. Thus, $y_3^* = 0$, and

$$\begin{aligned} 2y_1^* + 5y_2^* &= 9 \\ 2y_1^* + y_2^* &= 5 \end{aligned}$$

yielding $Y^* = (2,1,0)$. But this point is not a feasible solution to (4.5.9), as the reader may verify. Hence $(3,0,1,0)$ cannot be an optimal solution to (4.5.8).

Problem Set 4.5

1. Consider the linear programming problem of

$$\begin{aligned} &\text{Maximizing } x_1 + 2x_2 \\ &\text{subject to} \\ &2x_1 + x_2 \leq 3 \\ &x_1 + 2x_2 \leq 3 \\ &x_1, x_2 \geq 0 \end{aligned}$$

   (a) Determine the dual problem.
   (b) Show that $X^* = (1,1)$ and $Y^* = (0,1)$ are optimal solutions for the original and dual problems, respectively, by using the Complementary Slackness Theorem.
   (c) Note that at these solution points, both y_1^* and the slack in the corresponding first constraint of the max problem are zero.

2. Consider the linear programming problem of

$$\begin{aligned} &\text{Maximizing } 2x_1 + 2x_2 \\ &\text{subject to} \\ &x_1 + x_3 + x_4 \leq 1 \\ &x_2 + x_3 - x_4 \leq 1 \\ &x_1 + x_2 + 2x_3 \leq 3 \\ &x_1, x_2, x_3, x_4 \geq 0 \end{aligned}$$

(a) Determine the dual problem.

(b) Show that $X^* = (1,1,0,0)$ and $Y^* = (1,1,1)$ are feasible solutions to the original and dual problems, respectively.

(c) Show that for this pair of solutions, for each j, $x_j^* > 0$ implies that the slack in the corresponding dual constraint is zero.

(d) Show that Y^* is not an optimal solution to the dual.

(e) Does this contradict the Complementary Slackness Theorem?

3. Prove or disprove each of the following, using complementary slackness.

(a) $(1,1,0,0)$ is an optimal solution point to the maximization problem of Problem 2.

(b) $(0,4,0,2)$ is an optimal solution point to (4.5.8) on page 157.

(c) $(3,0,1,0,5)$ is an optimal solution point to the problem of

$$\begin{aligned}
&\text{Maximizing } 5x_1 + 16x_2 - 4x_3 - x_4 + 7x_5 \\
&\text{subject to} \\
&8x_1 - 2x_2 + 3x_3 - 2x_5 \le 18 \\
&2x_1 + 4x_2 - 7x_3 + 3x_4 + x_5 \le 4 \\
&x_1 + 3x_2 + x_3 - x_4 + 2x_5 \le 14 \\
&x_1, x_2, x_3, x_4, x_5 \ge 0
\end{aligned}$$

(d) $(1,0,1,0)$ is an optimal solution point to the problem of

$$\begin{aligned}
&\text{Minimizing } 5x_1 + 8x_2 + 4x_3 + 2x_4 \\
&\text{subject to} \\
&x_1 + 2x_2 - x_3 + x_4 \ge 0 \\
&2x_1 + 3x_2 + x_3 - x_4 \ge 3 \\
&x_1, x_2, x_3, x_4 \ge 0
\end{aligned}$$

(e) $(0,3,12)$ is an optimal solution point to the problem of

$$\begin{aligned}
&\text{Minimizing } 2y_1 - 5y_2 - 3y_3 \\
&\text{subject to} \\
&-3y_1 - 6y_2 + 2y_3 \ge 6 \\
&y_1 + 3y_2 + y_3 \ge 20 \\
&4y_1 + 7y_2 - 3y_3 \ge -15 \\
&y_1, y_2, y_3 \ge 0
\end{aligned}$$

(f) $(0,3,0,0,4)$ is an optimal solution point to the problem of

$$\begin{aligned}
&\text{Maximizing } 5x_1 + 4x_2 + 8x_3 + 9x_4 + 15x_5 \\
&\text{subject to} \\
&x_1 + x_2 + 2x_3 + x_4 + 2x_5 \le 11 \\
&x_1 - 2x_2 - x_3 + 2x_4 + 3x_5 \le 6 \\
&x_1, x_2, x_3, x_4, x_5 \ge 0
\end{aligned}$$

(g) $(0,8)$ is an optimal solution point to the problem of

$$\begin{aligned}
&\text{Maximizing } 3x_1 + 4x_2\\
&\text{subject to}\\
&x_1 + 2x_2 \leq 16\\
&x_1 + x_2 \leq 8\\
&x_1, x_2 \geq 0
\end{aligned}$$

(h) $(1,0,3)$ is an optimal solution point to the problem of

$$\begin{aligned}
&\text{Maximizing } 6x_1 + 9x_2 + 5x_3\\
&\text{subject to}\\
&3x_1 + 3x_2 + 2x_3 \leq 9\\
&x_1 + 2x_2 + x_3 \leq 4\\
&x_1, x_2, x_3 \geq 0
\end{aligned}$$

(i) $(2,1,0)$ is an optimal solution point to the problem of part (h).

(j) $(60,4,25)$ is an optimal solution point to the problem of

$$\begin{aligned}
&\text{Maximizing } 10x_1 + 3x_2 - 23x_3\\
&\text{subject to}\\
&7x_1 + 2x_2 - 16x_3 \leq 28\\
&3x_1 + x_2 - 7x_3 \leq 9\\
&x_1, x_2, x_3 \geq 0
\end{aligned}$$

CHAPTER 5

SENSITIVITY ANALYSIS

5.1 EXAMPLES IN SENSITIVITY ANALYSIS

In some applications of linear programming there may be a need not only to optimize a given function under specified conditions, but also to evaluate the effects changes in the conditions of the problem have on the optimal solution. For example, it could be that some of the coefficients a_{ij} of the coefficient matrix A are just approximations, and it would be desirable to know how their variance affects the optimal solution. Or it could be that the results of purchasing raw materials from other sources, yielding altered cost coefficients c_j, or of expanding one's storage capacities, yielding altered constants b_i, are to be measured. The study of techniques used to handle such problems is called *sensitivity analysis* or *postoptimality analysis*.

In this section we introduce, through three examples, some concepts involved in such an analysis. In the first example, we work with a linear programming problem with only two variables and two constraints. Our analysis of this problem is based on graphs in the plane, available because of the limited size of the problem, and the Duality Theorem. In the other examples, involving problems with more than two constraints, our analyses use only the Duality Theorem. Sensitivity analysis techniques for more general problems are developed in the subsequent sections of the chapter.

Example 5.1.1. Consider the problem of the poultry producer in Problem 7 of Section 2.2. The producer's stock requires daily at least 124 units of nutritional element A and 60 units of nutritional element B. Two feeds are available for use. One pound of Feed 1 costs 16 cents and contains 10 units of A and 3 units of B; 1 lb of Feed 2 costs 14 cents and contains 4 units of A and 5 units of B. Wishing to determine an adequate diet for the stock at minimal costs, the producer defines x_1 and x_2 to be the number of pounds of Feeds 1 and 2, respectively, used in the diet and formulates the following linear programming problem.

An Introduction to Linear Programming and Game Theory, Third Edition. By P. R. Thie and G. E. Keough.

Minimize $16x_1 + 14x_2$
subject to
$10x_1 + 4x_2 \geq 124$
$3x_1 + 5x_2 \geq 60$
$x_1, x_2 \geq 0$

The graph of the set of feasible solutions and one line from the family of lines $\{16x_1 + 14x_2 = c : c \text{ a constant}\}$ are sketched in Figure 5.1. From the geometry, it is clear that the minimal value of $16x_1 + 14x_2$ on the set of feasible solutions is attained at the point $(10,6)$. Thus the minimum daily cost for the poultry producer is \$2.44, attained by using 10 lb of Feed 1 and 6 lb of Feed 2.

Suppose, however, that the costs of the two feeds vary due to market conditions, weather patterns, labor negotiations, and the like, and the poultry producer would like some guidance on when to use what diet. We address this concern. As discussed in Section 2.2, using the slopes of the lines involved, the above result can be extended. If the costs of 1 lb of Feeds 1 and 2 are c_1 and c_2 cents ($c_1, c_2 > 0$), respectively, and thus the objective function is given by $c_1x_1 + c_2x_2$, the minimum-cost diet is attained at $(10,6)$ if $-\frac{5}{2} \leq -\frac{c_1}{c_2} \leq -\frac{3}{5}$, that is, if $\frac{3}{5} \leq \frac{c_1}{c_2} \leq \frac{5}{2}$, and at $(20,0)$ if $\frac{c_1}{c_2} \leq \frac{3}{5}$, and at $(0,31)$ if $\frac{5}{2} \leq \frac{c_1}{c_2}$. Thus, in terms of the ratio $\frac{c_1}{c_2}$, the minimum daily cost, in cents, of a feasible diet is

$$\begin{cases} 10c_1 + 6c_2, & \frac{3}{5} \leq \frac{c_1}{c_2} \leq \frac{5}{2} \\ 20c_1, & \frac{c_1}{c_2} \leq \frac{3}{5} \\ 31c_2, & \frac{5}{2} \leq \frac{c_1}{c_2} \end{cases} \tag{5.1.1}$$

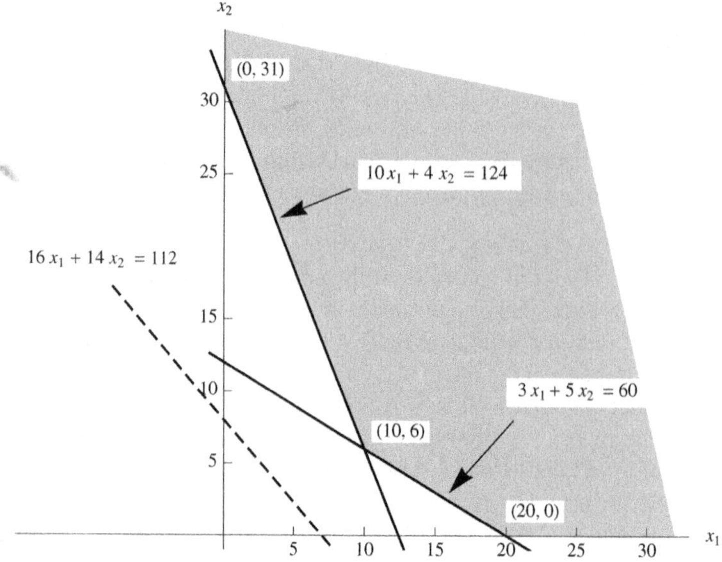

Figure 5.1

Using these results, we have, for example, that if tl increased by 50% to 21 cents while the cost of Feed producer's daily minimal cost would be \$2.86, but th doubled, the optimal diet would change and the dail reader may verify.

For this problem with only two variables, we have acterize the solution of the problem in terms of coeffic Suppose now that the poultry producer questions the e tional requirements on the minimal cost of an adequat units and 60 units required of elements A and B were estimates, and the producer would like to know what might be saved by decreasing these amounts. Or maybe the producer has discovered that by increasing the amounts of one or both of the nutrients, the stock has a higher market value, and wonders how this increased value compares with the increased feeding costs.

In responding to these queries, we cannot proceed directly as above; changing the constant terms of the constraints changes the set of feasible solutions and makes the sketch in Figure 5.1 obsolete. However, using duality, the queries can be addressed.

Assuming that the costs of the two feeds are 16 and 14 cents/lb, as originally stated, the dual to the original problem is to

$$\begin{aligned}
&\text{Maximize } 124y_1 + 60y_2\\
&\text{subject to}\\
&10y_1 + 3y_2 \leq 16\\
&\ \ 4y_1 + 5y_2 \leq 14\\
&y_1, y_2 \geq 0
\end{aligned}$$

From the graph in Figure 5.2, we see that the dual objective function $124y_1 + 60y_2$ on the set of feasible solutions to the dual is maximized at the point $(1,2)$, with a maximum value equal to the minimum daily cost, in cents, of 244, as guaranteed by the Duality Theorem. In fact, using the Duality Theorem and Figure 5.2, we can proceed just as above to estimate the effect of varying the required amounts of the nutritional elements on the minimum cost.

Let b_1 and b_2 denote the minimal amounts of elements A and B required daily, with the dual objective function equal to $b_1y_1 + b_2y_2$. Then, using slopes, we find that the optimal solution point to the dual is $(1,2)$ if $\frac{4}{5} \leq \frac{b_1}{b_2} \leq \frac{10}{3}$, and is $(0, 14/5)$ if $\frac{b_1}{b_2} \leq \frac{4}{5}$, and is $(8/5, 0)$ if $\frac{10}{3} \leq \frac{b_1}{b_2}$. Therefore, in terms of the ratio $\frac{b_1}{b_2}$, the maximum value of the dual objective function, and thus the producer's minimum daily cost, in cents, is

$$\begin{cases}
b_1 + 2b_2, & \frac{4}{5} \leq \frac{b_1}{b_2} \leq \frac{10}{3}\\
\frac{14}{5}b_2, & \frac{b_1}{b_2} \leq \frac{4}{5}\\
\frac{8}{5}b_1, & \frac{10}{3} \leq \frac{b_1}{b_2}
\end{cases}$$

In the second part of the above example, we were able to measure the consequence of variations in the daily nutritional requirements for the stock on the minimal

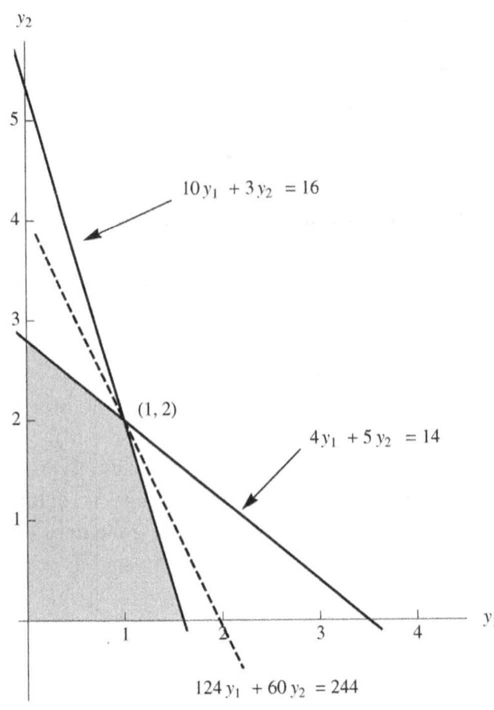

Figure 5.2

cost of an adequate diet; that is, we measured the effect of changes in the constant terms of the constraints of the problem on the optimal value of the objective function. We used the Duality Theorem along with the graph of the set of feasible solutions to the dual problem. Actually, for a given linear programming problem, even if the associated dual problem has more than two variables and graphing techniques are unavailable, we can still derive, using the Duality Theorem, useful information relating changes in the constant terms of the constraints of the original problem to the optimal value of the objective function. We illustrate with two examples.

Example 5.1.2. Consider the problem of the boat manufacturer in Example 2.3.1 on page 21. The problem is to determine a maximal profit production schedule in the manufacturing and selling of rowboats and canoes utilizing limited resources of aluminum, machine time, and labor. Letting R and C denote the number of rowboats and canoes, respectively, to be manufactured, the linear programming problem is to

$$\begin{aligned}&\text{Maximize } z = 50R + 60C \\ &\text{subject to} \\ &50R + 30C \leq 2000 \\ &\ \ 6R + \ \ 5C \leq \ \ 300 \\ &\ \ 3R + \ \ 5C \leq \ \ 200 \\ &R, C \geq 0\end{aligned} \tag{5.1.2}$$

The constant terms of the constraints come from the limits on the resources of aluminum (2000 lb), machine time (300 min), and labor (200 hr). As determined in Problem 1 of Section 2.3 (or see Section 8.1), the maximum possible profit is \$2750 earned in the production of 25 rowboats and 25 canoes.

Suppose that the boat manufacturer has the opportunity to purchase more aluminum. With additional aluminum available, (possibly) more boats can be produced. But the manufacturer needs to know how the profits from the additional sales would compare with the cost of the extra aluminum. In other words, how much should he be willing to pay for more aluminum?

In terms of the original problem of (5.1.2), what we need is a measure of the effect of a change in the constant of the first inequality on the maximum value of the objective function. To determine this, consider the dual to the problem of (5.1.2) and its solution. The dual is to

$$\begin{aligned}
&\text{Minimize } v = 2000y_1 + 300y_2 + 200y_3\\
&\text{subject to}\\
&50y_1 + 6y_2 + 3y_3 \geq 50\\
&30y_1 + 5y_2 + 5y_3 \geq 60\\
&y_1, y_2, y_3 \geq 0
\end{aligned}$$

The optimal value of v is 2750 attained at $y_1 = \frac{7}{16}$, $y_2 = 0$, $y_3 = \frac{75}{8}$ (see Problem 3 of Section 4.3, or verify the optimality of both the points (25, 25) and (7/16, 0, 75/8) using either Corollary 4.4.2 or Theorem 4.5.1). Note that the coefficients of the dual objective function v are, by definition, the constant terms from the original constraints. In fact, from the Duality Theorem, we have that

$$\text{Max } z = \text{Min } v = 2000\left(\tfrac{7}{16}\right) + 300(0) + 200\left(\tfrac{75}{8}\right)$$

It follows that as long as $(\frac{7}{16}, 0, \frac{75}{8})$ is an optimal solution point for the dual, the minimum of v and therefore the maximum profit will increase by $\$(\frac{7}{16}) \approx 44$ cents for each available pound of aluminum above the original 2000 lb. We can now answer the original question. Since the profit figures of \$50 for a rowboat and \$60 for a canoe would be determined by subtracting the cost of the required aluminum, machine time, and labor for each from the selling price, the manufacturer should be willing to pay for additional aluminum up to about 44 cents/lb more than was paid for the original 2000 lb of aluminum. For example, if 48 lb can be purchased for only \$15 more than the original cost of 48 lb of aluminum, sales can be increased by $\$(\frac{7}{16})48 = \$21$ and, therefore, a net gain of \$6 realized.

Obviously, we can extend this analysis to the other resources. Since $y_3 = \frac{75}{8} \approx 9.38$ in the optimal solution to the dual, each additional hour of finishing labor would increase profits by \$9.38. Similarly, $y_2 = 0$ implies that an increase in available machine time over the original 300 min will provide no increase in profits. Actually, this last fact is also apparent from the original optimal solution to (5.1.2). As can be easily calculated, the production schedule of $R = C = 25$ utilizes all 2000 lb of aluminum and 200 hr of labor but only 275 of the available 300 min of machine time. (The Complementary Slackness Theorem of Section 4.5 is relevant here.) Thus production is restricted by the limited amounts of aluminum and labor available. Machine time is an underutilized resource, so increasing its availability has no effect on profits.

In sum, the solution $(\frac{7}{16}, 0, \frac{75}{8})$ to the dual has provided estimates on the effects that changes in the constant terms of the problem have on the optimal value of the objective function. These numbers are sometimes called the *shadow prices or marginal values* of the constraints, since each provides some indication of the worth or cost on the optimal value of the objective function of a unit of the resource or demand generating the associated constraint. Certainly, however, there are limitations on their use. For example, the boat manufacturer would not be able to make unlimited profits even if an unlimited supply of aluminum were available, because the other two constraints would still restrict the total production. Thus the estimate of an increased profit of 44 cents/lb for each additional pound of aluminum available is accurate only to some upper limit. Once the change or changes in the constant terms of the original problem effect a change in the optimal solution point to the dual, these marginal values will change. In fact, in Example 5.1.1, by using geometry, we were able to state precisely, in terms of the ratio of the two constant terms, when a change in these constant terms would alter the optimal solution point to the dual. For a general problem this question can be much more involved. Techniques for dealing with it are developed in the subsequent sections of this chapter.

Example 5.1.3. An aluminum can company must produce monthly at least 2400 cases of a can of Type A and 2800 cases of a can of Type B. The company has three processes available for production. The first uses a special pure grade aluminum; the other two allow for some use of recycled aluminum. (Because of government regulations, the company must use 600 lb of recycled aluminum in its monthly production.) The input, output, and cost of 1 hr of operation of each process are as follows:

	Input	***Output***		
	Recycled Al (lb)	*Type A (cases)*	*Type B (cases)*	*Cost ($)*
Process 1	0	6	8	65
Process 2	2	12	12	150
Process 3	3	10	15	200

Obviously, the company manager would want to know the most economical operation of the plant. But other information, such as some estimate on the actual cost of meeting each of the two demands, and of utilizing the recycled aluminum, could be very helpful.

To respond to the first question, we formulate and solve the associated linear programming problem. Defining x_i to be the number of hours that Process i is used, $1 \leq i \leq 3$, the problem is to

$$\begin{aligned}
&\text{Minimize } 65x_1 + 150x_2 + 200x_3 \\
&\text{subject to} \\
&6x_1 + 12x_2 + 10x_3 \geq 2400 \\
&8x_1 + 12x_2 + 15x_3 \geq 2800 \\
&\qquad\quad 2x_2 + 3x_3 = 600 \\
&x_1, x_2, x_3 \geq 0
\end{aligned} \tag{5.1.3}$$

Subtracting slack variables x_4 and x_5 from the first two constraints, adding three artificial variables, and applying the simplex algorithm yields the reduced tableaux resolution in Table 5.1. (The first and last tableaux only are listed, and the artificial variable data are retained for future reference.) The company can meet its demands at a monthly cost of $41,250 by using Process 2 for 75 hr and Process 3 for 150 hr. An extra 350 cases of Type B cans would be produced.

Now to respond to the second set of questions, we consider the dual to (5.1.3):

$$\begin{aligned}
&\text{Maximize } v = 2400y_1 + 2800y_2 + 600y_3 \\
&\text{subject to} \\
&6y_1 + 8y_2 \leq 65 \\
&12y_1 + 12y_2 + 2y_3 \leq 150 \\
&10y_1 + 15y_2 + 3y_3 \leq 200 \\
&y_1, y_2 \geq 0,\ y_3 \text{ unrestricted}
\end{aligned} \tag{5.1.4}$$

From the data in the last row of Table 5.1, we see that $(3\frac{1}{8}, 0, 56\frac{1}{4})$ is an optimal solution to this problem. (See Section 4.4, or verify that this point satisfies the constraints of (5.1.4) and that v evaluated at the point is 41,250.) Thus,

$$\text{Min cost (in \$)} = \text{Max}\, v = 2400(3\tfrac{1}{8}) + 2800(0) + 600(56\tfrac{1}{4})$$

As long as this point remains an optimal solution point to (5.1.4), the shadow price for a case of Type A can is about $3.13 and $0 for a case of Type B (compare with the surplus production). The cost associated with using 1 lb of recycled aluminum is $56.25.

Table 5.1

	x_1	x_2	x_3	x_4	x_5	x_6	x_7	x_8	
x_6	6	12	10	-1	0	1	0	0	2,400
x_7	8	12	15	0	-1	0	1	0	2,800
x_8	0	2	3	0	0	0	0	1	600
	65	150	200	0	0	0	0	0	0
x_2	$\frac{9}{8}$	1	0	$-\frac{3}{16}$	0	$\frac{3}{16}$	0	$-\frac{5}{8}$	75
x_3	$-\frac{3}{4}$	0	1	$\frac{1}{8}$	0	$-\frac{1}{8}$	0	$\frac{3}{4}$	150
x_5	$-\frac{23}{4}$	0	0	$-\frac{3}{8}$	1	$\frac{3}{8}$	-1	$\frac{15}{4}$	350
	$46\frac{1}{4}$	0	0	$3\frac{1}{8}$	0	$-3\frac{1}{8}$	0	$-56\frac{1}{4}$	$-41,250$

Given this information, it would be obvious to the company manager that the major cost factor in the present operation is the use of the recycled aluminum. Production costs could be reduced, for example, by having the mandated amount of such aluminum to be used decreased or by investing in a can-producing process that utilizes more recycled aluminum.

The above examples demonstrate the practical value of both the solution of a linear programming problem and the solution of the associated dual problem. This suggests an obvious question. Can the solution of the dual always be determined from the solution of a given linear programming problem? We saw in Section 4.4 that the answer to that question is "yes" if the simplex algorithm is used to solve the initial problem and the tableau resolution of the problem is at hand. But what if Microsoft Excel's spreadsheet tool Solver is used? The answer here is also "yes," although the solution to the dual is not contained in the original spreadsheet. The solution to the dual (and much more) is available in the accompanying Sensitivity Report (see Appendix E). We demonstrate using Example 5.1.3.

A spreadsheet solution of the can company's problem of (5.1.3) is shown in Figure 5.3 and the associated Sensitivity Report is in Figure 5.4. The spreadsheet displays the original data of the problem, the optimal operation and associated minimal cost, and, with a little subtraction, the values of the slack variables, but no more.

The sensitivity report for the problem contains various related data. In particular, the solution point of $(3\frac{1}{8}, 0, 56\frac{1}{4})$ for the dual problem is in the lower half of the display, the Constraints half, in the Shadow Price column, as can be seen. Concerning the other columns in this half of the display, columns 1 and 2 (the Cell and Name columns) identify the rows of that part of the table, column 3 lists the final values of the left-hand sides of the constraints, and column 5 lists the initial constant terms.

	A	B	C	D	E	F	G
1	**Can Company**						
2				**Process**			**Required**
3			**1**	**2**	**3**		**Total**
4		**Output**: Type A (cases/hr)	6	12	10	≥	2400
5		**Output**: Type B (cases/hr)	8	12	15	≥	2800
6		**Input**: Al (lb/hr)	0	2	3	=	600
7		**Cost** (per hr of operation)	$65	$150	$200		
8							
9				**Variables**			
10		Process	**1**	**2**	**3**		
11		Hours to Be Used	0	75	150		
12							
13		Minimize Cost	$41,250				
14							
15		**Constraints**	LHS		RHS		
16		Minimum Type A cases	2400	≥	2400		
17		Minimum Type B cases	3150	≥	2800		
18		Aluminum Usage (lb)	600	=	600		

Figure 5.3

Can Company
Sensitivity Report

Adjustable Cells

Cell	Name	Final Value	Reduced Cost	Objective Coefficient	Allowable Increase	Allowable Decrease
C11	hours, Process 1	0	46.25	65	1E+30	46.25
D11	hours, Process 2	75	0	150	41.11	16.667
E11	hours, Process 3	150	0	200	25	61.667

Constraints

Cell	Name	Final Value	Shadow Price	Constraint R.H. Side	Allowable Increase	Allowable Decrease
C16	Type A cases LHS	2400	3.125	2400	1200	400
C17	Type B cases LHS	3150	0	2800	350	1E+30
C18	Aluminum LHS	600	56.25	600	120	93.333

Figure 5.4

These data can also be read off the spreadsheet. The last two columns, however, provide, as with the shadow price column, data of import.

We know that the use of the dual to provide estimates on the effect on the optimal value of the objective function if there is a change in a constraint's constant term is contingent upon the given solution point to the dual remaining optimal. The data in the table's last two columns address this concern. For example, the 120 in the Allowable Increase column and the $93\frac{1}{3}$ in the Allowable Decrease column of the aluminum constraint row are bounds on the range of values for the constant term 600 of the constraint to ensure that $(3\frac{1}{8}, 0, 56\frac{1}{4})$ remains the solution to the dual. In other words, as long as the required use of aluminum is between 506.667 and 720 lb, and *no* other data of the problem are changed, \$56.25/lb is the shadow price or marginal cost of meeting the aluminum requirement; increase the requirement to 700 lb, and minimal costs increase to $\$41,250 + 100(\$56.25) = \$46,875$, but decrease the requirement to 500 lb and costs decrease, but the exact amount cannot be determined from the data of the spreadsheet and sensitivity report. In Section 5.5 we will show how these bounds are determined from a final tableau (in this example, from Table 5.1). Actually, our perspective will be broader. In Section 5.5 we will consider the effects of any change in the data of the constant-term column on the final solution of a problem.

In Section 5.3 we will discuss the meaning and origins of the data of the upper half, the Adjustable Cells half, of the sensitivity report.

Problem Set 5.1

Problems 1–4 refer to Example 5.1.1, with costs and requirements as in the original problem..

1. (a) Show that if the ratio $\frac{c_1}{c_2} = \frac{3}{5}$, the two formulas of (5.1.1) eligible for calculating the daily minimal cost yield the same value.

(b) As in part (a), with $\frac{c_1}{c_2} = \frac{5}{2}$.

2. (a) Verify that if the element A requirement were increased by 2 units for every 1-unit decrease in the element B requirement, the producer's daily minimal cost remains unchanged. up to some upper limit.
   (b) Determine the upper limit.

3. Suppose that the value of the poultry producer's stock increases by 15 cents for each 10-unit increase in the daily requirement for either nutritional element A or element B. Should the producer increase one of the daily requirements, and if so, which one and up to what limit? Explain.

4. Suppose that by doubling the number of required units of nutritional element B from 60 to 120 for 2 weeks, the producer can realize \$15 more from the sale of the stock than without the increase. Is this worthwhile?

5. Suppose the minimal value of the objective function $c_1x_1 + \cdots + c_nx_n$ of a linear programming problem is attained at the point X_0 with the first m variables as basic variables. True or false:

   (a) If a c_j, $m+1 \leq j \leq n$, is increased, the minimal value is still attained at X_0.
   (b) If a c_j, $1 \leq j \leq m$, is decreased, the minimal value is still attained at X_0.

6. Consider the example in Problem 11 of Section 2.3. (See also Problem 8 of Section 3.5 and Problem 5 of Section 4.3.) Suppose the manager of the electronics firm wants an estimate on how altering the input of material and labor affects the maximum income earned from the sale of the circuits. Show that if b_1 and b_2 denote the number of units of material and labor, respectively, committed to the production of the circuits, then if $b_1/b_2 \leq \frac{3}{2}$, the maximum attainable income in dollars is $5b_1$; if $\frac{3}{2} \leq b_1/b_2 \leq 2$, the maximum income is $b_1 + 6b_2$; and if $b_1/b_2 \geq 2$, the maximum income is $8b_2$.

7. Using two raw materials, a firm can produce up to three different products. Inputs and profit per unit of production of each of the products are given in the following table.

	Input		
	Raw Material B_1	*Raw Material* B_1	*Profit (\$)*
Product 1	4	3	45
Product 2	2	1	17
Product 3	2	9	30

Suppose b_1 units of B_1 and b_2 units of B_2 are available.

   (a) Formulate the mathematical model for the problem of determining a production schedule that maximizes profit.
   (b) Write out the dual problem.
   (c) Using the dual problem, express as a function of the ratio of b_1 to b_2 the maximum profit function.

8. A bakery, using flour and sugar, makes cakes and pastries. Requirements and profits for making and selling a unit of each are as follows:

	Flour (lb)	*Sugar (lb)*	*Profit ($)*
Cake	10	15	40
Pastry	3	2	9

The bakery has available b_1 lb flour and b_2 lb of sugar. Assuming that all items made can be sold, express the maximum profit attainable as a function of the ratio of b_1 to b_2.

9. Determine the marginal values for the resources of fabric, wood, and labor in Problem 23 of Section 2.3. (Complementary slackness and the solution point given in part (b) of the problem could be used to solve the dual.)

10. Determine the shadow prices for the bluegrass and fescue seed requirements in Example 2.2.2 on page 14. In particular, estimate the effect on cost of a 1% increase in the bluegrass requirement for the final composition and also for the fescue requirement.

11. Determine the marginal costs for the manufacture of Products A and B of Problem 7 of Section 2.3.

12. For Problem 8 of Section 2.3, determine the marginal values of 1 hr of labor and 1 unit of plastic, and of the market restrictions on the sale of the elephants and donkeys.

13. Consider the transportation problem of Problem 2 of Section 2.4 and Example 4.3.3 on page 135.

(a) Show that

$$x_{11} = 300, x_{12} = 0, x_{13} = 0, x_{14} = 300$$
$$x_{21} = 0, x_{22} = 350, x_{23} = 400, x_{24} = 150$$

and

$$y_1 = 24, y_2 = 0, y_3 = 44, y_4 = 39, y_5 = 50, y_6 = 44$$

are optimal solutions for the original and dual problems. (You could use either Corollary 4.4.2 or the Complementary Slackness Theorem.)

(b) In the optimal shipping schedule, Outlet 1 is supplied by Warehouse 1 at a cost of 20 cents/case. But an increase in demand at Outlet 1 would increase the total shipping cost by more than 20 cents/case for each additional case required. Explain why this follows from the optimal solution to the dual problem.

14. Consider the linear programming problem of (2.3.1) on page 25.

(a) Show that $(19.55, 17.256, 9.132, 0)$ is a feasible solution.
(b) Determine the dual to the problem of (2.3.1).
(c) Show that $(20.483, 0, 0, 17.497, 8.86)$ is a feasible solution to the dual.

(d) Show that these solutions are, in fact, optimal solution points for their respective problems.
(e) Referring to the last paragraph of Section 2.3 on page 26, assist the division manager in responding to the vice president's questions.
(f) What does the marginal value for the first constraint of the original problem suggest about the use of overtime?

15. Combining raw materials M_1 and M_2 and labor, a company produces units of A, B, and C. The requirements and profit (excluding the cost of labor) for the production and sale of a unit of each are as follows:

	M_1 (lb)	*M_2 (lb)*	*Labor (hr)*	*Profit ($)*
A	6	16	2	105
B	12	25	3	165
C	4	7	1	60

For the next month, the company has available 1 ton of M_1, 2.5 tons of M_2, 500 hr of labor at $18/hr, and up to an additional 120 hr of overtime at $24/hr. (The company pays only for the labor used.)

To determine an optimal production schedule, the company manager defines x_1, x_2, and x_3 to be the number of A's, B's, and C's to be produced and x_4 to be the number of hours of overtime to be used and formulates the following model:

$$\begin{aligned} \text{Maximize } z &= 105x_1 + 165x_2 + 60x_3 - 18(2x_1 + 3x_2 + x_3) - 6x_4 \\ &= 69x_1 + 111x_2 + 42x_3 - 6x_4 \end{aligned}$$

subject to

$$\begin{aligned} 6x_1 + 12x_2 + 4x_3 &\leq 2000 \\ 16x_1 + 25x_2 + 7x_3 &\leq 5000 \\ 2x_1 + 3x_2 + x_3 &\leq 500 + x_4 \\ x_4 &\leq 120 \end{aligned}$$

$$x_1, x_2, x_3, x_4 \geq 0$$

Adding four slack variables and applying the simplex algorithm, the reduced tableaux resolution (initial and final tableaux only) are shown in Table 5.2.

(a) What is the optimal production schedule, and what profit does it yield?
(b) Write out the dual problem and determine an optimal solution point.
(c) Several employees offer to work additional hours of overtime (at the same $24/hr rate). Should the manager accept their offer?
(d) Suppose additional pounds of M_1 could be purchased, at a cost of $7.75/lb over what the company now pays for the raw material. Should more be purchased?

16. In Example 5.1.1 it was determined that $(1,2)$ is the optimal solution point to the dual of the poultry producer's diet problem as long as the ratio $\frac{b_1}{b_2}$ of the

Table 5.2

	x_1	x_2	x_3	x_4	x_5	x_6	x_7	x_8	
x_5	6	12	4	0	1	0	0	0	2000
x_6	16	25	7	0	0	1	0	0	5000
x_7	2	3	1	−1	0	0	1	0	500
x_8	0	0	0	1	0	0	0	1	120
	−69	−111	−42	6	0	0	0	0	0
x_6	0	4	0	0	1	1	−11	−11	180
x_4	0	0	0	1	0	0	0	1	120
x_1	1	0	0	0	$-\frac{1}{2}$	0	2	2	240
x_3	0	3	1	0	1	0	−3	−3	140
	0	15	0	0	$7\frac{1}{2}$	0	12	6	21,720

two nutritional requirements A and B is between $\frac{4}{5}$ and $\frac{10}{3}$. (Thus, for the original problem, with $\frac{b_1}{b_2} = \frac{16}{14}$, the marginal cost/shadow price of the element A requirement is 1 cent/unit and that of the element B requirement is 2 cents/unit.) Solver's Sensitivity Report for the original problem is shown in Figure 5.5. The data in the Shadow Price column confirms these marginal costs. We consider now the bounds listed in this part of the report.

(a) The entry of 95 in the bottom row of the Allowable Increase column suggests that shadow prices could change (i.e., the solution to the dual could change) if the nutritional requirement for element B increased by more than 95 units, with all other data (in particular, the 124-unit requirement for element A) remaining fixed. Using the results stated in the last paragraph of Example 5.1.1, verify that this is indeed the case.

(b) Proceeding as in part (a), verify Solver's Allowable Decrease entry of 22.8 units for the element B requirement.

Poultry Producer Problem
Sensitivity Report

Adjustable Cells

Cell	Name	Final Value	Reduced Cost	Objective Coefficient	Allowable Increase	Allowable Decrease
C12	Feed 1	10	0	16	19	7.6
D12	Feed 2	6	0	14	12.667	7.6

Constraints

Cell	Name	Final Value	Shadow Price	Constraint R.H. Side	Allowable Increase	Allowable Decrease
C18	Nutr. Element A LHS	124	1	124	76	76
C19	Nutr. Element B LHS	60	2	60	95	22.8

Figure 5.5

(c) Corroborate Solver's two bounds for the element A requirement.

(d) Construct an example in which both nutritional element requirements are changed, each within Solver's stated bound, but the solution to the dual changes.

17. You are production supervisor of a section of a large plant. Your section is responsible for delivering at least 100 units of A and 320 units of B each week for plant utilization. The B's can be bought externally for \$300/unit, and the A's and B's can be manufactured in your section using one or some combination of two different processes. Each process converts a rare metal and labor into A's and B's, with input and output for 1 hr of operation of each as follows:

	Input		***Output***	
	Rare Metal (lb)	*Labor (worker-hours)*	*A's*	*B's*
Process 1	11	6	3	7
Process 2	10	12	4	9

Each week , you can purchase up to 350 lb of rare metal at \$150/lb and have available up to 250 hr of labor at \$20/hr. (You pay only for what is used.)

To determine an optimal production schedule, you define x_i to be the number of hours to use Process i, $i = 1,2$, and x_3 to be the number of B's purchased externally and formulate the following model:

$$\begin{aligned} \text{Minimize } z &= 150(11x_1 + 10x_2) + 20(6x_1 + 12x_2) + 300x_3 \\ &= 1770x_1 + 1740x_2 + 300x_3 \end{aligned}$$

subject to

$$\begin{aligned} 11x_1 + 10x_2 \quad &\leq 350 \\ 6x_1 + 12x_2 \quad &\leq 250 \\ 3x_1 + 4x_2 \quad &\geq 100 \\ 7x_1 + 9x_2 + x_3 &\geq 320 \\ x_1, x_2, x_3 &\geq 0 \end{aligned}$$

Adding slack variables to each inequality and artificial variables to the last two, and applying the simplex algorithm, you obtain the reduced tableaux resolution shown in Table 5.3.

You are called to the plant manager's office. The manager would like an estimate on the present cost of supplying a unit of A and a unit of B, and the manager suggests that labor from another section of the plant (at the same \$20/hr pay rate) could be transferred to your section if this would help reduce your weekly production costs. With your printout in hand, how do you respond?

Table 5.3

	x_1	x_2	x_3	x_4	x_5	x_6	x_7	x_8	x_9	
x_4	11	10	0	1	0	0	0	0	0	350
x_5	6	12	0	0	1	0	0	0	0	250
x_8	3	4	0	0	0	-1	0	1	0	100
x_9	7	9	1	0	0	0	-1	0	1	320
	1770	1740	300	0	0	0	0	0	0	0
x_4	0	0	0	1	$\frac{7}{6}$	6	0	-6	0	$41\frac{2}{3}$
x_2	0	1	0	0	$\frac{1}{4}$	$\frac{1}{2}$	0	$-\frac{1}{2}$	0	$12\frac{1}{2}$
x_1	1	0	0	0	$-\frac{1}{3}$	-1	0	1	0	$16\frac{2}{3}$
x_3	0	0	1	0	$\frac{1}{12}$	$\frac{5}{2}$	-1	$-\frac{5}{2}$	1	$90\frac{5}{6}$
	0	0	0	0	130	150	300	-150	-300	$-78,500$

5.2 MATRIX REPRESENTATION OF THE SIMPLEX ALGORITHM

In this section, we will show that any tableau representation of a linear programming problem is completely determined once an ordered set of basic variables for the tableau is specified. In fact, we will develop formulas yielding these tableau data in terms of the original problem data. In the following sections, we will use these formulas to analyze the effects of an alteration in the original data of a linear programming problem on an already determined optimal solution to the problem. We will consider changes in the objective function coefficient vector c and in the constant term column vector b, and will consider additions of a new variable (an additional column in the coefficient matrix A) and a new constraint (an additional row in A). Our primary goal is not to provide a complete catalog of sensitivity analysis tools (notice, for example, that we will not consider the effects of a change in the entries of the coefficient matrix A); however, what follows should provide a deeper understanding of the questions that can be raised in postoptimality analysis and how the tools that we have at our disposal (and will have after this section and Section 5.6) can be used to answer these questions.

In matrix notation the standard form of the linear programming problem (Section 3.1) is to minimize $c \cdot X - z_0$ subject to $AX = b$, $X \geq 0$, where matrix A and vectors c, b, and X are defined in the obvious manner (in fact, just as in Section 4.2). Let $A^{(j)}$ denote the jth column of the coefficient matrix A for $j = 1, \ldots, n$. Then the matrix equation $AX = b$ is equivalent to the vector equation

$$\sum_{j=1}^{n} x_j A^{(j)} = b$$

in the m-dimensional column vectors $A^{(j)}$ and b. We have already discussed the equivalence in Example 3.2.3 on page 67.

We denote the initial tableau presentation of this problem in standard form as follows:

$$\begin{array}{c|c} A & b \\ \hline c & z_0 \end{array}$$

Suppose the pivot operation is applied several times, with the resulting equivalent linear programming problem being in canonical form and having the following tableau representation:

$$\begin{array}{c|c} A^* & b^* \\ \hline c^* & z_0^* \end{array}$$

Denote the basic variables in this representation by $x_{j(1)}, x_{j(2)}, \ldots, x_{j(m)}$, and define the $m \times m$ matrix $B = [A^{(j(1))}, A^{(j(2))}, \ldots, A^{(j(m))}]$ and the $1 \times m$ row vector $c_B = [c_{j(1)}, c_{j(2)}, \ldots, c_{j(m)}]$. Considering c and c^* as row vectors, in this section we will show that

$$\begin{array}{ll} A^* = B^{-1}A, & b^* = B^{-1}b \\ c^* = c - c_B B^{-1}A, & z_0^* = z_0 - c_B B^{-1}b \end{array}$$

Thus the above tableau representation can be calculated from the original tableau data and these auxiliary terms B and c_B, with the resulting tableau given by

$$\begin{array}{c|c} B^{-1}A & B^{-1}b \\ \hline c - c_B B^{-1}A & z_0 - c_B B^{-1}b \end{array}$$

Before these results are proved, we give an example.

Example 5.2.1. Consider the linear programming problem discussed in Sections 3.3 and 3.5. The problem is, from (3.3.4) on page 73, to

$$\begin{array}{l} \text{Minimize } z = 5x_1 + 3x_4 - 2x_5 + 21 \\ \text{subject to} \\ -6x_1 \qquad + x_3 - 2x_4 + 2x_5 = 6 \\ -3x_1 + x_2 \qquad + 6x_4 + 3x_5 = 15 \\ x_1, x_2, x_3, x_4, x_5 \geq 0 \end{array} \tag{5.2.1}$$

Applying the simplex algorithm generates, from Tables 3.1 and 3.3 of Section 3.5, the initial and final tableaux in Table 5.4.

Here $j(1) = 5$ and $j(2) = 1$ and so

$$B = [A^{(5)}, A^{(1)}] = \begin{bmatrix} 2 & -6 \\ 3 & -3 \end{bmatrix} \text{ and } c_B = [c_5, c_1] = [-2, 5].$$

Hence

$$B^{-1} = \tfrac{1}{12}\begin{bmatrix} -3 & 6 \\ -3 & 2 \end{bmatrix} = \begin{bmatrix} -\frac{1}{4} & \frac{1}{2} \\ -\frac{1}{4} & \frac{1}{6} \end{bmatrix}$$

Table 5.4

	x_1	x_2	x_3	x_4	x_5	
x_3	-6	0	1	-2	2	6
x_2	-3	1	0	6	3	15
	5	0	0	3	-2	-21
x_5	0	$\frac{1}{2}$	$-\frac{1}{4}$	$\frac{7}{2}$	1	6
x_1	1	$\frac{1}{6}$	$-\frac{1}{4}$	$\frac{3}{2}$	0	1
	0	$\frac{1}{6}$	$\frac{3}{4}$	$\frac{5}{2}$	0	-14

and

$$\begin{aligned}
B^{-1}A &= \begin{bmatrix} -\frac{1}{4} & \frac{1}{2} \\ -\frac{1}{4} & \frac{1}{6} \end{bmatrix} \begin{bmatrix} -6 & 0 & 1 & -2 & 2 \\ -3 & 1 & 0 & 6 & 3 \end{bmatrix} \\
&= \begin{bmatrix} 0 & \frac{1}{2} & -\frac{1}{4} & \frac{7}{2} & 1 \\ 1 & \frac{1}{6} & -\frac{1}{4} & \frac{3}{2} & 0 \end{bmatrix} \\
&= A^*
\end{aligned}$$

$$\begin{aligned}
B^{-1}b &= \begin{bmatrix} -\frac{1}{4} & \frac{1}{2} \\ -\frac{1}{4} & \frac{1}{6} \end{bmatrix} \begin{bmatrix} 6 \\ 15 \end{bmatrix} = \begin{bmatrix} 6 \\ 1 \end{bmatrix} = b^* \\
c - c_B B^{-1}A &= c - c_B A^* = [5,0,0,3,-2] - [-2,5]A^* \\
&= [5,0,0,3,-2] - [5,-\tfrac{1}{6},-\tfrac{3}{4},\tfrac{1}{2},-2] \\
&= [0,\tfrac{1}{6},\tfrac{3}{4},\tfrac{5}{2},0] = c^* \\
z_0 - c_B B^{-1}b &= z_0 - c_B b^* = -21 - (-7) = -14
\end{aligned}$$

Note how effective these formulas might be in sensitivity analysis on a linear programming problem. Suppose a given problem is solved using the simplex algorithm, but then changes are made on some of the original data (on A, b, c, or z_0). Using the formulas, we can determine the effect the changes would have on the previously calculated final tableaux, that is, on A^*, b^*, c^*, and z_0^*. (As we will see, this is especially easy if the changes do not involve B.) From this modified final tableau, we can determine if the former optimal solution remains optimal; if not, we can work from this tableau to resolve the modified problem.

We conclude this section with the verification of the formulas.

Definition 5.2.1. Let U_i denote the ith unit vector of $\mathbb{R}^m$.

Lemma 5.2.1. *The inverse of the matrix B always exists.*

Proof. Define $Y = [y_1, y_2, \ldots, y_m]^t$. For fixed i, $1 \leq i \leq m$, consider the system $BY = U_i$ of m equations in the m unknowns $y_1, y_2, \ldots, y_m$. By applying the same sequence of pivot steps to go from the original tableau

A	b
c	z_0

to the tableau

$$\begin{array}{c|c} A^* & b^* \\ \hline c^* & z_0^* \end{array}$$

to this system of equations, one obtains an equivalent system through this application of elementary row operations. (At each pivot step of the original simplex operation, multiples of one fixed row of the system of equations $AX = b$ were added to the remaining rows. Here we apply these same elementary row operations to the system of equations $BY = U_i$.) Since, in the final tableau, $x_{j(1)}, x_{j(2)}, \ldots, x_{j(m)}$ are the basic variables, this equivalent system will be in canonical form with the m basic variables $y_1, y_2, \ldots, y_m$. Thus this system of m equations and m unknowns has a solution, the results of these row operations on the column vector U_i. Denote this solution by the column vector $Y^{(i)}$ for each i, $1 \leq i \leq m$. Then

$$B[Y^{(1)}, Y^{(2)}, \ldots, Y^{(m)}] = [U_1, U_2, \ldots, U_m] = I$$

the $m \times m$ identity matrix. Therefore

$$B^{-1} = [Y^{(1)}, Y^{(2)}, \ldots, Y^{(m)}]$$

□

Example 5.2.2 (Continuation of Example 5.2.1)**.** Consider the system of equations $BY = U_1$; that is,

$$\begin{bmatrix} 2 & -6 \\ 3 & -3 \end{bmatrix} \begin{bmatrix} y_1 \\ y_2 \end{bmatrix} = \begin{bmatrix} 1 \\ 0 \end{bmatrix}$$

We can see from Table 5.4 that the sequence of row operations utilized to go from the initial to the final tableaux in that table takes the column vector $\begin{bmatrix} 2 \\ 3 \end{bmatrix}$ to $\begin{bmatrix} 1 \\ 0 \end{bmatrix}$ and the column vector $\begin{bmatrix} -6 \\ -3 \end{bmatrix}$ to $\begin{bmatrix} 0 \\ 1 \end{bmatrix}$. Thus, applying this sequence to the above system must generate a solution. In fact, in the standard *detached coefficients* notation, applying the sequence yields (as seen in the tables of Section 3.5)

$$\left[\begin{array}{cc|c} 2 & -6 & 1 \\ 3 & -3 & 0 \end{array}\right] \sim \left[\begin{array}{cc|c} 1 & -3 & \frac{1}{2} \\ 0 & 6 & -\frac{3}{2} \end{array}\right] \sim \left[\begin{array}{cc|c} 1 & 0 & -\frac{1}{4} \\ 0 & 1 & -\frac{1}{4} \end{array}\right]$$

Therefore $y_1 = -\frac{1}{4}$, $y_2 = -\frac{1}{4}$, and $Y^{(1)} = \begin{bmatrix} -\frac{1}{4} \\ -\frac{1}{4} \end{bmatrix}$.

The first formula that we prove is that $b^* = B^{-1}b$, that is, that $Bb^* = b$. In the example $Bb^* = b$ means that

$$\begin{bmatrix} 6 \\ 15 \end{bmatrix} = b = Bb^* = [A^{(5)}, A^{(1)}] \begin{bmatrix} 6 \\ 1 \end{bmatrix} = 6A^{(5)} + 1A^{(1)}$$

But this is equivalent to saying that the point $(1, 0, 0, 0, 6)$ is a solution to the constraints of (5.2.1), which we already know. The proof in the general case is just as above. It consists of observing that a basic feasible solution of a system of constraints is a solution to the system, corresponding to an expression for the constant-term column vector b as a linear combination of m of the column vectors from the coefficient matrix A.

Theorem 5.2.1. *$b^* = B^{-1}b$ or $Bb^* = b$.*

Proof. The two systems of equations $AX = b$ and $A^*X = b^*$ are equivalent and so have the same solution set. Define $X^* = [x_1^*, x_2^*, \ldots, x_n^*]^t$, a column vector in $\mathbb{R}^n$, by

$$x_j^* = \begin{cases} b_k^*, & x_j \text{ is the basic variable in the } k\text{th equation of the final tableau} \\ 0, & x_j \text{ is a nonbasic variable in the final tableau} \end{cases}$$

Then X^* is the basic feasible solution associated with the final tableau, and $A^*X = b^*$. Therefore

$$AX^* = b; \text{ that is, } \sum_{k=1}^{m} A^{(j(k))} b_k^* = b; \text{ that is, } Bb^* = b \qquad \square$$

Next, we show that $A^* = B^{-1}A$ or, equivalently, for each j, $1 \leq j \leq n$, $A^{*(j)} = B^{-1}A^{(j)}$. In our example, if we let $j = 4$, this would mean that

$$\begin{bmatrix} \frac{7}{2} \\ \frac{3}{2} \end{bmatrix} A^{*(4)} = B^{-1}A^{(4)} = B^{-1} \begin{bmatrix} -2 \\ 6 \end{bmatrix} \tag{5.2.2}$$

Suppose that in the example the original b was

$$\begin{bmatrix} -2 \\ 6 \end{bmatrix} = A^{(4)}$$

Then the b^* would be

$$\begin{bmatrix} \frac{7}{2} \\ \frac{3}{2} \end{bmatrix} = A^{*(4)}$$

as the result of our pivot operations on the column vector

$$\begin{bmatrix} -2 \\ 6 \end{bmatrix}$$

have been recorded in the fourth column of the tableaux. But from Theorem 5.2.1, we know that $b^* = B^{-1}b$, and applying that here would give us (5.2.2).

Theorem 5.2.2. *For any j, $1 \leq j \leq n$, $A^{*(j)} = B^{-1}A^{(j)}$.*

Proof. Fix j, $1 \leq j \leq n$. Suppose the initial column vector b of Theorem 5.2.1 is $A^{(j)}$. Then the resulting b^* would simply be $A^{*(j)}$, the result of the sequence of pivot steps on $A^{(j)}$. Thus Theorem 5.2.1 implies that $A^{*(j)} = B^{-1}A^{(j)}$. (Note that some of the entries of $A^{(j)}$ and $A^{*(j)}$ may be negative but that this has no effect on the proof of Theorem 5.2.1.) $\square$

Corollary 5.2.1. $A^* = B^{-1}A$.

Note that if $A^{(j)}$ is U_k, the kth unit vector in $\mathbb{R}^m$, then $B^{-1}A^{(j)} = B^{-1}U_k =$ the kth column of B^{-1}. Thus, if all the unit vectors $U_1, \ldots, U_m$ are represented in the initial coefficient matrix A and the matrix A^* is available, all the columns of the matrix B^{-1} can be read off immediately from the columns of A^*. In our example, $A^{(2)} = U_2$ and $A^{(3)} = U_1$. Hence the third column of A^* will be the first column of B^{-1}, and the second column of A^* will be the second column of B^{-1}. The reader should verify this.

Example 5.2.3. Similarly, using the data from the corresponding tables, the reader should confirm that B and B^{-1} for the final tableau of Example 3.5.2 on page 88 and Table 3.5 on page 88 are

$$B = \begin{bmatrix} 1 & 1 \\ -3 & 1 \end{bmatrix} \text{ and } B^{-1} = \frac{1}{4}\begin{bmatrix} 1 & -1 \\ 3 & 1 \end{bmatrix},$$

and for the final tableau of Example 3.5.3 and Table 3.6 are

$$B = \begin{bmatrix} -2 & 2 & -3 \\ -4 & 1 & -1 \\ 0 & 3 & -1 \end{bmatrix} \text{ and } B^{-1} = \frac{1}{24}\begin{bmatrix} 2 & -7 & 1 \\ -4 & 2 & 10 \\ -12 & 6 & 6 \end{bmatrix}.$$

To demonstrate the idea behind the proofs involving the data of the bottom rows of the tableaux, consider first the meaning of these data in our example. The third and sixth rows in Table 5.4 correspond to two expressions for the objective function of that example, namely,

$$z = 5x_1 + 3x_4 - 2x_5 + 21$$

and

$$z = \tfrac{1}{6}x_2 + \tfrac{3}{4}x_3 + \tfrac{5}{2}x_4 + 14$$

We know that these two expressions must deliver the same value when evaluated at any solution to the constraints of (5.2.1), and in particular, when evaluated at (1,0,0,0,6). Thus

$$5(1) + 3(0) - 2(6) + 21 = \tfrac{1}{6}(0) + \tfrac{3}{4}(0) + \tfrac{5}{2}(0) + 14$$

Clearing zeros, we have

$$\underbrace{5(1) - 2(6)}_{c_B b^*} + \underbrace{21}_{-z_0} = \underbrace{14}_{-z_0^*}$$

Theorem 5.2.3. $z_0^* = z_0 - c_B B^{-1} b = z_0 - c_B b^*$

Proof. The original linear programming problem is to minimize z with $z = -z_0 + c_1x_1 + \cdots + c_nx_n$ subject to $AX = b$, $X \geq 0$. After applying a sequence of pivot operations, we have the equivalent problem of minimizing z with $z = -z_0^* + c_1^*x_1 + \cdots + c_n^*x_n$ subject to $A^*X = b^*$, $X \geq 0$. These two expressions for the objective function, say $F(X) = -z_0 + c_1x_1 + \cdots + c_nx_n$ and $G(X) = -z_0^* + c_1^*x_1 + \cdots + c_n^*x_n$, although different expressions, have the same value at any point that satisfies the equivalent systems of equations $AX = b$ or $A^*X = b^*$ (see the discussion on the representation of the objective function in Section 3.2). Now X^*, as defined in the proof of Theorem 5.2.1, is a solution to the system of constraints, and so $F(X^*) = G(X^*)$. But $F(X^*) = -z_0 + c_Bb^*$, $G(X*) = -z_0^*$. Therefore, $-z_0^* = -z_0 + c_Bb^*$ or $z_0^* = z_0 - c_Bb^*$. □

One formula remains to be verified, namely, that $c^* = c - c_BB^{-1}A = c - c_BA^*$, or, equivalently, for each j, $1 \leq j \leq n$, $c_j^* = c_j - c_BA^{*(j)}$. Returning to our example, with $j = 4$, suppose, as before, that the original b were $A^{(4)}$, and also that the original z_0 were $3 = c_4$. Then not only would b^* of the final tableau be $A^{*(4)}$, as before, but z_0^* would be $\frac{5}{2} = c_4^*$, and Theorem 5.2.3, applied here, would yield

$$c_4^* = z_0^* = z_0 - c_Bb^* = c_4 - c_BA^{*(4)}$$

Theorem 5.2.4. *For any j, $1 \leq j \leq n$, $c_j^* = c_j - c_BB^{-1}A^{(j)} = c_j - c_BA^{*(j)}$.*

Proof. Fix j, $1 \leq j \leq n$. Suppose the initial column vector b in Theorem 5.2.3 is $A^{(j)}$, and the initial constant z_0 equals c_j. Then the resulting b^* would be $A^{*(j)}$, and the resulting z_0^* would be c_j^*. Thus Theorem 5.2.3, applied to the linear programming problem of minimizing z with $z = -c_j + c_1x_1 + \cdots + c_nx_n$ subject to $AX = A^{(j)}, X \geq 0$, implies that $c_j^* = c_j - c_BA^{*(j)}$. (Note again that negative entries in either $A^{(j)}$ or $A^{*(j)}$ will not affect the proof of Theorem 5.2.3.) □

Corollary 5.2.2. $c^* = c - c_BA^* = c - c_BB^{-1}A$.

The formulas in this section show that at any step of the simplex algorithm, all the relevant data can be calculated easily from the original data of the problem, as long as we know the corresponding matrix B^{-1}. This suggests another approach to the simplex process. Instead of using the full tableau, as we do to record the results of each step of the algorithm, suppose we simply record the original data along with the B^{-1} matrix. In fact, note that at each step of the simplex algorithm exactly one column of the B matrix is altered as exactly one variable in the basis is replaced by another variable. Rules can be given that prescribe the effects of this alteration on the B^{-1} matrix, and this leads to a modification of the simplex process known as the *revised simplex method*. For large problems, this technique for recording the simplex process has the advantage of requiring less computer time, calculation, and memory, and therefore enables one to handle larger problems with fixed computer facilities. We will not develop the revised simplex method in this text; refer to the books of Chvatal [11], Dantzig [7], or Hadley [12].

Problem Set 5.2

1. Consider the problem of Example 3.5.1 on page 87.

(a) Determine B, B^{-1}, and c_B for the final tableau. Note that B^{-1} is contained in the final tableau.
(b) Verify the formulas of this section for this tableau.

2. Consider the problem of Example 3.6.1 on page 95.

(a) Determine the matrix B^{-1} for the third tableau of Table 3.8.
(b) Determine B^{-1} for the final tableau of Table 3.9 on page 97, the completion of the problem of Table 3.8.
(c) What does this suggest about the data in the artificial variable columns of the coefficient matrix?

3. Consider the linear programming problem of

$$\begin{array}{l} \text{Minimizing } z = 3x_1 + 2x_2 + 5x_3 - 4x_4 \\ \text{subject to} \\ \quad x_1 - x_2 + 3x_3 + 2x_4 + 7x_5 = 31 \\ -2x_1 + 3x_2 - 6x_3 + 4x_4 - x_5 = 2 \\ x_1, x_2, x_3, x_4, x_5 \geq 0 \end{array}$$

(a) Determine the B, B^{-1}, and c_B for the tableau presentation of this problem with basic variables x_3 and x_4.
(b) Express $b = [31,2]^t$ as a linear combination of $A^{(3)} = [3,-6]^t$ and $A^{(4)} = [2,4]^t$ using Theorem 5.2.1.
(c) Show that this solution to the system of constraints is an optimal solution to the linear programming problem. (Compute c^* using Corollary 5.2.2.)

4. In the linear programming problem of Problem 3, suppose we select x_1 and x_3 as potential basic variables. Determine the corresponding matrix B. Does B have an inverse? Why does this not contradict Lemma 5.2.1?

5. For the linear programming problem of

$$\begin{array}{l} \text{Minimizing } z = -6x_1 + 2x_2 - 9x_3 + 12x_4 + 8x_5 \\ \text{subject to} \\ \quad x_1 + 6x_2 - 5x_3 + 2x_4 - 7x_5 = 15 \\ -x_1 - 4x_2 + 3x_3 - x_4 + 5x_5 = 25 \\ x_1, x_2, x_3, x_4, x_5 \geq 0 \end{array}$$

show, using the formulas in this section, that x_2 and x_5 can serve as basic variables in an optimal solution. What is this optimal solution point and the minimum of z?

6. For the linear programming problem of

$$\begin{aligned}
&\text{Minimizing } z = 4x_1 + 6x_2 + x_3 - 2x_4\\
&\text{subject to}\\
&3x_1 + 4x_2 - 3x_3 + x_4 = 60\\
&\ \ x_1 + 2x_2 + 2x_3 - x_4 = 25\\
&x_1, x_2, x_3, x_4 \geq 0
\end{aligned}$$

show that x_2 and x_4 can serve as basic variables for a basic feasible solution. Compute the associated $A^{(3)}$ and c_3^*. What is your conclusion?

7. Given the problem of maximizing $c \cdot X$ subject to $AX \leq b$, $X \geq 0$, with $b \geq 0$, suppose slack variables are added and the simplex algorithm is applied, leading to a finite optimal solution. Define B and c_B using the basic variables from the final tableau. Show that $-c_B B^{-1}$ is an optimal solution point to the dual problem. (*Hint*. Recall where in the final tableau the solution to the dual can be found.)

5.3 CHANGES IN THE OBJECTIVE FUNCTION

Suppose the simplex method has been used to solve the problem of minimizing z, $z = c \cdot X - z_0$, subject to $AX = b$, $X \geq 0$ with the final tableau given by

$$\begin{array}{c|c} A^* & b^* \\ \hline c^* & z_0^* \end{array}$$

and with X^* the associated optimal basic feasible solution, but that now one or more of the original coefficients of the objective function, the components of c, are changed. How does this affect the already computed solution to the problem?

From the formulas in the previous section, it follows that changes in c can induce changes only in c^* and z_0^*, and in fact, the new c^* can be determined using $c^* = c - c_B A^*$. If the modified c^* remains nonnegative, X^* would remain an optimal solution point; if not, more iterations of the simplex algorithm may be necessary to complete the modified problem. However, we could initiate the algorithm on the modified, previously final tableau. We illustrate.

Example 5.3.1. Consider the linear programming problem of

$$\begin{aligned}
&\text{Maximizing } z = 11x_1 + 4x_2 + x_3 + 15x_4 &(5.3.1)\\
&\text{subject to}\\
&3x_1 + x_2 + 2x_3 + 4x_4 \leq 28\\
&8x_1 + 2x_2 - x_3 + 7x_4 \leq 50\\
&x_1, x_2, x_3, x_4 \geq 0
\end{aligned}$$

Adding slack variables and then applying the simplex algorithm yields the reduced tableaux resolution shown in Table 5.5. The maximum value of z is 106, attained at the point $(0, 4, 0, 6)$.

Table 5.5

	x_1	x_2	x_3	x_4	x_5	x_6	
x_5	3	1	2	4	1	0	28
x_6	8	2	-1	7	0	1	50
	-11	-4	-1	-15	0	0	0
x_4	-2	0	5	1	2	-1	6
x_2	11	1	-18	0	-7	4	4
	3	0	2	0	2	1	106

Table 5.6

	x_1	x_2	x_3	x_4	x_5	x_6	
x_4	-2	0	5	1	2	-1	6
x_2	(11)	1	-18	0	-7	4	4
	-1	0	2	0	2	1	106
x_4	0	$\frac{2}{11}$	$\frac{19}{11}$	1	$\frac{8}{11}$	$-\frac{3}{11}$	$6\frac{8}{11}$
x_1	1	$\frac{1}{11}$	$-\frac{18}{11}$	0	$-\frac{7}{11}$	$\frac{4}{11}$	$\frac{4}{11}$
	0	$\frac{1}{11}$	$\frac{4}{11}$	0	$\frac{15}{11}$	$\frac{7}{11}$	$106\frac{4}{11}$

Suppose we now change the coefficient of x_1 in the objective function, the 11. Since x_1 is not a basic variable in the final tableau, a change in the 11 would not alter $c_B = [-15, -4]$. (Note that in this example the c_j's of the initial tableau are the negatives of the coefficients of the objective function of (5.3.1).) The only change would be in the first component of c and therefore in the first component of c^*. In particular, we presently have

$$3 = c_1^* = c_1 - c_B A^{*(1)}$$

and any increase or decrease in c_1 would generate the exact same change in c_1^*. Thus adding a constant λ to the 11 would decrease c_1 by λ and so decrease c_1^* by λ. c_1^* will remain nonnegative, and $(0,4,0,6)$ would remain the optimal solution point, as long as the 11 is not increased by more than 3 units. However, if the 11 is increased by more than 3 units, the modified c_1^* would become negative, and more iterations of the algorithm would be necessary to complete the modified problem. Computations could begin in the final tableau of Table 5.5, with c_1^* modified. For example, if the 11 is increased to 15, the maximum of z would then be $106\frac{4}{11}$, attained at $(\frac{4}{11}, 0, 0, 6\frac{8}{11})$ (see Table 5.6).

Clearly, this argument generalizes. Changes in the coefficients of the nonbasic variables translate directly into the corresponding changes in the appropriate components of c^*. In this example, with $c_3^* = 2$, c_3 can be decreased by up to 2 units, and the coefficient of x_3 in z increased by up to 2 units, before $(0,4,0,6)$ would no longer be optimal.

Suppose now that the only change in the objective function z is in the coefficient of x_4, from 15 to $15+\lambda$. Such a change would alter c_B, so we recalculate c^*.

$$\begin{aligned} c^* &= c - c_B A^* \\ &= [-11,-4,-1,-15-\lambda,0,0]-[-15-\lambda,-4]A^* \\ &= \underbrace{[-11,-4,-1,-15,0,0]-[-15,-4]A^*}_{\text{the former } c^*}+[0,0,0,-\lambda,0,0]-[-\lambda,0]A^* \\ &= [3,0,2,0,2,1]+[-2\lambda,0,5\lambda,0,2\lambda,-\lambda] \\ &= [3-2\lambda,0,2+5\lambda,0,2+2\lambda,1-\lambda] \end{aligned}$$

c^* will be nonnegative if $3-2\lambda \geq 0$, $2+5\lambda \geq 0$, $2+2\lambda \geq 0$, and $1-\lambda \geq 0$, that is, if $-\frac{2}{5} \leq \lambda \leq 1$. Thus, as long as λ is in this interval, the maximum of z will remain at $(0,4,0,6)$, with

$$\text{Max}\, z = 4(4)+(15+\lambda)6 = 106+6\lambda$$

Similarly, the coefficient of x_2 in z can range between $4-\frac{1}{4}$ and $4+\frac{1}{9}$ before $(0,4,0,6)$ would no longer be optimal, as the reader may verify (Problem 1).

Example 5.3.2. Consider the operation of the can-producing company of Example 5.1.3. The optimal production schedule calls for the use of Process 2 for 75 hr and Process 3 for 150 hr. Suppose, however, that the hourly cost of Process 3 may fluctuate because of the changing costs of the special materials used in this process, and management wants to know the limits on the range of change of this cost before the present $(0,75,150)$ operating schedule is no longer optimal. To respond, assume the operating cost of Process 3, in dollars/hr, is $c_3 = 200+\lambda$, and λ is a constant, not necessarily positive. Then, using the data of Table 5.1, we have

$$\begin{aligned} c^* &= c - c_B A^* \\ &= [65,150,200+\lambda,0,0]-[150,200+\lambda,0]A^* \\ &= [65,150,200,0,0]+[0,0,\lambda,0,0]-[150,200,0]A^*-\lambda[0,1,0]A^* \\ &= ([65,150,200,0,0]-[150,200,0]A^*)+[0,0,\lambda,0,0]-\lambda[0,1,0]A^* \\ &= \left[46\tfrac{1}{4},0,0,3\tfrac{1}{8},0\right]-\lambda\left[-\tfrac{3}{4},0,0,\tfrac{1}{8},0\right] \\ &= \left[46\tfrac{1}{4}+\tfrac{3\lambda}{4},0,0,3\tfrac{1}{8}-\tfrac{\lambda}{8},0\right] \end{aligned}$$

Thus the optimal production schedule remains $(0,75,150)$ as long as

$$46\tfrac{1}{4}+\tfrac{3\lambda}{4} \geq 0 \text{ and } 3\tfrac{1}{8}-\tfrac{\lambda}{8} \geq 0, \text{ that is }, -61\tfrac{2}{3} \leq \lambda \leq 25$$

The information determined above is also contained in Solver's Sensitivity Report for the can-producing problem, the report of Figure 5.4. We outline the upper half, the Adjustable Cells half, of the report. This half deals with the variables of the problem and their coefficients in the objective function. The first two columns of the section identify the rows. The third column lists the final values of the variables, and the fifth column the variables' initial coefficients in the objective function. The data in the last two columns are the most useful data in this half of the report. The values in a given row are bounds on how much the associated variable's coefficient in the objective function (the number in the fifth column) can change before the stated optimal solution changes, assuming no other changes in the data of the problem. In particular, in the x_3-Process 3 row in Figure 5.4 we see the bounds calculated above of $-61\frac{2}{3}$ and 25, thus verifying our calculations (or those of Microsoft Excel/Solver).

We can also easily confirm the data in the Reduced Cost column of the report. In general, the magnitude of the entries in this column indicates by how much the associated variable's coefficient in the objective function needs to change in order for the variable to enter the basis. In the solution of the can company's problem, Processes 2 and 3 are used in the optimal solution, hence the zeros in the corresponding x_2 and x_3 rows. Process 1 is not used in the optimal solution, and that certainly will not change if its hourly cost increases. However since x_1 is a nonbasic variable, decreasing its operating cost by $c_1^* = 46.25$ dollars (using Table 5.1) will allow x_1 into the basis, hence the (duplicate) information in the x_1 row.

The arithmetic associated with determining the bounds on allowable changes in the objective function coefficients can be reduced to routine calculations using the data in a final tableau resolution of a linear programming problem, as specified in Problem 7 of this section. Thus Solver's bounds can come immediately from a computer solution of the problem. But the formula for the bounds is dependent upon the fact that there is only one coefficient change in the objective function. If more than one coefficient were changed, Solver's data would not be immediately applicable and our computations in Example 5.3.2 would need revision. However, the theory that we have developed provides considerable flexibility. Using Corollary 5.2.2 we can simply calculate the new c^* and work from there.

Example 5.3.3. Considering again the operation of the can-producing company of the preceding example, suppose that labor is negotiating a new contract that would include a pay increase. While bargaining with labor, management needs to know how the tentative increase would influence the monthly production costs and if the present optimal production schedule would need to be modified. (Such a modification could necessitate some major one-time expenses.)

In particular, suppose that an increase of $\$\lambda$/hr in the pay of the workforce would change the cost in dollars of 1 hr of operation of Processes 1, 2, and 3 to $65+\lambda$, $150+4\lambda$, and $200+8\lambda$, respectively. (The amounts of labor required in the operation of the processes vary.) Using the data of Table 5.1, we have

$$\begin{aligned}
c^* &= c - c_B A^* \\
&= [65+\lambda, 150+4\lambda, 200+8\lambda, 0, 0] - [150+4\lambda, 200+8\lambda, 0]A^* \\
&= [46\tfrac{1}{4}, 0, 0, 3\tfrac{1}{8}, 0] + [\lambda, 4\lambda, 8\lambda, 0, 0] - [4\lambda, 8\lambda, 0]A^* \\
&= [46\tfrac{1}{4}, 0, 0, 3\tfrac{1}{8}, 0] + [\lambda, 4\lambda, 8\lambda, 0, 0] - \left[-\frac{3\lambda}{2}, 4\lambda, 8\lambda, \frac{\lambda}{4}, 0\right]
\end{aligned}$$

$$= \left[46\tfrac{1}{4} + \frac{5\lambda}{2}, 0, 0, 3\tfrac{1}{8} - \frac{\lambda}{4}, 0\right]$$

Thus, as long as the increase λ does not exceed $\frac{100}{8} = 12.50$, the optimal production schedule will remain 75 hr for Process 2 and 150 hr for Process 3 at a monthly cost in dollars of $75(150+4\lambda)+150(200+8\lambda) = 41,250+1500\lambda$.

Problem Set 5.3

1. Verify that $(0,4,0,6)$ is an optimal solution to (5.3.1) if the coefficient of x_2 in z is between $\frac{15}{4}$ and $\frac{37}{9}$ and all the other coefficients are as given in (5.3.1).

2. Suppose two changes are made in the coefficients of z in (5.3.1): the coefficient of x_1 is increased from 11 to 13 and the coefficient of x_4 is increased from 15 to $15\frac{3}{4}$. While each of these changes is within the corresponding allowable range for the coefficient as discussed in the text, show that $(0,4,0,6)$ is no longer optimal. Solve the modified problem.

3. Starting from the final tableau of Table 5.5, complete the problem of (5.3.1) if the objective function coefficient of
   (a) x_3 is increased from 1 to 4.
   (b) x_4 is increased from 15 to $16\frac{1}{2}$.
   (c) x_4 is decreased from 15 to 14 and the coefficient of x_3 is decreased from 1 to -2.

4. Consider the linear programming problem of Example 5.2.1 on page 176.
   (a) Determine the range on each of the coefficients in z such that if all the other coefficients of z remain fixed at the original values, the point $(1,0,0,0,6)$, as determined from the second tableau of Table 5.4, remains an optimal solution point.
   (b) Solve the modified problem, starting from the second tableau of Table 5.4, if z is changed as follows:
      (i) The coefficient of x_1 is increased to 7.
      (ii) The coefficient of x_1 is increased to 9 and the coefficient of x_5 is decreased to -3.
      (iii) The coefficients of both x_1 and x_5 are decreased by 2.

5. In the poultry producer's problem of Example 5.1.1 on page 161, with the initial cost/lb of Feeds 1 and 2 at 16 cents and 14 cents, respectively, the optimal diet is to use 10 lb of Feed 1 and 6 lb of Feed 2 daily. From (5.1.1), if the ratio $\frac{c_1}{c_2}$ of these costs remains between $\frac{3}{5}$ and $\frac{5}{2}$ and $(10,6)$ remains the optimal diet. From Solver's sensitivity report (Figure 5.5), if only one cost is changed from its original value, $(10,6)$ remains the optimal diet if the change is in c_1 and is between $-7\frac{3}{5}$ and 19, or if the change is in c_2 and is between $-7\frac{3}{5}$ and $12\frac{2}{3}$. Show that Solver's data can be determined by the bounds on the ratio $\frac{c_1}{c_2}$ from (5.1.1).

6. Consider the diet problem of Example 2.2.1 (page 10) and Example 4.3.1 (page 132). The shadow prices for the three nutritional requirements, from the final tableau of Table 4.2 (page 133), are 1 cent/unit for nutritional elements A and B and 0 cents/unit for C. Using the final tableau of Table 4.2, determine the range on the required number of units of A such that for any nutritional requirement of A within this range, with the requirements for B and C fixed at 84 and 72, respectively, the above shadow prices remain accurate. Similarly, determine the corresponding ranges for the requirements for B and C.

7. Suppose the simplex algorithm applied to the problem of minimizing $z = c \cdot X - z_0$ subject to $AX = b$, $X \geq 0$ leads to a finite optimal value attained at the point X^*, with x_s serving as the basic variable of the sth row of the final tableau (with data A^* and c^*). Show that if the coefficient c_s of x_s in z is changed to $c_s + \lambda$, then X^* remains optimal as long as

$$\text{Max}\{c_j^*/a_{rj}^* : a_{rj}^* < 0\} \leq \lambda \leq \text{Min}\{c_j^*/a_{rj}^* : a_{rj}^* > 0, j \neq s\}$$

8. Consider the operation of the aluminum can company of Example 5.1.3 on page 166.
   (a) Verify the bounds stated in the associated Sensitivity Report (Figure 5.4) on the allowable changes in c_2, the cost of 1 hr of operation of Process 2, before the optimal solution point would change.
   (b) As discussed in Example 5.3.3, suppose labor negotiates a \$15/hr increase in pay. Determine the new optimal production schedule.

9. Monthly profits are maximized for the company described in Problem 15 of Section 5.1 by the making and selling of 240 A's and 140 C's.
   (a) Using the data from Table 5.2, determine by how much the profit on the sale of a unit of A (the \$140) can vary before the optimal production schedule would change.
   (b) By how much do we need to increase the profit on the B's for them to be part of the optimal production schedule?
   (c) By how much can the profit on the sale of a unit of C (the \$60) vary before the optimal production schedule would change?
   (d) Suppose labor requests a pay raise, from \$18/hr to \$$(18 + \lambda)$/hr, with overtime pay to be \$$(24 + \lambda)$/hr. How large can λ be before the optimal production schedule would change?

10. The sensitivity report accompanying Microsoft Excel/Solver's solution of the problem considered above, that of Problem 15 of Section 5.1, is in Figure 5.6.
    (a) Use the report to verify your answers to parts (a), (b), and (c) of Problem 9.
    (b) Can you use the data of the report in responding to the question raised in part (d) of Problem 9? Why or why not?
    (c) Verify the entries in the shadow price column of the report.

11. Consider the situation described in Problem 17 of Section 5.1. You are called into the plant manager's office again. This time the manager wants to know how

Problem 15 of Section 5.1
Sensitivity Report

Adjustable Cells

Cell	Name	Final Value	Reduced Cost	Objective Coefficient	Allowable Increase	Allowable Decrease
\$C\$16	Units of A made	240	0	69	15	3
\$D\$16	Units of B made	0	-15	111	15	1E+30
\$E\$16	Units of C made	140	0	42	2	5
\$F\$16	Overtime hrs	120	0	-6	1E+30	6

Constraints

Cell	Name	Final Value	Shadow Price	Constraint R.H. Side	Allowable Increase	Allowable Decrease
\$C\$22	Material M1 LHS	2000	7.5	2000	480	140
\$C\$23	Material M2 LHS	4820	0	5000	1E+30	180
\$C\$24	Labor LHS	500	12	500	16.3636	120
\$C\$25	Overtime LHS	120	6	120	16.3636	120

Figure 5.6

contingent your optimal production schedule is on the price of the B's bought externally and on labor's pay rate. How do you respond?

12. (a) Suppose the coefficient of x_1 in the second constraint of the problem of (5.3.1) (the $a_{21} = 8$) is replaced by a constant α. Show that the optimal solution of the second tableau of Table 5.5 is not altered if $\alpha \geq 5$. (*Hint.* Compute c_1^* in terms of α.)

(b) Suppose the coefficient of x_2 in the second constraint of the problem (the $a_{22} = 2$) were changed. Would your analysis in this situation be any different than your approach in part (a)?

13. It is possible that Process 1, as described in Example 5.1.3 on page 166, could be modified to utilize some recycled aluminum, with no accompanying changes in output or operating cost. How much would need to be used for Process 1 to become part of the optimal production schedule?

14. In Problem 15 of Section 5.1, suppose the amount of raw material M_1 used in the manufacture of a unit of B can be reduced, causing no other changes in the data of the problem. By how much does the 12 lb/unit requirement need to be decreased to have the B's introduced into the optimal production schedule?

5.4 ADDITION OF A NEW VARIABLE

Suppose that the simplex process has been used to solve the problem of minimizing the objective function $z = c \cdot X - z_0$ subject to $AX = b$, $X \geq 0$, with the final tableau given by

$$\begin{array}{c|c} A^* & b^* \\ \hline c^* & z_0^* \end{array}$$

and with X^* the associated basic feasible solution, but that now we need to add another variable in the formulation of the original problem. For example, this variable could measure the amount of goods to be shipped by means of a newly opened shipping line, or it could measure the amount of production of a new product for which a market has just been developed. Let x_{n+1} be this new variable, with cost coefficient c_{n+1} and column vector of coefficients for the constraining equations $A^{(n+1)}$. Then the expanded, modified problem is to minimize z, with $z = c \cdot X + c_{n+1}x_{n+1}$ subject to

$$[A, A^{(n+1)}] \begin{bmatrix} X \\ x_{n+1} \end{bmatrix} = b, \quad X \geq 0, \quad x_{n+1} \geq 0 \tag{5.4.1}$$

X^* remains a basic feasible solution to (5.4.1) if we simply set the value of the nonbasic variable x_{n+1} equal to zero. Moreover, this point will provide an optimal solution if $c^*_{n+1} \geq 0$. And $c^*_{n+1} = c_{n+1} - c_B B^{-1} A^{(n+1)}$ from Theorem 5.2.4. Thus this quantity can be easily calculated. If $c_{n+1} \geq 0$, the original optimal solution remains optimal. If $c_{n+1} \leq 0$, the data for the new $(n+1)$ column in the system of constraints, $A^{*(n+1)}$ are needed to proceed with the simplex algorithm. But $A^{*(n+1)} = B^{-1}A^{(n+1)}$ from Theorem 5.2.2, so this information can also be easily calculated.

Example 5.4.1. Consider the problem of Example 5.3.1 in the previous section. Suppose we wish to introduce a new variable x_7, with the problem becoming the following:

$$\begin{aligned} &\text{Maximize } z = 11x_1 + 4x_2 + x_3 + 15x_4 + 12x_7 \\ &\text{subject to} \\ &3x_1 + \ x_2 + 2x_3 + 4x_4 + 3x_7 \leq 28 \\ &8x_1 + 2x_2 - \ x_3 + 7x_4 + 5x_7 \leq 50 \\ &x_1, x_2, x_3, x_4, x_7 \geq 0 \end{aligned} \tag{5.4.2}$$

The optimal solution point for the original problem is $(0,4,0,6)$, and so $x_2 = 4$, $x_4 = 6$, $x_1 = x_3 = x_7 = 0$ is a feasible solution to the expanded problem. This solution is optimal if $c^*_7 \geq 0$. Using the data of Table 5.5 on page 184 (note that the B^{-1} is contained in the second tableau), we have

$$\begin{aligned} c^*_7 &= c_7 - c_B B^{-1} A^{(7)} \\ &= -12 - [-15, -4] \begin{bmatrix} 2 & -1 \\ -7 & 4 \end{bmatrix} \begin{bmatrix} 3 \\ 5 \end{bmatrix} = -1 \end{aligned}$$

Thus this point is not an optimal solution point. To complete the problem, we determine $A^{*(7)}$,

$$A^{*(7)} = B^{-1}A^{(7)} = \begin{bmatrix} 2 & -1 \\ -7 & 4 \end{bmatrix} \begin{bmatrix} 3 \\ 5 \end{bmatrix} = \begin{bmatrix} 1 \\ -1 \end{bmatrix}$$

and expand the second tableau of Table 5.5 to include the new column. One iteration of the simplex algorithm (Table 5.7) shows that the maximum of z is now 112, attained at $x_2 = 10$, $x_7 = 6$, $x_1 = x_3 = x_4 = 0$.

Actually, we could have also used the dual to determine if the new variable x_7 is to enter the basis. The dual constraint corresponding to the new variable is

$$3y_1 + 5y_2 \geq 12 \tag{5.4.3}$$

Table 5.7

	x_1	x_2	x_3	x_4	x_5	x_6	x_7	
x_4	-2	0	5	1	2	-1	①	6
x_2	11	1	-18	0	-7	4	-1	4
	3	0	2	0	2	1	-1	106
x_7	-2	0	5	1	2	-1	1	6
x_2	9	1	-13	1	-5	3	0	10
	1	0	7	1	4	0	0	112

and the solution to the dual of the original problem is $(2,1)$, from the bottom row, slack variable columns, of Table 5.5. But the point $(2,1)$ does not satisfy this inequality, and so more iterations on the previously final tableau must be necessary to determine the optimal solution point to the dual of the expanded problem. In fact, the slack in this dual constraint (5.4.3) when evaluated at $(2,1)$ is equal to c_7^*, from Problem 9 of Section 4.4 (or from Theorem 5.2.4 and Problem 7 of Section 5.2).

In terms of marginal values, this is all quite plausible. Suppose the problem of (5.4.2) models a problem of maximizing profits through an operation consisting of up to five activities (the x_i's) using two limited resources (the constraints). Allowing the use of only the first four activities gives marginal values to the first and second resources of 2 and 1. If the new activity can more profitably utilize these resources (it delivers a return of 12 on an investment of $11 = 3 \cdot 2 + 5 \cdot 1$), it will be incorporated into the optimal production schedule.

Example 5.4.2. Suppose the aluminum can company of Example 5.1.3 on page 166 could invest in a fourth can-producing process. Operation of this process for 1 hr would produce 14 cases of the Type A can and 16 cases of the Type B can, use 1 lb of recycled aluminum, and cost \$110. To determine if this process would be utilized, consider the total marginal value of 1 hr of operation,

$$14(3\tfrac{1}{8}) + 16(0) + 1(56\tfrac{1}{4}) = \$100$$

This value is \$10 less than the cost of 1 hr of operation; the process would not be used. Only if the hourly cost could be reduced to less than \$100 would the process be included in the operation of the plant (see Problem 4).

Problem Set 5.4

1. Solve the problem of (5.3.1) of Section 5.3, starting from the second tableau of Table 5.5, if a variable $x_7 \geq 0$ is added to the problem as stated in (5.3.1), with:

   (a) $A^{(7)} = [2,5]'$, coefficient of x_7 in $z = 9$
   (b) $A^{(7)} = [2,5]'$, coefficient of x_7 in $z = 10$
   (c) $A^{(7)} = [-6,-11]'$, coefficient of x_7 in $z = 15$
   (d) $A^{(7)} = [7,12]'$, coefficient of x_7 in $z = 27$

2. Solve the problem of (5.2.1) of Section 5.2, using the data of Table 5.4, if a variable $x_6 \geq 0$ is added to the problem as stated in (5.2.1), with:
   (a) $A^{(6)} = [2,1]'$, coefficient of x_6 in $z = -1$
   (b) $A^{(6)} = [2,1]'$, coefficient of x_6 in $z = -2$
   (c) $A^{(6)} = [8,6]'$, coefficient of x_6 in $z = -9$
   (d) $A^{(6)} = [8,6]'$, coefficient of x_6 in $z = -11$

3. Suppose a variable x_5 is added to the problem of Example 3.6.1 on page 95, with the problem becoming

$$\begin{aligned}&\text{Minimize } 2x_1 - 3x_2 + x_3 + x_4 + 5x_5\\ &\text{subject to}\\ &x_1 - 2x_2 - 3x_3 - 2x_4 + 4x_5 = 3\\ &x_1 - x_2 + 2x_3 + x_4 + 3x_5 = 11\\ &x_1, x_2, x_3, x_4, x_5 \geq 0\end{aligned}$$

   Starting from the second tableau of Table 3.9, solve the problem.

4. (a) Using $c_j^* = c_j - c_B B^{-1} A^{(j)}$, verify that the c_j^* for the final tableau of Table 5.1 corresponding to the new process described in Example 5.4.2 is $+10$.
   (b) Suppose the cost of the new process could be reduced to \$90/hr. Determine the new monthly operating schedule and cost.

5. Reconsider Problem 13 of Section 5.3 using marginal values.

6. The company in Problem 15 of Section 5.1 could also make D's, with a unit of D requiring 8 lb of M_1, 10 lb of M_2, and 1.5 hr of labor. What minimum profit, excluding labor costs (i.e., selling price less cost of raw materials), is necessary before the company would produce and sell D's?

7. You, the production supervisor in Problem 17 of Section 5.1, have access to a Process 3, which turns out in 1 hr 6 A's and 10 B's using 16 lb of rare metal and 9 hr of labor. Should you use the process?

5.5 CHANGES IN THE CONSTANT-TERM COLUMN VECTOR

Suppose that the minimum of z, $z = c \cdot X - z_0$ subject to $AX = b$, $X \geq 0$ has been determined, but that now the constant term column vector b must be altered. Changing the original b will affect b^* (and z_0^*) of the final tableau but not c^* (and A^*). The modified $b^* = B^{-1}b$ can be calculated. If the entries remain nonnegative, since $c^* \geq 0$, the optimal solution point to the modified problem will have the same basic variables as the solution point to the original problem, with values given by the adjusted b^*, and evaluating z at this solution point will give the adjusted optimal value for the objective function.

Example 5.5.1. Suppose $b_2 = 50$ in the problem of (5.3.1) on page 183 is changed to 53. Using the data of Table 5.5, we have

$$b^* = B^{-1}b = \begin{bmatrix} 2 & -1 \\ -7 & 2 \end{bmatrix} \begin{bmatrix} 28 \\ 53 \end{bmatrix} = \begin{bmatrix} 3 \\ 16 \end{bmatrix} \geq 0$$

Thus the second tableau of Table 5.5, with b^* now given by $[3, 16]'$ (and the z_0^* value to be corrected), provides the final tableau for the modified problem. The optimal value of the objective function is attained at the point $(0, 16, 0, 3)$. Evaluating z at this point gives

$$\text{Max}\, z = 4 \cdot 16 + 15 \cdot 3 = 109$$

(and so the corrected $z_0^* = 109$).

In fact, we can generalize. Suppose $b_2 = 50$ is to be changed to $50 + \lambda$, λ a constant. Then

$$\begin{aligned} b^* &= B^{-1}b = B^{-1} \begin{bmatrix} 28 \\ 50 + \lambda \end{bmatrix} \\ &= \underbrace{B^{-1} \begin{bmatrix} 28 \\ 50 \end{bmatrix}}_{\text{the former } b^*} + B^{-1} \begin{bmatrix} 0 \\ \lambda \end{bmatrix} = \begin{bmatrix} 6 \\ 4 \end{bmatrix} + \begin{bmatrix} -\lambda \\ 4\lambda \end{bmatrix} = \begin{bmatrix} 6 - \lambda \\ 4 + 4\lambda \end{bmatrix} \end{aligned}$$

As long as $6 - \lambda \geq 0$ and $4 + 4\lambda \geq 0$, that is, as long as $-1 \leq \lambda \leq 6$, the optimal solution point to the more general problem is $(0, 4 + 4\lambda, 0, 6 - \lambda)$, with

$$\text{Max}\, z = 4(4 + 4\lambda) + 15(6 - \lambda) = 106 + \lambda$$

Note that the coefficient of λ here, the $+1$, is the marginal value for the second constraint. Of course, this is to be expected. As long as x_4 and x_2 remain the basic variables in the optimal solution, the second tableau of Table 5.5 functions as the final tableau, the optimal solution point to the dual remains $(2, 1)$, and

$$\begin{aligned} \text{Max}\, z &= \text{Min of dual objective function} \\ &= 2(28) + 1(50 + \lambda) = 106 + \lambda \end{aligned}$$

Finally, suppose $b_2 = 50$ is increased to 57. Then b^* would be $(-1, 32)'$ and the point $(0, 32, 0, -1)$ would be a solution to the modified constraints. But it is not a feasible solution, and it is not at all clear how to proceed at this point. Since the simplex algorithm as we have developed it must move from feasible solution to feasible solution, it would seem that here we would be forced to initiate the algorithm on the original problem with the constant term column vector now $(28, 57)'$ and that we could not make use of the data in the final tableau of Table 5.5. However, there is a variation of the simplex algorithm that can be used with negative entries in the constant-term column and that, in this case, would save considerable effort. In the next section we will develop the algorithm; then, with this new tool at our disposal, we will come back to the above problem.

In Section 5.1 we noted that for a linear programming problem, the components of the solution to the corresponding dual problem provide useful information relating the effects on the optimal value of the original objective function to changes in the constant terms of the constraints. When used in this context, we introduced the labels *shadow price* or *marginal value* or *cost* for the components. However, it was also noted that their use as shadow prices or marginal values/costs was restricted. If the changes in the original problem's constant-term data (and therefore the changes in the coefficients of the dual objective function) brought about a change in the optimal solution point to the dual, shadow prices could change.

We can now be more precise on the range of validity for the original set of shadow prices. As we have seen in this section, if the changes in the constant-term column b do not introduce any negative entries in the adjusted column vector b^*, the previously final tableau, with the modified b^* now in the constant-term column, serves as the final tableau for the modified problem. Thus the final tableaux for the original problem and for the modified problem are identical except for the constant-term columns. They share the same bottom row and hence have the same solution to their respective dual problems. Shadow prices have not changed. On the other hand, if changes in b introduce negative entries in the adjusted column vector b^*, additional action is required to solve the modified problem, leaving us with no information on the solution to its dual.

Example 5.5.2. Solver's Sensitivity Report of Figure 5.4 states that the shadow prices for the can company problem of Example 5.1.3 on page 166 remain unchanged if the demand for Type A cases does not decrease from the original 2400 cases by more than 400 cases or increase by more than 1200 cases, with no other data of the problem changing. To verify this claim, let the demand for a Type A case be $2400+\lambda$ cases. Then $b=[2400+\lambda, 2800, 600]'$, and using the data of Table 5.1, we have

$$b^* = B^{-1}b = B^{-1}\begin{bmatrix} 2400 \\ 2800 \\ 600 \end{bmatrix} + B^{-1}\begin{bmatrix} \lambda \\ 0 \\ 0 \end{bmatrix} = \begin{bmatrix} 75 \\ 150 \\ 350 \end{bmatrix} + \lambda \begin{bmatrix} \frac{3}{16} \\ -\frac{1}{8} \\ \frac{3}{8} \end{bmatrix}.$$

Thus $b^* \geq 0$ if $75+\frac{3\lambda}{16} \geq 0$, $150-\frac{\lambda}{8} \geq 0$, and $350+\frac{3\lambda}{8} \geq 0$, that is, if $-400 \leq \lambda \leq 1200$, as expected.

(Note that for λ in this range, the optimal operating schedule is $x_1=0$, $x_2=b_1^*=75+\frac{3\lambda}{16}$, $x_3=b_2^*=150-\frac{\lambda}{8}$, and the value of the objective function at this point is

$$65(0)+150\left(75+\frac{3\lambda}{16}\right)+200\left(150-\frac{\lambda}{8}\right) = \underbrace{41,250}_{\text{original optimal value}} + \underbrace{3\tfrac{1}{8}}_{\text{shadow price of Type A can}} \lambda,$$

as it should be.)

Problem Set 5.5

1. Consider the problem of Example 3.6.1 on page 95. Determine both the minimal value of z and a point at which this value is attained if:

   (a) b_1 is increased from 3 to 7 and b_2 is decreased from 11 to 8.
   (b) b_1 is increased from 3 to 8 and b_2 is decreased from 11 to 8.

2. Consider the linear programming problem of Example 3.5.1 on page 87. Determine the maximum value of the objective function and a point at which this value is attained if

   (a) b_2 is increased from 10 to 30 units, b_1 and b_3 remaining unchanged.
   (b) b_1, b_2, and b_3 are each decreased by 10 units from their original values.

3. For the problem of (5.3.1) in Section 5.3 on page 183, suppose $b_1 = 28$ is changed to $28 + \lambda$ (with b_2 fixed at 50). Determine the range on λ so that x_4 and x_2 remain as basic variables in an optimal solution. What would the optimal solution point be and the maximum of z be?

4. Consider the problem of (5.2.1) in Section 5.2 on page 176.

   (a) Determine the range on λ_1 such that if λ_1 is added to $b_1 = 6$ (b_2 remaining at 15), x_5 and x_1 still serve as basic variables in an optimal solution. What is this optimal solution, and what is the minimum of z?
   (b) Make a similar analysis assuming λ_2 is added to $b_2 = 15$, with b_1 remaining at 6.
   (c) Make a similar analysis with $b = [6 + \lambda_3, 15 + \lambda_3]'$.

5. Consider the operation of the aluminum can company of Example 5.1.3 on page 166.

   (a) Determine the optimal production schedule if the monthly requirement for Type B cans is reduced to 2500 cases.
   (b) Determine the optimal production schedule if the monthly requirement for each type of can is increased by 400.
   (c) The company, wishing to offset the present surplus production of the Type B can (with requirements at the original 2400 and 2800 cases), negotiates a contract with another cannery. The cannery would need λ cases of the Type A can and 3λ cases of the Type B can monthly, but the λ would vary from month to month. The can company agrees to the contract, but limits the amount λ to no more than the maximum allowed, with the (modified) second tableau of Table 5.1 still the final tableau. Determine this limit and the revised optimal production schedule (as a function of λ).
   (d) Suppose the monthly can requirements are fixed at the original 2400 and 2800 values, but the amount of recycled aluminum that must be used can vary. Determine the range on this requirement so that the (modified) second tableau of Table 5.1 can serve as the final tableau, and thus confirm the corresponding data in Solver's Sensitivity Report in Figure 5.4.

6. For the poultry producer's problem of Example 5.1.1 on page 161, use the results developed in that example relating the solution to the dual and the ratio $\frac{b_1}{b_2}$ of the daily required amounts of elements A and B to verify the bounds in Solver's Sensitivity Report (Figure 5.5 on page 173) concerning the shadow prices and the allowable

   (a) deviation from the original 124 units in element A's daily requirement
   (b) deviation from the original 60 units in element B's daily requirement

7. Consider the situation described in Problem 15 of Section 5.1.

   (a) Suppose $(500 + \lambda)$ hr of labor at \$18/hr can be used for production, $-16 \leq \lambda \leq 16$, with no other change in the data of the problem. Determine the optimal production schedule and profit.
   (b) Suppose the amount of available raw material M_1 is somewhat flexible. With 1 ton available, its marginal value is \$7.50/lb. For what interval around 2000 lb does M_1 retain this marginal value (with all other data as in the original problem)?
   (c) Use Solver's Sensitivity Report for Problem 15 of Section 5.1 (see Figure 5.6 on page 189) to verify your answer to part (b).
   (d) Verify the validity of the data in the sensitivity report on the bounds on changes in the available hours of labor and labor's marginal value of \$12/hr.
   (e) As in part (d), but for the data on changes in the available amount of M_2.

8. The plant manager in Problem 17 of Section 5.1 was impressed with your estimate on the present cost of supplying a unit of A. However, the plant's use of the A's may vary, and the manager wants to know by how much this weekly requirement of 100 units can change before your estimate on the cost of production of a unit of A would need to be revised. How do you respond?

5.6 THE DUAL SIMPLEX ALGORITHM

In the previous section, we saw that changes in the initial values of the constant-term column vector b can bring about a linear programming problem in the following form. The system of constraints is in canonical form with a specified set of basic variables, the objective function is expressed in terms of the nonbasic variables, and the corresponding coefficients c_j are nonnegative, but the associated basic solution is not feasible; that is, the constant-term column contains negative entries. With the simplex method as we have developed it, the only way to handle such a problem would be to multiply the equations with negative constant terms by (-1), add artificial variables if necessary to put the problem into canonical form, and proceed with the two-stage simplex process.

Example 5.6.1.

$$\begin{aligned}&\text{Minimize } z = 10x_1 + 5x_2 + 4x_3 \\ &\text{subject to} \\ &3x_1 + 2x_2 - 3x_3 \geq 3 \\ &4x_1 \quad\quad + 2x_3 \geq 10 \\ &x_1, x_2, x_3 \geq 0\end{aligned} \tag{5.6.1}$$

Adding slack variables x_4 and x_5, we have

$$\begin{aligned}3x_1 + 2x_2 - 3x_3 - x_4 \quad &= 3 \\ 4x_1 \quad + 2x_3 \quad - x_5 &= 10 \\ 10x_1 + 5x_2 + 4x_3 \quad &= z\end{aligned}$$

To apply the simplex process to this problem, we would now add two artificial variables and proceed. On the other hand, by multiplying the two constraints by (-1), we have the following:

$$\begin{aligned}-3x_1 - 2x_2 + 3x_3 + x_4 \quad &= -3 \\ -4x_1 \quad - 2x_3 \quad + x_5 &= -10 \\ 10x_1 + 5x_2 + 4x_3 \quad &= z\end{aligned} \tag{5.6.2}$$

The problem expressed in this way is in the form described above – the system of constraints is in canonical form with basic variables x_4 and x_5; the objective function is expressed in terms of the nonbasic variables x_1, x_2, and x_3; and the associated coefficients 10, 5, and 4 are nonnegative, but the associated basic solution $x_1 = x_2 = x_3 = 0$, $x_4 = -3$, $x_5 = -10$ is not feasible.

In this section, we will develop an algorithm for resolving problems in this form. The algorithm, called the *Dual Simplex Algorithm*, is intimately related to the dual problem of the linear programming problem under consideration. We will develop this relationship after describing the steps of the algorithm by means of the above example.

The basic step of the Dual Simplex Algorithm is the pivot operation that we have already seen. However, this algorithm differs from the standard simplex process by the rules used to determine the pivot term at each step. In this algorithm, at each step, first the row in which to pivot is determined and then the column is determined. Thus here we determine first what variable to extract from the basis and then what variable to enter into the basis.

Example 5.6.2 (Continuation of Example 5.6.1)**.** Consider the tableau presentation in Table 5.8 of the problem as stated in (5.6.2). To apply the Dual Simplex Algorithm, we determine first the row in which to pivot. According to the algorithm, the pivot term can be in any row with a negative constant term. In this tableau, $b_1 = -3$ and $b_2 = -10$; therefore the pivot term can come from either row. An arbitrary rule to use in such a case is to pivot in the row with the smallest b_i term; so, here, we pivot in the second row, extracting x_5 from the basis.

Table 5.8

	x_1	x_2	x_3	x_4	x_5	
x_4	−3	−2	3	1	0	−3
x_5	−4	0	−2	0	1	−10
	10	5	4	0	0	0

Table 5.9

	x_1	x_2	x_3	x_4	x_5	
x_4	−3	−2	3	1	0	−3
x_5	−4	0	(−2)	0	1	−10
	10	5	4	0	0	0
x_4	−9	−2	0	1	$\frac{3}{2}$	−18
x_3	2	0	1	0	$-\frac{1}{2}$	5
	2	5	0	0	2	−20

Next, we determine the column in which to pivot. The algorithm dictates that the pivot term be at a negative a_{ij} entry; so, here, the pivot term will be either at $a_{21} = -4$ or $a_{23} = -2$. To determine at which entry we pivot, the ratios c_j/a_{rj} must be considered for those $a_{rj} < 0$ (where r is the pivoting row), and the pivot term must be in that column, say column s, for which

$$\frac{c_s}{a_{rs}} = \text{Max}\left\{\frac{c_j}{a_{rj}} : a_{rj} < 0\right\}$$

In this case, we compare $c_1/a_{21} = \frac{10}{-4} = -\frac{5}{2}$ with $c_3/a_{23} = \frac{4}{-2} = -2$. The maximum occurs in the third column, and therefore we pivot at $a_{23} = -2$. (Note that here we are comparing two nonpositive ratios and seeking the maximum, and therefore are actually seeking the ratio of minimum absolute value. By the nature of the algorithm, this will always be the case.)

Pivoting here, we have the tableaux of Table 5.9. Notice that the c_j^* entries, here 2, 5, 0, 0, and 2, have remained nonnegative. Our choice of pivoting column guarantees this. In the second tableau, $b_1^* = -18$ is the only negative constant term, so we must pivot in the first row. Comparing those ratios corresponding to negative a_{rj}^* terms, we have $c_1^*/a_{11}^* = -\frac{2}{9} > c_2^*/a_{12}^* = -\frac{5}{2}$, and so we pivot at the $a_{11}^* = -9$ term. The resulting tableau is in Table 5.10.

Again, after this step, the constant-term column entries are nonnegative. In fact, with the original problem presented in this form, we have reached the solution of the problem, as Theorem 3.4.1 applies. The minimum value of the objective function is 24 and is attained at the point $(2,0,1,0,0)$.

Table 5.10

	x_1	x_2	x_3	x_4	x_5	
x_4	(-9)	-2	0	1	$\frac{3}{2}$	-18
x_3	2	0	1	0	$-\frac{1}{2}$	5
	2	5	0	0	2	-20
x_1	1	$\frac{2}{9}$	0	$-\frac{1}{9}$	$-\frac{1}{6}$	2
x_3	0	$-\frac{4}{9}$	1	$\frac{2}{9}$	$-\frac{1}{6}$	1
	0	$\frac{41}{9}$	0	$\frac{2}{9}$	$\frac{7}{3}$	-24

We summarize the steps of the Dual Simplex Algorithm. Consider the linear programming problem of minimizing $z = c \cdot X - z_0$ subject to $AX = b$, $X \geq 0$ (with b not necessarily ≥ 0).

0. Assume that
   (a) The system of constraints is in canonical form with a specified set of basic variables.
   (b) The objective function z is expressed in terms of the nonbasic variables only, and the corresponding coefficients c_j are all nonnegative.

1. If all $b_i \geq 0$, the minimum value of the objective function has been attained (Theorem 3.4.1 applies).

2. If there exists an r such that $b_r < 0$ and $a_{rj} \geq 0$ for all j, the system of constraints has no feasible solutions.

3. Otherwise, pivot. To determine the pivot term:
   (a) Pivot in any row with a negative b_i term. If there are several negative b_i terms, pivoting in the row with the smallest b_i may reduce the total number of steps necessary to complete the problem. Assume we pivot in row r.
   (b) To determine the column of the pivot term, find that column, say column s, such that

$$\frac{c_s}{a_{rs}} = \text{Max}\left\{\frac{c_j}{a_{rj}} : a_{rj} < 0\right\}$$

4. After pivoting, the problem will remain in the form described in Step 0. Now return to Step 1.

There are some obvious questions associated with the algorithm. We list some of them here. (See also Problems 1 and 2.)

1. If the problem initially is not in the form described in Step 0, is there a systematic way of driving the problem into this form?
2. If the pivoting term is chosen as in Step 3, why will the problem remain in the form described in Step 0?
3. Will this algorithm always terminate, that is, will we always reach a point where either Step 1 or Step 2 applies?

The answer to the first question is "yes." However, for our purposes, we will need and use the algorithm only for problems already in the form described in Step 0, and we will not develop here any of the techniques for putting a problem into the desired form. Refer to Dantzig [7] and Lemke [13].

The answers to the other two questions follow from the relationship between the Dual Simplex Algorithm and the standard simplex algorithm applied to the dual of the original problem. To see this, consider the linear programming problem in the form described above in Step 0. To determine its dual, we do the following:

1. Consider the basic variables of the system of constraints as slack variables with coefficients $+1$. Drop them from the problem, replacing the equations of the constraints with $(\leq)$ inequalities.
2. Multiply each constraint by (-1), creating a minimization problem with $(\geq)$ inequalities for constraints.

The dual of this equivalent problem is readily determined. The dual would be a maximization problem with $(\leq)$ inequalities as constraints, with the coefficients of the objective function corresponding to the negative of the b_i terms of the original problem and the constant terms of the dual corresponding to the c_j terms of the original problem.

To apply the simplex process to this dual problem, it must first be put into canonical form. The slack variables added to the $(\leq)$ inequalities in the system of constraints can serve as the basic variables, and the associated basic solution is feasible, as the coefficients c_j of the original objective function were assumed to be nonnegative. Finally, multiplication of the coefficient of the dual objective function by (-1) produces the required minimization problem.

The simplex algorithm can now be applied. Notice that the coefficients of the nonbasic variables of the objective function are precisely the b_i terms of the original problem, so that determining the pivoting column corresponds directly to Step 3(a) of the Dual Simplex Algorithm. And the a_{ij} entries in this pivoting column correspond to the negative of the a_{ij} entries in the pivoting row determined by Step 3(a) of the Dual Simplex Algorithm. Thus determination of the pivoting row here corresponds directly to Step 3(b) of the Dual Simplex Algorithm.

Instead of attempting to write out in precise terms the relationship between the algorithms for a general linear programming problem and its dual, we will demonstrate the relationships by means of the example of this section.

Example 5.6.3 (Continuation of Example 5.6.1). The problem, as stated in (5.6.2), is to minimize z with

$$\begin{aligned} -3x_1 - 2x_2 + 3x_3 + x_4 \qquad &= -3 \\ -4x_1 \qquad - 2x_3 \qquad + x_5 &= -10 \\ 10x_1 + 5x_2 + 4x_3 \qquad &= z \end{aligned}$$

Dropping the basic variables, the constraints become

$$\begin{aligned} -3x_1 - 2x_2 + 3x_3 &\leq -3 \\ -4x_1 \qquad - 2x_3 &\leq -10 \end{aligned}$$

Table 5.11

	y_1	y_2	y_3	y_4	y_5	
y_3	3	4	1	0	0	10
y_4	2	0	0	1	0	5
y_5	-3	(2)	0	0	1	4
	-3	-10	0	0	0	0
y_3	(9)	0	1	0	-2	2
y_4	2	0	0	1	0	5
y_2	$-\frac{3}{2}$	1	0	0	$\frac{1}{2}$	2
	-18	0	0	0	5	20
y_1	1	0	$\frac{1}{9}$	0	$-\frac{2}{9}$	$\frac{2}{9}$
y_4	0	0	$-\frac{2}{9}$	1	$\frac{4}{9}$	$\frac{41}{9}$
y_2	0	1	$\frac{1}{6}$	0	$\frac{1}{6}$	$\frac{7}{3}$
	0	0	2	0	1	24

Multiplying each by (-1), we have the equivalent problem of minimizing z with

$$\begin{aligned} 3x_1 + 2x_2 - 3x_3 &\geq 3 \\ 4x_1 \qquad\quad + 2x_3 &\geq 10 \\ 10x_1 + 5x_2 + 4x_3 &= z \end{aligned}$$

(Note that this is simply the problem in (5.6.1).) The dual problem is to maximize v with

$$\begin{aligned} 3y_1 + 4y_2 &\leq 10 \\ 2y_1 \qquad &\leq 5 \\ -3y_1 + 2y_2 &\leq 4 \\ 3y_1 + 10y_2 &= v \\ y_1, y_2 &\geq 0 \end{aligned}$$

The equivalent problem in canonical form is to minimize $-v$ with

$$\begin{aligned} 3y_1 + 4y_2 + y_3 \qquad\qquad &= 10 \\ 2y_1 \qquad\qquad + y_4 \qquad &= 5 \\ -3y_1 + 2y_2 \qquad\qquad + y_5 &= 4 \\ -3y_1 - 10y_2 \qquad\qquad\qquad &= (-v) \\ y_1, y_2, y_3, y_4, y_5 \geq 0 \end{aligned}$$

Now compare the steps of the simplex algorithm applied to this problem (Table 5.11) with the steps of the Dual Simplex Algorithm applied to the original problem (Tables 5.9 and 5.10).

We conclude this section with some applications of the algorithm.

Table 5.12

	x_1	x_2	x_3	x_4	x_5	x_6	
x_4	-2	0	5	1	2	$\textcircled{-1}$	-1
x_2	11	1	-18	0	-7	4	32
	3	0	2	0	2	1	113
x_6	2	0	-5	-1	-2	1	1
x_2	3	1	2	4	1	0	28
	1	0	7	1	4	0	112

Example 5.6.4. In Example 5.5.1, the final change proposed for the $b_2 = 50$ in the problem of (5.3.1) of Section 5.3 was an increase to 57. This change generated a negative b_1^* in the modified final tableau for the original problem (the second tableau of Table 5.5 on page 184). Now we can easily handle the proposed change. Modifying the previously final tableau (now $b^* = (-1, 32)^t$ and $z_0^* = 4(32) + 15(-1) = 113$) and applying the Dual Simplex Algorithm, we see, from Table 5.12, that the maximum of z is 112, attained at $(0, 28, 0, 0)$.

Example 5.6.5. New government regulations require that the aluminum can company of Example 5.1.3 on page 166 increase its monthly use of recycled aluminum by 28%, from 600 lb to 768 lb, and the company manager wants to know how this increase affects the company's operation and the marginal costs of producing the cans and using the recycled aluminum.

To respond, we first calculate the modified b^* for the second tableau of Table 5.1. (Note that B^{-1} is contained in the artificial variable column data of the tableau.) We have

$$\begin{aligned} b^* = B^{-1}b = B^{-1}\begin{bmatrix} 2400 \\ 2800 \\ 600 \end{bmatrix} + B^{-1}\begin{bmatrix} 0 \\ 0 \\ 168 \end{bmatrix} \\ = \begin{bmatrix} 75 \\ 150 \\ 350 \end{bmatrix} + \begin{bmatrix} \frac{3}{16} & 0 & -\frac{5}{8} \\ -\frac{1}{8} & 0 & \frac{3}{4} \\ \frac{3}{8} & -1 & \frac{15}{4} \end{bmatrix}\begin{bmatrix} 0 \\ 0 \\ 168 \end{bmatrix} \\ = \begin{bmatrix} 75 \\ 150 \\ 350 \end{bmatrix} + \begin{bmatrix} -105 \\ 126 \\ 630 \end{bmatrix} = \begin{bmatrix} -30 \\ 276 \\ 980 \end{bmatrix} \end{aligned}$$

Now b_1^* is negative. Thus we modify the constant-term column of the second tableau of Table 5.1 ($z_0^* = -(150(-30) + 200(176)) = -50,700$) and apply the Dual Simplex Algorithm (Table 5.13). Monthly demands can now be met by using Process 3 for 256 hr. An extra 160 cases of Type A cans and 1040 cases of Type B cans are produced. The marginal cost for meeting each of the requirements is zero; and the marginal cost of using the recycled aluminum is increased to \$66.67/lb.

Table 5.13

	x_1	x_2	x_3	x_4	x_5	x_6	x_7	x_8	
x_2	$\frac{9}{8}$	1	0	$\left(-\frac{3}{16}\right)$	0	$\frac{3}{16}$	0	$-\frac{5}{8}$	−30
x_3	$-\frac{3}{4}$	0	1	$\frac{1}{8}$	0	$-\frac{1}{8}$	0	$\frac{3}{4}$	276
x_5	$-\frac{23}{4}$	0	0	$-\frac{3}{8}$	1	$\frac{3}{8}$	−1	$\frac{15}{4}$	980
	$46\frac{1}{4}$	0	0	$3\frac{1}{8}$	0	$-3\frac{1}{8}$	0	$-56\frac{1}{4}$	−50,700
x_4	−6	$-\frac{16}{3}$	0	1	0	−1	0	$\frac{10}{3}$	160
x_3	0	$\frac{2}{3}$	1	0	0	0	0	$\frac{1}{3}$	256
x_5	−8	−2	0	0	1	0	−1	5	1,040
	65	$16\frac{2}{3}$	0	0	0	0	0	$-66\frac{2}{3}$	−51,200

Problem Set 5.6

1. Explain the conclusion of Step 2 in the Dual Simplex Algorithm.

2. If a linear programming problem is presented in the form described in Step 0 of the Dual Simplex Algorithm, there cannot exist a set of feasible solutions to the system of constraints on which the objective function is unbounded below. Prove this.

3. Using the second tableau of Table 5.5, find the optimal value of the objective function and a point at which the value is attained if $b = [28, 50]^t$ in the problem of (5.3.1) of Section 5.3 is changed to

   (a) $[28, 49]^t$.
   (b) $[29, 50]^t$.
   (c) $[28, -23]^t$.

4. Solve the problem of Example 3.5.3 on page 88 if $b = [20, 10, 60]^t$ is changed to

   (a) $[18, 13, 60]^t$.
   (b) $[18, 14, 60]^t$.
   (c) $[20 - \lambda, 10 + 2\lambda, 60]^t$ for $-15/2 \leq \lambda \leq 20$.

5. Solve the problem of Example 3.6.1 on page 95 if $b = [3, 11]^t$ is changed to $[9, 8]^t$.

6. In Example 4.3.1 on page 132, the shadow prices for the three nutritional requirements in the diet problem of Example 2.2.1 on page 10 were determined to be 1 cent/unit for nutritional elements A and B, and 0 cents/unit for C. Suppose that the cost per pound of each feed has been increased by 10 cents, so that Feed 1 now costs 20 cents/lb and Feed 2 costs 14 cents/lb. Determine the new shadow prices for the constraints.

7. The aluminum can company of Example 5.1.3 on page 166 has just signed a contract calling for the delivery of an additional 1800 cases of the Type A can per month (with all other data as stated in the original example). Determine

the revised optimal operating schedule and monthly costs, and the new marginal costs for the constraints.

8. The company in Problem 15 of Section 5.1 has access to an additional 600 lb of raw material M_1 How should the optimal monthly production schedule and profit be revised, and how much would additional hours of overtime be worth to the company now?

9. Accepting your estimate that new workers for your section would save the plant weekly operating costs of $130 for each additional hour available, the plant manager in Problem 17 of Section 5.1 provides you with two new workers (a total of 80 additional hr/week) at the standard pay rate of $20/hr. You revise the operating schedule of your section accordingly (which becomes?), but note that your weekly costs (which now are?) are reduced by less than $130(80) = $10,400. How do you explain this to the plant manager?

5.7 ADDITION OF A CONSTRAINT

Suppose that after the optimal value for a linear programming problem has been found by means of the simplex method, we wish to alter the original problem by adding a new constraint. It could be that the optimal solution found previously satisfies the new constraint. If this is the case, the solution is also optimal for the expanded problem, because clearly, by the addition of a constraint, we have not changed the objective function or increased the set of feasible solutions to the system of constraints. On the other hand, if the previous optimal solution does not satisfy the new constraint, we must find a new optimal solution. Under certain circumstances, however, this problem may be resolved quite easily by the application of the Dual Simplex Algorithm to data determined from the final tableau solution to the original problem.

Example 5.7.1. Consider the problem of Example 5.3.1 on page 183. Using the second tableau of Table 5.5 on page 184, the problem, with slack variables x_5 and x_6 added, is to minimize $(-z)$ (originally to maximize z) with

$$\begin{aligned}
-2x_1 \quad\quad + 5x_3 + x_4 + 2x_5 - x_6 &= 6 \\
11x_1 + x_2 - 18x_3 \quad\quad - 7x_5 + 4x_6 &= 4 \\
3x_1 \quad\quad + 2x_3 \quad\quad + 2x_5 + x_6 &= 106 + (-z) \\
x_1, x_2, x_3, x_4, x_5, x_6 \geq 0 &
\end{aligned} \tag{5.7.1}$$

The maximum value of z is 106, attained at the point $(0,4,0,6,0,0)$.

Suppose now that we also demand that the solution satisfy the $(\leq)$ inequality

$$3x_1 + x_2 + 3x_4 \leq 20 \tag{5.7.2}$$

The point $(0,4,0,6,0,0)$ does not satisfy this constraint. Thus we add the constraint to the problem and attempt to put this expanded problem into a form to which we can apply a solution algorithm. Adding the slack variable x_7 to the new constraint gives the equation

$$3x_1 + x_2 + 3x_4 + x_7 = 20$$

Next, we seek a set of basic variables for this system of three equations and seven unknowns. An obvious choice, as long as the added constraint introduces a slack variable, is this slack variable along with the basic variables from the final tableau of the original problem. Here then we can use as basic variables x_4, x_2, and x_7. The only operation that must be considered is the removal of the basic variables of the previous problem from the new constraint; clearly, this can always be done by simply adding appropriate multiples of the original constraints to the new constraint. Here we subtract the second equation and three times the first equation from the new constraint,

$$\begin{array}{rrrrrrrrl}
3x_1 & + x_2 & & + 3x_4 & & & + x_7 & = & 20 \\
-(11x_1 & + x_2 & - 18x_3 & & - 7x_5 & + 4x_6 & & = & 4) \\
-3(-2x_1 & & + 5x_3 & + x_4 & + 2x_5 & - x_6 & & = & 6) \\
\hline
-2x_1 & & + 3x_3 & & + x_5 & - x_6 & + x_7 & = & -2
\end{array} \tag{5.7.3}$$

yielding an equivalent system of equations in canonical form. Moreover, the expression for the objective function from the last tableau still contains no basic variables, since the added basic variable x_7 is a slack variable with a zero cost coefficient; and the associated coefficients of the objective function are nonnegative, corresponding to a terminating tableau in the simplex algorithm. The modified problem, with the equation of (5.7.3) added to the constraints of (5.7.1), is in a form to which we can apply the Dual Simplex Algorithm (Table 5.14). The maximum of z is now $100\frac{2}{3}$ attained at $(0,0,\frac{2}{3},6\frac{2}{3},0,4,0)$.

Example 5.7.2 (Continuation of Example 5.7.1)**.** Suppose that instead of the inequality (5.7.2) we add to the problem of (5.7.1) the $(\geq)$ inequality

$$4x_1 + x_2 + 4x_4 \geq 29 \tag{5.7.4}$$

Again, the point $(0,4,0,6,0,0)$ does not satisfy the additional constraint. Introducing the slack variable x_7 to (5.7.4) gives

$$4x_1 + x_2 + 4x_4 - x_7 = 29 \tag{5.7.5}$$

To use x_7 along with x_2 and x_4 as basic variables, we multiply (5.7.5) by (-1), then add to the result the second constraint of (5.7.1) and four times the first constraint. This yields the equation

$$-x_1 + 2x_3 + x_5 + x_7 = -1 \tag{5.7.6}$$

The Dual Simplex Algorithm applied to the problem of (5.7.1), with the additional constraint (5.7.6), is given in Table 5.15. The problem now has no feasible solutions.

Table 5.14

	x_1	x_2	x_3	x_4	x_5	x_6	x_7	
x_4	-2	0	5	1	2	-1	0	6
x_2	11	1	-18	0	-7	4	0	4
x_7	-2	0	3	0	1	(-1)	1	-2
	3	0	2	0	2	1	0	106
x_4	0	0	2	1	1	0	-1	8
x_2	3	1	(-6)	0	-3	0	4	-4
x_6	2	0	-3	0	-1	1	-1	2
	1	0	5	0	3	0	1	104
x_4	1	$\frac{1}{3}$	0	1	0	0	$\frac{1}{3}$	$6\frac{2}{3}$
x_3	$-\frac{1}{2}$	$-\frac{1}{6}$	1	0	$\frac{1}{2}$	0	$-\frac{2}{3}$	$\frac{2}{3}$
x_6	$\frac{1}{2}$	$-\frac{1}{2}$	0	0	$\frac{1}{2}$	1	-3	4
	$\frac{7}{2}$	$\frac{5}{6}$	0	0	$\frac{1}{2}$	0	$\frac{13}{3}$	$100\frac{2}{3}$

Table 5.15

	x_1	x_2	x_3	x_4	x_5	x_6	x_7	
x_4	-2	0	5	1	2	-1	0	6
x_2	11	1	-18	0	-7	4	0	4
x_7	(-1)	0	2	0	1	0	1	-1
	3	0	2	0	2	1	0	106
x_4	0	0	1	1	0	-1	-2	8
x_2	0	1	4	0	4	4	11	-7
x_1	1	0	-2	0	-1	0	-1	1
	0	0	8	0	5	1	3	103

Example 5.7.3 (Continuation of the previous two examples). Suppose we add to the problem of (5.7.1) the equality constraint

$$x_1 + 3x_3 + x_4 - 4x_5 - x_6 = -18$$

a constraint also not satisfied by the point $(0,4,0,6,0,0)$. When we add to the final tableau of the original problem the above equality, the problem is completed in three pivot steps (Table 5.16). The first pivot step renews the status of x_4 as a basic variable, removing the variable from the new constraint. The purpose of the second pivot step is to introduce a basic variable into the new constraint row, necessary here. as no slack variable came with the equality constraint. To determine the pivoting term for this row, we use the pivoting rules for the Dual Simplex Algorithm. Thus we pivot at the -6 term ($\frac{2}{-6} > \frac{2}{-2}$ or, using absolute values, $\frac{1}{3} < 1$).

Table 5.16

	x_1	x_2	x_3	x_4	x_5	x_6	
x	-2	0	5	(1)	2	-1	6
x_2	11	1	-18	0	-7	4	4
x	1	0	3	1	-4	-1	-18
	3	0	2	0	2	1	106
x_4	-2	0	5	1	2	-1	6
x_2	11	1	-18	0	-7	4	4
x	3	0	-2	0	(-6)	0	-24
	3	0	2	0	2	1	106
x_4	-1	0	$\frac{13}{3}$	1	0	(-1)	-2
x_2	$\frac{15}{2}$	1	$-\frac{47}{3}$	0	0	4	32
x_5	$-\frac{1}{2}$	0	$\frac{1}{3}$	0	1	0	4
	4	0	$\frac{4}{3}$	0	0	1	98
x_6	1	0	$-\frac{13}{3}$	-1	0	1	2
x_2	$\frac{7}{2}$	1	$\frac{5}{3}$	4	0	0	24
x_5	$-\frac{1}{2}$	0	$\frac{1}{3}$	0	1	0	4
	3	0	$\frac{17}{3}$	1	0	0	96

The Dual Simplex Algorithm can now be properly applied; the bottom row coefficients have remained nonnegative, and the system of constraints is in canonical form. One more iteration of the algorithm completes the problem. The maximum of z is 96, attained at $(0,24,0,0,4,2)$.

Example 5.7.4. The directors of the aluminum can company of Example 5.1.3 on page 166 find the optimal operating schedule suggested in that example unacceptable; they want each of their processes to be used at least 16 hr/month.

To satisfy this requirement, we append to the previously completed problem, for the time being, the single constraint $x_1 \geq 16$. (In the original solution, the other two processes were each used for far more than 16 hr/month.) Introducing a slack variable x_9 and multiplying by (-1), we have $-x_1 + x_9 = -16$. Adding this equation to the second tableau of Table 5.1 and applying the Dual Simplex Algorithm generates the tableaux of Table 5.17. The corresponding optimal solution point calls for using Process 1 for 16 hr, Process 2 for 57 hr, and Process 3 for 162 hr. This production schedule uses all three processes for at least 16 hr and thus is an optimal operating schedule.

Table 5.17

	x_1	x_2	x_3	x_4	x_5	x_9	x_6	x_7	x_8	
x_2	$\frac{9}{8}$	1	0	$-\frac{3}{16}$	0	0	$\frac{3}{16}$	0	$-\frac{5}{8}$	75
x_3	$-\frac{3}{4}$	0	1	$\frac{1}{8}$	0	0	$-\frac{1}{8}$	0	$\frac{3}{4}$	150
x_5	$-\frac{23}{4}$	0	0	$-\frac{3}{8}$	1	0	$\frac{3}{8}$	-1	$\frac{15}{4}$	350
x_9	(-1)	0	0	0	0	1	0	0	0	-16
	$46\frac{1}{4}$	0	0	$3\frac{1}{8}$	0	0	$-3\frac{1}{8}$	0	$-56\frac{1}{4}$	$-41,250$
x_2	0	1	0	$-\frac{3}{16}$	0	$\frac{9}{8}$	$\frac{3}{16}$	0	$-\frac{5}{8}$	57
x_3	0	0	1	$\frac{1}{8}$	0	$-\frac{3}{4}$	$-\frac{1}{8}$	0	$\frac{3}{4}$	162
x_5	0	0	0	$-\frac{3}{8}$	1	$-\frac{23}{4}$	$\frac{3}{8}$	-1	$\frac{15}{4}$	442
x_1	1	0	0	0	0	-1	0	0	0	16
	0	0	0	$3\frac{1}{8}$	0	$46\frac{1}{4}$	$-3\frac{1}{8}$	0	$-56\frac{1}{4}$	$-41,990$

Problem Set 5.7

1. In Example 5.7.1, a ($\leq$) inequality was added to the set of constraints of a linear programming problem. The solution point to the original problem did not satisfy this new constraint, so the inequality was modified by the addition of a slack variable (x_7) and the elimination of the basic variables (x_2 and x_4). This resulted in a negative constant term (here -2), and so the Dual Simplex Algorithm could be used on the expanded problem. What would we have done if this resulting constant term were nonnegative?

2. Solve the problem of Example 3.6.1 on page 95 after adding the constraint

   (a) $x_1 + x_2 + 2x_3 + 2x_4 \leq 37$
   (b) $x_1 + x_2 + 2x_3 + 2x_4 \leq 17$
   (c) $x_1 - x_2 \leq 7$
   (d) $x_1 + 2x_2 + 20x_3 - 2x_4 \geq 38$

3. Starting with the data from the second tableau of Table 5.4, solve the problem of (5.2.1) of Section 5.2 with the following constraints added:

   (a) $x_4 + x_5 = 5$
   (b) $x_4 + x_5 = 7$

   In each, follow the steps used in Example 5.7.3. (Note that in the second pivot in that example, the fact that the constant term in the pivoting row, the -24, was negative was not relevant at that point.)

4. Solve the problem of Example 3.5.1 on page 87 after expanding the problem by adding the constraint

   (a) $2x_1 - x_2 - 5x_3 \geq 12$
   (b) $x_3 \geq 6$
   (c) $2x_1 + x_2 \geq 39$
   (d) $x_1 + x_2 + x_3 = 20$

(e) $x_1 + x_2 + x_3 = 27$

5. Consider the situation described in Example 5.7.4

   (a) What is the marginal cost (in \$/hr required) of using each process at least 16 hr/month?

   (b) By how much can this monthly minimum use requirement increase before the marginal cost in part (a) changes?

   (c) Determine the optimal production schedule if each process must be used at least 40 hr/month. What is the marginal cost of the monthly minimum use requirement now?

6. (a) To maintain its visibility in the marketplace, the company in Problem 15 of Section 5.1 considers producing and selling each month at least 25 units of each of its three products. What would the corresponding optimal production schedule be, and what would this visibility factor cost?

   (b) The company decides against setting minimal sales requirements for its products. But now the Sales Department claims it can sell at most 200 A's in any month. Determine the revised optimal production schedule.

7. The plant manager in Problem 17 of Section 5.1 wants the dependence on the external source of the B's to be reduced to no more than 70/week. (From Table 5.3, the present operating schedule calls for the weekly purchase of $90\frac{5}{6}$ units.)

   (a) After doing some calculations based on the second tableau of Table 5.3, what do you tell the plant manager?

   (b) A compromise is offered – no more than 80/week. Do you accept the restriction, and if so, how do you implement it?

CHAPTER 6

INTEGER PROGRAMMING

6.1 INTRODUCTION TO INTEGER PROGRAMMING

The term *integer programming* (or *mixed integer programming*) refers to the study of linear programming problems for which the domains of all (or some) of the variables of the problem are restricted to be integral. Models leading to such problems readily occur. For example, suppose we are considering a transportation problem, where the units to be shipped are automobiles, refrigerators, or soldiers, or a production problem, where the units to be produced are homes, swimming pools, or submarines. Certainly the optimal solution to such a problem cannot contain fractional values. But the use of integer programming is not restricted to obvious situations such as these. Many other optimization problems involving, for example, fixed charges, favorable objective function shifts, alternative constraints, "on" or "off" variables, and so on, can be modeled using integer programming techniques. We will see examples of such problems in Section 6.2.

In Sections 6.3 and 6.4, we will develop two different algorithms that can be used to calculate optimal solutions to integrally restricted problems. The first algorithm uses a cutting plane method of solution; the second uses a branch and bound approach. Both algorithms make use of the Dual Simplex Algorithm. In Section 6.5 we demonstrate the use of Solver in resolving integer programming problems.

In this section, we illustrate some of the difficulties involved in determining the solution to an integrally restricted linear programming problem. The most obvious possible solution technique for such a problem would be initially to ignore the integral restrictions, solve the corresponding linear programming problem using whatever algorithm or means is suitable, and round off this calculated solution to integral values. However, this method fails, as the following example demonstrates.

Example 6.1.1. Consider the problem of

$$\begin{aligned}
&\text{Maximizing } f(x_1,x_2) = 3x_1 + 13x_2\\
&\text{subject to}\\
&\ \ 2x_1 + 9x_2 \le 40\\
&11x_1 - 8x_2 \le 82\\
&x_1,x_2 \ge 0 \text{ and integral}
\end{aligned}$$

Table 6.1

	x_1	x_2	x_3	x_4	
x_3	2	(9)	1	0	40
x_4	11	-8	0	1	82
	-3	-13	0	0	0
x_2	$\frac{2}{9}$	1	$\frac{1}{9}$	0	$4\frac{4}{9}$
x_4	($\frac{115}{9}$)	0	$\frac{8}{9}$	1	$117\frac{5}{9}$
	$-\frac{1}{9}$	0	$\frac{13}{9}$	0	$57\frac{7}{9}$
x_2	0	1	$\frac{11}{115}$	$-\frac{2}{115}$	$2\frac{2}{5}$
x_1	1	0	$\frac{8}{115}$	$\frac{9}{115}$	$9\frac{1}{5}$
	0	0	$\frac{167}{115}$	$\frac{1}{115}$	$58\frac{4}{5}$

If we ignore the integral restrictions, we can apply the simplex algorithm (after adding slack variables x_3 and x_4). From the tableaux of Table 6.1, the maximum value of f for the simple linear programming problem is $58\frac{4}{5}$ and is attained at the point $(9\frac{1}{5}, 2\frac{2}{5})$. As for the original problem with the integrally restricted variables, it would seem reasonable that we should simply round off the above solution point $(9\frac{1}{5}, 2\frac{2}{5})$ to $(9,2)$ or maybe $(10,2)$, $(10,3)$, or $(9,3)$. However, none of these four points is feasible; the first three do not satisfy the second inequality, and the last two do not satisfy the first. Thus the simplex algorithm has provided us with no useful information, and it is not at all clear how one could proceed, at least in general.

Actually, for this simple example with only two variables, we can graph the solution set to the system of constraints and work from there. This graph is sketched in Figure 6.1. There are 36 lattice points (points with both coordinates integral) in the region bounded by the constraints. Since the coefficients of the objective func-

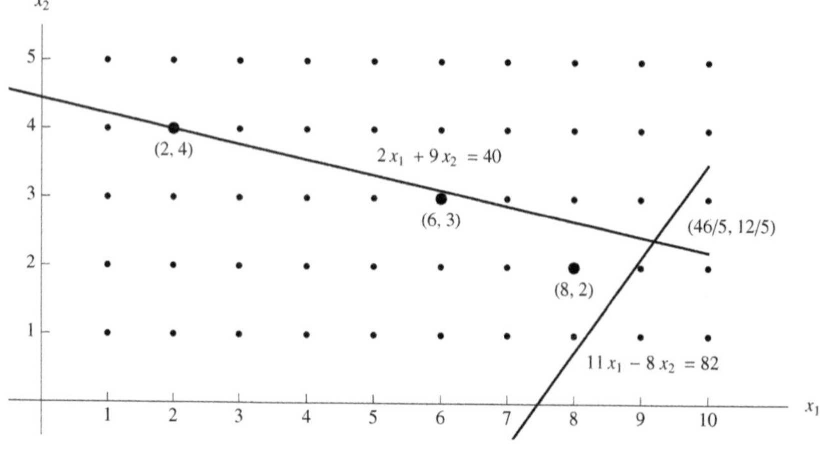

Figure 6.1

tion are positive, there can be no lattice points in the feasible region that are to the right or above the point at which the maximal value of the objective function is attained, and so the optimal value of f must occur at either $(2,4)$, $(6,3)$, or $(8,2)$. Now $f(2,4) = 58$, $f(6,3) = 57$, and $f(8,2) = 50$, so the maximal value of f is 58 and is attained at the point $(2,4)$.

Note the distance between the integer programming solution point $(2,4)$ and the linear programming solution point $(9\frac{1}{5}, 2\frac{2}{5})$ and contrast this distance to the proximity of the feasible lattice point $(8,2)$ to $(9\frac{1}{5}, 2\frac{2}{5})$. Compare also the value of f of 50 at this closest lattice point to its value of 58, a 16% increase, at $(2,4)$. Clearly, integer programming problems may require special techniques.

Problem Set 6.1

1. Consider the problem of

$$\begin{aligned} &\text{Maximizing } 5x_1 + 2x_2 \\ &\text{subject to} \\ &\quad 6x_1 + 2x_2 \leq 13 \\ &-6x_1 + 7x_2 \leq 14 \\ &x_1, x_2 \geq 0 \text{ and integral} \end{aligned}$$

   (a) Using the simplex algorithm, show that the optimal value of the objective function of the problem with the integral restrictions ignored is $11\frac{5}{6}$ and is attained at the point $(1\frac{1}{6}, 3)$.

   (b) Show graphically that the optimal solution to the original restricted variable problem is at $(2,0)$.

   Conclusion. Given an integer programming problem, the fact that some (but not all of the restricted) coordinates of the solution point of the corresponding linear programming problem are integral is of little value in solving the integer programming problem.

2. Consider the problem of

$$\begin{aligned} &\text{Maximizing } 2x_1 + 9x_2 \\ &\text{subject to} \\ &2x_1 + x_2 \leq 20 \\ &\ x_1 + 5x_2 \leq 24 \\ &x_1, x_2 \geq 0 \text{ and integral} \end{aligned}$$

   (a) Using the simplex algorithm, show that the optimal value of the problem without the integral constraints is attained at $(8\frac{4}{9}, 3\frac{1}{9})$.

   (b) Show that the lattice point $(8,3)$ is a feasible solution to the system of constraints.

(c) Show graphically (or algebraically) that the optimal solution to the original integrally restricted variable problem is at $(4,4)$.

Conclusion. Given an integer programming problem, the fact that the coordinates of the solution point of the corresponding linear programming problem round off to a feasible solution to the system of constraints is of little value in solving the original problem.

3. Consider the problem of

$$\begin{aligned}&\text{Maximizing } x_1+x_2\\&\text{subject to}\\&3x_2 \geq 1\\&3x_2 \leq 2\\&x_1,x_2 \geq 0 \text{ and integral}\end{aligned}$$

(a) Solve the problem, ignoring the integral restrictions.
(b) Solve the problem with the integral restrictions.
(c) Can you demonstrate any greater a difference between the solutions to an integrally restricted problem and the corresponding nonintegrally restricted problem?

4. Given an integer programming problem, suppose the optimal value of the corresponding linear programming problem is attained at a point with all coordinates integral. Is this point the solution point to the original problem?

6.2 Models with Integer Programming Formulations

In this section, we will discuss various problem situations that can be formulated as integer (or mixed integer) programming models. Three general areas will be discussed: the allocation of discrete commodities, modifications of the objective function, and problems with alternative constraints. Many other application areas exist. These applications show that restricting the domains of some of the variables of a problem to a discrete set can be an effective tool in the formulation of mathematical models. However, solution algorithms for integer programming problems may require many computationally heavy iterations. The size and type of a given problem are critical factors in determining if the computer system and the software at hand have the capacity to solve the problem. Integer programming is an active area in theoretical and computational research.

The Allocation of Discrete Commodities

Suppose we have a fixed amount of a resource, such as capital, space, or machine time, that we wish to utilize in a way that maximizes profit or gain. Suppose also that the alternative methods through which our resource can be utilized are such that only multiples of fixed-size lots can be allocated to each. By introducing integrally restricted variables to correspond to the allotment made to each of the alternatives, such problems can be formulated as integer programming problems.

Example 6.2.1. An investment firm, wishing to maximize profit, has \$100,000 to invest in a construction project requiring an investment of \$48,000 and providing a profit of \$2900; in any number of units of a portfolio of stocks requiring an investment of \$19,000/unit and yielding a profit of \$1,100/unit; and in any number of shares of a certain stock costing \$1750/share and yielding a profit of \$95/share.

To formulate a mathematical model, introduce three variables x_1, x_2, and x_3, defined as follows: x_1 will be 1 if the construction project is utilized and 0 if not, x_2 will be the number of units of the portfolio utilized, and x_3 will be the number of units of the stock utilized. The integer programming model is

$$\begin{aligned}
&\text{Maximize } 2900x_1 + 1100x_2 + 95x_3\\
&\text{subject to}\\
&48,000x_1 + 19,000x_2 + 1750x_3 \leq 100,000\\
&0 \leq x_1 \leq 1, x_2, x_3 \geq 0 \text{ and } x_1, x_2, x_3 \text{ integral}
\end{aligned}$$

Example 6.2.2. An airplane can carry up to W lb of extra cargo on a scheduled flight. There are n different items that could be transported, with Item i weighing a_i lb and providing a profit of c_i dollars if transported. What items should be shipped so as to maximize profit?

Define n variables x_i,

$$x_i = \begin{cases} 1 \text{ if the } i\text{th item is to be shipped} \\ 0 \text{ if not} \end{cases}$$

The corresponding integer programming problem is

$$\begin{aligned}
&\text{Maximize } \sum_{i=1}^{n} c_i x_i\\
&\text{subject to}\\
&\sum_{i=1}^{n} a_i x_i \leq W\\
&0 \leq x_i \leq 1 \text{ and integral, } i = 1, \ldots, n
\end{aligned}$$

Suppose that two restrictions are now added to the problem: Item 2 cannot be shipped unless Item 1 is also shipped, and Items 3 and 4 cannot be shipped together. These restrictions can be easily expressed in terms of the integrally restricted variables; we add to the constraints the inequalities

$$x_2 \leq x_1 \text{ and } x_3 + x_4 \leq 1$$

Note that the final models in these two examples are simple integer programming problems. This is our goal for each of the examples of this section, that is, to translate the problem at hand into a linear programming problem with integral restrictions on (possibly some of) the variables of the problem so that the problem can be solved by the use of available integer programming solution techniques. Remember that the introduction and interpretation of the variables for the problem is just an intermediate (but often the most difficult) step, and that it is contingent on the problem formulator's ascertaining whether the final integer programming problem adequately reflects the situation at hand and, in particular, the desired interpretation of the variables.

Modifications of the Objective Function

A simple example of a model that requires modification of the objective function is the *Fixed Charge Problem.* Consider an operation in which the costs or profits associated can involve lump sum quantities, depending on whether or not certain processes, facilities, or whatever are utilized, that is, situations in which there is a fixed charge or cost for the use of a process only if that process is used. For example, if a machine is to be used in the manufacture of a product, there could be a setup cost to prepare the equipment for the run. Or if a new market is to be opened for selling a product, there could be an initial market development cost totally independent of the number of units eventually sold in that market. In a transportation problem, there could be rental costs for the warehouses utilized.

To formulate models for such operations, we introduce, for each fixed charge, an auxiliary variable, restricted to either 0 or 1, with the interpretation that the variable will be 1 only if the corresponding fixed charge is to be assessed. Then the product of the fixed charge times the variable is added to the objective function of the model, and inequalities are added to the constraints so that this "on or off" role of the auxiliary variable is maintained.

Example 6.2.3. Suppose a machine shop has three processes that it can use to manufacture two different parts, with each process combining various amounts of raw material and labor to produce different quantities of the two parts. Suppose the shop has weekly limits on its raw material and labor and must meet weekly fixed demands for the two parts. Assume that with each process there is both a setup and a maintenance charge d_i if the process is utilized, and also a cost c_i for every hour of its operation. We wish to determine the weekly operation of the shop that minimizes total cost while meeting the given demands for the parts.

Define variables:

$$x_i = \text{number of hours per week that Process } i \text{ is used}$$

$$y_i = \begin{cases} 0, & x_1 = 0 \\ 1, & x_1 > 0 \end{cases}$$

Then the restrictions on the available raw material and labor would be reflected in ($\leq$) inequalities involving the x_i's, the weekly demands for the parts would correspond to ($\geq$) inequalities in the x_i's, and the objective function for the model would be

$$c_1x_1 + c_2x_2 + c_3x_3 + d_1y_1 + d_2y_2 + d_3y_3$$

Now all we need do is add constraints to regulate the y_i's. Suppose that there are known upper bounds on the x_i's independent of the final operation of the plant. In this problem, let us assume that each x_i must be less than or equal to 40. Then, to the constraints we add the inequalities

$$y_i \geq \frac{x_i}{40}$$
$$0 \leq y_i \leq 1 \text{ and integral}, i = 1,2,3$$

Notice that if the optimal operation calls for the use of Process i, the restrictions on y_i force it to be 1, since $x_i/40$ will be greater than 0, whereas if Process i is not to be used, the constraints would permit y_i to be 0 or 1, but the minimization of the objective function would lead to $y_i = 0$. Thus the desired role of these auxiliary variables is achieved.

Example 6.2.4. A distributor supplies n retail outlets, with Outlet j requiring d_j units monthly. The distributor can rent storage facilities in up to m warehouses, with Warehouse i having a storage capacity of s_i units and a monthly rental fee of r_i dollars. There is a cost of c_{ij} dollars to ship 1 unit from Warehouse i to Outlet j. Determine what warehouses are to be utilized in the implementation of a feasible shipping schedule that minimizes total costs. (This problem would be a standard transportation problem if it were not for the rental fees of the occupied warehouses.)

Define variables:

$$x_{ij} = \text{number of units shipped monthly from Warehouse } i \text{ to Outlet } j$$
$$y_i = \begin{cases} 0 & \text{if Warehouse } i \text{ is not utilized} \\ 1 & \text{if Warehouse } i \text{ is utilized} \end{cases}$$

Then an integer programming model for this problem is to

$$\text{Minimize} \sum_{i,j} c_{ij}x_{ij} + \sum_i r_i y_i \tag{6.2.1}$$

subject to

$$\sum_j x_{ij} \leq s_i, i = 1, \ldots, m$$
$$\sum_i x_{ij} \geq d_j, j = 1, \ldots, n$$
$$y_i \geq \sum_j \frac{x_{ij}}{s_i}, i = 1, \ldots, m$$
$$0 \leq x_{ij}, 0 \leq y_i \leq 1 \text{ and integral}$$

In a solution to this problem, the y_i's would indicate what warehouses are to be used, and the x_{ij}'s would provide a corresponding minimal-cost shipping schedule.

If the "fixed charge" is in fact a rebate, that is, a benefit to be provided as long as certain conditions are fulfilled, the problem can be handled in the same fashion as the above, only now the associated auxiliary variable would need to be forced "off" if the required conditions are not fulfilled.

Example 6.2.5 (Continuation of Example 6.2.4)**.** Suppose the distributor of Example 6.2.4 is entitled to a monthly rent reduction of $\$R$ if both Warehouses 1 and 2 are utilized.

Define

$$z = \begin{cases} 0, & y_1 + y_2 < 2 \\ 1, & y_1 + y_2 = 2 \end{cases}$$

Then we add to the objective function of (6.2.1) the term $(-Rz)$ and to the constraints the restrictions

$$\begin{gathered} z \leq (y_1 + y_2)/2 \\ 0 \leq z \leq 1 \text{ and integral} \end{gathered}$$

Now minimization of the objective function would require z to be 1 whenever possible, but the restrictions on z constrain it to be 0 unless both y_1 and y_2 are 1.

Another situation that can require modifications in the objective function is when there are sliding or changing costs, depending on how much of a quantity is used. The modeling problems of Chapter 2 that involved the use of labor, with overtime at a premium price, are examples of this type.

Example 6.2.6. Processing a material M through a machine, a company can produce A's and B's. The requirements and selling price of a unit of each are as follows:

	M (units)	*Machine Time (min)*	*Selling Price (\$)*
A	5	2	25
B	8	4	45

The company has available 110 min of machine time weekly at no cost. The material M, however, must be purchased from an outside vendor. The company can purchase weekly 150 units at \$2/unit and an additional 100 units at \$3/unit.

Assuming that there are no other expenses associated with the production and sale of the A's and B's, one possible model for this problem is the following. Define x_1 and x_2 to be the number of A's and B's, respectively, to be produced and sold weekly, x_3 the number of units of M purchased at \$2/unit, and x_4 the number purchased at \$3/unit. The problem then is to

$$\begin{aligned} &\text{Maximize } 25x_1 + 45x_2 - 2x_3 - 3x_4 \\ &\text{subject to} \\ &5x_1 + 8x_2 = x_3 + x_4 \\ &2x_1 + 4x_2 \leq 110 \\ &\qquad\quad x_3 \leq 150 \\ &\qquad\quad x_4 \leq 100 \\ &x_1, x_2, x_3, x_4 \geq 0 \end{aligned} \tag{6.2.2}$$

For this problem, as stated, we have used no auxiliary discrete variables. By the nature of the objective function, we are guaranteed that in any optimal solution, x_4 will be positive only if x_3 equals 150.

Now we alter the circumstances of the problem.

Example 6.2.7 (Continuation of Example 6.2.6). Suppose that new sources of material M become available to the outside vendor, and so the vendor changes the price of M as follows: still \$2/unit for the first 150 units but only \$1.75/unit for the next 100 units.

Letting x_4 now be the number of units purchased at \$1.75/unit, one approach here might be to simply change the coefficient of x_4 in the objective function of (6.2.2) from -3 to -1.75. However, that alone would not suffice; nothing in the model would force the use of the first 150 units of M before the use of the less expensive last 100 units. We need to introduce an auxiliary variable.

Define

$$y_1 = \begin{cases} 0, & x_3 < 150 \\ 1, & x_3 = 150 \end{cases}$$

and, in the problem of (6.2.2), change the $-3x_4$ term of the objective function to $-1.75x_4$ and add the constraints

$$\begin{aligned} &y_1 \leq x_3/150 \\ &x_4 \leq 100y_1 \\ &0 \leq y_1 \leq 1 \text{ and integral} \end{aligned}$$

Then x_4 can be positive only if the discrete variable y_1 is 1, and y_1 can equal 1 only when $x_3 = 150$.

We can extend this technique to handle various shifts in cost.

Example 6.2.8 (Continuation of Example 6.2.7). Suppose the vendor offers an unlimited amount of material M to the company, priced at \$2/unit for the first 150 units, \$1.75/unit for the next 100, and only \$1/unit for any number of units purchased over the first 250.

Define x_3, x_4, and x_5 to be the number of units of M purchased at \$2, \$1.75, and \$1/unit, respectively, and define x_1 and x_2 as before. Introduce

$$y_1 = \begin{cases} 0, & x_3 < 150 \\ 1, & x_3 = 150 \end{cases} \qquad y_2 = \begin{cases} 0, & x_4 < 100 \\ 1, & x_4 = 100 \end{cases}$$

The problem:

$$
\begin{aligned}
&\text{Maximize } 25x_1 + 45x_2 - 2x_3 - 1.75x_4 - 1x_5 \\
&\text{subject to} \\
&5x_1 + 8x_2 = x_3 + x_4 + x_5 \\
&2x_1 + 4x_2 \leq 110 \\
&\qquad\quad x_3 \leq 150 \\
&y_1 \leq x_3/150 \qquad y_2 \leq x_4/100 \\
&x_4 \leq 100y_1 \qquad x_5 \leq 245y_2 \\
&x_1, x_2, x_3, x_4, x_5 \geq 0 \\
&0 \leq y_1, y_2 \leq 1 \text{ and integral}
\end{aligned}
$$

(The constraint $x_5 \leq 245y_2$, which restricts the x_5 to 0 until y_2 is 1 and x_4 is 100, requires an upper bound on the x_5. Here $2x_1 + 4x_2 \leq 110$ implies $x_1 \leq 55$ and $x_2 \leq 27.5$, and so, in any feasible production schedule, $5x_1 + 8x_2 \leq 5(55) + 8(27.5) = 495$ (see Problem 4).)

The Problem of Alternative Constraints

Consider a situation in which the amounts of the quantities involved, represented by the variables of the problem, must satisfy one (or more) set(s) from two (or more) alternative sets of constraints. For example, in the manufacture of a certain product, it may be economically feasible to produce either none at all or an amount exceeding some minimal batch size. Or, in the bidding for various contracts, those bids submitted must exceed certain minimal operating costs, and so the bids are either 0 or greater than or equal to some lower bound. Or, in the utilization of the resources of a plant, one of several methods may be employed, and associated with each method is a corresponding system of constraints reflecting the limited supplies and required demands.

To formulate models for problems of this type, we again introduce auxiliary on or off variables equal to either 0 or 1 and use the variables to modify the system of constraints.

Example 6.2.9. Suppose a variable x_i, representing the amount of an item to be produced, must be either 0 or greater than or equal to a minimal batch size b_i, and assume also that in any feasible solution, x_i is bounded above by m_i. To model this, we introduce a discrete variable y_i, adding to the constraints the restrictions

$$
\begin{aligned}
&x_i \leq m_i y_i \\
&x_i \geq b_i y_i \\
&0 \leq y_i \leq 1 \text{ and integral}
\end{aligned}
$$

Then y_i can equal only 0 or 1. If $y_i = 0$, the two inequalities force $x_i = 0$, and if $y_i = 1$, the first inequality has no significance but the second demands $x_i \geq b_i$. Thus x_i is properly restricted.

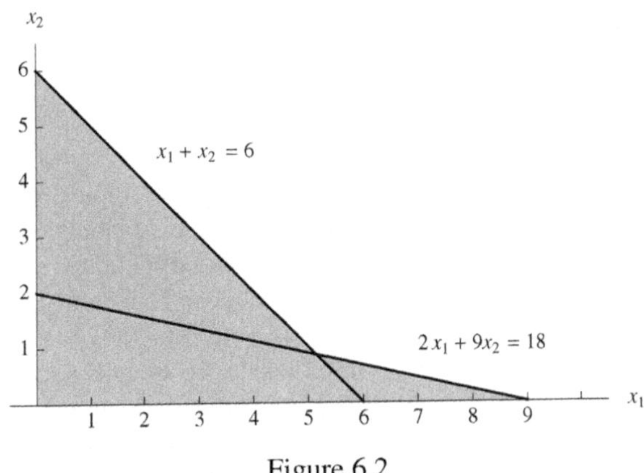

Figure 6.2

Example 6.2.10. Suppose nonnegative variables x_1 and x_2 must satisfy either $2x_1 + 9x_2 \leq 18$ or $x_1 + x_2 \leq 6$. The graph of the set of feasible points (x_1, x_2) is illustrated in Figure 6.2. (Notice that this set is not convex.) Introduce variables y_1 and y_2, with

$$y_1 = \begin{cases} 0 \text{ if the first inequality is not satisfied} \\ 1 \text{ if the first inequality is satisfied} \end{cases}$$

and similarly for y_2 and the second inequality.

To formulate an integer programming model to reflect these interpretations, we need to determine upper bounds to use in the constraints in order to render the associated inequalities nonrestrictive when they are off. Here, if (x_1, x_2) is a feasible solution to the either/or constraints, we must have $x_1 \leq 9$ and $x_2 \leq 6$ (see Figure 6.2). Thus, for any feasible point (x_1, x_2),

$$2x_1 + 9x_2 - 18 \leq 18 + 54 - 18 = 54$$

and

$$x_1 + x_2 - 6 \leq 9 + 6 - 6 = 9$$

Using these bounds, one can express the either/or restrictions as follows:

$$\begin{aligned} 2x_1 + 9x_2 - 18 &\leq 54(1 - y_1) \\ x_1 + x_2 - 6 &\leq 9(1 - y_2) \\ y_1 + y_2 &\geq 1 \\ x_1, x_2 &\geq 0 \\ 0 \leq y_1, y_2 &\leq 1 \text{ and integral} \end{aligned} \tag{6.2.3}$$

To ascertain that this system properly reflects the original conditions of the problem, observe first that the inequality $y_1 + y_2 \geq 1$ demands that at least one of the y_i's will be 1. If $y_1 = 1$, the first inequality reduces to the original $2x_1 + 9x_2 \leq 18$, and so, if (x_1, x_2, y_1, y_2) is a solution to (6.2.3) with $y_1 = 1$, the point (x_1, x_2) must satisfy this inequality. Similarly, if $y_2 = 1$, a solution point must satisfy the original second inequality. Finally, when $y_1 = 0$, the first constraint places no meaningful restrictions on the solution set, because all feasible points (x_1, x_2) must satisfy $2x_1 + 9x_2 - 18 \leq 54$. Similarly when $y_2 = 0$. (See Problem 1.)

Notice the nature of the end product in the above example. We have a simple collection of linear inequalities involving nonnegative variables, with some of the variables restricted to be integral. Thus, unless other difficulties were present, integer programming techniques could be used with these constraints. Of course, if the optimization problem were simply to maximize, say $f(x_1, x_2) = 4x_1 + x_2$ subject to either $x_1 + x_2 \leq 6$ or $2x_1 + 9x_2 \leq 18$, with x_1 and x_2 nonnegative, one could consider each constraint separately, solve the corresponding linear programming problem, and compare the two optimal values. The larger value would be the solution to the problem with the either/or constraints. However, the techniques of the example do show some of the range of integer programming and also generalize quite easily to encompass more complicated situations.

Example 6.2.11. Consider the problem of finding all points (x_1, x_2, x_3) satisfying

$$x_1 + 2x_2 + 3x_3 \leq 600, \quad x_1, x_2, x_3 \geq 0$$

and at least three of the following five alternative sets of constraints:

$$\{5x_1 + 10x_2 \leq 500\}, \quad \{x_1 - x_2 + x_3 \leq 450\}, \quad \{x_2 + x_3 \geq 100\},$$
$$\{2x_1 - 3x_2 \geq 100 \text{ and } x_3 \geq 50\}, \quad \{x_1 + x_3 = 150\}$$

Introduce discrete variables y_i, with

$$y_i = \begin{cases} 0, & \text{the } i\text{th set of constraints is not satisfied} \\ 1, & \text{the } i\text{th set is satisfied} \end{cases}$$

To determine bounds to render the constraints redundant when the y_i's equal 0, note first that $x_1 + 2x_2 + 3x_3 \leq 600$ implies that $x_1 \leq 600$, $x_2 \leq 300$, $x_3 \leq 200$ for any solution point. Thus, for any feasible point (x_1, x_2, x_3), we have

$$\begin{aligned}
5x_1 + 10x_2 - 500 &\leq 5(600) + 10(300) - 500 = 5500 \\
x_1 - x_2 + x_3 - 450 &\leq 600 - 0 + 200 - 450 = 350 \\
x_2 + x_3 - 100 &\geq 0 + 0 - 100 = -100 \\
2x_1 - 3x_2 - 100 &\geq 0 - 900 - 100 = -1000 \\
x_3 - 50 &\geq 0 - 50 = -50 \\
x_1 + x_3 - 150 &\leq 600 + 200 - 150 = 650 \\
x_1 + x_3 - 150 &\geq 0 + 0 - 150 = -150
\end{aligned}$$

The problem then becomes one of determining all (x_1, x_2, x_3) and $(y_1, y_2, y_3, y_4, y_5)$ satisfying

$$
\begin{array}{l}
x_1 + 2x_2 + 3x_3 \leq 600 \\
5x_1 + 10x_2 - 500 \leq 5500(1-y_1) \\
x_1 - x_2 + x_3 - 450 \leq 350(1-y_2) \\
x_2 + x_3 - 100 \geq -100(1-y_3) \\
2x_1 - 3x_2 - 100 \geq -1000(1-y_4) \\
x_3 - 50 \geq -50(1-y_4) \\
x_1 + x_3 - 150 \leq 650(1-y_5) \\
x_1 + x_3 - 150 \geq -150(1-y_5) \\
y_1+y_2+y_3+y_4+y_5 \geq 3 \\
0 \leq y_1,y_2,y_3,y_4,y_5 \leq 1 \text{ and integral} \\
x_1,x_2,x_3 \geq 0
\end{array}
$$

Problem Set 6.2

1. In working with auxiliary variables in problems with alternative constraints, we used upper and lower bounds in the inequalities associated with these variables so that these inequalities were rendered redundant when the corresponding auxiliary variables were off (i.e., usually equal to 0). To show the need for these bounds, consider Example 6.2.10. Suppose we attempt to replace the either/or constraints of that example with the following:

$$
\begin{array}{l}
2x_1 + 9x_2 - 18 \leq 1-y_1 \\
x_1 + x_2 - 6 \leq 1-y_2 \\
y_1+y_2 \geq 1 \\
x_1,x_2 \geq 0 \\
0 \leq y_1,y_2 \leq 1 \text{ and integral}
\end{array}
\tag{6.2.4}
$$

(This is simply (6.2.3), with the upper bounds of 54 and 9 omitted from the first two inequalities.)

   (a) Show that if (x_1,x_2,y_1,y_2) satisfies (6.2.4), then either $2x_1+9x_2 \leq 18$ or $x_1+x_2 \leq 6$.

   (b) Show that for some (x_1,x_2) that satisfy either $2x_1+9x_2 \leq 18$ or $x_1+x_2 \leq 6$, we cannot find y_1 and y_2 such that (x_1,x_2,y_1,y_2) is a solution to (6.2.4). (Try finding y_1 and y_2 for the point $(3,3)$, for example.)

   (c) In fact, show that if (x_1,x_2,y_1,y_2) is a solution to (6.2.4), then (x_1,x_2) satisfies

$$
\left\{ \begin{array}{l} 2x_1+9x_2 \leq 18 \\ x_1+x_2 \leq 7 \end{array} \right\} \quad \text{or} \quad \left\{ \begin{array}{l} 2x_1+9x_2 \leq 19 \\ x_1+x_2 \leq 6 \end{array} \right\}
$$

2. (a) Show that replacing the inequality $y_1+y_2 \geq 1$ in (6.2.3) with the equality $y_1+y_2 = 1$ yields an equivalent system of constraints (with respect to (x_1,x_2)).

   (b) Express the either/or constraints of Example 6.2.10 with linear restrictions using only one discrete variable.

3. Show that the optimal production schedules for the problems of Examples 6.2.6, 6.2.7, and 6.2.8 are all distinct.

   (*Hint.* The simplex algorithm can be used in 6.2.6. The solution suggests that in any optimal solution to 6.2.7, at least 150 units of M will be used, and with this assumption, 6.2.7 can be formulated as a linear programming problem without discrete variables.)

4. Show that in Example 6.2.8, the 495 upper bound on the number of M's used could be replaced by 275. (*Hint.* $5x_1 + 8x_2 = 2(2x_1 + 4x_2) + x_1$.)

5. For each of the shaded regions in Figures 6.3, 6.4, and 6.5, determine an equivalent system of linear constraints.

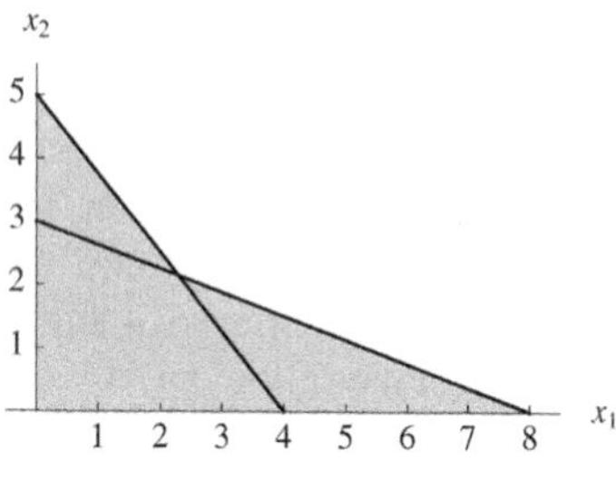

Figure 6.3

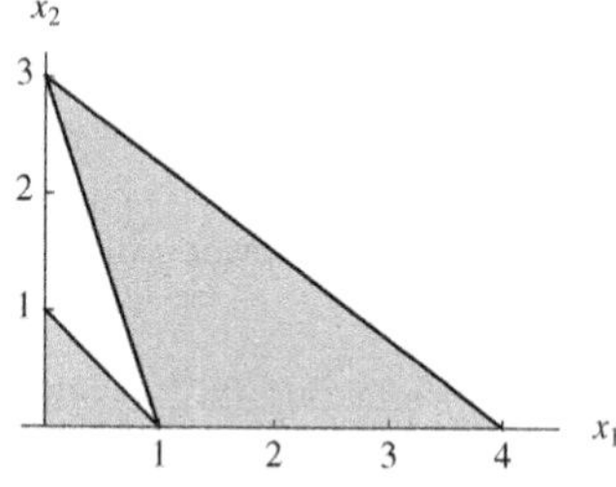

Figure 6.4

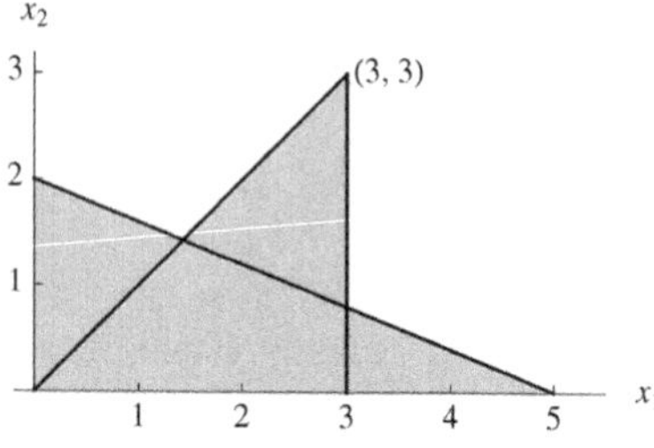

Figure 6.5

Formulate integer (or mixed integer) programming models for the following. (Do not attempt to solve the problems.)

6. An investment firm, wanting to maximize profit, has \$500,000 to invest in the following:

- A development project, requiring a lump sum investment of \$390,000 and yielding a return of 6.7%.
- A construction project, requiring a lump sum investment of \$220,000 and yielding a return of 6.5%.
- Any number of units in a portfolio of stocks, costing \$25,000/unit and yielding a return of 6.3%.
- Any number of units of a certain stock, costing \$1300/share and yielding a return of 5.7%.

7. A construction firm has available M dollars in capital to be used for the development of up to n different sites, with Site i costing a_i dollars for development and returning an expected profit of c_i dollars upon completion. However, $\sum a_i > M$. What sites should be selected to optimize the expected profit?

8. (a) (*The Knapsack Problem*) A backpacker's knapsack has a volume of V in.^{3} and can hold up to W lb of gear. The backpacker has a choice of n items to carry in it, with the ith item requiring a_i in.^{3} of space, weighing w_i lb, and providing c_i units of value for the trip. What items should be taken in the knapsack?

(b) Refine part (a) to include the following considerations: Item 1, a can of tuna fish, Item 2, a can of corn, and Item 3, a can of stew, have no value unless Item 4, the can opener, is taken; and only one snack, either Item 5, potato chips (light but bulky), or Item 6, unpopped popcorn (small but heavy), is to go. Of course Items 2, 3, and 6 all use Item 7, the cooking pot.

9. A road construction firm seeks to assign its force optimally over the next 28-week period. They can be assigned to any combination of the following:

- For any number of 10-week periods, working for the state, and earning a profit of \$3200/week.
- For any number of 6-week periods, working for the county, and earning a profit of \$2900/week.
- For any number of 3-week periods, working for a private land developer, and earning \$2750/week.
- For any number of weeks, working on parking lot construction, and earning \$2550/week.
- However, if the firm does any work at all for either the state or county (or both), it is expected to contribute \$7500 to the campaign fund of a certain anonymous political figure.

10. (a) A manufacturer supplies six outlets and has a choice of renting space in up to three warehouses to maintain stocks to deliver to the outlets. Determine the minimal-cost renting and shipping schedule using the following data:

Shipping Costs per Unit		*Outlets* 1	2	3	4	5	6	*Storage Capacity*	*Monthly Rent*
	1	12	9	16	13	11	23	150	300
Warehouse	2	10	13	12	7	12	26	200	500
	3	13	12	14	10	17	21	300	700
Monthly Demands		70	45	35	50	75	60		

(b) As above, but with the added condition that the carrier that would be used to deliver units from Warehouse 2 to Outlets 4, 5, and 6 offers a shipping cost reduction of 1/unit for each unit shipped after the first 80.

(c) As in part (b), except that the carrier is more generous, offering the cost reduction on all units shipped if this number exceeds 80.

11. (a) A firm has M units of a new product to be sold in up to n different new market areas. To develop the ith market for sales, there is an initial research and advertising cost of d_i dollars. Once opened, the ith market can sell up to u_i units at a profit of c_i dollars/unit. What areas should be developed and how many units should each of these areas receive so as to maximize profit?

(b) Suppose also that because of personnel limitations, at most k market areas $(k < n)$ can be developed.

(c) Suppose also that if both Markets 1 and 2 are developed, the firm must pay a tax of T dollars; but if both Markets 3 and 4 are developed, the firm receives tax credits worth R dollars.

12. (a) A small division of an automobile plant manufactures two parts to meet the monthly demands of the major assembly plant. Three different machines can be used in the process, each having varying input and output capacities and setup and maintenance costs if used. Determine the most economical monthly operation of the division, using the following data:

	One Hour of Operation — ***Input*** Labor (hr)	Raw Material (lb)	***Output*** Part A (units)	Part B (units)	*Monthly Maintenance*
Machine 1	12	95	16	15	\$1350
Machine 2	9	70	10	12	\$ 950
Machine 3	14	75	5	17	\$1575

Monthly demands: Part A – 400 units
Part B – 500 units
Monthly supplies: Labor – 550 hours
Raw material – unlimited
Costs: Labor – \$19.50/hr
Raw material – \$12.50/lb

(b) Assume in addition that the division has available 55 hr of overtime labor at \$29.25/hr, and 1000 lb of raw material at \$10/lb.

13. An ice cream plant can make up to 28 different flavors of ice cream each month. Flavor i requires a_i lb of sugar/gal and earns c_i dollars/gal sold, but at least u_i gal must be made per month if the flavor is to sell. With M lb of sugar available for the month, what flavors should be made to optimize profit?

14. Maximize $5x_1 + 12x_2$ subject to

$$7x_1 + 3x_2 \leq 16 \quad \text{or} \quad 3x_1 + 10x_2 \leq 20 \quad \text{with} \quad x_1, x_2 \geq 0$$

15. Maximize $3x_1 + 5x_2 + 7x_3$ subject to

$5x_1 + 4x_2 + 2x_3 \leq 300, x_1, x_2, x_3 \geq 0$, and

$$\left\{ \begin{array}{r} x_1 + x_3 \leq 100 \\ x_1 - x_2 \geq \quad 0 \end{array} \right\} \quad \text{or} \quad \left\{ \begin{array}{r} 2x_1 - 4x_2 + 5x_3 \leq 250 \\ x_2 - 2x_3 \geq \quad 50 \end{array} \right\}$$

16. Maximize $9x_1 + 8x_2 + 7x_3$ subject to

$$\begin{aligned} x_1 + x_2 + x_3 &\leq 500 \\ x_1, x_2, x_3 &\geq 0 \end{aligned}$$

and such that (x_1, x_2, x_3) satisfies at least two of the following three constraints:

$$3x_1 - 3x_2 + 4x_3 \leq 1000, \quad x_1 - 2x_3 \geq 200, \quad \text{and} \quad x_1 + x_2 = 300.$$

17. (a) In spring, in preparation for the summer trade, a shop makes outdoor tables. Four types can be made, with input (wood and labor) and selling price for a table of each type as follows:

Type	*Wood (units)*	*Labor (hr)*	*Selling Price ($)*
A	7	3	90
B	5	2	68
C	4	4	85
D	8	12	175

For the project, the shop has available up to 2000 hr of labor at $8/hr. The wood is purchased from a mill and costs $5/unit for the first 1600 units, $4.75/unit for the next 900 units, and $4.50/unit for any number above 2500. Assuming that all tables made now will be sold in the coming summer, how many of each type should be made to maximize profit?

(b) It is next spring, and the mill has changed its prices: $5/unit of wood if the shop purchases no more than 2500 units and $4.65/unit for all units purchased if the shop wants more than 2500.

18. A company must produce weekly either 1500 *A*'s and 1000 *B*'s or 1000 *A*'s and 1500 *B*'s. Three different processes can be used in production, with input (labor and raw material *M*) and output (*A*'s and *B*'s) of 1 hr of operation of each as follows:

	Input		*Output*	
	Labor (hr)	*M's (units)*	*A's (units)*	*B's (units)*
Process 1	20	35	40	42
Process 2	12	12	45	35
Process 3	25	28	36	44

An unlimited number of M's are available weekly at \$15/unit and up to 600 hr of labor at \$12/hr. How many A's and B's should be made, using what production schedule, to minimize weekly costs?

19. A microbrewery can brew three different beers: a light, a dark, and an ale. The requirements (grain and hops) and revenue (selling price less normal costs) for a batch of each are as follows:

	Grain (lb)	*Hops (lb)*	*Revenue (\$)*
Light	25	10	395
Dark	40	8	440
Ale	30	16	475

The brewery has available each week 800 lb of grain and 250 lb of hops. It can make a total of 20 batches weekly with its regular labor force and up to an additional 4 batches using overtime. (Batches made on overtime cost an additional \$95/batch.)

The brewery wants to market exactly two types and make at least five batches of each. There is also a weekly equipment preparation/conversion cost that is a function of the two beers being brewed.

Products	*Cost (\$)*
Light and Dark	250
Light and Ale	150
Dark and Ale	225

What beers should the brewery make, and how many batches of each should be made to maximize net revenue?

6.3 Gomory's Cutting Plane Algorithm

There are various algorithms available for the solution of integer programming problems. The reason for this abundance is that no one algorithm has proved to be computationally efficient for all problems, and thus the search continues for more effective algorithms. To introduce some of the methods used for solving these problems, we will present in this section and the next two integer programming algorithms. Specifically, in this section we will develop one version of *Gomory's Cutting Plane*

Algorithm. This algorithm was one of the first of its kind, published in 1958 ([14]), and is still an effective tool for solving certain integer programming problems.

Consider a pure integer programming problem, that is, a standard linear programming problem with integral restrictions on all of the variables. The fundamental idea underlying Gomory's Cutting Plane Algorithm is to add constraints to the problem one at a time so that we eventually have a linear programming problem with an optimal solution with integral coordinates. The algorithm works as follows. First, we solve the original linear programming problem, ignoring the integral restrictions. Then, if this solution has all integral coordinates, it is also an optimal solution to the integer programming problem, and we are done. If not, we generate a new constraint to be added to the problem. This constraint will have two fundamental properties: first, the nonintegral optimal solution to the original linear programming problem will not satisfy this constraint; and second, all integral feasible solutions to the original problem will satisfy the new constraint. Thus this constraint essentially cuts off a subset of the set of feasible solutions to the linear programming problem, but a subset that contains no feasible integral solutions. We add this constraint to the problem and proceed to solve the expanded problem as before, first ignoring the integral restrictions and continuing on. We demonstrate the algorithm first by an example.

Example 6.3.1. Consider the problem of

$$\begin{aligned}
&\text{Minimizing } x_1 - 3x_2 \\
&\text{subject to} \\
&\ \ x_1 - \ \ x_2 \leq \ 2 \\
&2x_1 + 4x_2 \leq 15 \\
&x_1, x_2 \geq 0 \text{ and integral}
\end{aligned}$$

Adding integrally restricted slack variables x_3 and x_4 and using the simplex algorithm on the associated linear programming problem, we have the tableaux of Table 6.2. The minimal value of the objective function, ignoring the integral constraints, is attained at $x_1 = 0$, $x_2 = \frac{15}{4} = 3\frac{3}{4}$ (and $x_3 = 5\frac{3}{4}$, $x_4 = 0$). Since this point has nonintegral coordinates, we need to generate a new constraint. To do this, we can work with any constraining equation from the final tableau that has a nonintegral constant

Table 6.2

	x_1	x_2	x_3	x_4	
x_3	1	-1	1	0	2
x_4	2	(4)	0	1	15
	1	-3	0	0	0
x_3	$\frac{3}{2}$	0	1	$\frac{1}{4}$	$\frac{23}{4}$
x_2	$\frac{1}{2}$	1	0	$\frac{1}{4}$	$\frac{15}{4}$
	$\frac{5}{2}$	0	0	$\frac{3}{4}$	$\frac{45}{4}$

term, and so, in this case, we could use either equation. Let us consider the equation defined by the first row of the final tableau:

$$\tfrac{3}{2}x_1 + x_3 + \tfrac{1}{4}x_4 = \tfrac{23}{4}$$

Separating all constants into their integral and fractional parts, we have

$$(1 + \tfrac{1}{2})x_1 + x_3 + \tfrac{1}{4}x_4 = 5 + \tfrac{3}{4}$$

Therefore we must have

$$\tfrac{1}{2}x_1 + \tfrac{1}{4}x_4 - \tfrac{3}{4} = 5 - x_1 - x_3$$

Since we want only integral solutions, the right side and therefore the left side of this equation must be integral. And, since all variables are nonnegative, $\frac{1}{2}x_1 + \frac{1}{4}x_4 \geq 0$, and so the smallest integer the left side can equal is 0. Thus we want solutions such that

$$\tfrac{1}{2}x_1 + \tfrac{1}{4}x_4 - \tfrac{3}{4}$$

is a nonnegative integer, say x_5. It is this constraint that we add to the two original constraints, giving us the expanded problem of

$$\begin{array}{l}
\text{Minimizing } \tfrac{5}{2}x_1 + \tfrac{3}{4}x_4 - \tfrac{45}{4} \\
\text{subject to} \\
\begin{array}{lllllll}
\tfrac{3}{2}x_1 & & + x_3 & + \tfrac{1}{4}x_4 & & = & \tfrac{23}{4} \\
\tfrac{1}{2}x_1 & + x_2 & & + \tfrac{1}{4}x_4 & & = & \tfrac{15}{4} \\
\tfrac{1}{2}x_1 & & & + \tfrac{1}{4}x_4 & - x_5 & = & \tfrac{3}{4}
\end{array} \\
x_1, x_2, x_3, x_4, x_5 \geq 0 \text{ and integral}
\end{array}$$

Here we have used the final tableau data of Table 6.2 in expressing the original problem. Notice that the optimal solution found above, $x_1 = x_4 = 0$, $x_2 = \frac{15}{4}$, $x_3 = \frac{23}{4}$, does not satisfy the new constraint.

Now we proceed as before, solving the corresponding linear programming problem. To do this, since we have simply added a constraint to a completed problem, we can use the Dual Simplex Algorithm (Sections 5.6 and 5.7). After multiplying the new constraint by (-1), we have as basic variables for the first tableau the basic variables of the previous final tableau (x_3 and x_2) along with the new variable, x_5. The tableaux are presented in Table 6.3. The solution point here has all integral coordinates, and therefore we have the solution to the original integer programming problem. The minimal value of the objective function is -9 and is attained at the point $x_1 = 0$, $x_2 = 3$.

The geometry of this example explains the name of the algorithm. The feasible points for the original system of constraints are the lattice points of the shaded region in Figure 6.6. The constraint added can be expressed as $\frac{1}{2}x_1 + \frac{1}{4}x_4 - \frac{3}{4} \geq 0$. Using $x_4 = 15 - 2x_1 - 4x_2$, this inequality reduces to

$$2x_1 + (15 - 2x_1 - 4x_2) \geq 3 \quad \text{or} \quad 4x_2 \leq 12 \quad \text{or} \quad x_2 \leq 3$$

Table 6.3

	x_1	x_2	x_3	x_4	x_5	
x_3	$\frac{3}{2}$	0	1	$\frac{1}{4}$	0	$\frac{23}{4}$
x_2	$\frac{1}{2}$	1	0	$\frac{1}{4}$	0	$\frac{15}{4}$
x_5	$-\frac{1}{2}$	0	0	$\boxed{-\frac{1}{4}}$	1	$-\frac{3}{4}$
	$\frac{5}{2}$	0	0	$\frac{3}{4}$	0	$\frac{45}{4}$
x_3	1	0	1	0	1	5
x_2	0	1	0	0	1	3
x_4	2	0	0	1	−4	3
	1	0	0	0	3	9

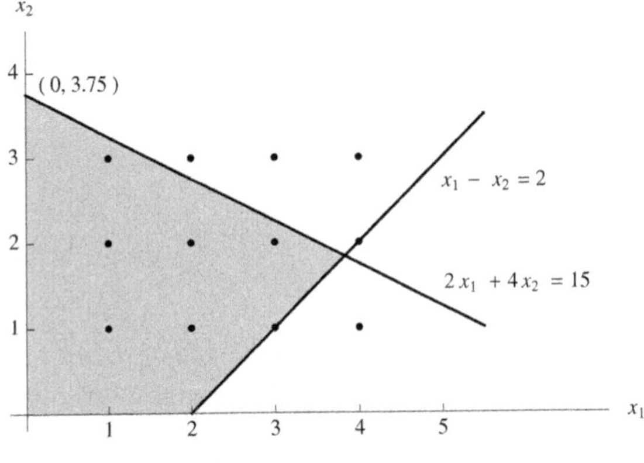

Figure 6.6

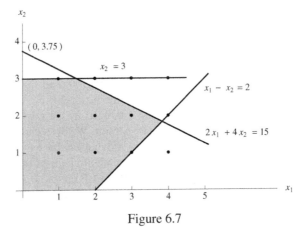

Figure 6.7

As can be seen in Figure 6.7, the new constraint is equivalent to an inequality that cuts off from the feasible set the original nonintegral optimal solution $(0, 3\frac{3}{4})$ but does not exclude from consideration any feasible lattice points.

We now describe in detail how these new constraints are generated. Suppose, after solving the associated linear programming problem, the constant term of the ith row of the final tableau is not an integer. Then, in the optimal basic solution corresponding to this final tableau, the value of the basic variable isolated in the ith row will not be integral, and so we need to add a new constraint.

Attaching back the variables, suppose this ith constraint is

$$\sum_j a_{ij}x_j = b_i \tag{6.3.1}$$

Letting $[a]$ denote the greatest integer in a (i.e., the greatest integer less than or equal to a, and so $[3\frac{4}{5}] = 3$, $[1] = 1$, $[-3\frac{2}{5}] = -4$), define the *fractional part* of any number a to be $a - [a]$. Thus the fractional part of $3\frac{4}{5}$ is $\frac{4}{5}$, that of 1 is 0, and that of $-3\frac{2}{5}$ is $-3\frac{2}{5} - (-4) = \frac{3}{5}$. Note that the fractional part of a number must be nonnegative and less than 1. Let f_{ij} and f_i denote the fractional parts of a_{ij} and b_i, respectively, that is,

$$f_{ij} = a_{ij} - [a_{ij}] \text{ and } f_i = b_i - [b_i]$$

Then we can rewrite (6.3.1) as

$$\sum_j ([a_{ij}] + f_{ij})x_j = [b_i] + f_i \quad \text{or} \quad \sum_j f_{ij}x_j - f_i = [b_i] - \sum_j [a_{ij}]x_j \tag{6.3.2}$$

Notice that all the constant terms on the right side of (6.3.2) are integral. Thus, for any integral solution to the original system of constraints, the right side, and therefore the left side of (6.3.2), must be integral. Moreover, since all variables are nonnegative and f_i is less than 1, the left side of (6.3.2) must be greater than or equal to the integer 0. Hence the new constraint:

$$\sum_j f_{ij}x_j - f_i \geq 0 \text{ and integral} \tag{6.3.3}$$

We have developed (6.3.3) so that any feasible integral solution to the original system of constraints will satisfy this new constraint, and so that by adding this constraint, we still have a pure integer programming problem. Furthermore, the optimal basic feasible solution from the final tableau of the corresponding linear programming problem does not satisfy this constraint. We have chosen i such that b_i was not an integer, and so $f_i > 0$. Now the only variables x_j that can appear in (6.3.3) are the nonbasic variables of the final tableau; the coefficients of the basic variables are either 0 or 1, and so have fractional part 0. (In Example 6.3.1, the variables of the added constraint were x_1 and x_4, the nonbasic variables of the final tableau solution of the original problem.) Hence the corresponding basic feasible solution, with $x_j = 0$ for all nonbasic variables x_j, does not satisfy (6.3.3).

If, in the final tableau for the corresponding linear programming problem, several of the constant terms b_i are not integral, we have a choice of what row to use to generate the new constraint. In fact, rules governing the choice of row to use can be given (Gomory [15] or Hadley [16]) that will guarantee in theory at least the convergence of the algorithm to an optimal integral solution in a finite number of steps. However, after many iterations and several hours of computer time with no feasible solution in sight, a vice-president of computer affairs would probably be somewhat unimpressed with theoretical convergence arguments. In practice, the simple rule of using the row containing the constant term b_i with the largest fractional value is easy to apply and is usually quite effective.

We summarize the steps of Gomory's Cutting Plane Algorithm for pure integer programming problems. Consider the integer programming problem of optimizing $c \cdot X$ subject to $AX = b$, $X \geq 0$ and integral.

1. Solve the corresponding linear programming problem, ignoring the integral restrictions on X. If this solution has all integral coordinates, then it is an optimal solution to the original problem.
2. Otherwise, a new constraint is added to the problem.
   (a) To construct this constraint, select any row from the final optimal tableau solution of the linear programming problem with a nonintegral constant term b_i. (Using the row containing the constant term with the largest fractional value may reduce the total number of iterations necessary to complete the problem.)
   (b) Suppose the ith row is selected and the corresponding equation is

$$\sum_j a_{ij} x_j = b_i$$

   Then form the constraint

$$-\sum_j f_{ij} x_j + x = -f_i$$

   where

$$\begin{aligned} f_{ij} &= a_{ij} - \lfloor a_{ij} \rfloor = \text{fractional part of } a_{ij} \\ f_i &= b_i - \lfloor b_i \rfloor = \text{fractional part of } b_i \\ x &= \text{a new slack variable, restricted to be nonnegative and integral} \end{aligned}$$

   (c) Add this constraint to the problem and return to Step 1. Note that now, when solving the corresponding linear programming problem, the Dual Simplex Algorithm can be used.

Example 6.3.2. Consider the programming problem to

$$\begin{aligned} &\text{Minimize } x_1 - 2x_2 \\ &\text{subject to} \\ &\quad 2x_1 + \ \ x_2 \leq 5 \\ &-4x_1 + 4x_2 \leq 5 \\ &x_1, x_2 \geq 0 \text{ and integral} \end{aligned}$$

Table 6.4

	x_1	x_2	x_3	x_4	
x_3	2	1	1	0	5
x_4	-4	(4)	0	1	5
	1	-2	0	0	0
x_3	(3)	0	1	$-\frac{1}{4}$	$\frac{15}{4}$
x_2	-1	1	0	$\frac{1}{4}$	$\frac{5}{4}$
	-1	0	0	$\frac{1}{2}$	$\frac{5}{2}$
x_1	1	0	$\frac{1}{3}$	$-\frac{1}{12}$	$\frac{5}{4}$
x_2	0	1	$\frac{1}{3}$	$\frac{1}{6}$	$\frac{5}{2}$
	0	0	$\frac{1}{3}$	$\frac{5}{12}$	$\frac{15}{4}$

Table 6.5

	x_1	x_2	x_3	x_4	x_5	
x_1	1	0	$\frac{1}{3}$	$-\frac{1}{12}$	0	$\frac{5}{4}$
x_2	0	1	$\frac{1}{3}$	$\frac{1}{6}$	0	$\frac{5}{2}$
x_5	0	0	($-\frac{1}{3}$)	$-\frac{1}{6}$	1	$-\frac{1}{2}$
	0	0	$\frac{1}{3}$	$\frac{5}{12}$	0	$\frac{15}{4}$
x_1	1	0	0	$-\frac{1}{4}$	1	$\frac{3}{4}$
x_2	0	1	0	0	1	2
x_3	0	0	1	$\frac{1}{2}$	-3	$\frac{3}{2}$
	0	0	0	$\frac{1}{4}$	1	$\frac{13}{4}$

Adding integrally restricted slack variables x_3 and x_4 and applying the simplex algorithm, we have the tableaux of Table 6.4.

Both constant terms of the final tableau are nonintegral, but $\frac{5}{2}$ has the larger fractional value. The second row of this tableau generates the constraint

$$-\tfrac{1}{3}x_3 - \tfrac{1}{6}x_4 + x_5 = -\tfrac{1}{2}$$

where x_5 is a new slack variable. Adding this equation and using the Dual Simplex Algorithm leads to the tableaux of Table 6.5. The first row of the final tableau of this table generates the constraint

$$-\tfrac{3}{4}x_4 + x_6 = -\tfrac{3}{4}$$

where x_6 is a new slack variable. (Note that $-\frac{1}{4} - [-\frac{1}{4}] = -\frac{1}{4} - (-1) = \frac{3}{4}$. In general, be careful when working with the fractional part of a negative number.) Adding this constraint, we have the tableaux of Table 6.6. Thus the minimal value of the objective function is -3 and is attained at $x_1 = 1, x_2 = 2$.

Table 6.6

	x_1	x_2	x_3	x_4	x_5	x_6	
x_1	1	0	0	$-\frac{1}{4}$	1	0	$\frac{3}{4}$
x_2	0	1	0	0	1	0	2
x_3	0	0	1	$\frac{1}{2}$	-3	0	$\frac{3}{2}$
x_6	0	0	0	$-\frac{3}{4}$ (circled)	0	1	$-\frac{3}{4}$
	0	0	0	$\frac{1}{4}$	1	0	$\frac{13}{4}$
x_1	1	0	0	0	1	$-\frac{1}{3}$	1
x_2	0	1	0	0	1	0	2
x_3	0	0	1	0	-3	$\frac{2}{3}$	1
x_4	0	0	0	1	0	$-\frac{4}{3}$	1
	0	0	0	0	1	$\frac{1}{3}$	3

With the problem complete, we illustrate the action of the cutting planes geometrically. The first constraint added corresponds to

$$\tfrac{1}{3}x_3 + \tfrac{1}{6}x_4 - \tfrac{1}{2} \geq 0$$

that is,

$$\tfrac{1}{3}(5 - 2x_1 - x_2) + \tfrac{1}{6}(5 + 4x_1 - 4x_2) - \tfrac{1}{2} \geq 0$$

This reduces to $x_2 \leq 2$. Similarly, the second additional constraint reduces from $\frac{3}{4}x_4 - \frac{3}{4} \geq 0$ to $-x_1 + x_2 \leq 1$. The graph is sketched in Figure 6.8.

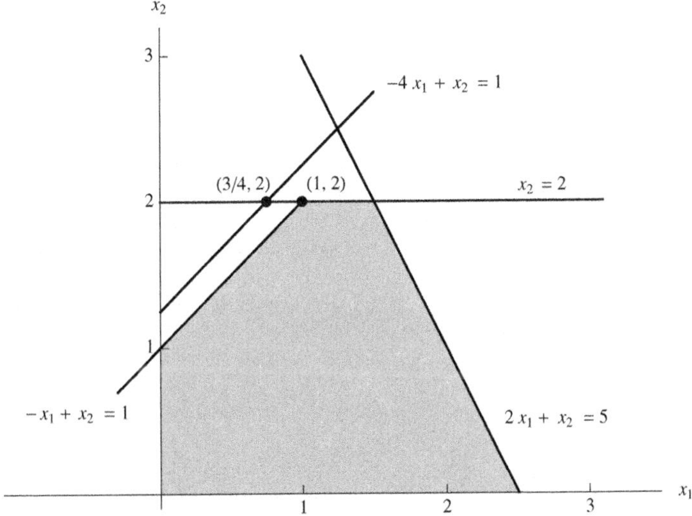

Figure 6.8

Problem Set 6.3

1. Solve each of the following problems using the Cutting Plane Algorithm and then sketch the graph of the original feasible region and the cutting plane. (In each, it should be necessary to add only one additional constraint before reaching an optimal integral solution.)

   (a) Minimize $x_1 - x_2$
   subject to
   $3x_1 + 4x_2 \leq 6$
   $x_1 - x_2 \leq 1$
   $x_1, x_2 \geq 0$ and integral

   (b) Maximize $x_1 + 2x_2$
   subject to
   $x_1 + 3x_2 \leq 13$
   $2x_1 - x_2 \leq 6$
   $x_1, x_2 \geq 0$ and integral

2. Solve the following using the Cutting Plane Algorithm.

   (a) Maximize $x_1 + 3x_2$
   subject to
   $-x_1 + 3x_2 \leq 6$
   $2x_1 + x_2 \leq 12$
   $x_1, x_2 \geq 0$ and integral

   (b) Maximize $3x_1 + 8x_2$
   subject to
   $x_1 + 2x_2 \leq 9$
   $2x_2 \leq 5$
   $x_1, x_2 \geq 0$ and integral

   (c) Maximize $2x_1 - 4x_2 + x_3$
   subject to
   $x_1 - x_2 \leq 12$
   $2x_2 + 3x_3 \leq 28$
   $x_1, x_2, x_3 \geq 0$ and integral

   (d) Maximize $2x_1 - 4x_2 + x_3$
   subject to
   $x_1 - x_2 = 12$
   $2x_2 + 3x_3 = 28$
   $x_1, x_2, x_3 \geq 0$ and integral

3. Is it possible, after several iterations of the Cutting Plane Algorithm, to arrive at an optimal solution to the corresponding linear programming problem with the property that only the slack variables defined by the appended constraints assume nonintegral values and all the variables of the original problem assume integer values?

4. What do you suppose would happen if the Cutting Plane Algorithm were applied to an integer programming problem with the property that the corresponding linear programming problem had feasible nonintegral solutions but no feasible integral solutions?

5. Test your answer to Problem 4 on the following:

   Maximize $x_1 + 2x_2$
   subject to

   $$\begin{aligned} 3x_1 + 3x_2 &\leq 2 \\ 3x_1 \quad\quad &\geq 1 \end{aligned}$$

   $x_1, x_2 \geq 0$ and integral

6. Solve the following problem using the Cutting Plane Algorithm:

   Maximize $x_1 + 2x_2 + x_3$
   subject to

   $$\begin{aligned} x_1 + 4x_2 + 2x_3 &\leq 7 \\ -x_1 + 3x_2 \quad\quad &\geq 4 \end{aligned}$$

   $x_1, x_2, x_3 \geq 0$ and integral

6.4 A Branch and Bound Algorithm

In the previous section, we developed a version of Gomory's Cutting Plane Algorithm. In this section, we will demonstrate the basic idea underlying another important class of algorithms, *branch and bound algorithms*, used to solve integer programming problems. These algorithms originate from the work of Land and Doig [17], published in 1960, and the version we demonstrate is a modification by Dakin [18]. Branch and bound algorithms, along with their refinements and extensions, form a constructive set of solution techniques for integer programming problems.

Given an integer programming problem, the first step in the branch and bound approach is to ignore the integral restrictions and solve the corresponding linear programming problem. If this problem does not have an integral optimal solution, then, as in the Cutting Plane Algorithm, new constraints are generated to cut off this optimal nonintegral solution. But here, instead of expanding the original problem by the addition of a single constraint, we create two distinct problems, each coming about by the addition of a new constraint to the original set of constraints.

These two new constraints are generated from the nonintegral optimal solution to the original problem as follows. Select a variable, say x_j, that assumes a nonintegral value in this optimal solution. Suppose $x_j = b_i$ in this solution. Then the two new problems are created by adding to the original constraint set for one problem the constraint $x_j \leq [b_i]$ and for the other the constraint $x_j \geq [b_i]+1$. For example, if in the original problem the optimal solution has the variable $x_3 = 8\frac{1}{4}$, the constraint sets for the two new problems would contain the original set of constraints plus, for one, the constraint $x_3 \leq 8$ and, for the other, the constraint $x_3 \geq 9$.

Note that the original nonintegral optimal solution is not a feasible solution to either new problem, but that any integral feasible solution to the original problem would be a solution to one of these new problems. However, in contrast to the Cutting Plane Algorithm, there are now two integer programming problems to deal with, and the integral optimal solution to the original problem could be contained in either problem. We continue, considering the two new problems just as before. For each, we initially ignore the integral restrictions, solve, and, if the problem has a nonintegral optimal solution, we again branch from that problem, formulating two new problems using the above method.

It may seem that with this branching process we are compounding our difficulties by continually expanding the set of problems to be solved. However, this is not quite the case for two reasons. First, some of the newly formed problems may have no feasible solutions as a result of the increased restrictions from the additional constraints. Second, some of these problems may have integral optimal solutions. Such a solution would certainly satisfy the constraints of the original problem and would provide a bound for the optimal value to the original integer programming problem. This bound would allow us to eliminate from consideration any problems generated through the branching process from a problem with an optimal value not better than this bound. By making use of such arguments based on bounds, we can eliminate problems to consider and will eventually be left with the optimal integral solution to the original problem.

We illustrate with an example.

Example 6.4.1. Consider the integer programming problem of Example 6.3.2. The problem is to

$$\begin{aligned}
&\text{Minimize } x_1 - 2x_2\\
&\text{subject to}\\
&\quad 2x_1 + \ \ x_2 \leq 5\\
&-4x_1 + 4x_2 \leq 5\\
&x_1, x_2 \geq 0 \text{ and integral}
\end{aligned}$$

The solution found in the previous section to the problem with the integral restrictions ignored is $\text{Min}\, z = -3\frac{3}{4}$, attained at $x_1 = 1\frac{1}{4}$, $x_2 = 2\frac{1}{2}$. Using the branch and bound algorithm, we formulate two new problems here by restricting either the x_1 or the x_2 variable. We arbitrarily select the x_1 variable and use the constraints $x_1 \leq 1$ and $x_1 \geq 2$ to form two new problems. We denote this branching process in Figure 6.9.

Box 1 in the figure corresponds to the original problem: its solution, with the integral restrictions ignored, is inside. Boxes 2A and 2B correspond to the two new problems, with the new constraints indicated on the branches leading to the respective boxes. The solutions to these problems, again with the integral restrictions ig-

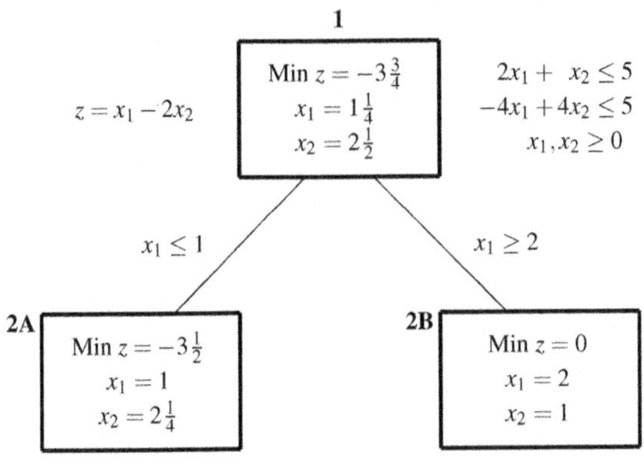

Figure 6.9

nored, could be determined by using the Dual Simplex Algorithm in conjunction with the final tableau solution to the original problem, because in each we have simply added a constraint to a completed problem. The solutions are listed in the appropriate boxes. (Finding these solutions can require some effort. But note that LP Assistant does allow for the duplication and expansion of an existing (final) tableau, a useful tool when implementing a branch and bound algorithm.) Problem 2A has an optimal solution of Min $z = -3\frac{1}{2}$ at $x_1 = 1, x_2 = 2\frac{1}{4}$. The optimal value for this problem is greater than the optimal value for Problem 1, as would be expected, because we have added a constraint and therefore reduced the solution set of feasible points on which to minimize the objective function. Problem 2B provides a feasible integral solution $x_1 = 2, x_2 = 1$ to the constraints of the original problem and an upper bound of 0 for the optimal value of the integrally restricted problem. Since better integral solutions may be contained in Problem 2A, we branch again off Problem 2A using the x_2 variable, as illustrated in Figure 6.10.

Problem 3B, to

$$
\begin{aligned}
&\text{Minimize } x_1 - 2x_2 \\
&\text{subject to} \\
&\quad 2x_1 + \ \ x_2 \leq 5 \\
&-4x_1 + 4x_2 \leq 5 \\
&\quad\ x_1 \qquad\ \leq 1 \\
&\qquad\qquad x_2 \geq 3 \\
&x_1, x_2 \geq 0 \text{ and integral}
\end{aligned}
$$

has no feasible solutions, as the reader may verify (consider the bounds on x_1 and x_2 and the second constraint), and so this branch terminates. However, we must continue by branching at Problem 3A. See Figure 6.11.

Problem 4A yields the integral feasible solution of $x_1 = 1, x_2 = 2$ with the optimal value of $z = -3$. The optimal value for Problem 4B is $-2\frac{1}{2}$, and so further branching here can lead only to solutions with a value greater than $-2\frac{1}{2}$. Since we already have an integral solution to the constraints of the original problem at which the value

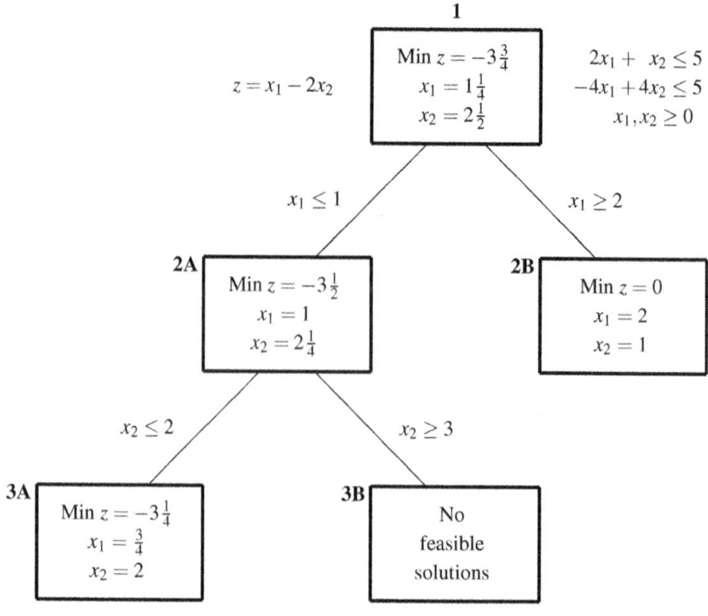

Figure 6.10

of the objective function is -3, there is no need to branch at Problem 4B. No other problems remain to be considered, so we are done. The original integer programming problem must have an optimal value of -3 attained at $x_1 = 1$, $x_2 = 2$, the optimal solution to Problem 4A. Note that here, in fact, we could have ceased calculations once the solution to Problem 4A had been found. Because the objective function has integral coefficients, it follows from the optimal solution to Problem 1 that the smallest value the objective function can possibly attain at a feasible integral solution is -3, and the solution to Problem 4A provides a lattice point at which this value is attained.

In using this branch and bound algorithm, there are three possible reasons for not continuing a branch at a particular problem. The problem under consideration may have no feasible solutions, it may have an optimal integral solution, or bounds from previously determined integral solutions may render further consideration of the problem unnecessary. In fact, one advantage of this algorithm over the Cutting Plane Algorithm in the previous section is that this algorithm generates feasible integral solutions to the constraints of the problem as it proceeds, along with estimates on how close to the optimal value these solutions might be. Thus, in a large and complicated problem, if we are unable to complete the problem using the branch and bound algorithm, maybe because of limited computer capacity, the algorithm may still provide an integral feasible solution not necessarily optimal but adequate for our purposes.

In the quest for increased efficiency of the algorithm, questions such as what problem to examine next if two or more branches remain open and what variable to

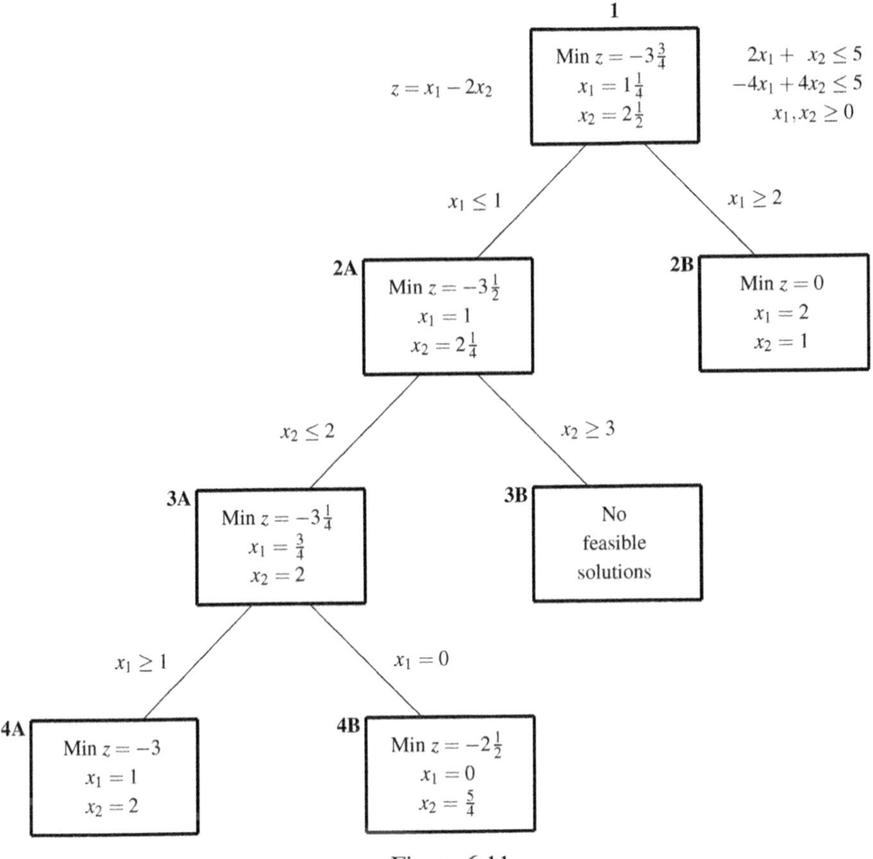

Figure 6.11

use when defining the new constraints if more than one variable takes on nonintegral values in the optimal solution to the problem at hand must be considered. There are no universally accepted answers to these questions. The rules to follow when selecting the variable to use in defining the new constraints can be rather involved. When selecting the problem from which to branch, one possible rule is to select the problem with the most favorable optimal value; another frequently used rule is to select the problem most recently generated. (See Example 6.4.3 for an application of this first rule.)

We conclude this section with two more examples.

Example 6.4.2. Consider the problem of

$$\begin{aligned}
&\text{Maximizing } z = 8x_1 + 15x_2 \\
&\text{subject to} \\
&10x_1 + 21x_2 \leq 156 \\
&\ \ 2x_1 + \quad x_2 \leq 22 \\
&x_1, x_2 \geq 0 \text{ and integral}
\end{aligned}$$

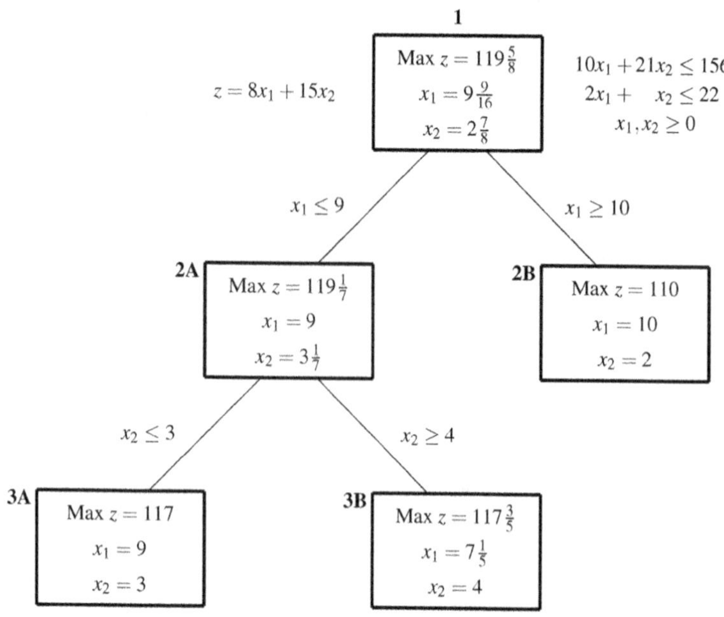

Figure 6.12

Figure 6.12 contains the completed branch and bound diagram. Upon solving the original problem, Problem 1 in the diagram, we know that the optimal value of z restricted to integral feasible solutions is at most 119. Arbitrarily selecting the x_1 variable, we create Problems 2A and 2B. Problem 2B provides a feasible integral solution to the original constraints and a lower bound of 110 for the final maximal value of z.

Restrictions on x_2 lead to Problems 3A and 3B from Problem 2A. The integral solution to Problem 3A yields the improved lower bound of 117 for the optimal value of z. The optimal value for Problem 3B exceeds 117, but only by a fraction, and so the value of z at any integral solution to the constraints of Problem 3B cannot exceed 117. Thus the algorithm terminates. The optimal value for the objective function of the integrally restricted problem is 117, and one point at which this value is attained is $x_1 = 9, x_2 = 3$.

Example 6.4.3. Consider the problem of Example 6.1.1. The problem is to

$$\begin{aligned}
&\text{Maximize } z = 3x_1 + 13x_2 \\
&\text{subject to} \\
&\ \ 2x_1 + 9x_2 \leq 40 \\
&11x_1 - 8x_2 \leq 82 \\
&x_1, x_2 \geq 0 \text{ and integral}
\end{aligned}$$

Figure 6.13 contains the completed branch and bound diagram. Note that after branching from Problem 1 to Problems 2A and 2B and solving these problems, we

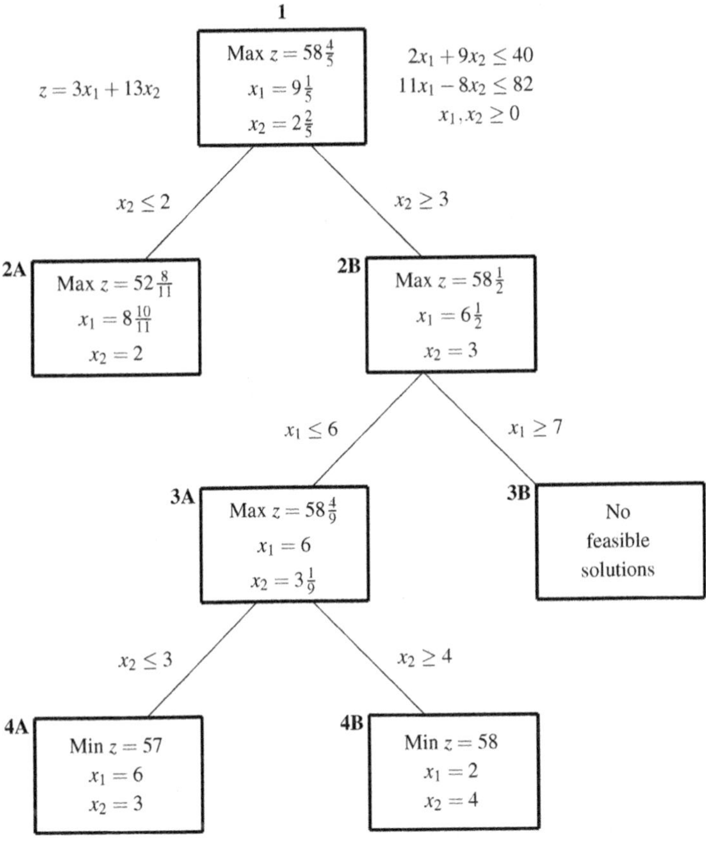

Figure 6.13

have an option of which problem to investigate. Here, however, the choice seems obvious. The maximal value of the objective function under the constraints of Problem 2B is $58\frac{1}{2}$, but under the constraints of Problem 2A it is only $52\frac{8}{11}$. Thus, at this time, we work from Problem 2B and hold Problem 2A in abeyance for future consideration. But then the optimal value for z of 58 in Problem 4B, attained at the feasible integral solution $x_1 = 2, x_2 = 4$, makes further consideration of Problem 2A unnecessary, and the algorithm terminates with this solution as the optimal solution to the integer programming problem.

Problem Set 6.4

1. Solve the integer programming problems of the following using the branch and bound algorithm.

   (a) Section 6.1, Problem 1.
   (b) Section 6.1, Problem 2.
   (c) Section 6.3, Problem 2(a).

(d) Section 6.3, Problem 2(c).
(e) Section 6.3, Problem 6.

2. Solve the following using the branch and bound approach.

(a) Maximize $4x_1 + 5x_2 + 3x_3$
subject to
$$3x_1 - 2x_2 + x_3 \le 15$$
$$x_1 + 2x_2 + x_3 \le 8$$
$x_1, x_2, x_3 \ge 0$ and integral

(b) Maximize $9x_1 + 2x_2 + 3x_3$
subject to
$$x_1 + x_2 - x_3 \le 5$$
$$2x_1 - x_2 + 3x_3 \le 8$$
$x_1, x_2, x_3 \ge 0$ and integral

(c) Maximize $32x_1 + 21x_2 + 12x_3$
subject to
$$3x_1 + 7x_2 + 3x_3 \le 14$$
$$9x_1 + 5x_2 + 3x_3 \le 37$$
$x_1, x_2, x_3 \ge 0$
x_1, x_2 integral

3. Construct a flow chart for the branch and bound algorithm.

6.5 Spreadsheet Solution of an Integer Programming Problem

Microsoft Excel's Solver uses a branch and bound method for integer programming problems. We illustrate it in action with several examples.

Example 6.5.1. Company Zeta of Example 3.10.1 on page 115 produces Products 1, 2, and 3 combining units of the component materials A, B, C, and D. An optimal production schedule was determined for the given month using Solver. The solution appears in Figure 3.5 on page 116.

Now, 1 month later, Company Zeta is faced with the same problem of optimizing profits. But the available supplies of the component materials for this month have changed; the four quantities have been rounded off to the nearest 100. The associated mathematical model now is to

	A	B	C	D	E	F
1	**Company Zeta**					
2				**Product**		
3		Component Material	**1**	**2**	**3**	Supply (units)
4		A	16	30	28	1600
5		B	24	40	36	2000
6		C	30	50	32	2400
7		D	10	20	15	1000
8		**Profit** ($/unit)	$78	$136	$104	
9						
10				**Variables**		
11		Product	**1**	**2**	**3**	
12		Units to Be Made	0	43.08	7.69	
13						
14	**Linear Programming Solution**					
15		Maximize Profit	$6,658			
16						
17		**Constraints**	LHS		RHS	
18		Component A	1507.69	≤	1600	
19		Component B	2000	≤	2000	
20		Component C	2400	≤	2400	
21		Component D	976.92	≤	1000	

Figure 6.14

$$\text{Maximize profit } z \text{ (in \$)}, \; z = 78x_1 + 136x_2 + 104x_3 \tag{6.5.1}$$

subject to

$$16x_1 + 30x_2 + 28x_3 \le 1600$$
$$24x_1 + 40x_2 + 36x_3 \le 2000$$
$$30x_1 + 50x_2 + 32x_3 \le 2400$$
$$10x_1 + 20x_2 + 15x_3 \le 1000$$
$$x_1, x_2, x_3 \ge 0$$

Using Solver on this linear programming problem generates the solution in Figure 6.14, but here the optimal production schedule of $(0, 43.08, 7.69)$ is not integral and thus is not practical. An integral solution is required.

To achieve this, the variables in Solver's "constraints" frame are set to be *integer* (*int*), and in the "options" frame the tolerance is set to 0. (The tolerance is the percentage by which the value of the objective function at a feasible integral solution can differ from the optimal value for the corresponding unrestricted linear programming problem and still be acceptable.)

Solver now generates a viable optimal production schedule (see Figure 6.15). Maximum profit this month is $6642, attained by producing 3 units of Product 1, 41 units of Product 2, and 8 units of Product 3. There remain surplus units of materials A, C, and D. (There is no sensitivity report available for integer programming problems, as there is no single last tableau — just think of the branch and bound displays for the elementary problems in the previous section.)

	A	B	C	D	E	F
1	**Company Zeta**					
2				**Product**		
3		Component Material	**1**	**2**	**3**	Supply (units)
4		A	16	30	28	1600
5		B	24	40	36	2000
6		C	30	50	32	2400
7		D	10	20	15	1000
8		**Profit** ($/unit)	$78	$136	$104	
9						
10				**Variables**		
11		Product	**1**	**2**	**3**	
12		Units to Be Made	3	41	8	
13						
14	**Integer Programming Solution**					
15		Maximize Profit	$6,642			
16						
17		**Constraints**	LHS		RHS	
18		Component A	1502	≤	1600	
19		Component B	2000	≤	2000	
20		Component C	2396	≤	2400	
21		Component D	970	≤	1000	

Figure 6.15

Example 6.5.2. Consider the problem of (6.5.1), except that now fixed charges are to be included in the model. In particular, assume that a fixed cost of $75 is assessed if any units of Product 1 are produced; likewise, any units of Products 2 or 3 produced have fixed costs of $350 and $120, respectively.

To formulate a mathematical model, we define binary variables

$$y_i = \begin{cases} 0, & x_i = 0 \\ 1, & x_i > 0 \end{cases}, \quad i = 1,2,3$$

and establish upper bounds for the x_i's. We have

$$\text{Max}\, x_1 = 80 = \text{Min}\{1600/16, 2000/24, 2400/30, 1000/10\},$$

and similarly $\text{Max}\, x_2 = 48$ and $\text{Max}\, x_3 = 56$. Using these variables and bounds, our problem here is to

Maximize profit z (in \$), (6.5.2)

$$z = 78x_1 + 136x_2 + 104x_3 - 75y_1 - 350y_2 - 120y_3$$

subject to

$$16x_1 + 30x_2 + 28x_3 \leq 1600$$
$$24x_1 + 40x_2 + 36x_3 \leq 2000$$
$$30x_1 + 50x_2 + 32x_3 \leq 2400$$
$$10x_1 + 20x_2 + 15x_3 \leq 1000$$
$$y_1 \geq \frac{x_1}{80}, \quad y_2 \geq \frac{x_2}{48}, \quad y_3 \geq \frac{x_3}{56} \quad \left(\text{or simply } y_i \geq \frac{x_i}{80},\ i = 1,2,3\right)$$
$$x_1, x_2, x_3 \geq 0, \quad y_1, y_2, y_3 \text{ binary}$$

	A	B	C	D	E	F
1	**Company Zeta** (w/ fixed charges)					
2				**Product**		
3		Component Material	**1**	**2**	**3**	Supply (units)
4		A	16	30	28	1600
5		B	24	40	36	2000
6		C	30	50	32	2400
7		D	10	20	15	1000
8		**Profit** ($/unit)	$78	$136	$104	
9						
10		Fixed Charge If Used	$75	$350	$120	
11		Bound on Units Produced	80	48	56	
12						
13				**Variables**		
14		Product	**1**	**2**	**3**	
15		Units to Be Made	0	48	0	
16		1 If Units Are Made, Else 0	0	1	0	<-- binary y's
17						
18	**Integer Programming Resoultion**					
19		Maximize Profit	$6,178			
20						
21		**Constraints**	LHS		RHS	
22		Component A	1440	≤	1600	
23		Component B	1920	≤	2000	
24		Component C	2400	≤	2400	
25		Component D	960	≤	1000	
26		y1 ≥ x1/max	0	≥	0	
27		y2 ≥ x2/max	1	≥	1	
28		y3 ≥ x3/max	0	≥	0	

Figure 6.16

The spreadsheet for the model is displayed in Figure 6.16. (The y_i's, the variables of cells C16:E16, are designated as *binary* (*bin*) in Solver's "constraints" frame.) Now only Product 2 is utilized. Producing 48 units realizes an income of $6528 and a $350 fixed charge for a net profit of $6178.

Interestingly, if the fixed charge for Product 3 production is reduced by $5 from $120 to $115, the optimal production schedule changes dramatically: from producing 48 units of Product 2 to producing 71 units of Product 1 and 8 units of Product 3, as the reader with Solver at hand may verify (see Problem 1). Of course, there is no reason to expect continuity in the optimal solution point of an integer programming problem.

Example 6.5.3. In Problem 19 of Section 6.2, a microbrewery that can brew up to three different beers, a light, a dark, and an ale, needs to determine what beers, and how many batches of each, to make in order to maximize weekly net revenue. Production is restricted by the limited resources of grain and hops. There are also restrictions on batch size and the number of types of beer brewed, as well as costs involving overtime and equipment conversion.

To formulate a mathematical model, we define the variables:

$$x_1 = \text{number of batches of beer of type 1 (light) brewed}$$
$$x_2 = \text{number of batches of beer of type 2 (dark) brewed and}$$
$$x_3 = \text{number of batches of beer of type 3 (ale) brewed}$$
$$y_i = \begin{cases} 1, & x_i > 0 \\ 0, & x_i = 0 \end{cases} \quad \text{for } i = 1,2,3$$
$$v = \text{the number of batches brewed with overtime}$$

The mircobrewery's problem, then, in terms of these variables, is to

Maximize profit z (in \$),

$$\begin{aligned} z = \; & 395x_1 + 440x_2 + 475x_3 \\ & - [250(1-y_3) + 150(1-y_2) + 225(1-y_1) + 95v] \end{aligned}$$

subject to

$$\begin{aligned}
&25x_1 + 40x_2 + 30x_3 \le 800 && \text{(grain restriction)}\\
&10x_1 + 8x_2 + 16x_3 \le 250 && \text{(hops restriction)}\\
&x_1 + x_2 + x_3 \le 20 + v,\ v \le 4 && \text{(batch and overtime restriction)}\\
&5y_1 \le x_1 \le 25y_1 &&\\
&5y_2 \le x_2 \le 20y_2 && \text{(y_i's on/off; minimum batch size)}\\
&5y_3 \le x_3 \le 15y_3 &&\\
&y_1 + y_2 + y_3 = 2 && \text{(number of types)}\\
&x_1, x_2, x_3 \ge 0 \text{ and integral, } y_1, y_2, y_3 \text{ binary} &&
\end{aligned}$$

A spreadsheet representation of the problem, and Solver's solution, is in Figure 6.17. Optimal net profit is \$9580, attained by brewing 13 batches of dark beer and 9 batches of ale. Two batches are brewed using overtime.

We conclude with a toast to Solver.

Problem Set 6.5

1. (a) Change the coefficient of y_3 in the objective function of (6.5.2) from 120 to 115, run Solver on the modified problem, and verify that $(71,0,8)$ is the optimal production schedule, as discussed in the last paragraph of Example 6.5.2. By how much does the optimal profit increase over the profit in the original fixed charge problem?

   (b) On your spreadsheet for the modified problem of part (a), set the six variable cells equal to 0 and the tolerance in Solver's "options" frame to 5 and run Solver. Do you get the correct answer?

	A	B	C	D	E	F
1	Microbrewery					
2				Beer Type		Available
3		**Requirements+Revenue**	1 (Light)	2 (Dark)	3 (Ale)	Supply (lb)
4		Grain (lb)	25	40	30	800
5		Hops (lb)	10	8	16	250
6		Revenue/Batch	$395	$440	$475	
7		(Max Batches Possible)	25	20	15	
8						
9			Light & Dark	Light & Ale	Dark & Ale	
10		**Prep./Conv. Cost**	$250	$150	$225	
11						
12		**Other Restrictions**	Regular	Overtime		
13		Labor Available (hr)	20	4		
14		Min Batch Size	5			
15		Overtime per Batch	$95			
16		Beers Brewed ≤	2			
17						
18				**Variables**		
19		Beer Type	1 (Light)	2 (Dark)	3 (Ale)	
20		Number of Batches Made	0	13	9	
21		1 If Batch Is Made, Else 0	0	1	1	
22		(Batches Using Overtime)	2	0	0	
23						
24		Maximize Profit	$9,580			
25						
26		**Constraints**	LHS		RHS	
27		Grain (lb)	790	≤	800	
28		Hops (lb)	248	≤	250	
29		Overtime Batches	2	≤	2	
30		Overtime Bound	2	≤	4	
31		x1 ≤ 25 y1	0	≤	0	
32		x2 ≤ 20 y2	13	≤	20	
33		x3 ≤ 15 y3	9	≤	15	
34		5 y1 ≤ x1	0	≤	0	
35		5 y2 ≤ x2	5	≤	13	
36		5 y3 ≤ x3	5	≤	9	
37		Number of Beers Brewed	2	=	2	

Figure 6.17

For each of the following, formulate a mathematical model for the problem, and then solve the problem using Solver.

2. Using Feeds 1, 2, and 3, a blend which contains at least 1375 units of component A and 1800 units of component B is desired; the data are as follows:

Components	*Feeds (units/lb)* *1*	*2*	*3*	*Units Required*
A	5	10	20	1375
B	8	6	24	1800
Cost (cents/lb)	60	80	200	

(a) Verify that a minimal-cost blend is attained by using 75 lb of Feed 1 and 50 lb of Feed 3 and costs $145.

(b) For each of the following, determine a minimal-cost blend with the stated restrictions added (the parts are independent).

 (i) Feed 1 can be bought only in 20-lb bags; Feed 2 in 10-lb bags; Feed 3 in 12-lb bags.

 (ii) There is a change in cost: the 60 cents/lb cost of Feed 1 is valid only for the first 40 lb; the cost thereafter is 45 cents/lb.

 (iii) The cost changes again: cost of the first 40 lb of Feed 1 is 60 cents/lb and that of the next 50 lb is 45 cents/lb; cost thereafter is 30 cents/lb.

3. Problem 9 of Set 6.2

4. (a) Problem 12(a) of Set 6.2
 (b) Problem 12(b) of Set 6.2

5. Problem 17(a) of Set 6.2

CHAPTER 7

THE TRANSPORTATION PROBLEM

7.1 A DISTRIBUTION PROBLEM

In Section 2.4 we formulated a mathematical model for the standard transportation problem, the problem of determining a minimal-cost shipping schedule between sources and destinations. The model is a linear programming problem, and so the simplex method of Chapter 3 can be used to solve it. However, one would have a formidable linear programming problem in trying to determine an optimal shipping schedule between, say, 100 warehouses and 300 retail outlets. There would be 30,000 variables and 400 constraints. Fortunately, other more efficient algorithms exist for solving the transportation problem – algorithms that can solve large-scale problems with only moderate computer effort. We will develop one such algorithm in Section 7.2. Then, in Section 7.3, we will consider other optimization models to which we can apply the transportation problem algorithm.

In this section, we will develop an algorithm to solve the following *distribution problem*. Suppose a single commodity is produced in varying amounts at a set of plants or origins and is in demand at a set of markets or destinations, with varying demands at the different destinations. To meet the demands, the commodity must be shipped through links connecting the sources with the destinations, but there is an upper limit on the amount of the commodity that can be shipped through each link. Is it possible to meet the demands of the destinations with the supplies at the sources using a shipping schedule that does not exceed the capacities of the links? Although this problem is of interest in its own right, the computational technique we develop here will be used in the transportation problem algorithm we develop in Section 7.2, and it is primarily for this reason that we consider the problem.

Specifically, suppose there are m origins or plants and n destinations or markets. Let a_i denote the supply at the ith origin, b_j the demand at the jth destination, and k_{ij} the maximum amount that can be shipped from the i-th origin to the jth destination, where k_{ij} is a nonnegative integer, $1 \leq i \leq m$, $1 \leq j \leq n$. Letting x_{ij} denote the amount to be shipped from the ith origin to the jth destination, we want to determine if there exists a shipping flow $\{x_{ij}\}$ satisfying

An Introduction to Linear Programming and Game Theory, Third Edition. By P. R. Thie and G. E. Keough.

$$\sum_{j=1}^{n} x_{ij} \leq a_i, \text{for each } i, 1 \leq i \leq m$$
$$\sum_{i=1}^{m} x_{ij} \geq b_j, \text{for each } j, 1 \leq j \leq n$$
$$0 \leq x_{ij} \leq k_{ij}, \text{for } 1 \leq i \leq m, 1 \leq j \leq n$$

and if such a flow exists, a method for finding it. The first set of inequalities requires that the flow out of each origin does not exceed its supply, and the second set requires that the demand at each destination is met. Clearly, to have a *feasible solution* to the problem, we must have

$$\sum_{i=1}^{n} a_i \geq \sum_{j=1}^{m} b_j$$

Note that this distribution problem is not an optimization problem; there is no function to be optimized. The question to be answered here is whether a feasible flow exists, and if so, how to find it. The approach we use to answer the question is straightforward. We begin at the first origin and assign flow values x_{ij} within the link capacities and meeting the demands, as much as possible, at the destinations. Then we move on to the next origin, and so on. If the initial assignments do not meet all the demands, and they probably will not, then we simply rearrange the flow, "returning" units to origins from destinations that can be supplied by surplus from other origins, and then shipping these returned units to points where there is still demand. We illustrate with a simple example.

Example 7.1.1. Consider the distribution problem with three origins, four destinations, and the following data:

	Destinations				
	1	*2*	*3*	*4*	*Origin Supplies*
Origin 1	4	1	5	1	5
Origin 2	2	2	6	5	7
Origin 3	5	1	4	0	7
Destination Demands	3	3	9	4	

Here the entries in the right column represent the supplies available at each origin, the entries in the bottom row the demands at each destination, and the entry in the main body of the table for Origin i and Destination j the shipping capacity k_{ij}.

Throughout this chapter, we propose using a more compact data presentation for such a problem by eliminating the labeling; thus we will summarize the relevant data for supplies, demands, and shipping capacities as

4	1	5	1	5
2	2	6	5	7
5	1	4	0	7
3	3	9	4	

In this problem, the total supply equals the total demand, equal to 19, and so it is possible that feasible solutions exist.

Starting with the first row and moving from left to right, we ship units to the destinations, the amounts shipped limited only by the destination demands, the link capacities, and the supply. Thus, from the first origin we send 3 units to the first destination, 1 to the second, and 1 to the third. We now expand the columns of our previously reduced table to include these initial x_{ij} values alongside the corresponding k_{ij} as follows:

4	3	1	1	5	1	1		5
2		2		6		5		7
5		1		4		0		7
3		3		9		4		

We continue in this manner, shipping the 7 units from the second row and 3 units from the third, again keeping in mind that the amounts shipped are limited by the destination demands, link capacities, and supply available. This brings us to the following situation:

4	3	1	1	5	1	1		5	
2		2	2	6	5	5		7	
5		1		4	3	0		7	*S*
3		3		9		4	4		

In this simple manner we have constructed an initial shipping flow. However, the flow is not feasible, since a demand for 4 units remains in the fourth column. We denote this unmet demand with the auxiliary entry in the bottom row. This demand cannot be met directly with the surplus from the third row, the only row with a surplus. We label this row with an "*S*."

We now begin the more complicated part of the solution process, readjusting this flow in an attempt to increase it. For example, 1 unit of the demand for 3 units in the second column could be met by the third row. This would free the 1 unit presently assigned to the second column from the first row, and this unit could then be reassigned to the fourth column. We illustrate as follows:

4	3	1	~~1~~0	5	1	1	~~0~~1	5	
2		2	2	6	5	5		7	
5		1	~~0~~1	4	3	0		7	*S*
3		3		9		4	~~4~~3		

Table 7.1

4	3	1	0	5	1	1	1	5	
2		2	2	6	~~5~~4	5	~~0~~1	7	
5		1	1	4	~~3~~4	0		7	S
3		3		9		4	~~3~~2		

Table 7.2

4	~~3~~1	1		5	~~1~~3	1	1	5	
2		2	2	6	~~4~~2	5	~~1~~3	7	
5	~~0~~2	1	1	4	4	0		7	~~S~~
3		3		9		4	~~2~~0		

Similarly, we increase the flow to the fourth column through the third row, the third column, and the second row. See Table 7.1.

Unfortunately, we still have an unmet demand for 2 units in the fourth column and a surplus of 2 units in the third row. However, the 2 units in the third row can be shipped only to the first column, since all the other links from this source are used to full capacity. Shipping these units to the first column frees 2 units in the first row. Although these units cannot be shipped directly to the fourth column, they can be shipped to the third column, freeing 2 units from the second row that can be used to meet the fourth-column demand. Thus we have the following chain:

$$\text{row } 3 \rightarrow \text{column } 1 \rightarrow \text{row } 1 \rightarrow \text{column } 3 \rightarrow \text{row } 2 \rightarrow \text{column } 4$$

With this adjustment, illustrated in Table 7.2, we have constructed a feasible shipping schedule, and the problem is solved.

In larger problems, the construction of such chains that enable the flow to be readjusted and thereby increased might be rather complicated. Thus we will introduce a *labeling scheme* for the rows and columns of the table to facilitate the finding and tracing back of these chains. When units can be shipped to a column, we will record in that column the row number from which the units can come. Similarly, if a row has units that can be reassigned, we will record in that row the column number that allows the reassignment. Then, if we reach a column with an unmet demand, we can use these numbers to trace the chain back to the source with a surplus.

We illustrate this procedure with the above example. Consider the problem with the flow given in Table 7.1 (see Table 7.3). The surplus in row 3 can be sent to column 1, and so we place a 3 at the bottom of column 1, indicating the source of

Table 7.3

4	3	1		5	1	1	1	5	
2		2	2	6	4	5	1	7	
5		1	1	4	4	0		7	S
3		3		9		4	2		

Table 7.4

4	3	1		5	1	1	1	5	1
2		2	2	6	4	5	1	7	
5		1	1	4	4	0		7	S
3		3		9		4	2		
3									

Table 7.5

4	3	1		5	1	1	1	5	1
2		2	2	6	4	5	1	7	2 or 3
5		1	1	4	4	0		7	S
3		3		9		4	2		
3		1		1		2			

supply. Continuing, column 1 is receiving units from row 1, so these units in row 1 can be reassigned. We label row 1 with a 1, indicating that from this row, units can be reassigned, and that this is possible because of a readjustment in column 1. The augmented table is illustrated in Table 7.4.

Now the units of row 1 could be sent to columns 2 or 3, so we append columns 2 and 3 with a 1. From either of these columns row 2 units can be reassigned, so we append row 2 with a 2 or 3. But now column 4 could be sent reassigned units from row 2, so column 4 is labeled with a 2. See Table 7.5. We have labeled a column with an unmet demand, so we readjust and increase the flow. We can use the row and column labels to trace back from column 4 the route of supply.

$$\text{column } 4 \leftarrow \text{row } 2 \leftarrow \text{columns } 2 \text{ or } 3 \leftarrow \text{row } 1 \leftarrow \text{column } 1 \leftarrow \text{row } 3$$

If we use column 3, we see that we can adjust the flow by 2 units, and thus we can solve the problem, as previously done.

The numbers used in the labeling scheme play a double role. The fact that a row is labeled indicates that there are free units in that row that can be assigned to any column for which the corresponding shipping link is not being used to full capacity. And the source of these free units is the column number used as the label. Similarly, a labeled column indicates that any row shipping units to that column can reassign these units elsewhere, and the alternate row source for this column is the number used as the label. This scheme is a compromise between simplicity and completeness. We could introduce into the scheme a second index indicating at each step the maximum number of units that could flow to that point through the chain already constructed. For our textbook problems, such an index is not necessary. However, if this algorithm were to be programmed to handle large-scale problems, this second index should be incorporated into the program in order to systematize the determination of the flow increase and the readjustment of the flow. For a discussion of this expanded scheme, see Ford and Fulkerson [19].

We summarize the steps of the *distribution problem algorithm.* Consider a distribution problem with m sources, n destinations, supplies a_i, demands b_j, capacities k_{ij}, and $\sum a_i \geq \sum b_j$.

1. Construct an initial flow $\{x_{ij}\}$ by shipping as much as possible without exceeding demands or capacities first from row 1, then row 2, and so on.
2. Calculate the unmet demand $b_j - \sum_i x_{ij}$ for each column. If this is zero for all columns, we have a feasible flow and the problem is solved.
3. Otherwise, initiate the labeling procedure.
   (a) Label all rows with surplus units, that is, rows with $\sum_j x_{ij} < a_i$, with an "S." Let I denote this set of rows.
   (b) For each $i \in I$, determine all unlabeled columns j for which $x_{ij} < k_{ij}$. Label these columns with the corresponding row number $i \in I$. Let J denote this set of columns.
   (c) For each $j \in J$, determine all unlabeled rows i for which $x_{ij} > 0$. Label these rows with the corresponding column number $j \in J$. Let I denote this new set of labeled rows. Return to Step 3(b).
4. Continue this labeling procedure, moving from rows to columns to rows to columns, and so on, until either:
   (a) In Step 3(b), a column with an unmet demand is labeled. Then increase the flow into this column by readjusting the flow values x_{ij}. Erase all labels and return to Step 2.

   or

   (b) In either Step 3(b) or 3(c), no previously unlabeled column or row, respectively, is labeled. Then the problem has no feasible solution.

Before discussing the convergence of the algorithm and the claim of nonfeasibility made in Step 4(b), we illustrate the algorithm with two additional examples.

Example 7.1.2. Consider the distribution problem having the following data:

5	0	4	4	0	6
4	4	4	0	1	9
3	3	4	5	7	6
7	5	0	4	3	17
8	5	6	8	10	

The total supply of 38 exceeds the total demand for 37, so we begin the algorithm by constructing the initial flow and determining the unmet demands. See Table 7.6. At this point, there are various routes that can be used to ship units from the surplus in row 4 to help meet the unmet demands in columns 4 and 5. For example, we can ship 2 units through row 4 $\to$ column 1 $\to$ row 1 $\to$ column 4 (meeting completely the unmet demand in column 4), 1 unit through row 4 $\to$ column 1 $\to$ row 2 $\to$ column 5, and 1 unit through row 4 $\to$ column 2 $\to$ row 3 $\to$ column 4. We record these changes in Table 7.7, and since an unmet demand remains in column 5 (and since

Table 7.6

5	5	0		4	1	4		0		6	
4	3	4	4	4	2	0		1		9	
3		3	1	4	3	5	2	7		6	
7		5		0		4	4	3	3	17	S
8		5		6		8	2	10	7		

Table 7.7

5	~~5~~3	0		4	1	4	~~0~~2	0		6	1
4	~~3~~2	4	4	4	2	0		1	~~0~~1	9	1 or 2
3		3	~~1~~0	4	3	5	2	7	~~0~~1	6	3 or 4
7	~~0~~3	5	~~0~~1	0		4	4	3	3	17	S
8		5		6		8		10	5		
4		4		1 or 2		1		3			

Table 7.8

5	~~3~~0	0		4	~~1~~4	4	2	0		6	3
4	2	4	4	4	2	0		1	1	9	1 or 2
3		3		4	~~3~~0	5	2	7	~~1~~4	6	4
7	~~3~~6	5	1	0		4	4	3	3	17	S
8		5		6		8		10	2		
4		4		2		1		3			

there are no other "short" routes from row 4 to column 5), we initiate the labeling procedure. Column 5 is labeled, and working backward, we can construct the chain

$$\text{column } 5 \leftarrow \text{row } 3 \leftarrow \text{column } 3 \leftarrow \text{row } 1 \leftarrow \text{column } 1 \leftarrow \text{row } 4$$

A maximum of 3 units can be shipped through this chain, the limit coming from $x_{11} = k_{13} - x_{13} = x_{33} = 3$. The adjusted flow is represented in Table 7.8. (Other chains from row 4 to column 5 could be constructed from Table 7.7, such as

$$\text{column } 5 \leftarrow \text{row } 3 \leftarrow \text{column } 3 \leftarrow \text{row } 2 \leftarrow \text{column } 2 \leftarrow \text{row } 4,$$

with capacity $2 = k_{23} - x_{23}$. When several routes exist, augmenting the flow with one of greater capacity may lead to a quicker solution of the problem.)

The flow of Table 7.8 leaves an unmet demand for 2 units in column 5, and so we again label. Several chains from the surplus to column 5 can be constructed. In particular, the route

$$\text{column } 5 \leftarrow \text{row } 3 \leftarrow \text{column } 4 \leftarrow \text{row } 1 \leftarrow \text{column } 3 \leftarrow \text{row } 2 \leftarrow \text{column } 2 \leftarrow \text{row } 4$$

allows a flow increase of 2 units. The resulting feasible flow is shown in Table 7.9.

Table 7.9

5		0		4	~~4~~2	4	~~2~~4	0		6
4	2	4	~~4~~2	4	~~2~~4	0		1	1	9
3		3		4		5	~~2~~0	7	~~4~~6	6
7	6	5	~~1~~3	0		4	4	3	3	17
8		5		6		8		10		

Example 7.1.3. Consider the following distribution problem.

4	3	1	4	0	8
5	3	2	0	4	6
3	0	1	10	0	13
0	2	4	6	4	5
6	6	6	9	5	

Supply equals demand, and so we construct the initial flow, determine unmet demands, and label (Table 7.10). Thus 2 units can flow through row 3 → column 1 → row 2 → column 4. Recording this adjustment, deleting old labels, and labeling anew, we are led to Table 7.11. Now, however, we reach a point where Step 4(b) of the algorithm applies. The surplus in row 3 can be sent to columns 1 or 4, and so the units assigned to column 1 from row 1 can be reassigned. But the only link out of row 1 not being used to capacity is to column 4, an already labeled column. Thus our labeling procedure terminates and, since we have been unable to label the

Table 7.10

4	4	3	3	1	1	4		0		8	1
5	2	3	3	2	1	0		4		6	1
3		0		1	1	10	9	0		13	S
0		2		4	3	6		4	2	5	
6		6		6		9		5	3		
3				2		3		2			

Table 7.11

4	4	3	3	1	1	4		0		8	1
5	0	3	3	2	1	0		4	2	6	
3	2	0		1	1	10	9	0		13	S
0		2		4	3	6		4	2	5	
6		6		6		9		5	1		
3						3					

column with the unmet demand, we can conclude that the problem has no feasible solution.

Actually, we do not need to apply the algorithm to demonstrate that this problem has no feasible solution. Consider the total demand of columns 2, 3, and 5 (the unlabeled columns), that is, 17. The demand can be partially met with the supply from rows 2 and 4 (the unlabeled rows). However, the total supply of these two rows is 11, and so, in any feasible solution to the problem, at least $6 = 17 - 11$ units must flow from rows 1 and 3 (the labeled rows) to columns 2, 3, and 5. But the sum of the capacities of the six links connecting these two rows with these three columns is only $5 = 3+0+1+1+0+0$. Thus there can be no feasible solutions.

It is precisely the generalization of this argument that we will use to prove the nonfeasibility claim of Step 4(b) of the algorithm. Clearly, however, the rows and columns to consider will come from the labeling procedure. Before we do this, we will prove a theorem stated primarily for its application to the transportation problem algorithm of the next section. The theoretical questions of convergence and nonfeasibility of the distribution problem algorithm will then be discussed.

Theorem 7.1.1. *Suppose the distribution problem algorithm is applied to a distribution problem, generating a flow $\{x_{ij}\}$, but that, with this flow, Step 4(b) of the algorithm is reached. Let R denote the set of all the labeled rows and C the set of all the labeled columns. Then*

(a) $i \in R$, $j \notin C$ implies that $x_{ij} = k_{ij}$.

(b) For any $j \in C$, $x_{ij} > 0$ implies that $i \in R$.

(c) $\sum_{j \in C} b_j < \sum_{i \in R} a_i$.

Proof. To prove part (a), suppose that $i \in R$. Then either row i has a surplus or units in row i can be reassigned. If $x_{ij} < k_{ij}$, this shipping link could be used to send units from row i to column j, and j would be in C. Thus $j \notin C$ implies that $x_{ij} = k_{ij}$.

For part (b), let $j \in C$. Then column j has an alternate source of supply, and units assigned to column j can be reassigned. Thus, if $x_{ij} > 0$, row i has such units and so $i \in R$.

Finally, for any $j \in C$, $\sum_{i=1}^{m} x_{ij} = b_j$; otherwise, Step 4(a) of the algorithm would have been implemented. Therefore

$$\begin{aligned}
\sum_{j \in C} b_j &= \sum_{j \in C}\left(\sum_{i=1}^{m} x_{ij}\right) = \sum_{j \in C}\left(\sum_{i \in R} x_{ij}\right) \qquad \text{(from part (b))} \\
&= \sum_{i \in R}\left(\sum_{j \in C} x_{ij}\right) \leq \sum_{i \in R}\left(\sum_{j \in C} x_{ij}\right) + \sum_{i \in R}\left(\sum_{j \notin C} x_{ij}\right) \\
&= \sum_{i \in R}\left(\sum_{j=1}^{n} x_{ij}\right)
\end{aligned}$$

But for any i, $\sum_{j=1}^{n} x_{ij} \leq a_i$, and there is at least one row in R with a surplus, that is, at least one $i \in R$ with $\sum_{j=1}^{n} x_{ij} < a_i$. Therefore $\sum_{i \in R}(\sum_{j=1}^{n} x_{ij}) < \sum_{i \in R} a_i$. □

Theorem 7.1.2. *Let R and C be as defined in Theorem 7.1.1, and let R' and C' denote their complements. Then*

$$\sum_{j \in C'} b_j > \sum_{i \in R'} a_i + \sum_{j \in C'} \left(\sum_{i \in R} k_{ij} \right)$$

That is, the total demand in the C' columns is strictly greater than the total supply in the R' rows plus the sum of the capacities of the links from the remaining rows to the C' columns.

Proof. For any j, $\sum_{i=1}^{m} x_{ij} \leq b_j$, and for at least one $j \in C'$, $\sum_{i=1}^{m} x_{ij} < b_j$. Also, from part (b) of Theorem 7.1.1, $i \in R'$ and $j \in C$ implies that $x_{ij} = 0$. Thus

$$\begin{aligned}
\sum_{j \in C'} b_j > \sum_{j \in C'} \left(\sum_{i=1}^{m} x_{ij} \right) &= \sum_{j \in C'} \sum_{i \in R'} x_{ij} + \sum_{j \in C'} \sum_{i \in R} x_{ij} \\
&= \sum_{i \in R'} \sum_{j \in C'} x_{ij} + \sum_{j \in C'} \sum_{i \in R} k_{ij} \\
&= \sum_{i \in R'} \sum_{j=1}^{n} x_{ij} + \sum_{j \in C'} \sum_{i \in R} k_{ij} \\
&= \sum_{i \in R'} a_i \quad + \sum_{j \in C'} \sum_{i \in R} k_{ij}
\end{aligned}$$

since $i \in R'$ implies that row i has no surplus. □

Corollary 7.1.1. *If Step 4(b) is reached in the application of the distribution problem algorithm, the associated problem has no feasible solutions.*

Corollary 7.1.2. *If the distribution problem algorithm is applied to a distribution problem with integral supplies, demands, and capacities, the solution process must terminate in a finite number of iterations, either at Step 2 with the construction of a feasible integral solution or at Step 4(b) with the determination of nonfeasibility.*

Proof. Notice first that the initial flow and all subsequent modified flows will be integral. In fact, every readjustment of the flow must increase the total flow by at least 1 unit. Since the total demand is finite, the algorithm must therefore terminate after a finite number of iterations. □

Problem Set 7.1

1. For each of the following distribution problems, determine either a feasible shipping schedule or the set of rows R' and columns C' of Theorem 7.1.2, and verify that these rows and columns satisfy the inequality of that theorem. (A capacity of ∞ means that there is no limit on the number of units that can be shipped through the corresponding link.)

(a)

20	5	8	0	12
15	1	8	4	23
2	12	10	5	15
10	13	21	6	

(b)

3	2	0	1	5	4
1	0	3	2	2	6
1	2	3	3	0	7
0	1	4	4	1	8
4	3	6	5	6	

(c)

8	0	5	2	10
6	18	10	4	20
10	15	3	20	22
10	1	20	2	28
14	26	18	22	

(d)

∞	0	∞	0	0	∞	31
0	∞	0	∞	∞	0	23
0	∞	∞	0	∞	∞	17
∞	0	0	∞	0	0	29
21	12	24	9	18	15	

(e)

∞	∞	0	∞	0	∞	31
∞	0	∞	∞	∞	∞	23
0	∞	∞	0	∞	0	17
∞	∞	0	∞	0	0	29
21	12	24	9	18	15	

(f)

4	0	10	10	20	15
8	7	10	8	15	20
10	9	0	6	20	40
5	10	15	20	22	

(g)

5	0	6	2	1	7
2	4	5	0	2	9
8	7	2	3	2	17
3	0	8	3	9	9
6	5	13	5	12	

(h)

∞	∞	0	0	∞	∞	10
∞	∞	∞	∞	0	0	5
0	0	∞	∞	0	∞	14
∞	∞	0	∞	∞	∞	5
4	3	6	7	12	2	

(i)

10	8	10	20	8	20
10	4	15	10	5	30
10	20	4	12	10	40
20	2	25	5	3	50
15	25	35	45	20	

(j)

∞	∞	0	∞	0	0	9
∞	∞	∞	0	∞	∞	10
∞	∞	0	∞	0	0	7
0	∞	∞	∞	∞	∞	12
5	6	8	4	6	9	

(k)

∞	0	∞	∞	0	∞	∞	75
0	∞	0	∞	0	∞	0	103
∞	0	0	0	∞	0	∞	86
∞	∞	0	∞	0	∞	0	200
0	∞	∞	0	∞	0	∞	136
91	43	112	54	120	75	95	

(l)

10	10	1	0	4	8	8	23
3	4	9	15	0	6	5	19
0	5	5	10	5	5	5	18
10	5	10	0	3	3	0	30
8	10	5	4	2	2	2	26
21	14	10	20	12	19	20	

2. Suppose Step 4(b) is reached in the application of the distribution problem algorithm to a problem for which each k_{ij} is equal to either 0 or ∞. Let R and C be the labeled rows and columns. Explain why $i \in R$ and $j \notin C$ implies that $k_{ij} = 0$.

3. Suppose there is a column j of a distribution problem for which $\sum_{i=1}^{m} k_{ij} < b_j$. Then the problem has no feasible solution. Determine sets R' and C' for which the inequality of Theorem 7.1.2 applies.

4. Prove that the distribution problem algorithm will terminate after a finite number of steps if applied to a problem with all data rational and not necessarily integral.

5. Consider the following assignment problem. Suppose there are n jobs to be assigned to n individuals, but each individual is capable of doing only some of the jobs. The problem is to determine if there is an assignment of individuals to jobs so that all individuals are assigned jobs for which they are qualified.
   (a) Formulate this assignment problem as a distribution problem. (*Hint.* Associate individuals with sources and jobs with destinations, and let each of the a_i's and b_j's equal 1.)
   (b) Apply the distribution problem algorithm to determine if proper assignments can be made for the following assignment problems. (An "x" in the ijth entry indicates that individual I_i is qualified for job J_j.)

(i)

	J_1	J_2	J_3	J_4	J_5	J_6	J_7	J_8
I_1	x	x					x	
I_2		x		x				x
I_3			x					x
I_4			x			x	x	
I_5	x			x	x			
I_6	x					x		
I_7		x		x	x		x	
I_8	x					x		

(ii)

	J_1	J_2	J_3	J_4	J_5	J_6	J_7	J_8
I_1	x			x		x		
I_2		x	x			x		x
I_3				x		x		
I_4	x		x		x		x	
I_5	x	x		x	x		x	x
I_6	x					x		
I_7		x	x		x			x
I_8	x			x		x		

(c) For any set I of individuals, let $J(I)$ denote the set of all jobs J_j for which at least one individual in I is qualified. Prove that a proper assignment of individuals and jobs exists if and only if for every set I, the number of elements in $J(I)$ is greater than or equal to the number of elements in I. (*Hint*. To prove that the condition is sufficient, use Corollary 7.1.2, Theorem 7.1.1, and Problem 2.)

6. Determine a way of using the distribution problem algorithm to solve the problem of assigning to all individuals jobs for which they are qualified as described in Problem 5, but under the assumption that there are more jobs than individuals.

7.2 The Transportation Problem

The transportation problem is one of determining a minimal-cost shipping schedule of a commodity between plants or sources or origins and markets or distribution centers or destinations. Specifically, suppose that there are m origins and n destinations, and that a_i units of the commodity are available at the ith origin and a demand for b_j units is to be met at the jth destination. Suppose the cost of shipping a unit from origin i to destination j is c_{ij}. Then a mathematical model of the problem is to

$$\begin{aligned}
&\text{Minimize} \sum_{i=1}^{m}\sum_{j=1}^{n} c_{ij}x_{ij} \\
&\text{subject to} \\
&\sum_{j=1}^{n} x_{ij} \le a_i, i = 1,\ldots,m \qquad (7.2.1)\\
&\sum_{i=1}^{m} x_{ij} = b_j, j = 1,\ldots,n \\
&x_{ij} \ge 0, 1 \le i \le m, 1 \le j \le n
\end{aligned}$$

where x_{ij} denotes the number of units to be shipped from origin i to destination j.

Clearly, this problem would have no feasible solutions if the total supply $\sum_i a_i$ were less than the total demand $\sum_j b_j$. In fact, in the rest of this section, we will assume that $\sum_i a_i = \sum_j b_j$ and, with this assumption, the above problem is equivalent to the problem of

$$\text{Minimizing} \sum_{i=1}^{m}\sum_{j=1}^{n} c_{ij}x_{ij}$$

subject to

$$\sum_{j=1}^{n} x_{ij} = a_i, i = 1,\ldots,m \tag{7.2.2}$$

$$\sum_{i=1}^{m} x_{ij} = b_j, j = 1,\ldots,n$$

$$x_{ij} \geq 0$$

We lose no generality here by assuming equality of supply and demand; if we encounter a problem with $\sum_i a_i > \sum_j b_j$, we can simply create an additional destination, with demand $\sum_i a_i - \sum_j b_j$ and shipping costs of zero from each origin. The formulation of this expanded problem would then have total supply equal to total demand, and any solution to the problem would give a solution to the original problem, with the interpretation that all units scheduled to be sent to the additional destination are surplus units that would remain at their respective origins. We emphasize here in passing that with the assumption that $\sum_i a_i = \sum_j b_j$, since there are no capacity restrictions on the shipping links, all of our transportation problems have feasible solutions, and the problem is solely to determine a minimal-cost feasible solution. This is in contrast to the distribution problem of the previous section, where the existence of feasible solutions to the problem was the primary consideration.

The algorithm developed in this section to solve the transportation problem of (7.2.2) is due essentially to Ford and Fulkerson ([20] or [19]). It is based on the dual to the problem of (7.2.2). Recall that in each iteration of the standard simplex algorithm developed in Chapter 3, we moved from feasible solution to feasible solution of the problem, attempting to improve the value of the objective function at each step. In the algorithm that we develop now, in each iteration we will move from feasible solution to feasible solution of the dual problem to (7.2.2), improving at each step the dual objective function.

(In fact, the algorithm is a *primal-dual algorithm*. At each iteration, associated with this (not necessarily basic) feasible solution to the dual of (7.2.2) will be a flow $\{x_{ij}\}$, a partial solution to the constraints of (7.2.2). While moving from feasible solution to feasible solution of the dual, we will also be building upon these flows. Moreover, at each iteration, the feasible solution to the dual and the associated flow will maintain complementary slackness, that is, an x_{ij} will be allowed to be positive only if the slack in the corresponding dual constraint is zero.)

The dual to the problem of (7.2.2), from the definition developed in Section 4.2, is a maximization problem with $m+n$ unrestricted variables, corresponding to the $m+n$ equality constraints of (7.2.2), and mn ($\leq$) inequality constraints, corresponding to the mn x_{ij}'s. To formulate the problem, we introduce dual variables $u_1,\ldots,u_m$ to associate with the m origin constraints of (7.2.2) and variables $v_1, \dots, v_n$ to associate with the n destination constraints. With these variables the dual of (7.2.2) is to

$$\begin{aligned}
&\text{Maximize} \sum_{i=1}^{m} a_i u_i + \sum_{j=1}^{n} b_j v_j \\
&\text{subject to} \qquad\qquad\qquad\qquad\qquad\qquad\qquad\qquad (7.2.3) \\
&u_i + v_j \leq c_{ij}, 1 \leq i \leq m, 1 \leq j \leq n \\
&u_i, v_j \text{ unrestricted}
\end{aligned}$$

The mn constraints of (7.2.3) are especially simple in form. Each x_{ij} of (7.2.2) appears with coefficient 1 in only two constraints of (7.2.2), the ith origin constraint and the jth destination constraint.

Example 7.2.1. Consider the transportation problem with two origins, three destinations, and data given as follows:

3	5	7	30
4	7	11	50
20	25	35	

As in Section 7.1, the entries in the right column represent the supplies and the entries of the bottom row the demands but now, the ij-th entry in the body of the table is the cost c_{ij}. Supply equals demand, and the associated linear programming problem is

$$\begin{aligned}
&\text{Minimizing } 3x_{11} + 5x_{12} + 7x_{13} + 4x_{21} + 7x_{22} + 11x_{23} \\
&\text{subject to} \\
&x_{11} + x_{12} + x_{13} = 30 \\
&x_{21} + x_{22} + x_{23} = 50 \\
&x_{11} + x_{21} = 20 \\
&x_{12} + x_{22} = 25 \\
&x_{13} + x_{23} = 35 \\
&x_{ij} \geq 0
\end{aligned}$$

The dual is to

$$\begin{aligned}
&\text{Maximize } 30u_1 + 50u_2 + 20v_1 + 25v_2 + 35v_3 \\
&\text{subject to} \\
&u_1 + v_1 \leq 3 \\
&u_1 + v_2 \leq 5 \\
&u_1 + v_3 \leq 7 \\
&u_2 + v_1 \leq 4 \\
&u_2 + v_2 \leq 7 \\
&u_2 + v_3 \leq 11 \\
&u_1, u_2, v_1, v_2, v_3 \text{ unrestricted}
\end{aligned}$$

The fundamental step in the solution algorithm to the transportation problem of (7.2.2) involves a distribution problem. In the algorithm, we start with a feasible solution $\{u_i, v_j\}$ to (7.2.3), the dual of (7.2.2). Then a distribution problem between the given origins and destinations is defined as follows: supplies and demands as given in the original transportation problem, and capacities k_{ij} defined to be either infinite or 0, depending on whether $u_i + v_j = c_{ij}$ or $u_i + v_j < c_{ij}$. If this distribution problem has a feasible solution, such a solution is also a minimal-cost solution to the transportation problem, as we will prove in Theorem 7.2.2 later in this section. If the distribution problem is not feasible, a better solution to the dual problem (7.2.3) is constructed by changing the values of some of the u_i's and v_j's. The variables with value changes correspond to the labeled rows and columns of Theorem 7.1.1 of the previous section. With this new solution to the dual, we begin again, considering the associated distribution problem, and so on. We outline the steps of the algorithm.

The Transportation Problem Algorithm

Consider a transportation problem as in (7.2.2) with m origins, n destinations, supplies a_i, demands b_j, transportation costs c_{ij}, and total supply $\sum_i a_i$ equal to total demand $\sum_j b_j$.

1. Construct an initial solution to the dual problem (7.2.3) by defining

$$u_i = \operatorname*{Min}_j \{c_{ij}\}, i = 1, \ldots, m$$

and then

$$v_j = \operatorname*{Min}_i \{c_{ij} - u_j\}, j = 1, \ldots, n$$

2. Associate with $\{u_i, v_j\}$ a distribution problem: origins and their supplies and destinations and their demands as in the transportation problem, and link capacities

$$k_{ij} = \begin{cases} \infty, & u_i + v_j = c_{ij} \\ 0, & u_i + v_j < c_{ij} \end{cases}$$

3. Attempt to solve this distribution problem.
   (a) If there are feasible solutions to the distribution problem, then any such solution is a minimal-cost shipping schedule for the original transportation problem.
   (b) If the distribution problem has no feasible solutions, determine the labeled rows R and columns C of Theorem 7.1.1.
4. Define a new solution $\{u'_i, v'_j\}$ to the dual. Let

$$d = \operatorname*{Min}_{\substack{i \in R \\ j \notin C}} \{c_{ij} - (u_i + v_j)\}$$

and define

$$u'_i = \begin{cases} u_i + d, & i \in R \\ u_i, & i \notin R \end{cases} \quad \text{and} \quad v'_j = \begin{cases} v_j - d, & j \in C \\ v_j, & j \notin C \end{cases}$$

Return to Step 2, now using this solution $\{u'_i, v'_j\}$ to the dual.

Before discussing convergence and proving the claims of the algorithm, namely, that in Step 3(a), a feasible solution to the associated distribution problem will be a minimal-cost solution to the transportation problem, and that in Step 4, a better feasible solution to the dual is defined, we illustrate the algorithm with two examples. The table that we use for recording the steps of the algorithm will be an expansion on the table suggested in Example 7.2.1. To record the present values of the dual variables, we introduce a new left-hand column for the u_i values and a new top row for the v_j values. Since the capacities for the associated distribution problem can be only 0 or ∞, we can denote the existence of a shipping link from an origin i to a destination j by simply placing a circle next to the cost c_{ij} in the main body of the table. Then the x_{ij} values calculated when attempting to solve the corresponding distribution problem can be placed in these circles.

Example 7.2.2 (Continuation of Example 7.2.1). The data are

3	5	7	30
4	7	11	50
20	25	35	

We first construct an initial solution to the dual, letting $u_1 = 3$, the minimal cost in the first row, and $u_2 = 4$, the minimal cost in the second row. Since we need $u_i + v_j \leq c_{ij}$ for all i and j, the largest value v_1 could have is

$$\text{Min}\{c_{11} - u_1, c_{21} - u_2\} = \text{Min}\{0, 0\} = 0$$

and similarly

$$v_2 = \text{Min}\{5 - 3, 7 - 4\} = 2$$
$$v_3 = \text{Min}\{7 - 3, 11 - 4\} = 4$$

See Table 7.12. Next, we indicate with circles the usable shipping links for the associated distribution problem, that is, those entries for which $u_i + v_j = c_{ij}$ (Table 7.13). Note that in attempting to solve this distribution problem, the only data from the tableau that are used are the supplies 30 and 50, the demands 20, 25, and 35, and the presence of the circles. This particular distribution problem is not feasible, but a partial solution found by using the algorithm in the previous section is indicated in Table 7.14. There is a surplus in the second row, and these units can be sent to the

Table 7.12

	$v_1 = 0$	$v_2 = 2$	$v_3 = 4$	
$u_1 = 3$	3	5	7	30
$u_2 = 4$	4	7	11	50
	20	25	35	

Table 7.13

	0	2	4	
3	3 ○	5 ○	7 ○	30
4	4 ○	7	11	50
	20	25	35	

Table 7.14

	0	2	4		
3	3 ○	5 (25)	7 (5)	30	
4	4 (20)	7	11	50	S
	20	25	35 30		
	2				

Table 7.15

	~~0~~ −1	2	4	
3	3	5	7	30
5 ~~4~~	4	7	11	50
	20	25	35	

first column only (as $k_{22} = k_{23} = 0$ and $x_{21} = 20 < \infty = k_{21}$). But the first column is not receiving units from any other row, and so the flow cannot be increased. Thus row 2 is the labeled row and column 1 is the labeled column; that is, $R = \{2\}$ and $C = \{1\}$.

To construct a better solution to the dual, calculate

$$d = \operatorname*{Min}_{\substack{i=2\\ j=2,3}} \{c_{ij} - (u_i + v_j)\}$$
$$= \text{Min}\{7 - (2+4), 11 - (4+4)\} = 1$$

and so increase u_2 by 1 and decrease v_1 by 1, leaving u_1, v_2, and v_3 unchanged. See Table 7.15. Now we proceed just as before, circling the existing shipping links and attempting to solve the associated distribution problem (Table 7.16). Again the distribution problem is not feasible, and we are led to the sets $R = \{2\}$, corresponding to the row with surplus, and $C = \{1,2\}$, corresponding to the columns from which units could be reassigned. Note that in the application of the transportation problem algorithm, when Step 3(b) is encountered, not only do we determine that the associated distribution problem is not feasible, but we must also determine the sets of labeled rows R and columns C to which Theorem 7.1.1 applies.

Table 7.16

	−1	2	4		
3	3	5 ( )	7 (30)	30	
5	4 (20)	7 (25)	11	50	S
	20	25	35 5		
	2	2			

Table 7.17

	~~−1~~ −3	~~2~~ 0	4	
3	3	5	7	30
7 ~~5~~	4	7	11	50
	20	25	35	

Table 7.18

	−3	0	4		
3	3	5	7 (30)	30	
7	4 (20)	7 (25)	11 (5)	50	S
	20	25	35		

For the problem of Table 7.16,

$$d = \underset{\substack{i=2 \\ j=3}}{\text{Min}}\{c_{ij} - (u_i + v_j)\}$$

$$= \text{Min}\{11 - (4+5)\} = 2$$

and so we increase u_2 by 2, decrease v_1 and v_2 by 2, and leave u_1 and v_3 unchanged (Table 7.17). Circling and considering the associated distribution problem, we have Table 7.18.

But now the distribution problem is feasible, and so the indicated solution, $x_{11} = x_{12} = 0$, $x_{13} = 30$, $x_{21} = 20$, $x_{22} = 25$, $x_{23} = 5$, is a minimal-cost shipping schedule for the original transportation problem. The actual shipping cost would be $520 = 7 \cdot 30 + 4 \cdot 20 + 7 \cdot 25 + 11 \cdot 5$.

Example 7.2.3. Consider the transportation problem with four origins, six destinations, and data as follows:

7	6	5	8	7	8	16
2	5	6	7	4	6	12
2	2	1	3	3	1	10
1	3	4	3	2	5	18
5	11	3	13	7	17	

Constructing the initial solution to the dual and circling to indicate the associated distribution problem, we have Table 7.19. Attempting to solve the distribution problem leads to Table 7.20. From that table, we have $R = \{1,2\}$, $C = \{1,2,3\}$, and $d = 1$. We record the new solution to the dual and the associated distribution problem in Table 7.21.

Notice that while we have gained three new shipping links (row 1 to columns 4 and 5 and row 2 to column 5), we have lost three from the previous distribution problem (row 3 to columns 2 and 3 and row 4 to column 1). However, none of these lost links were being used in the final attempted solution to the first distribution problem (and this is always the case – see Problem 13), and so, in attempting to solve the new distribution problem, we need not start with Step 1 in the distribution problem algorithm of the previous section but, instead, can use the flow already constructed from the first iteration as the initial flow here.

Table 7.19

	0	1	0	2	1	0	
5	7	6 ○	5 ○	8	7	8	16
2	2 ○	5	6	7	4	6	12
1	2	2 ○	1 ○	3 ○	3	1 ○	10
1	1 ○	3	4	3 ○	2 ○	5	18
	5	11	3	13	7	17	

Table 7.20

	0	1	0	2	1	0		
5	7	6 (11)	5 (3)	8	7	8	16	S
2	2 (5)	5	6	7	4	6	12	S
1	2	2 ○	1 ○	3 ○	3	1 (10)	10	
1	1 ○	3	4	3 (13)	2 (5)	5	18	
	5	11	3	13	7 2	17 7		
	2	1	1					

Table 7.21

	~~0~~ −1	~~1~~0	~~0~~ −1	2	1	0	
6~~5~~	7	6 ○	5 ○	8 ○	7 ○	8	16
3~~2~~	2 ○	5	6	7	4 ○	6	12
1	2	2	1	3 ○	3	1 ○	10
1	1	3	4	3 ○	2 ○	5	18
	5	11	3	13	7	17	

Table 7.22

	−1	0	−1	2	1	0		
6	7	6 (11)	5 (3)	8 (2)	7 ○	8	16	4
3	2 (5)	5	6	7	4 (7)	6	12	5
1	2	2	1	3 ○	3	1 (10)	10	
1	1	3	4	3 (11)	2 ○	5	18	S
	5	11	3	13	7	17 7		
	2	1	1	4	4			

Table 7.23

	~~−1~~ −3	~~0~~ −2	~~−1~~ −3	~~2~~0	~~1~~ −1	0	
8~~6~~	7	6 (11)	5 (3)	8 (2)	7 ○	8 ○	16
5~~3~~	2 (5)	5	6	7	4 (7)	6	12
1	2	2	1	3	3	1 (10)	10
3~~1~~	1	3	4	3 (11)	2 ○	5	18
	5	11	3	13	7	17	

Building on this flow, we are led to the flow in Table 7.22. Here $R = \{1,2,4\}$, $C = \{1,2,3,4,5\}$, and $d = 2$. Table 7.23 records the new solution to the dual along with the last flow. Using the surplus of the fourth row to increase the flow, we are led to Table 7.24. Here $R = \{2,4\}$, $C = \{1,4,5\}$, and $d = 1$, leading to Table 7.25. The last flow has been retained. But now the flow can be increased and, in fact, the given distribution problem has a feasible solution. Table 7.26, the final table, provides a minimal-cost flow.

We now consider the theoretical questions raised after the statement of the transportation problem algorithm.

Table 7.24

	−3	−2	−3	0	−1	0		
8	7	6 (11)	5 (3)	8 ( )	7 ( )	8 (2)	16	
5	2 (5)	5	6	7	4 (7)	6	12	5
1	2	2	1	3	3	1 (10)	10	
3	1	3	4	3 (13)	2 ( )	5	18	S
	5	11	3	13	7	17 5		
	2			4	4			

Table 7.25

	~~−3~~ −4	−2	−3	~~0~~ −1	~~−1~~ −2	0		
8	7	6 (11)	5 (3)	8	7	8 (2)	16	
6 ~~5~~	2 (5)	5	6	7	4 (7)	6 ( )	12	5
1	2	2	1	3	3	1 (10)	10	
4 ~~3~~	1	3	4	3 (13)	2 ( )	5	18	S
	5	11	3	13	7	17 5		
	2			4	4	2		

Table 7.26

	−4	−2	−3	−1	−2	0	
8	7	6 (11)	5 (3)	8	7	8 (2)	16
6	2 (5)	5	6	7	4 (2)	6 (5)	12
1	2	2	1	3	3	1 (10)	10
4	1	3	4	3 (13)	2 (5)	5	18
	5	11	3	13	7	17	

Theorem 7.2.1. *When Step 4 of the transportation problem algorithm is implemented, the $\{u'_i, v'_j\}$ constructed is a new, feasible solution to the dual problem (7.2.3), and the value of the objective function of (7.2.3) at this point is strictly greater than the value of the function at the $\{u_i, v_j\}$ solution point.*

Proof. First, from Problem 2 of Section 7.1, for all $i \in R$ and $j \notin C$, $c_{ij} > u_i + v_j$, and so $d > 0$. Thus the point $\{u'_i, v'_j\}$ is distinct from the point $\{u_i, v_j\}$.

Next, we can see that $u'_i + v'_j \leq c_{ij}$ for all i and j by considering each of the four possibilities:

(a) If $i \in R$, $j \in C$, $u'_i + v'_j = (u_i + d) + (v_j - d) = u_i + v_j \leq c_{ij}$.

(b) If $i \notin R$, $j \notin C$, $u_i' + v_j' = u_i + v_j \le c_{ij}$.

(c) If $i \notin R$, $j \in C$, $u_i' + v_j' = u_i + (v_j - d) < u_i + v_j \le c_{ij}$.

(d) If $i \in R$, $j \notin C$, $u_i' + v_j' = (u_i + d) + v_j \le c_{ij}$ by the choice of d.

Thus $\{u_i', v_j'\}$ is a feasible solution to (7.2.3).

Finally,

$$\begin{aligned}
\sum_{i=1}^{m} a_i u_i' + \sum_{j=1}^{n} b_j v_j' &= \sum_{i\in R} a_i u_i' + \sum_{i\notin R} a_i u_i' + \sum_{j\in C} b_j v_j' + \sum_{j\notin C} b_j v_j' \\
&= \sum_{i\in R} a_i(u_i + d) + \sum_{i\notin R} a_i u_i + \sum_{j\in C} b_j(v_j - d) + \sum_{j\notin C} b_j v_j \\
&= \sum_{i=1}^{m} a_i u_i + \sum_{j=1}^{n} b_j v_j + d\left(\sum_{i\in R} a_i - \sum_{j\in C} b_j\right) \\
&> \sum_{i=1}^{m} a_i u_i + \sum_{j=1}^{n} b_j v_j
\end{aligned}$$

(This last inequality is guaranteed by Theorem 7.1.1, which showed that $\sum_{i\in R} a_i - \sum_{j\in C} b_j > 0$.) □

Theorem 7.2.2. *Suppose that in the implementation of the transportation problem algorithm, the distribution problem associated with a solution $\{u_i, v_j\}$ of (7.2.3) has a feasible flow solution $\{x_{ij}\}$. Then this flow $\{x_{ij}\}$ is an optimal solution to (7.2.2).*

Proof. Since the problems of (7.2.2) and (7.2.3) are dual, from Corollary 4.4.2, all we need show is that the value of the objective function of (7.2.2) at $\{x_{ij}\}$ equals the value of the objective function of (7.2.3) at $\{u_i, v_j\}$. And, using the fact that x_{ij} can be nonzero only if $c_{ij} = u_i + v_j$, we have

$$\begin{aligned}
\sum_{i=1}^{m}\sum_{j=1}^{n} c_{ij}x_{ij} &= \sum_{i=1}^{m}\sum_{j=1}^{n} (u_i + v_j)x_{ij} \\
&= \sum_{i=1}^{m}\sum_{j=1}^{n} u_i x_{ij} + \sum_{i=1}^{m}\sum_{j=1}^{n} v_j x_{ij} \\
&= \sum_{i=1}^{m} u_i\left(\sum_{j=1}^{n} x_{ij}\right) + \sum_{j=1}^{n} v_j\left(\sum_{i=1}^{m} x_{ij}\right)
\end{aligned}$$

But the flow $\{x_{ij}\}$ is a feasible solution to the distribution problem, and so $\sum_{i=1}^{m} x_{ij} = b_j$ for each j and, since $\sum_{i=1}^{m} a_i = \sum_{j=1}^{n} b_j$, $\sum_{j=1}^{n} x_{ij} = a_i$ for each i. Thus

$$\sum_{i=1}^{m}\sum_{j=1}^{n} c_{ij}x_{ij} = \sum_{i=1}^{m} a_i u_i + \sum_{j=1}^{n} b_j v_j$$ □

Theorem 7.2.3. *Suppose the data for the transportation problem of (7.2.2) are all integral. Then the transportation problem algorithm applied to (7.2.2) will terminate after a finite number of steps to a minimal-cost flow with all integral values.*

Proof. The objective function of (7.2.2), $\sum_i \sum_j c_{ij} x_{ij}$, is bounded below by $c \sum_i a_i$ where $c = \text{Min}_{i,j}\{c_{ij}\}$, and so (7.2.2) and therefore its dual (7.2.3) have finite optimal solutions (Duality Theorem 4.4.2). Now, in each iteration of the transportation problem algorithm, the dual objective function is increased by the quantity $d(\sum_{i \in R} a_i - \sum_{j \in C} b_j)$ (Proof of Theorem 7.2.1), and this quantity must be at least 1 if the data are integral. Since the dual objective function is bounded, the algorithm must therefore eventually terminate at Step 3(a), at a distribution problem with a feasible solution. And this distribution problem will have integral solutions $\{x_{ij}\}$, from Corollary 7.1.2. □

The proof of Theorem 7.2.2 shows that when the transportation problem algorithm terminates, the associated $\{u_i, v_j\}$ point is an optimal solution to the dual problem. This can provide useful information about the original transportation problem.

Example 7.2.4. Consider the transportation problem with three origins, four destinations, and the following data:

13	18	32	27	25
15	20	35	28	25
11	14	29	30	25
10	15	20	25	

The total demand of 70 is five less than the total supply of 75, so before we apply the transportation problem algorithm, we add a fifth column with demand 5 and costs 0.

The four iterations necessary to solve the modified problem are presented in Tables 7.27–7.30. The optimal shipping schedule of Table 7.30 leaves the surplus 5 units at row 2. The minimum shipping cost is

$$1645 = 13 \cdot 10 + 32 \cdot 10 + 27 \cdot 5 + 28 \cdot 20 + 14 \cdot 15 + 29 \cdot 10$$

Table 7.27

	11	14	29	27	0	
0	13	18	32	27 (25)	0 ○	25
0	15	20	35	28	0 (5)	25
0	11 (10)	14 (15)	29 ○	30	0 ○	25
	10	15	20	25	5	

$R = \{2\}, C = \{5\}, d = 1$

Table 7.28

	11	14	29	27	−1	
0	13	18	32	27 (25)	0	25
1	15	20	35	28 ( )	0 (5)	25
0	11 (10)	14 (15)	29 ( )	30	0	25
	10	15	20	25	5	

$R=\{1,2\}, C=\{4,5\}, d=2$

Table 7.29

	11	14	29	25	−3	
2	13 (10)	18	32	27 (15)	0	25
3	15	20	35	28 (10)	0 (5)	25
0	11 (0)	14 (15)	29 (10)	30	0	25
	10	15	20	25	5	

$R=\{1,2\}, C=\{1,4,5\}, d=1$

Table 7.30

	10	14	29	24	−4	
3	13 (10)	18	32 (10)	27 (5)	0	25
4	15	20	35	28 (20)	0 (5)	25
0	11	14 (15)	29 (10)	30	0	25
	10	15	20	25	5	

This is also the value of the dual objective function at the associated dual solution point, that is,

$$1645 = 3\cdot 25 + 4\cdot 25 + 0\cdot 25 + 10\cdot 10 + 14\cdot 15 + 29\cdot 20 + 24\cdot 25 + (-4)5 \quad (7.2.4)$$

In fact, from duality, these calculations certify the correctness of our solution, since the $\{x_{ij}\}$ flow and the $\{u_i, v_j\}$ point of Table 7.30 are feasible solutions to their respective problems. (We do not even need to calculate the values of the corresponding objective functions. The Complementary Slackness Theorem guarantees optimality as long as $u_i + v_j \leq c_{ij}$ for all i and j, the associated flow $\{x_{ij}\}$ is feasible, and the x_{ij}'s are positive only where $u_i + v_j = c_{ij}$.)

The reader may also verify that the flow of the final table is the only feasible flow that can be constructed using the circled links of that table. From this it follows that the optimal flow for this transportation problem is unique. (See Problem 16.)

Now, returning to the original problem, suppose we wish to increase the demand in column 2 by 3 units, from 15 to 18. Obviously, this new problem has feasible solutions. For example, we could modify the flow of Table 7.30 by shipping 3 units from the surplus of 5 in row 2 directly to column 2, at an additional cost of 60. However, the optimal solution to the dual from Table 7.30 suggests that we can do better. We are changing the value of b_2 of that table from 15 to 18 and the value of b_5 from 5 to 2. The net effect of these changes on the dual objective function, evaluated in (7.2.4), would be

$$14\underbrace{(+3)}_{15\to 18} + (-4)\underbrace{(-3)}_{5\to 2} = 54$$

Thus, as long as the $\{u_i, v_j\}$ point remains an optimal solution to the dual, the altered problem has an optimal flow with a cost increase of only 54. (In fact, such a flow exists. To find it, modify the flow of Table 7.30 using the circled links.)

Problem Set 7.2

1. Solve the transportation problems with the following data tables:

(a)

5	6	7	5	12
9	10	9	6	14
3	4	4	2	10
9	8	11	8	

(b)

4	3	5	2	2	15
6	7	5	5	4	15
5	5	7	3	4	10
8	4	6	10	12	

(c)

8	10	9	8	7	8
9	5	4	6	9	10
10	8	5	8	8	12
5	6	4	8	7	

(d)

12	15	21	8	17	50
10	13	20	8	16	60
13	16	25	12	20	75
15	15	19	13	18	40
25	35	45	55	65	

(e)

4	12	6	5	2	10	20
9	13	7	7	3	10	26
2	10	1	3	1	9	28
5	7	5	6	2	5	22
13	17	15	19	21	11	

2. (a) Solve the following transportation problem. (Note that supply exceeds demand, and so before the transportation problem algorithm can be properly applied, a fifth column for the surplus must be added.)

1	4	3	4	10
5	5	5	5	10
4	4	2	1	10
6	6	6	6	

(b) Apply the transportation problem algorithm directly to the data of the above table without first adding a column for the surplus. Compare the cost of the final flow with your answer in part (a).

(c) True or false: The assumption that total supply equals total demand in the transportation problem algorithm is critical.

3. Solve the transportation problems defined by the following tables. ($c_{ij} = \infty$ indicates the impossibility of shipping between the corresponding origin and destination, and $c_{ij} < 0$ indicates the effects of government subsidies.)

(a)

5	6	7	5	15
9	10	9	6	10
3	4	4	2	15
9	8	11	8	

(b)

13	∞	18	-3	9
17	12	∞	-2	11
19	14	23	-4	13
6	7	9	7	

(c)

5	-2	-7	6	4	20
14	1	-6	16	∞	10
∞	-1	0	∞	13	25
12	-2	-9	14	12	20
10	9	15	20	11	

4. Suppose the demand in column 1 in the transportation problem of 3(a) is increased from 9 to $9+\lambda$, for $0 \leq \lambda \leq 4$.

(a) Estimate the corresponding change on the total shipping cost using the solution to the dual from Problem 3(a).
(b) Determine an optimal flow when $\lambda = 2$. What is the increase in cost?
(c) Determine an optimal flow when $\lambda = 3$. What is the increase in cost now?

5. (a) Solve the following transportation problem:

5	6	3	6	25
12	15	10	14	75
3	5	7	4	100
20	40	60	80	

(b) Suppose now the demand at column 3 is to be increased to 65, and to meet it, 5 additional units are to be added to the supply at one of the rows. Estimate to which row the 5 units should be added so that the increase in shipping cost would be minimized, basing your estimate on:
(i) what row is presently supplying column 3
(ii) the costs c_{13}, c_{23}, and c_{33}.
(iii) the optimal solution to the dual.
(c) Determine in fact where the 5 units should be placed so that the increase in cost is minimized.

6. (a) Using the optimal solution to the dual from Table 7.18, estimate the change in shipping costs if, for the transportation problem of Example 7.2.1, the supply in row 1 is increased to $30+\lambda$ and the demand in column 1 is increased to $20+\lambda$.
(b) For what range on $\lambda \geq 0$ is this estimate in fact precise?

7. (a) Solve the following transportation problem:

29	31	32	28	30	50
10	18	35	12	19	75
22	31	37	20	30	100
11	16	34	17	13	125
45	85	55	65	75	

(b) Estimate using the solution to the dual the net increase in shipping cost if the demand in column 5 is increased to 90.
(c) Verify the accuracy of your estimate in part (b).
(d) Solve the problem of part (a) under the assumption that no surplus can remain in row 3. Start with your final table of part (a) after adjusting c_{36} (and v_6).

8. Solve the transportation problem of 3(a) under the added condition that no surplus can remain in row 2.

9. (a) Solve the following transportation problem using the transportation problem algorithm.

5	1	5	3
5	5	1	3
1	5	5	2
2	2	2	

(b) Suppose the demand at column 1 is increased by 1 unit. Estimate, using the above solution to the dual, the effect on the total shipping cost.

(c) Determine the actual change in cost when the demand at column 1 is increased by 1.

(d) Explain.

10. Determine the value of the minimal-cost shipping schedule for the transportation problem given by

(a)

2	12	5
12	2	25
15	15	

(b)

2	12	15
12	2	25
15	25	

Note that in part (b), we are simply shipping 10 more units through the same network. Is this what is meant by the term *economies of scale*?

11. True or false: At each iteration of the transportation problem algorithm, the flow is increased. (*Hint.* Look at Example 7.2.4, or apply the algorithm to the following problem.)

1	5	5	5	1	1
5	1	2	3	4	2
5	5	1	5	5	1
5	5	5	1	5	1
1	1	1	1	1	

12. Prove that under the assumption that $\sum_i a_i = \sum_j b_j$, any feasible solution to the problem of (7.2.1) is also a feasible solution to the problem of (7.2.2); that is, prove that the problems of (7.2.1) and (7.2.2) are equivalent.

13. Prove that when going from Step 4 to Step 2 in the transportation problem algorithm, any shipping link that was being used (i.e., the $x_{ij} > 0$) in the attempted solution to the first distribution problem will remain open in the distribution problem of the next iteration (i.e., the $u'_i + v'_j = c_{ij}$). (*Hint.* Use part (b) of Theorem 7.1.1).

14. Verify for the first two examples in this section that at the final iteration when a distribution problem with feasible solutions is constructed, the value of the cost function $\sum_{i,j} c_{ij}x_{ij}$ at the feasible flow $\{x_{ij}\}$ equals the value of the dual objective function at the corresponding solution point $\{u_i, v_j\}$ to the dual.

15. Prove Theorem 7.2.2 using the Complementary Slackness Theorem.

16. Given a transportation problem, let $\{u_i, v_j\}$ be any optimal feasible solution to its dual, the problem of (7.2.3). Show then that if $\{x_{ij}\}$ is any minimal-cost solution to the transportation problem, $x_{ij} > 0$ implies that $u_i + v_j = c_{ij}$. In other words, suppose the transportation problem algorithm is used to solve a transportation problem. Then any minimal-cost flow solution to the problem, no matter how it is found, can use only the circled links from the final table of the algorithm solution to the problem. (*Hint.* Use the Complementary Slackness Theorem.)

17. True or false: If $\{u_i, v_j\}$ is an optimal solution to the problem of (7.2.3), then so also is the point $\{u_i + \lambda, v_j - \lambda\}$ for any constant λ.

18. Prove that the transportation problem algorithm will lead to an optimal solution after a finite number of steps if the algorithm is applied to a problem with all data rational.

19. Suppose a minimal-cost shipping flow $\{x_{ij}\}$ is determined for a standard transportation problem with total supply equal to total demand. Does this solution remain optimal if

(a) All the costs c_{ij} from an origin i are altered by a fixed amount q?
(b) All the costs c_{ij} to a destination j are altered by a fixed amount q?

Suppose in the initial transportation problem $\sum_i a_i > \sum_j b_j$. Would your answers to the above be the same?

20. Consider Problem 2 of Section 2.4.

(a) Solve the problem using the transportation problem algorithm.
(b) Suppose shipping costs from the East Coast Warehouse to Outlet 3 could be reduced from 30 to $30 - \lambda$ cents/case. How large does λ need to be in order to reduce total shipping costs?
(c) Let λ equal 7 in part (b) and solve the modified problem.

21. Consider Problem 3 of Section 2.4.

(a) Solve the problem using the transportation problem algorithm.
(b) A private shipper offers to transport cases from Plant 1 to Outlet 2. At what cost/case shipped would the beverage company be interested?

22. Solve Problem 5 of Section 2.4.

23. Consider Problem 11 of Section 2.6.

(a) Solve the problem using the transportation problem algorithm.

(b) Suppose the supply at Plant 2 can be increased from 600 to $600+\lambda$ units. Estimate, using the solution to the dual found above, the effect this increase would have on total costs. For what upper bound on λ is this estimate in fact precise?

(c) Solve the original problem (supply at Plant 2 equal to 600), but with the added condition that because of contractual obligations, the firm must buy 300 units from the outside source.

24. Solve each of the following problems using the transportation problem algorithm.

(a) The transportation problem with data given by

5	4	1	3	4	5
4	7	2	5	6	8
6	6	2	4	5	7
6	6	5	4	3	

has total supply of 20 and total demand of 24. Determine a minimal-cost shipping schedule that distributes the entire 20 units among the five destinations so that no demand is exceeded.

(b) As above, but suppose now that there is also a cost of q_j per unit of unmet demand at the jth destination, with $q_1 = 4$, $q_2 = q_4 = 5$, $q_3 = 1$, and $q_5 = 7$.

25. Solve the following problems using the transportation problem algorithm.

(a) Suppose that in the data table in Problem 24(a), the a_i's represent the available supplies, the b_j's represent the number of units that can be sold at the jth destination, and the c_{ij}'s represent the net profit from the sale of 1 unit shipped from the ith origin and sold at the jth destination. How should the supply be distributed so that profit is maximized?

(b) As above, using the data in Problem 1(c).

7.3 APPLICATIONS

In each of the examples we consider in this section, under a suitable interpretation or structural modification, we will be able to formulate the problem at hand as a standard transportation problem (the problem of (7.2.2) on page 265) and therefore solve the problem using the transportation problem algorithm. However, the reader should consider the presentation as primarily a demonstration of how the mathematical model of a particular situation or problem may be adjusted to make use of an already available solution technique. We do not mean to imply that the transportation problem algorithm is the only technique for solving the types of problems that follow. In several of the examples the problems have a special structure, and more efficient algorithms exist to take advantage of that structure. Indeed, the study of

Table 7.31

8	6	5	7	20
12	4	3	9	30
6	7	4	6	50
15	20	25	35	

Table 7.32

8	6	5	7	0	20
12	4	3	9	0	30
6	7	4	6	0	12
6	7	4	∞	0	38
15	20	25	35	5	

network flow problems, of which the transportation problem can be considered an example, is a major area of mathematical programming.

Example 7.3.1 (A Capacitated Transportation Problem)**.** Consider the transportation problem with three origins, four destinations, and the data given in Table 7.31. Suppose also that the shipping link between Origin 3 and Destination 4 has an upper bound capacity of 12 units, that is, at most 12 units can flow from Origin 3 to Destination 4. With this restriction, we cannot apply the transportation problem algorithm directly to the problem; the final optimal flow may (in fact, would) have x_{34} greater than 12.

However, if we divide the supply at Origin 3 into two parts, one part with 12 units and shipping links to all the destinations and the other part with 38 units but no shipping link to Destination 4, the algorithm can be applied. The initial data, so modified, are in Table 7.32. The flow out of Origin 3 corresponds to the entries in rows 3 and 4, and $c_{44} = \infty$ limits Origin 3's flow to Destination 4 to the supply of row 3. The computation of an optimal flow is left as an exercise (Problem 1(a)).

This technique can be extended. Other minor modifications of a simple transportation problem sometimes can be handled by the addition of another row and/or column, with suitable associated costs. Here, for example, if the capacity of the shipping link from Origin 3 to Destination 3 were also restricted to, say, 15 units, we could divide Origin 3's supply of 50 into three parts (see Problem 1(b)). Clearly, however, if all 12 shipping links were capacitated, other, more general solution techniques would be necessary.

Example 7.3.2 (A Transshipment Problem)**.** Suppose two sources supply three destinations, with supplies, demands, and direct shipping costs as follows:

20	30	12	50
15	28	13	100
25	50	75	

Suppose units may also be shipped from the sources to the destinations through two transshipment nodes. A transshipment node has no supply or demand, but units may be shipped from the sources to the destinations through these nodes. (For example, suppose goods can be shipped not only from Washington and New York directly to Portland, Seattle, and Spokane, but also from the East Coast cities to St. Louis and Chicago, and then from St. Louis and Chicago to the West Coast cities.) Suppose also that the number of units that can flow through each transshipment node is bounded (e.g., St. Louis and Chicago have limited transfer and/or storage facilities). Here suppose transshipment Node 1 has a capacity of 60 units and Node 2 a capacity of 70 units, with shipping costs given as follows:

		Nodes	
		1	*2*
Origins	*1*	3	6
	2	4	5

		Destinations		
		1	2	*3*
Nodes	*1*	16	25	8
	2	12	20	6

We wish to determine a minimal-cost shipping schedule.

If the transshipment nodes did not have capacity restrictions, we might simply determine, for each source and destination, the least expensive way of shipping units via the three routes (directly, through Node 1, and through Node 2) and use these data in a standard transportation problem model. However, we can still use the transportation problem algorithm here if we include in both our sources and destinations the two transshipment nodes, setting supplies and demands equal to capacities and allowing no shipping between the two nodes. The following would be the modified data for such an application of the algorithm:

		Nodes		***Destinations***			
		1	*2*	*1*	*2*	*3*	*Supply*
Origins	*1*	3	6	20	30	12	50
	2	4	5	15	28	13	100
Nodes	*1*	0	∞	16	25	8	60
	2	∞	0	12	20	6	70
Demand		60	70	25	50	75	

The first two rows and last three columns correspond to the two sources and three destinations, with the direct flow between them to be noted in the upper right corner of the table. Rows 3 and 4 and columns 1 and 2 provide the auxiliary sources and destinations associated with the transshipment nodes. Entries in the upper left corner of the table would correspond to units shipped from the sources to the nodes, and entries in the lower right corner to units shipped from the nodes to the destinations. These later quantities are properly limited by the row supplies of 60 and 70, and the availability of these 60 and 70 units for the last three columns is contingent upon how much of the 60-unit and 70-unit demands of the first two columns is met by shipment from the two sources through rows 1 and 2. The x_{31} and x_{42} entries would correspond to unused node capacity at Nodes 1 and 2, respectively. Completion of the problem is left to the reader (Problem 2).

Example 7.3.3 (A Dynamic Scheduling Problem). An automobile company makes transmissions at two plants to meet the needs at three of its assembly centers. For the next 2 months, the output at each plant and the requirements at each center are as follows:

	Output		***Requirements***		
	Plant 1	*Plant 2*	*Center 1*	*Center 2*	*Center 3*
Month 1	225	275	150	200	100
Month 2	260	240	175	225	125

Delivery costs, which do not vary from month to month, are (in $/unit shipped):

	Centers		
	1	*2*	*3*
Plant 1	12	13	10
Plant 2	21	19	18

The monthly requirements of the assembly centers must be met exactly. Any transmission made at a plant but not delivered may be stored at that plant for delivery next month, with a storage cost of $4/unit at Plant 1 and $5/unit at Plant 2, or such units may be sold to local parts companies, at a profit of $27 at Plant 1 and $20 at Plant 2.

The translation of the problem of determining a distribution plan that minimizes net expenses into a standard transportation problem is straightforward. We have in fact four sources (the output of each plant over each month) and six primary destinations (the monthly requirements of each assembly center). There are shipping restrictions: units produced in Month 2 cannot be used to meet the Month 1 requirements. There are also some additional shipping costs: units produced in Month 1 and used to meet Month 2 requirements incur storage costs. But these factors can all be accounted for with appropriate c_{ij}'s. See Table 7.33 (and Problem 3).

Example 7.3.4 (An Assignment Problem). Suppose a plant manager has six different jobs to be performed daily and six different machines to do the jobs. Suppose also that for each machine and job, there is a known cost to be incurred if the given machine is assigned to perform the given job. These cost factors could include setup time expenses, production costs, expected costs due to breakdowns, and so on. Obviously, the manager seeks an assignment of machines to jobs that would minimize the total daily costs.

Table 7.33

		Center 1 in Month		*Center 2* in Month		*Center 3* in Month			
		1	2	*1*	2	*1*	2	*Surplus*	*Output*
Plant 1	*in Month 1*	12	16	13	17	10	14	−27	225
	in Month 2	∞	12	∞	13	∞	10	−27	260
Plant 2	*in Month 1*	21	26	19	24	18	23	−20	275
	in Month 2	∞	21	∞	19	∞	18	−20	240
	Requirements	150	175	200	225	100	125	25	

To formulate a mathematical model of this assignment problem, let c_{ij} be the cost to be incurred if machine M_i is assigned to job J_j, and introduce 36 variables x_{ij}, $1 \leq i, j \leq 6$, with the interpretation that

$$x_{ij} = \begin{cases} 1, & M_i \text{ is assigned } J_j \\ 0, & M_i \text{ is not assigned } J_j \end{cases}$$

Now consider the integer programming problem of

$$\text{Minimizing } \sum_{i=1}^{6} \sum_{j=1}^{6} c_{ij} x_{ij} \tag{7.3.1}$$

subject to

$$\sum_{i=1}^{6} x_{ij} = 1, j = 1, \ldots, 6$$

$$\sum_{j=1}^{6} x_{ij} = 1, i = 1, \ldots, 6$$

$$0 \leq x_{ij} \leq 1 \text{ and integral}, 1 \leq i, j \leq 6$$

Each variable x_{ij} can be only 0 or 1. The first set of constraints in (7.3.1) demand that for each j, there is exactly one i for which $x_{ij} = 1$, that is, for each job there is exactly one machine assigned the job. Similarly, the second set of constraints in (7.3.1) demand that for each i, there is exactly one j for which $x_{ij} = 1$, that is, each machine is assigned exactly one job. Our desired interpretation of the x_{ij} variables as designating an assignment of machines to jobs is accomplished. Moreover, the objective function of (7.3.1) properly measures the cost associated with an assignment $\{x_{ij}\}$, and so we have an appropriate mathematical model for the original assignment problem.

This integer programming problem closely resembles a transportation problem with six sources (with each $a_i = 1$) and six destinations (with each $b_j = 1$). But (7.3.1) differs from a transportation problem in that the x_{ij} variables are restricted to be both integral and not greater than 1. The latter restriction, that the $x_{ij} \leq 1$,

Table 7.34

	J_1	J_2	J_3	J_4	J_5	J_6
M_1	14	13	12	17	15	10
M_2	8	12	9	11	12	9
M_3	3	4	2	6	7	5
M_4	7	8	6	9	11	6
M_5	11	17	14	15	16	∞
M_6	6	7	8	7	9	6

Table 7.35

	~~0~~ − 1	1	~~0~~ − 1	1	3	~~0~~ − 1	
11 ~~10~~	14	13	12	17	15	10 (1)	1
9 ~~8~~	8 ○	12	9	11	12 (1)	9	1
3 ~~2~~	3	4 ○	2 (1)	6	7	5	1
7 ~~6~~	7	8 (1)	6 ○	9	11	6 ○	1
12 ~~11~~	11 (1)	17	14	15	16	∞	1
6	6	7 ○	8	7 (1)	9 ○	6	1
	1	1	1	1	1	1	

can be ignored; the equations of (7.3.1) and the nonnegativity of the x_{ij}'s make the upper bound constraints redundant. Let us now assume that the c_{ij}'s are all integral. Then Theorem 7.2.3 in the previous section guarantees that if we consider (7.3.1) a transportation problem and apply the transportation problem algorithm, we will construct an optimal flow with all x_{ij}'s integral. Thus the transportation problem algorithm applied to this assignment problem will generate an optimal assignment $\{x_{ij}\}$.

For example, if the c_{ij}'s are as in Table 7.34, the application of the transportation problem algorithm would lead to the solution of the problem in two iterations (Table 7.35). One possible optimal assignment is M_1 to J_6, M_2 to J_5, M_3 to J_3, M_4 to J_2, M_5 to J_1, and M_6 to J_4, at the minimal cost of $50 = 10+12+2+8+11+7$.

Other, more general assignment problems can be solved in the same manner. If the number of machines and the number of jobs are not equal, either additional machines or jobs can be added to the problem to provide the necessary equality, associating with these auxiliary rows or columns cost factors of zero. If the c_{ij}'s are rational but not all integral, there exists a constant $c \neq 0$ such that cc_{ij} is integral for all i and j, and the assignment problem with costs cc_{ij} would be equivalent to the original. If some c_{ij}'s are irrational, they can be approximated by rationals. Other variations are left to the problems.

Problem Set 7.3

All problems in this set that ask for the determination of some optimal scheme can be modeled as standard transportation problems and solved using the transportation problem algorithm.

1. Solve the problem of Example 7.3.1
   (a) as stated (Table 7.32).
   (b) with the additional restriction that the capacity of the shipping link from Origin 3 to Destination 3 is 15.
2. (a) Determine a minimal-cost shipping schedule for the problem of Example 7.3.2.
   (b) Using the associated solution to the dual found in part (a), estimate the change in total shipping cost if the capacity of
      (i) Node 1 is reduced to $60 - \lambda$.
      (ii) Node 2 is reduced to $70 - \lambda$.
3. (a) Solve the problem of Example 7.3.3 (Table 7.33).
   (b) Suppose the production of Plant 1 could be increased for either Month 1 or Month 2, but not for both months. Using the solution to the dual found above, determine in which month production should be increased to achieve the greater savings and estimate this savings.
4. Consider the transportation problem with the following data:

6	7	3	2	20
12	14	10	12	20
5	3	2	2	20
12	14	16	18	

   (a) Solve the problem as stated.
   (b) Solve the problem assuming that the capacity of the shipping link from Origin 1 to Destination 4 is 8.
   (c) Solve the problem assuming that any number of units can be shipped from Origin 1 to Destination 4, but that the first 8 shipped cost 2/unit and any over 8 cost 4/unit.
   (d) Solve the original problem, with the one additional condition that at least one-half of the supply at Origin 1 must be shipped to Destination 2.
   (e) Solve the original problem, with the one additional condition that at least one-half of the supply at Origin 1 must be shipped to the first two destinations.
5. Compute a minimal-cost shipping schedule for Problem 6 of Section 2.4.
6. Consider a transportation problem having three origins and four destinations in which production at both Origins 1 and 2 is fixed at 60 units but surplus units

may be sold at each of these origins, at 6/unit above cost at Origin 1 and at 16/unit above cost at Origin 2. Also suppose that there is no market for surplus at Origin 3 — and although the cost of any surplus production here would have to be absorbed at a significant loss, production capacity is somewhat flexible. At least 50 units must be produced at Origin 3, but no more than 65. Complete cost data are as follows:

32	24	35	28	60
33	22	20	25	60
28	16	30	20	50 - 65
65	25	60	30	

How many units should be made at Origin 3, and how should the output of the three origins be distributed so that net costs are minimized?

7. For the transportation problem with three origins, five destinations, and the following data

15	17	18	16	21	25
18	19	16	20	22	15 - 25
16	18	15	15	17	15 - 25
8	10	12	14	16	

suppose production at Origin 1 is fixed at 25 and surplus units at this origin have a net value of zero. Production at Origins 2 and 3 can be adjusted between 15 and 25 units each so that production and distribution costs are minimized. A surplus at these two origins, however, must be avoided. Determine how many units should be produced at Origins 2 and 3, and how the total demand should be met, to minimize costs.

8. Solve the transportation problem of Problem 7(a) of Section 7.2, but with the added condition that the combined flow from Origins 1 and 2 to Destination 5 be at least 40 units.

9. Solve the transportation problem of Problem 1(d) of Section 7.2, but with the added requirement that the combined flow from Origins 1 and 2 to Destinations 1 and 2 be at least 20 units. (*Hint.* Add an additional row and column, with supply and demand each 20, and with shipping links from Origins 1 and 2 to the additional column and from the additional row to Destinations 1 and 2. Use the fact that $c_{11} - c_{21} = c_{12} - c_{22}$ in setting costs.)

10. Solve the transportation problem of Problem 1(e) of Section 7.2, but with the added requirement that the combined flow from Origins 1 and 2 to Destinations 2 and 3 be at least 12 units.

11. (a) Determine an optimal shipping schedule for the transshipment problem with two sources (supplies of 80 and 120), three destinations (demands of 45, 60, and 75), one transshipment node (capacity of 55), and shipping costs as follows:

		To Destination			
		Node	*1*	*2*	*3*
	Source 1	7	21	28	16
From	*Source 2*	6	17	20	12
	Node		10	15	7

(b) Suppose the capacity of the transshipment node could be increased to $55 + \lambda$. Estimate the effect this would have on total shipping costs using the solution to the dual found above.

(c) For what range on $\lambda \geq 0$ is your estimate in part (b) accurate?

(d) Solve the original problem, but with the following change in status of the transshipment node: namely, that instead of having an upper bound capacity of 55, it is required that in any solution exactly 65 units flow through the node.

12. (a) Determine an optimal shipping schedule for the transshipment problem with two sources, four destinations, three transshipment nodes, and data as follows:

Shipping Costs From	*Supplies*	To Node *1*	*2*	*3*	To Destination *1*	*2*	*3*	*4*
Source 1	120	8	4	5	10	12	31	15
Source 2	200	5	6	3	8	13	33	20

Shipping Costs From	*Capacities*	To Destination *1*	*2*	*3*	*4*
Node 1	100	1	6	25	11
Node 2	50	5	7	28	14
Node 3	75	4	4	27	15

Demands by Destination			
1	*2*	*3*	*4*
65	85	50	90

(b) Suppose the capacity of one of the transshipment nodes could be increased by a small amount. Using the solution to the dual found above, estimate at which node the increase should be made to achieve the greatest savings.

13. Determine an optimal shipping schedule for the transshipment problem of Example 7.3.2 with the modification that each transshipment node must be used to its full capacity.

14. A firm with two plants must supply three outlets over the next three time periods. The supplies and demands over the three periods and the shipping costs for any period are given in the following tables.

	Plant Supplies		*Outlet Demands*		
Period	*1*	*2*	*1*	*2*	*3*
1	15	25	10	5	20
2	15	25	10	10	10
3	10	20	10	15	10

Shipping Costs	*To Outlet* 1	2	3
From Plant 1	7	9	12
From Plant 2	10	11	16

The period demands at each outlet must be met exactly. Any units produced but undelivered at a plant may be stored at the plant for later delivery with a storage cost of 3/unit/period at Plant 1 and 2/unit/period at Plant 2, or such units may be sold at a profit of 5/unit at Plant 1 and 8/unit at Plant 2. How should the units be distributed so that net expenses are minimized?

15. Reconsider Problem 14 using the following data:

	Plant Supplies		*Outlet Demands*		
Period	*1*	*2*	*1*	*2*	*3*
1	15	15	5	6	2
2	15	10	8	10	4
3	10	5	10	7	6

Shipping Costs	*To Outlet* 1	2	3
From Plant 1	2	1	3
From Plant 2	6	4	5

	Plant 1	*Plant 2*
Storage Cost/Unit/Period	3	1
Profit/Unit Sold	10	8

16. Two plants supply four outlets weekly, with supplies, demands, and production and transportation costs as follows:

53	50	43	44	200
50	47	42	44	400
250	200	150	75	

The total supply of 600, which must be distributed, represents output using regular time. Plant 1 can produce up to another 40 units weekly using overtime at a cost of 5/unit over the costs given in the above table; Plant 2 can produce up to another 80 units weekly using overtime, at an increase in cost of 7/unit. Determine a minimal-cost production and delivery schedule.

17. (a) Solve Problem 9 of Section 2.6.

(b) Suppose the monthly demand for differentials is increased from 225 to $225 + \lambda$ units. Using the solution to the dual, estimate the effect this increase would have on costs.

18. (a) Solve Problem 8 of Section 2.5.

(b) Suppose 25 more engines need to be produced, but we are free to choose the quarter in which they are delivered. Using the solution to the dual, determine in which quarter they should be delivered so as to minimize the increase in production costs and estimate this increase.

19. (a) Solve Problem 5 of Section 2.5.

(b) Suppose the capacity of the storage facility is reduced from 300 to $300 - \lambda$ units. Estimate, using the solution to the dual, the effect on profits.

20. (a) Solve Problem 3 of Section 2.5.

(b) Using the associated solution to the dual, estimate:

(i) the effect on profits if the capacity of the storage facility is reduced from 45 to $45 - \lambda$ units.

(ii) in which month it would be most profitable to increase sales potential, and the effect this increase would have on the total profit (storage capacity for the original 45 units).

21. *The Caterer Problem* (Jacobs [21]). A caterer must supply 110 napkins on Monday, 90 on Tuesday, 130 on Wednesday, and 170 on Thursday. The caterer initially has no napkins on hand. New napkins can be bought for 7 cents each. Used napkins can be laundered for use the next day at 4 cents/napkin or laundered for use in 2 days or more at 2 cents/napkin. At the end of the week, all used napkins have no value. How can the caterer meet these demands at minimal cost? (*Hint.* Consider this as a transportation problem with four sources – the new-napkin outlet and the first 3 days' collections of used napkins.)

22. Reconsider Problem 21, with demands of 70 for Monday, 60 for Tuesday, 80 for Wednesday, 100 for Thursday, and 90 for Friday, and new-napkins costs of 12 cents, next-day laundry service of 5 cents, 2-day laundry service of 3 cents, and 3-day or more service of 2 cents. Furthermore, assume that all used napkins are worth 1 cent at the end of the week.

23. Determine optimal assignments and total minimal costs for the assignment problems defined by the following rating matrices:

(a)

17	27	20
23	25	19
11	18	13

(b)

6	6	7	8	3
2	4	2	1	3
6	9	8	9	7
1	3	2	3	0

(c)

12	16	14	10	5	12	18	13
8	6	7	9	8	11	10	12
13	18	16	14	9	11	14	17
20	18	17	19	12	13	15	14
13	15	12	10	6	18	13	13
6	5	8	9	8	7	4	7
1	4	7	6	3	3	2	5
11	9	10	12	7	5	7	11

24. (a) Is the optimal assignment of Problem 23(b) unique? (See Problem 16 of Section 7.2.)

(b) Solve Problem 23(b) with the added condition that the job corresponding to column 2 of the matrix must be assigned.

25. (a) Determine an optimal assignment (which leaves one job unassigned) for the assignment problem with the following rating matrix:

8	9	12	11	8
4	3	6	7	5
13	20	17	18	12
23	26	25	33	20

(b) Is the above solution unique?

(c) Solve this assignment problem with the restriction that the jobs corresponding to columns 3 and 4 must be assigned.

26. (a) Determine an optimal assignment for the assignment problem with the following rating matrix:

4	8	7	2	5
7	9	10	3	7
5	10	11	6	6
2	6	9	1	8
9	8	7	5	9
3	8	6	4	8

(b) Determine an optimal assignment for these six machines to the five jobs, with the added condition that there is also a savings to be earned, the amount of which depends upon which machine is unassigned. In particular, if either M_1 or M_2 is unassigned, 2 units are saved; M_3 or M_4, 3 units; and M_5 or M_6, 5 units.

27. Solve the following assignment problems, but assume that the c_{ij}'s represent profit and so the objective function $\sum_{i,j} c_{ij}x_{ij}$ is to be maximized.

(a)

7	8	3	10
6	5	0	11
5	6	2	12
6	5	1	9

(b)

6	9	5	7	4
3	8	4	5	7
2	7	3	4	6
5	6	5	7	5

28. Each of six individuals is to be assigned to one of six different jobs, and all six jobs must be completed. The individuals have ranked the jobs in order of preference, giving a 1 to the most desirable job, and so on. These rankings are given in the following table:

	J_1	J_2	J_3	J_4	J_5	J_6
I_1	3	2	1	6	4	4
I_2	1	2	2	4	6	5
I_3	3	1	2	∞	∞	4
I_4	2	1	4	3	5	6
I_5	4	4	1	2	6	3
I_6	1	3	2	6	4	5

Note that the rankings include some ties and that I_3 is not qualified for two jobs. How should the assignments be made?

29. Two jobs, the first requiring three workers and the second two workers, must be completed. There are seven workers who are qualified for the jobs. The wages a worker would receive depend on both the worker and the job assigned and are as follows:

	W_1	W_2	W_3	W_4	W_5	W_6	W_7
J_1	105	90	85	95	80	65	80
J_2	125	105	135	115	100	105	95

Furthermore, each of the first four workers must still be paid 50 units if unassigned, whereas each of the last three workers receives only 25 units if unassigned. Determine a minimal-cost assignment.

30. (a) The shop of Problem 29 is given an order to be completed in 2 weeks. The order requires 9 man-weeks of labor for a J_1-type job and 4 man-weeks of labor for a J_2-type job. Assuming that the wages listed in Problem 29 represent wages per week for the workers, how should the seven workers be assigned over the 2-week period so that the order is completed and the costs are minimized?

(b) As in part (a), but assume now that the shop has 3 weeks to complete the order.

31. A machine shop is given six orders. However, the shop has only five machines and therefore cannot fulfill one order. The profits realized from assigning a machine to an order, and the penalty costs incurred if an order is not completed, are as follows:

		Orders					
		1	*2*	*3*	*4*	*5*	*6*
Machines (profit)	*1*	75	80	85	70	70	x
	2	56	55	65	50	48	48
	3	20	35	30	x	24	x
	4	50	46	42	38	42	32
	5	65	62	48	55	52	45
Penalty Cost		20	25	5	15	24	18

An "x" in the table indicates that a machine is not suited for the associated job. Determine an assignment that optimizes net profit.

32. *The Tanker Scheduling Problem* (Dantzig and Fulkerson [22]). A shipping company has contractual obligations to provide oil tankers for service over the following routes:

Route 1, from Port A to Port B, two tankers daily
Route 2, from Port C to Port B, three tankers daily
Route 3, from Port C to Port D, one tanker daily
Route 4, from Port B to Port E, four tankers daily

The time in days for a tanker (laden or empty) to travel between ports is as follows:

	B	C	D	E
A	23	12	30	28
B		16	10	8
C			19	20
D				5

A tanker requires 1 day in port to load and 1 day to unload. Once unloaded, a tanker is reassigned. If the next route does not originate at the present port of delivery, the tanker sails empty to the originating port of the next route. Determine how the 10 tankers completing assignments daily should be reassigned so that the total number of tankers necessary to meet these requirements is minimized.

(*Hint.* The daily requirements of 10 ships are at four "destinations," the originating ports of the four routes, with demands of 2, 3, 1, and 4; and the "sources," the terminal ports of the four routes, provide the supplies of 2, 3, 1, and 4. To determine a c_{ij}, suppose a ship completing Route 1 is assigned to Route 3. Then a total of 41 ships would be required to maintain this part of the steady-state flow: 1 ship loading in Port A, 23 in transit to Port B, 1 unloading in Port B, and 16 empty ships in transit to Port C.)

33. (a) Reconsider Problem 32, now with an additional port and the following route requirements:

Route	*From*	*To*	*Daily Requirement*
1	Port A	Port E	3 tankers
2	B	C	5
3	C	A	1
4	D	F	1
5	B	F	1

Travel time between Port F and the other ports is:

	To A	*To B*	*To C*	*To D*	*To E*
From F	21	18	7	22	25

(b) Using the solution to the dual, estimate the number of new ships needed to accommodate doubling Route 3's requirement to two ships daily. Is this estimate accurate?

34. True or false: In the final tableau corresponding to the solution of an assignment problem, the sum of the dual variables, $\sum_i u_i + \sum_j v_j$, equals the total cost of an optimal assignment.

35. Given an assignment problem, prove that for any optimal assignment, at least one machine is assigned to a job for which the machine's cost factor is minimal, that is, at least one machine is assigned its best job.

36. Suppose a rating matrix of an assignment problem is altered by the addition of a fixed constant to all the entries of either a row or a column of the matrix. Show that an assignment is optimal for the original problem if and only if it is optimal for the altered problem.

37. Given an assignment problem with an $n \times n$ rating matrix (c_{ij}), show that there are n numbers $r_1, \ldots, r_n$ such that the assignment problem with rating matrix $(c_{ij} + r_i)$ (i.e., r_i is added to the ith row of (c_{ij}) for each i) has the property that the cost of an optimal assignment is $\sum_{j=1}^{n} \text{Min}_i\{c_{ij} + r_i\}$ (i.e., the sum of the minimal entries from each column). (*Hint.* Make use of the u_i's from the final table of a solution to the original problem.)

38. Given an assignment problem with rating matrix (c_{ij}), show that there are numbers r_i and s_j such that the assignment problem with ranking matrix (c'_{ij}), where $c'_{ij} = c_{ij} + r_i + s_j$ (i.e., r_i is added to the ith row of (c_{ij}) and s_j to the jth column), has the property that all the c_{ij}'s are nonnegative and the cost of an optimal assignment is zero.

39. From Problem 16 of Section 7.2, it follows that if the transportation problem algorithm is used to solve an assignment problem, then for any optimal assignment, no matter how it is found, if M_i is assigned J_j, then $u_i + v_j = c_{ij}$, where u_i and v_j are the dual variables of the final table corresponding to the original solution of the problem. Provide an alternate proof of this using Problems 36–38.

40. A private high school with 368 students has classified its aid packages to students as low, mid-range, or high, and using these categories, the following cross-classification table was constructed.

	Class				
Aid Package	*Freshman*	*Sophomore*	*Junior*	*Senior*	*Row Sums*
Low	34	52	3	38	127
Mid-range	12	43	34	27	116
High	46	15	24	40	125
Column sums	92	110	61	105	368

The school's recruitment officer would like to tell prospective students and their parents the total number of students receiving aid in each of the three categories. However, the size of each class is public information, and the school fears that with both the row and column sums of the above table known, the specific number of students in a given category might be determined, or at least closely approximated, compromising individual privacy. The school is especially concerned when the number in a cross-classification category is low, such as in the juniors/low-aid category.

Focusing on this cell, the question then is: In the set of all possible ways of reconstructing a cross-classification table from the above row and column sums alone, what are the lower and upper bounds for the entries in the junior/low-aid cell? If the difference in the bounds is small, privacy could be compromised; if the difference is large, fears on this issue for this category are unfounded.

In fact, the two bounds can be easily determined using linear programming and, in particular, using the transportation problem algorithm. Consider the transportation problem with three sources with supplies equal to the row sums of the table, four destinations with demands the table's corresponding column sums, and with all costs $c_{ij} = 0$ except $c_{13} = -1$ (notation as in (7.2.1)). The following is a solution of the problem:

	0	0	-1	0	
0	0 (66)	0 ( )	-1 (61)	0 ( )	127
0	0 (26)	0 (90)	0	0 ( )	116
0	0 (0)	0 (20)	0	0 (105)	125
	92	110	61	105	

We see that the minimal cost is -61, that is, the minimum of $-x_{13}$ and therefore the maximum of x_{13} is 61, with this value attained at the indicated shipping flow. In part (b) below, you are asked to verify that 0 is the attainable lower bound for x_{13} in a reconstructed cross-classification table with these specified row and column sums. With the difference in bounds of 61, we can conclude that the privacy of the junior/low-aid individuals is well protected if the row and column sums data are known.

(a) What modification of the above transportation problem could be used to determine the lower bound for feasible values of x_{13}?

(b) Use your answer to part (a) to verify that the lower bound for x_{13} is 0.

(c) The fact that 52 sophomores are receiving only a low amount of aid may, if known, upset some members of the sophomore class. Determine if this is a realizable fear if the table's row and column sums are public knowledge.

Note. In most applications, practitioners would want assurances that the data in each cell would remain reasonably protected before the row and column sum data were made public. For an $m \times n$ table, this would mean that $2mn$ linear programming problems, each with mn variables and $m+n$ constraints, would need to be solved. But since in fact the problems can be formulated as quickly-solved simple transportation problems, the procedure outlined above remains a practical tool even when m and n are large.

CHAPTER 8

OTHER TOPICS IN LINEAR PROGRAMMING

8.1 AN EXAMPLE INVOLVING UNCERTAINTY

In the first three sections of this chapter, we consider by means of examples variations in the decision problems with which we have been working and how these variations might be handled using the machinery of linear programming. In Section 8.4, an application of linear programming in a setting quite distinct from that of the general optimization problem model is presented. The four sections are independent of each other and may be read in any order.

Our first example involves the element of probability. In Chapter 5 we developed techniques for measuring the effects changes in the constants of a linear programming problem had on an optimal solution to the problem. In this section, we consider a way of working with uncertainty in the values of these constant terms. As a secondary result, in the solution of the problem in this section, a special technique that can be used to solve linear programming problems with upper bounds on many of its variables is illustrated.

For many applications, it is more realistic to assume that the terms of the problem are not fixed but, instead, can range over sets of values subject to estimated probability distributions. For example, consider the problem of a manufacturer of goods to be sold in the market. In general, the manufacturer seeks to determine a production schedule that maximizes profits and/or minimizes costs. But first, the problem of estimating the demand for the product must be addressed. It may be that the manufacturer cannot assume that everything produced will be sold, but that unpredictable conditions such as weather, strength of the competition, whims of the public, and so on, influence the salability of the product. However, it could be that from past records, a probability distribution for the number of units sold can be estimated with some degree of accuracy.

The example that follows demonstrates one possible technique by which information such as this can be incorporated into a linear programming problem. The basic idea behind the approach is set forth in a paper by A. Ferguson and G. Dantzig [23] (see also Chapter 28 of Dantzig's book [7]), in which the realistic problem of the allocation of commercial aircraft to meet uncertain demands is discussed.

An Introduction to Linear Programming and Game Theory, Third Edition. By P. R. Thie and G. E. Keough.

Example 8.1.1. Consider the problem of the boat manufacturer described in Example 2.3.1 on page 21 (see also Problem 3 of Section 4.3 and Example 5.1.2). The manufacturer produces two types of small boats, a rowboat and a canoe, with the total number produced restricted by the availability of aluminum, machine time, and finishing labor. With a profit of \$50 on the sale of a rowboat and \$60 on the sale of a canoe, the specific linear programming problem is concerned with optimizing profits, and is to

$$\begin{aligned} &\text{Maximize } 50R + 60C \\ &\text{subject to} \\ &50R + 30C \leq 2000 \\ &\ \ 6R + \ \ 5C \leq \ \ 300 \\ &\ \ 3R + \ \ 5C \leq \ \ 200 \end{aligned} \tag{8.1.1}$$

where $R \geq 0$ and $C \geq 0$ are the number of rowboats and canoes produced, respectively. Of course, in this formulation of the problem, we are assuming that all boats produced are sold. With this assumption, an optimal production schedule can be easily determined geometrically. The graph of the set of feasible solutions is sketched in Figure 8.1. The value of the objective function $50R + 60C$ can be computed at the four vertices of this set, and the maximum value, \$2750, attained at the point (25, 25) easily determined. Thus, if all boats produced can be sold, the manufacturer realizes a maximal profit of \$2750 by making 25 boats of each type.

Let us suppose now that the market for the boats is not fixed but is contingent on various factors, the primary one being the weather conditions in early summer.

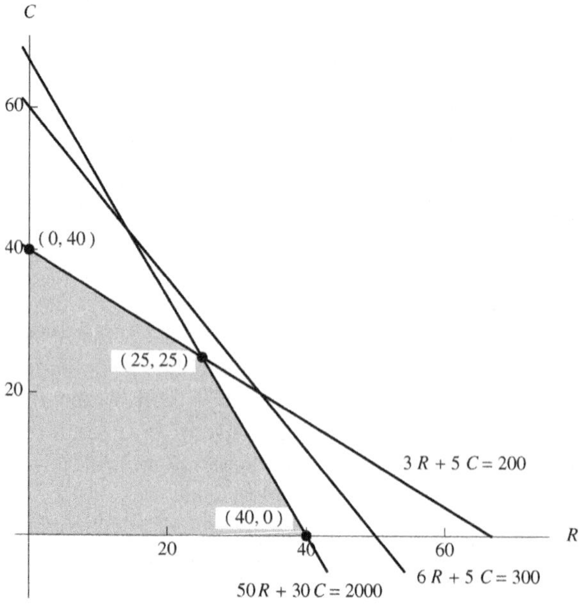

Figure 8.1

Also assume that the manufacturer is attempting to meet these summer demands by spring production and thus must estimate from past experience the demands for boats. It is known that the family rowboat market is more variable than the sports canoe market. More specifically, assume it is equiprobable that there will be either 20 potential rowboat buyers or 50 potential rowboat buyers, and that there will be either 20, 30, or 40 people wanting canoes, with probabilities $\frac{1}{4}$, $\frac{1}{2}$, and $\frac{1}{4}$, respectively. Our problem is to develop a model that takes this new information into account.

One somewhat simplistic approach would be to restrict the number of rowboats and canoes produced to the average or expected number of potential buyers. (In a finite probability space, the expected value of an event is defined to be the sum of the values of each of the possible outcomes of the event times the probability that the outcome occurs.) The expected number of rowboat buyers is $20(\frac{1}{2})+50(\frac{1}{2})=35$, and the expected number of canoe buyers is $20(\frac{1}{4})+30(\frac{1}{2})+40(\frac{1}{4})=30$. Thus, using this approach, the manufacturer would simply add two new constraints to the set in (8.1.1), that $R \leq 35$ and $C \leq 30$. Since the optimal value of the objective function of (8.1.1) is attained at a point that also satisfies these new constraints, it follows that the point $(25,25)$ also provides the solution to this modified problem. However, the profit expected from this production schedule is not the \$2750 of the original problem, but is less because it is no longer certain that all boats produced will be sold. In fact, we are certain that only the first 20 of each type of boat will be sold. Since the probability that 50 people will desire rowboats is $\frac{1}{2}$, the probability that the last five rowboats produced will be sold is $\frac{1}{2}$. Similarly, the probability that the last five canoes produced will be sold is $\frac{3}{4}$ (the probability that there will be at least 30 buyers is $\frac{1}{2}+\frac{1}{4}$). Thus the expected profit is

$$\$50\cdot 20+\$50\cdot 5\cdot \tfrac{1}{2}+\$60\cdot 20+\$60\cdot 5\cdot \tfrac{3}{4}=\$2550$$

This computation of the expected profit suggests a way of refining our model that may lead to another production schedule with a higher expected profit. Our model should take into consideration the differences in expected profit associated with the sale of the first 20 rowboats and the next 30, and the differences associated with the sale of the first 20 canoes, the next 10, and the next 10. We can do this as follows.

Consider the total number R of rowboats produced as divided into two increments, the number between 0 and 20, denoted by R_1, and the number between 20 and 50, denoted by R_2. Note that no more than 50 such boats can be sold, and so we restrict our attention to only these profitable increments. Similarly, consider the number C of canoes produced to be divided into three increments, C_1, C_2, and C_3, where C_1 denotes the number between 0 and 20, C_2 the number between 20 and 30, and C_3 the number between 30 and 40. For example, corresponding to the above $(25,25)$ production schedule, we have $R_1=20$, $R_2=5$, $C_1=20$, $C_2=5$, and $C_3=0$. For these five new variables we have the following constraints.

$$\begin{array}{ll} R_1+R_2=R & 0\leq C_1\leq 20 \\ 0\leq R_1\leq 20 & 0\leq C_2\leq 10 \\ 0\leq R_2\leq 30 & 0\leq C_3\leq 10 \\ & C_1+C_2+C_3=C \end{array}$$

R and C must still satisfy the original three production constraints given in (8.1.1). However, the expected profit function can now be expressed precisely using the new variables. The expected profit associated with the sale of $R = R_1 + R_2$ rowboats is $\$50 \cdot R_1 + \$50 \cdot R_2 \cdot \frac{1}{2}$, since the probability of selling the first 20 boats is 1 and the probability of selling the next 30 is $\frac{1}{2}$. Similarly, the sale of $C = C_1 + C_2 + C_3$ canoes, with the C_i's restricted as above, will realize an expected profit of $\$60 \cdot C_1 + \$60 \cdot C_2 \cdot \frac{3}{4} + \$60 \cdot C_3 \cdot \frac{1}{4}$, since the probability of selling the first 20 canoes is 1, selling the next 10 is $\frac{3}{4}$, and selling the last 10 is $\frac{1}{4}$. Combining all of this, we have the following linear programming problem:

$$
\begin{aligned}
&\text{Maximize } 50R_1 + 25R_2 + 60C_1 + 45C_2 + 15C_3 \\
&\text{subject to} \\
&50R + 30C \le 2000 \\
&\ \ 6R + \ \ 5C \le \ \ 300 \\
&\ \ 3R + \ \ 5C \le \ \ 200 \\
&R_1 + R_2 = R \\
&C_1 + C_2 + C_3 = C \\
&R_1 \le 20 \\
&R_2 \le 30 \\
&C_1 \le 20 \\
&C_2 \le 10 \\
&C_3 \le 10 \\
&R, C, R_1, R_2, C_1, C_2, C_3 \ge 0
\end{aligned}
\tag{8.1.2}
$$

One problem that may have occurred to the reader is that if our interpretation of the variables R_1 and R_2 and of C_1, C_2, and C_3 is to be valid, we should consider only those solutions to the above problem for which $R_2 = 0$ whenever $R_1 < 20$, $C_2 = C_3 = 0$ whenever $C_1 < 20$, and $C_3 = 0$ whenever $C_1 = 20$ and $C_2 < 10$. However, the objective function of (8.1.2) forces any optimal solution to have this property — the coefficient of R_1 in the objective function is greater than the coefficient of R_2, and a similar relationship holds for the coefficients of C_1, C_2, and C_3.

We now proceed to solve the problem of (8.1.2). Although the simplex method could be applied to the problem as it stands, frequently problems such as this, with upper bounds on many of the variables, can be solved without the introduction of a full set of slack variables. We will solve this problem in such a manner.

First, note that from the graph in Figure 8.1 of the solution set to the constraints of (8.1.1), it is obvious that the inequality $6R + 5C \le 300$ is satisfied by any point satisfying the other two constraints of (8.1.1), and so this inequality can be dropped from (8.1.2). Introducing slack variables X_1 and X_2 in the first and third inequalities of (8.1.2), we have

$$
\begin{aligned}
5R + 3C + X_1 \qquad &= 200 \\
3R + 5C \qquad + X_2 &= 200
\end{aligned}
\tag{8.1.3}
$$

Pivoting at the R term of the first equation and then at the C term of the second, (8.1.3) is equivalent to

$$\begin{aligned} R \quad + \tfrac{5}{16}X_1 - \tfrac{3}{16}X_2 &= 25 \\ C - \tfrac{3}{16}X_1 + \tfrac{5}{16}X_2 &= 25 \end{aligned}$$

Using these two equations and eliminating the variables R and C from the constraints in (8.1.2), we see that the problem of (8.1.2) is equivalent to the problem of

$$\begin{aligned} &\text{Maximizing } 50R_1 + 25R_2 + 60C_1 + 45C_2 + 15C_3 = z \\ &\text{subject to} \\ &R_1 + R_2 \qquad\qquad\qquad + \tfrac{5}{16}X_1 - \tfrac{3}{16}X_2 = 25 \\ &\qquad\qquad C_1 + C_2 + C_3 - \tfrac{3}{16}X_1 + \tfrac{5}{16}X_2 = 25 \\ &0 \le R_1 \le 20 \qquad 0 \le C_1 \le 20 \\ &0 \le R_2 \le 30 \qquad 0 \le C_2 \le 10 \\ &X_1 \ge 0 \qquad\quad\;\; 0 \le C_3 \le 10 \\ &X_2 \ge 0 \end{aligned} \tag{8.1.4}$$

We know from our previous work that $R_1 = 20, R_2 = 5, C_1 = 20, C_2 = 5, C_3 = X_1 = X_2 = 0$ is a solution to the constraints of (8.1.4) with $z = 2550$. Notice that the values of the variables R_2 and C_2 in this solution are strictly between their upper and lower bounds. Thus we use the two equations of (8.1.4) to eliminate R_2 and C_2 from the equation in (8.1.4) defining z. Subtracting 25 times the first equation plus 45 times the second, we have

$$25R_1 + 15C_1 - 30C_3 + \tfrac{5}{8}X_1 - \tfrac{75}{8}X_2 = -1750 + z \tag{8.1.5}$$

In (8.1.5), the coefficients of R_1, C_1, and X_1 are positive and the coefficients of C_3 and X_2 are negative. This suggests that the value of z can be increased by moving to another solution of the constraints in (8.1.4) for which either R_1, C_1, or X_1 is larger or C_3 or X_2 is smaller. But in the $R_1 = 20, R_2 = 5, C_1 = 20, C_2 = 5, C_3 = X_1 = X_2 = 0$ solution, the values of R_1 and C_1 are at their maximum and the values of C_3 and X_2 are at their minimum. Consider, however, X_1. Letting $R_1 = 20, C_1 = 20, C_3 = X_2 = 0$, the equations of (8.1.4) become

$$\begin{aligned} 20 + R_2 + \tfrac{5}{16}X_1 &= 25 \\ 20 + C_2 - \tfrac{3}{16}X_1 &= 25 \end{aligned} \quad \text{or} \quad \begin{aligned} R_2 &= 5 - \tfrac{5}{16}X_1 \\ C_2 &= 5 + \tfrac{3}{16}X_1 \end{aligned} \tag{8.1.6}$$

Since $R_2 \ge 0$, the first equality implies that $X_1 \le 16$. Since $C_2 \le 10$, the second equality implies that $X_1 \le \frac{80}{3}$. Thus, under these conditions, the largest possible value for X_1 is 16. Using (8.1.6) to solve for R_2 and C_2, the corresponding solution to the constraints of (8.1.4) is

$$R_1 = 20, R_2 = 0, C_1 = 20, C_2 = 8, C_3 = 0, X_1 = 16, X_2 = 0$$

with $z = 2560$ at this point. Moreover, in this solution, only the values of the variables C_2 and X_1 are not equal to one of the limits of their bounds. Thus we eliminate these two variables from the expression for the objective function. Using the first constraining equation in (8.1.4) and equation (8.1.5), we have

$$23R_1 - 2R_2 + 15C_1 - 30C_3 - 9X_2 = -1800 + z \tag{8.1.7}$$

From (8.1.7), we see that the maximum value of z will be attained when R_1 and C_1 assume their maximum values and R_2, C_3, and X_2 assume their minimum values. But this is precisely the situation in the above solution.

Thus the maximum expected profit for the boat manufacturer is \$2560 and is realized by producing 20 rowboats and 28 canoes. Note that this approach has led us to a solution with an expected profit \$10 greater than the expected profit corresponding to the solution that used only expected values for the demands.

Problem Set 8.1

1. In the example of this section, show that if the probabilities of there being 20 or 50 rowboat buyers are changed to $\frac{2}{5}$ and $\frac{3}{5}$, respectively, the maximum expected profit is attained at the $R = C = 25$ solution.

2. Suppose that in this example the probabilities of potential boat buyers are given by the following table:

Rowboat Buyers	*Probability*	*Canoe Buyers*	*Probability*
20	$\frac{1}{6}$	15	$\frac{4}{5}$
30	$\frac{1}{3}$	30	$\frac{1}{10}$
40	$\frac{1}{2}$	50	$\frac{1}{10}$

   (a) Compute the optimal production schedule using only expected values for the demands.

   (b) Compute the optimal production schedule using the approach of this section.

   (c) Do the two above answers agree?

3. Consider the situation of the dealer of home heating oil described in Example 2.5.1 on page 39. Reformulate the problem using the approach of this section to incorporate the following information on the probability distributions for the demands of oil for the three time periods.

	Gallons of Oil That Can Be Sold	*Probability*
First Month	4000	$\frac{1}{3}$
	6000	$\frac{1}{3}$
	8000	$\frac{1}{3}$
Second Month	6000	$\frac{1}{2}$
	8000	$\frac{1}{2}$
Third Month	8000	1

4. Consider the linear programming problem of

$$
\begin{aligned}
&\text{Maximizing } z = 3x_2 - 5x_3 - 2x_4 + 8x_5 \\
&\text{subject to} \\
&x_1 + x_2 - x_3 + 3x_5 = 16 \\
&2x_1 + 2x_3 + x_4 - x_5 = 24 \\
&0 \le x_1 \le 10,\ 0 \le x_2 \le 10,\ 0 \le x_3 \le 5,\ 0 \le x_4 \le 5,\ 0 \le x_5
\end{aligned}
$$

(a) Show that $x_1 = 10$, $x_3 = 0$, $x_5 = 0$, $x_2 = 6$, $x_4 = 4$ is a feasible solution.

(b) In this solution, only the variables x_2 and x_4 are strictly between their respective bounds. Use the two equations to express the objective function in terms of the remaining variables x_1, x_3, and x_5. Note that this resulting expression suggests that we try to increase the value of either x_1 or x_3, or decrease the value of x_5. Thus we work with x_3. Why?

(c) For $x_1 = 10$ and $x_5 = 0$, show that the two equations and the bounds force $x_3 \le 2$, leading to the solution $x_1 = 10$, $x_4 = 0$, $x_5 = 0$, $x_2 = 8$, $x_3 = 2$.

(d) In this solution x_2 and x_3 are strictly between their respective bounds. Put the system of two equations into canonical form with these as basic variables and use these equations to express z in terms of x_1, x_4, and x_5.

(e) Note that the resulting expression for z suggests that x_1 enter the basis. With $x_4 = x_5 = 0$, determine the lower bound value for x_1 and the corresponding solution point.

(f) At this solution point, what variables are strictly between their bounds? Expressing z in terms of the remaining variables, show that the optimal solution point has been attained and that the maximum value of z is 15.

5. Using the solution technique outlined in this section (and in the above problem), solve the following linear programming problems.

(a) Maximize $4x_1 + x_2 + x_3$

subject to

$$
\begin{aligned}
&3x_1 + x_2 - x_3 = 14 \\
&2x_1 + 2x_3 + x_4 = 10 \\
&0 \le x_1, x_2 \le 4,\ 0 \le x_3, x_4 \le 6
\end{aligned}
$$

Start with the solution $(4,2,0,2)$.

(b) Minimize $x_1 + x_2 + 3x_3 - 2x_4 + 7x_5$

subject to

$$
\begin{aligned}
&x_1 + 2x_3 - x_4 + 8x_5 = 10 \\
&x_2 - x_3 + 2x_4 - 3x_5 = 20 \\
&0 \le x_1,\ 0 \le x_2, x_3 \le 8,\ 0 \le x_4, x_5 \le 12
\end{aligned}
$$

Start with the solution $(6,4,8,12,0)$.

8.2 AN EXAMPLE WITH MULTIPLE GOALS

Our general linear programming model allows for the optimization of a single linear function, the objective function of the problem. However, a decision maker, in attempting to determine the policy of operation for a complex system, may have a multitude of objectives to consider. These objectives may be conflicting (e.g., increasing profits versus reducing pollution) and incommensurate (e.g., maintaining a stable workforce versus reducing losses during periods of market stagnation). One tool that has been useful in such situations is *goal programming*. In this approach the objectives of the project at hand are defined, and specific goals are established for each. Auxiliary variables are introduced to measure the deviation of each objective from its stated goal, and then a plan of operation is determined that minimizes in some sense these deviations. The function to be minimized could be a weighted sum of the deviations, for example, or, as in the method we demonstrate in this section, a set of objectives ordered by priority. (For further discussion see the texts of Ignizio [24] or Lee [25].)

Example 8.2.1. Consider again the situation of the boat manufacturer described in Example 2.3.1 on page 21 and the example in the previous section. Using aluminum, machine time, and labor, the company produces rowboats and canoes. The data are as follows.

	Aluminum (lb)	*Machine Time (min)*	*Labor (hr)*	*Profit (\$)*
Rowboat	50	6	3	50
Canoe	30	5	5	60

Suppose the company is having difficulty obtaining aluminum of sufficient quantity and quality for its overall operation, and because of this, management is considering reducing its commitment to the small boat division. In particular, suppose that for the next quarter only 1500 lb of usable aluminum are available to the division. In this situation, management questions whether the company can also reduce the machine time devoted to boat production to 215 min and the labor to 160 hr and still maintain a viable operation, which they define as meeting the needs of their long-term customers (this requires 10 rowboats and 20 canoes) and generating a profit of at least \$2500. Thus the question: does the following set of inequalities have a feasible solution (where R and C are the number of rowboats and canoes to be produced)?

$$\begin{aligned} 50R + 30C &\leq 1500 \\ 6R + 5C &\leq 215 \\ 3R + 5C &\leq 160 \\ R \quad &\geq 10 \\ C &\geq 20 \\ 50R + 60C &\geq 2500 \\ R, C &\geq 0 \end{aligned} \tag{8.2.1}$$

The answer to this question is "no"; (8.2.1) has no feasible solutions (as we will see shortly). Given this information, all management knows is that all the prescribed goals cannot be attained simultaneously. But some of the constraints are fixed (e.g., the supply of aluminum), some are probably somewhat flexible (e.g., the profit margin), and others may be of low priority (e.g., the labor restrictions). Management asks: is it possible that some adjustment of these goals could lead to an acceptable plan of operation?

To respond to this question, management considers the restrictions and goals that have been set. The supply of aluminum is limited by external factors; the company has no control over it. However, all the other considerations are internal, and so management can establish an order of priority for them. Suppose it is decided that the goal of highest priority is the restriction of machine time to 215 min, because this machine is very heavily used in other operations. Management sets as its second goal meeting the needs of its traditional customers and, as its third goal, maintaining a profit margin of \$2500. Since it may be possible to shift other workers to the boat division if necessary, the goal of lowest priority is the restriction of labor to 160 hr.

Now, to measure deviations from the goals (and to allow flexibility in meeting a goal of lower priority in order to meet a goal of higher priority), we introduce for each goal a pair of nonnegative variables, one (u) to measure underachievement and one (v) to measure overachievement. For example, we can state the restriction on machine time as

$$6R + 5C + u_1 - v_1 = 215$$

The value u_1 plays the role of the traditional slack variable, and v_1 allows and measures machine time above 215 min. Our first priority would be to reduce v_1 to zero, that is, to minimize v_1.

With these variables then, management's second question can be stated as the problem of

$$\begin{array}{lll}
\text{Minimizing } \{v_1; u_2 + u_3; u_4; v_5\} & & \\
\text{subject to} & & \\
50R + 30C + u_0 = 1500 & \text{(aluminum)} & \\
6R + 5C + u_1 - v_1 = 215 & \text{(machine time)} & \\
R + u_2 - v_2 = 10 & \text{(production requirements)} & \\
C + u_3 - v_3 = 20 & & \\
50R + 60C + u_4 - v_4 = 2500 & \text{(profit)} & \\
3R + 5C + u_5 - v_5 = 160 & \text{(labor)} & \\
R, C, u_i, v_i \geq 0 & &
\end{array} \tag{8.2.2}$$

The expression for the objective function defines and displays the priority levels. It is interpreted as follows: we first minimize v_1; then, with our variables set to maintain this minimum, we minimize $u_2 + u_3$; and so on. (The goals of producing 10 rowboats and 20 canoes have been placed at the same priority level.)

Several solution techniques exist for working with an ordered set of objective functions such as this. In our example, we will simply consider the objective functions one by one. This approach has the advantage that at each step we have a standard linear programming problem, which can be solved using our already developed techniques.

To begin the solution process, we must consider the function v_1 subject to the constraints of (8.2.2). However, only the first two constraints are relevant here; without limits on the u_i and v_i, $2 \le i \le 5$, the last four equations place no restrictions on the R and C (and the u_1 and v_1). Thus our first problem:

$$\begin{aligned}
&\text{Minimize } v_1\\
&\text{subject to}\\
&50R + 30C + u_0 \qquad\quad = 1500\\
&\;\;6R + \;\;5C + u_1 - v_1 = \;\;215\\
&R, C, u_0, u_1, v_1 \ge 0
\end{aligned}$$

This problem is trivial. The minimum of v_1 is 0 and is attained at many points (e.g., $v_1 = R = C = 0$, $u_0 = 1500$, $u_1 = 215$). Thus our first goal can be attained, and we set $v_1 = 0$.

Next, consider the problem corresponding to the second level of priority. Letting $v_1 = 0$ in the relevant constraints of (8.2.2), we have the problem of

$$\begin{aligned}
&\text{Minimizing } u_2 + u_3\\
&\text{subject to}\\
&50R + 30C + u_0 \qquad\quad = 1500\\
&\;\;6R + \;\;5C + u_1 \qquad\quad = \;\;215\\
&\;\;\;R \qquad\qquad + u_2 - v_2 = \;\;\;10\\
&\qquad\qquad\;\; C + u_3 - v_3 = \;\;\;20\\
&R, C, u_i, v_i \ge 0
\end{aligned} \tag{8.2.3}$$

To solve this problem, we apply the simplex algorithm. The variables u_0, u_1, u_2, and u_3 can serve as the initial basic variables for the constraints, and subtracting the last two equations from the expression for the objective function eliminates the u_2 and u_3 from this form (see Table 8.1). The minimum of $u_2 + u_3$ is 0, attained at $R = 10$, $C = 20$, and $u_2 = u_3 = 0$. The second goal is achieved, and to maintain it, we set u_2 and u_3 equal to 0.

The third level of priority is the restriction on the profit margin. From (8.2.2), omitting the last constraint and setting $v_1 = u_2 = u_3 = 0$ (alternatively, by building upon (8.2.3)), we have the problem of

Table 8.1

	R	C	u_0	u_1	u_2	v_2	u_3	v_3	
u_0	50	30	1	0	0	0	0	0	1500
u_1	6	5	0	1	0	0	0	0	215
u_2	1	0	0	0	1	−1	0	0	10
u_3	0	(1)	0	0	0	0	1	−1	20
	−1	−1	0	0	0	1	0	1	−30
u_0	50	0	1	0	0	0	−30	30	900
u_1	6	0	0	1	0	0	−5	5	115
u_2	(1)	0	0	0	1	−1	0	0	10
C	0	1	0	0	0	0	1	−1	20
	-1	0	0	0	0	1	1	0	−10
u_0	0	0	1	0	−50	50	−30	30	400
u_1	0	0	0	1	−6	6	−5	5	55
R	1	0	0	0	1	−1	0	0	10
C	0	1	0	0	0	0	1	−1	20
	0	0	0	0	1	0	1	0	0

$$
\begin{aligned}
&\text{Minimizing } u_4 \\
&\text{subject to} \\
&50R + 30C + u_0 = 1500 \\
&6R + 5C + u_1 = 215 \\
&R - v_2 = 10 \\
&C - v_3 = 20 \\
&50R + 60C + u_4 - v_4 = 2500 \\
&R, C, u_i, v_i \geq 0
\end{aligned}
\tag{8.2.4}
$$

To determine a solution, we can make use of our previous work. The first four constraints of (8.2.4) are the constraints of (8.2.3), and therefore equivalent to the constraints of the last tableau of Table 8.1 (with $u_2 = u_3 = 0$). Thus we add the last equation of (8.2.4) to this system, eliminate the variables R and C from the equation using multiples of the equations isolating R and C, and then extract the variable u_4 from the expression for the objective function. This equivalent problem is in canonical form with basic variables u_0, u_1, R, C, and u_4. One iteration of the simplex algorithm completes the problem (Table 8.2). The minimum of u_4 is 140, attained at $R = 10$, $C = 31$. Thus, given the limits on the supply of aluminum and the restrictions established by meeting the first two goals, a profit margin of \$2500 cannot be attained; the maximum that can be earned is \$2360.

Moreover, from the last row of data in Table 8.2,

$$u_4 = 140 + 12u_1 + 22v_2 + v_4$$

Table 8.2

	R	C	u_0	u_1	u_2	v_2	u_3	v_3	
u_0	0	0	1	0	50	30	0	0	400
u_1	0	0	0	1	6	(5)	0	0	55
R	1	0	0	0	-1	0	0	0	10
C	0	1	0	0	0	-1	0	0	20
u_4	0	0	0	0	50	60	1	-1	800
	0	0	0	0	-50	-60	0	1	-800
u_0	0	0	1	-6	14	0	0	0	70
v_3	0	0	0	$\frac{1}{5}$	$\frac{6}{5}$	1	0	0	11
R	1	0	0	0	-1	0	0	0	10
C	0	1	0	$\frac{1}{5}$	$\frac{6}{5}$	0	0	0	31
u_4	0	0	0	-12	-22	0	1	-1	140
	0	0	0	12	22	0	0	1	-140

and so the minimum of u_4 can be achieved only with $u_1 = v_2 = v_4 = 0$. To come within \$140 of the third goal, then, we must have

$$\begin{array}{llllll} R = 10, & u_0 = 70, & u_1 = 0, & u_2 = 0, & u_3 = 0, & u_4 = 140 \\ C = 31, & & v_1 = 0, & v_2 = 0, & v_3 = 11, & v_4 = 0 \end{array}$$

Hence the last goal is also unattainable. Letting $R = 10$ and $C = 31$ in the last equation in (8.2.2) yields $u_5 = 0$, $v_5 = 25$. The boat division requires an additional 25 hr of labor to meet its first two goals and maintain an (unacceptable) profit margin of \$2360.

Given this information, suppose management proposes some adjustments in the operation of the boat division. In particular, they estimate that they can shift up to 20 hr more of labor into the division, but definitely no more (now making the restriction in labor to 180 hr their third priority); and they lower their profit margin expectation, now their last priority, to \$2250. But before the determination of a feasible operation can be made, the Sales Department comes in with a discouraging report. The market for canoes is soft, and the profit return for the next quarter on any canoes sold after the first 24 must be reduced from \$60 to \$40.

To formulate this modified problem in terms of goal programming, we first need to be able to measure the amount of canoe production over 24. The variable v_3 of (8.2.2) already measures production over 20, and so, in keeping with the spirit of our notation, we introduce here u_6 and v_6 and the constraint

$$v_3 + u_6 - v_6 = 4$$

The goal programming problem then is to

$$\begin{aligned}
&\text{Minimize } \{v_1; u_2+u_3; v_5; u_4\} \qquad (8.2.5)\\
&\text{subject to}\\
&50R + 30C \quad + u_0 = 1500\\
&6R + 5C \quad + u_1 - v_1 = 215\\
&R \quad + u_2 - v_2 = 10\\
&C \quad + u_3 - v_3 = 20\\
&3R + 5C \quad + u_5 - v_5 = 160\\
&v_3 + u_6 - v_6 = 4\\
&50R + 60C - 20v_6 + u_4 - v_4 = 2250\\
&R, C, u_i, v_i \geq 0
\end{aligned}$$

As the reader is invited to show (Problem 1), the first three goals of (8.2.5) are attainable, and we can come within 20 of the last. The corresponding production schedule is $R = 15$, $C = 25$.

Problem Set 8.2

1. Show that the solution to the problem of (8.2.5) is as stated.
2. Solve the following goal programming problems.

(a) Minimize $\{v_1; u_2; u_3; v_4\}$ subject to

$$\begin{aligned}
x \quad + u_1 - v_1 &= 6\\
y + u_2 - v_2 &= 8\\
x - y + u_3 - v_3 &= 1\\
x + 2y + u_4 - v_4 &= 20\\
x, y, u_i, v_i \geq 0,\ 1 \leq i \leq 4&
\end{aligned}$$

(b) Minimize $\{v_1; v_2; u_3; v_4\}$ subject to

$$\begin{aligned}
x + 2y + u_1 - v_1 &= 20\\
x - y + u_2 - v_2 &= 5\\
x + y + u_3 - v_3 &= 15\\
x \quad + u_4 - v_4 &= 9\\
x, y, u_i, v_i \geq 0,\ 1 \leq i \leq 4&
\end{aligned}$$

(c) Minimize $\{v_1; v_2; u_3; u_4\}$
subject to

$$\begin{aligned} x + 2y + u_1 - v_1 &= 20 \\ x - y + u_2 - v_2 &= 5 \\ x + y + u_3 - v_3 &= 13 \\ y + u_4 - v_4 &= 7 \\ x, y, u_i, v_i \geq 0,\ 1 \leq i &\leq 4 \end{aligned}$$

(d) Minimize $\{v_1 + v_2; u_3; v_3 + 4u_4\}$
subject to

$$\begin{aligned} x + 2y + u_1 - v_1 &= 20 \\ x - y + u_2 - v_2 &= 5 \\ 2x + y + u_3 - v_3 &= 16 \\ x + u_4 - v_4 &= 8 \\ x, y, u_i, v_i \geq 0,\ 1 \leq i &\leq 4 \end{aligned}$$

3. Formulate as a goal programming problem and then solve each of the following variations of Problem 7(a) of Section 2.2 (and Example 5.1.1).
   (a) The poultry producer establishes the following goals, listed in order of priority:
      (i) Meet the element B nutritional requirement (at least 60 units/day).
      (ii) Spend no more than \$2.06 daily (with costs of 16 cents/lb and 14 cents/lb for Feeds 1 and 2, respectively).
      (iii) Use at least 4 lb of Feed 1 daily.
      (iv) Meet the element A nutritional requirement (at least 124 units/day).
   (b) The producer sets two goals:
      (i) Spend no more than \$2.06 daily.
      (ii) Meet the two nutritional requirements, but, if this is not possible, weigh the deviations so that the deviation from the element A requirement of 124 units carries twice the penalty that the deviation from the element B requirement of 60 units carries.

4. Formulate the following problem variations as goal programming problems. The goals for each are listed in order of priority. (Do not attempt to solve the problems.)
   (a) Problem 10 of Section 2.3, with a fixed constraint that the wood supply cannot exceed 1032 units and with these goals:
      (i) Use all the available labor (1750 hr).
      (ii) Purchase no more than 175 frames from the local mill.
      (iii) Maintain a profit margin of \$9675.
   (b) Problem 15 of Section 2.3, with fixed constraints being the stated upper bounds on flowers available from the local wholesaler and the distant dealer, and with these goals:
      (i) Make at least 50 arrangements of each type.

   (ii) Use all the flowers available from the local wholesaler.
   (iii) Maintain a profit margin of \$350 (incorporating the following price change: the first 200 Type A arrangements sell for \$2.75, but any over 200 sell for \$2.25).
   (iv) Purchase no carnations from the distant dealer.

(c) Example 2.2.2 on page 14, with fixed data being the composition and cost of the two grass seed blends and the following goals:
   (i) Make 1000 lb of a combination seed that is at least 30% fescue.
   (ii) Do not exceed a budget of \$675.
   (iii) Maintain the percentage of bluegrass in the combination at between 25% and 28%. (Assume that deviation below 25% weighs the same as deviation above 28%.)

5. (a) Another approach to problems with multiple goals is to consider minimizing the maximum deviation. For example, for the problem with goals

$$\begin{aligned} 2x_1 + x_2 - x_3 &= 10 \\ x_1 + x_3 &\geq 15 \\ x_2 + 3x_3 &\leq 12 \\ x_1, x_2, x_3 &\geq 0 \end{aligned}$$

we could

$$\begin{aligned} &\text{Minimize the maximum of } \{u_1, v_1, u_2, v_3\} \\ &\text{subject to} \\ &2x_1 + x_2 - x_3 + u_1 - v_1 = 10 \\ &x_1 + x_3 + u_2 - v_2 = 15 \\ &x_2 + 3x_3 + u_3 - v_3 = 12 \\ &x_1, x_2, x_3, u_1, u_2, u_3, v_1, v_2, v_3 \geq 0 \end{aligned}$$

Show that this problem is equivalent to the linear programming problem of

$$\begin{aligned} &\text{Minimizing } y \\ &\text{subject to} \\ &2x_1 + x_2 - x_3 + u_1 - v_1 = 10 \\ &x_1 + x_3 + u_2 - v_2 = 15 \\ &x_2 + 3x_3 + u_3 - v_3 = 12 \\ &y \geq u_1, y \geq v_1, y \geq u_2, y \geq v_3 \\ &x_1, x_2, x_3, u_1, u_2, u_3, v_1, v_2, v_3, y \geq 0 \end{aligned}$$

(b) Solve using the simplex algorithm:

$$\begin{aligned} &\text{Minimize the maximum of } \{u, v\} \\ &\text{subject to} \\ &3x_1 + 2x_2 + 4x_3 + u = 9 \\ &2x_1 + x_2 + 3x_3 - v = 4 \\ &x_1, x_2, x_3, u, v \geq 0 \end{aligned}$$

8.3 An Example Using Decomposition

In theory, the simplex algorithm can be used to solve any linear programming problem. In practice, however, complications may arise. For example, the size of the problem to be solved could exceed the capacity of the computer system being used to implement the algorithm. There are various techniques, though, which can be used in working with large problems that may allow or at least facilitate their solution.

One such technique is the *decomposition principle*. This approach is especially applicable if the primary problem has a special structure, such as a constraint set that can be divided into independent or partially independent subsystems, leading to subproblems that are easily solved. In practice, such a problem could occur in the management of a collection of divisions of a corporation. Each division independently produces goods for profit utilizing its own labor and production facilities, along with the corporation's capital, raw materials, and marketing network. The sum of the division's profits is to be maximized, subject to the global constraints on capital, raw materials, and sales and the local constraints for each division on labor and production capacities (the subproblems). One benefit in applying the decomposition technique in this situation is the generation of important pricing factors for the global commodities. These prices allow the division managers to work quite independently of each other and of the corporate directors. The decomposition principle was first set out by G. Dantzig and P. Wolfe [26]. (See also Chapter 23 of Dantzig's book [7].)

Example 8.3.1. Two plants are under the control of Company Z. Plant X, using raw materials M_1 and M_2, labor, and a stamping machine, produces two products. Data for the manufacture and sale of a unit of each are as follows:

	M_1 *(units)*	M_2 *(units)*	*Labor (hr)*	*Machine Time (min)*	*Profit ($)*
Product χ_1	1	1	3	5	2
Product χ_2	3	2	5	1	4

The plant has available weekly 400 hr of labor and 300 min of machine time.

Plant Y, using these same raw materials and its own labor, milling machine, and transportation network, also produces two products, with data as follows:

	M_1 *(units)*	M_2 *(units)*	*Labor (hr)*	*Machine Time (min)*	*Transportation (units)*	*Profit ($)*
Product ψ_1	4	1	1	1	8	3
Product ψ_2	0	1	4	2	3	1

Plant Y has available weekly 260 hr of labor, 140 min for milling, and 600 units of transportation.

A total of 456 units of M_1 and 260 units of M_2 are at Company Z's disposal each week for distribution to the two plants. How should these supplies be divided, and what should the production schedule for each plant be, in order to maximize the company's total profits?

To formulate a mathematical model for this problem, let x_i be the number of units of χ_i produced by Plant X weekly and y_i the number of units of ψ_i produced by Plant Y, $i = 1, 2$. Then the company's problem is to

$$\begin{array}{ll} \text{Maximize } 2x_1 + 4x_2 + 3y_1 + y_2 & \\ \text{subject to} & \\ x_1 + 3x_2 + 4y_1 \leq 456 & \\ x_1 + 2x_2 + y_1 + y_2 \leq 260 & \\ 3x_1 + 5x_2 \leq 400 & y_1 + 4y_2 \leq 260 \\ 5x_1 + x_2 \leq 300 & y_1 + 2y_2 \leq 140 \\ x_1, x_2 \geq 0 & 8y_1 + 3y_2 \leq 600 \\ & y_1, y_2 \geq 0 \end{array} \tag{8.3.1}$$

Notice the special structure of the constraints. We have two global constraints involving the x's and y's and two sets of independent constraints, one in the x's and the other in the y's. In the decomposition algorithm, we consider these two subproblems independently, then use the optimal points generated in the global problem. Consideration of the global problem will generate prices for the raw materials, which we then use in reconsidering the local problems. Before we demonstrate the algorithm, we formulate a linear programming problem equivalent to (8.3.1), called the *master problem*. Understanding the decomposition algorithm is much easier with this problem at hand.

Suppose the manager of Plant X ignores, for the time being, the global restrictions on the raw materials. Then Manager X has the following optimization problem to solve:

$$\begin{array}{l} \text{Maximize } 2x_1 + 4x_2 \\ \text{subject to} \\ 3x_1 + 5x_2 \leq 400 \\ 5x_1 + x_2 \leq 300 \\ x_1, x_2 \geq 0 \end{array} \tag{8.3.2}$$

The problem involves only two variables and can be solved geometrically. In Figure 8.2, we see that the convex set S_X of feasible solutions has four vertices: $X_0^* = (0,0)$, $X_1^* = (0,80)$, $X_2^* = (50,50)$, and $X_4^* = (60,0)$. Moreover, each point of S_X can be expressed as a convex combination of these vertices. In fact, we have

$$\begin{aligned} S_X &= \left\{ \lambda_0 X_0^* + \lambda_1 X_1^* + \lambda_2 X_2^* + \lambda_3 X_3^* : \sum_{i=0}^{3} \lambda_i = 1, \lambda_i \geq 0 \right\} \\ &= \left\{ \lambda_1 X_1^* + \lambda_2 X_2^* + \lambda_3 X_3^* : \sum_{i=1}^{3} \lambda_i \leq 1, \lambda_i \geq 0 \right\} \end{aligned}$$

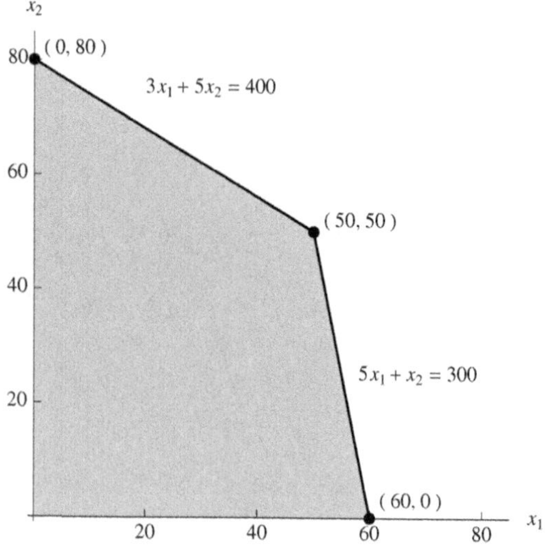

Figure 8.2

This follows from the general theory of convex sets, although the reader should be able to verify the validity of this set equality in the plane using vector geometry. (Start with Problem 1 of Section 3.9.)

Thus all feasible solutions to the constraints of (8.3.2) can be represented using X_1^*, X_2^*, and X_3^*. For future reference, we also note here the amounts of the two raw materials required by these three production schedules and the profit that each yields: X_1^* uses 240 units of M_1 and 160 units of M_2 and delivers a profit of \$320; X_2^*, 200 units of M_1 and 150 units of M_2, with a profit of \$300; and X_3^*, 60 units of M_1, 60 units of M_2, and a profit of \$120. (If we define

$$M_X = \begin{bmatrix} 1 & 3 \\ 1 & 2 \end{bmatrix} \text{ and } c_X = [2,4]$$

then, considering the X_i^*'s as column vectors, these quantities are given simply by the matrix products $M_X X_i^*$ and $c_X X_i^*$.)

Similarly, the manager of Plant Y ignores the restrictions on the raw materials and considers the following optimization problem:

$$\begin{aligned} &\text{Maximize } 3y_1 + y_2 \\ &\text{subject to} \\ &\quad y_1 + 4y_2 \leq 260 \\ &\quad y_1 + 2y_2 \leq 140 \\ &8y_1 + 3y_2 \leq 600 \\ &y_1, y_2 \geq 0 \end{aligned} \tag{8.3.3}$$

The set S_Y of feasible solutions to (8.3.3) has five vertices: $Y_0^* = (0,0)$, $Y_1^* = (75,0)$, $Y_2^* = (0,65)$, $Y_3^* = (60,40)$, and $Y_4^* = (20,60)$ (see Figure 8.3). Hence

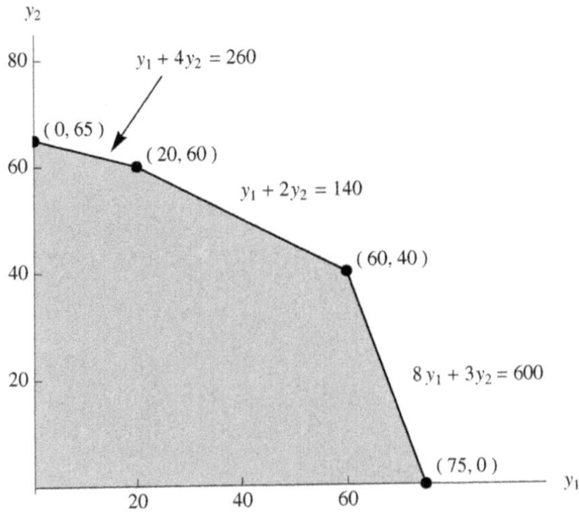

Figure 8.3

$$S_Y = \left\{ \mu_1 Y_1^* + \mu_2 Y_2^* + \mu_3 Y_3^* + \mu_4 Y_4^* : \sum_{j=1}^{4} \mu_j \leq 1, \mu_j \geq 0 \right\}$$

Raw material requirements (in units) and profits (in dollars) are as follows:

	Y_1^*	Y_2^*	Y_3^*	Y_4^*
Units of M_1	300	0	240	80
Units of M_2	75	65	100	80
Profit (\$)	225	65	220	120

(Again, if we define

$$M_Y = \begin{bmatrix} 4 & 0 \\ 1 & 1 \end{bmatrix} \text{ and } c_Y = [3, 1]$$

and consider the Y_j^*'s as column vectors, these quantities are simply the products $M_Y Y_j^*$ and $c_Y Y_j^*$.)

Combining this information, we can reformulate the problem of (8.3.1) as follows:

$$\begin{aligned}
&\text{Maximize } 320\lambda_1 + 300\lambda_2 + 120\lambda_3 + 225\mu_1 + 65\mu_2 + 220\mu_3 + 120\mu_4 \qquad (8.3.4)\\
&\text{subject to}\\
&240\lambda_1 + 200\lambda_2 + 150\lambda_1 + 300\mu_1 \qquad\qquad + 240\mu_3 + 80\mu_4 \leq 456\\
&160\lambda_1 + 150\lambda_2 + 60\lambda_3 + 75\mu_1 + 65\mu_2 + 100\mu_3 + 80\mu_4 \leq 260\\
&\lambda_1 + \lambda_2 + \lambda_3 \leq 1,\ \lambda_i \geq 0\\
&\mu_1 + \mu_2 + \mu_3 + \mu_4 \leq 1, \mu_j \geq 0
\end{aligned}$$

This master problem can be stated in terms of the X_i^*'s and Y_j^*'s as

$$\text{Maximize} \sum_{i=1}^{3} \lambda_i(c_X X_i^*) + \sum_{j=1}^{4} \mu_j(c_y Y_j^*) \tag{8.3.5}$$

subject to

$$\sum_{i=1}^{3} \lambda_i(M_X X_i^*) + \sum_{j=1}^{4} \mu_j(M_Y Y_j^*) \leq [456, 260]^t$$

$$\sum_{i=1}^{3} \lambda_i \leq 1, \lambda_i \geq 0$$

$$\sum_{j=1}^{4} \mu_j \leq 1, \mu_j \geq 0$$

The roles of the X_i^*'s and Y_j^*'s in the master problem are transparent in (8.3.5). We are seeking the convex combination of the X_i^*'s and the convex combination of the Y_j^*'s that maintains feasibility in the raw materials (the first two inequalities in (8.3.4) and (8.3.5)) and maximizes total profit. Given an optimal solution (λ_i^*, μ_j^*) to (8.3.4), the company director would instruct Manager X to follow production scheme $\sum \lambda_i^* X_i^*$ and Manager Y the scheme $\sum \mu_j^* Y_j^*$.

The dual of (8.3.4) will also be of use. To define it, we introduce variables σ_1 and σ_2 to correspond to the two raw material constraints and σ_X and σ_Y to correspond to the upper bound constraints in the λ's and μ's. The dual then is to

$$\begin{array}{llllll} \text{Minimize } 456\sigma_1 + 260\sigma_2 + \sigma_X + \sigma_Y \\ \text{subject to} \\ 240\sigma_1 & + & 160\sigma_2 & + \sigma_X & & \geq 320 \\ 200\sigma_1 & + & 150\sigma_2 & + \sigma_X & & \geq 300 \\ 150\sigma_1 & + & 60\sigma_2 & + \sigma_X & & \geq 120 \\ 300\sigma_1 & + & 75\sigma_2 & & + \sigma_Y & \geq 225 \\ & & 65\sigma_2 & & + \sigma_Y & \geq 65 \\ 240\sigma_1 & + & 100\sigma_2 & & + \sigma_Y & \geq 220 \\ 80\sigma_1 & + & 80\sigma_2 & & + \sigma_Y & \geq 120 \\ \sigma_1, \sigma_2, \sigma_X, \sigma_Y \geq 0 \end{array} \tag{8.3.6}$$

or, in terms of the X_i^*'s and Y_j^*'s,

$$\begin{array}{l} \text{Minimize } 456\sigma_1 + 260\sigma_2 + \sigma_X + \sigma_Y \\ \text{subject to} \\ (M_X X_i^*)[\sigma_1, \sigma_2]^t + \sigma_X \geq c_X X_i^*, \ 1 \leq i \leq 3 \\ (M_Y Y_j^*)[\sigma_1, \sigma_2]^t + \sigma_Y \geq c_Y Y_j^*, \ 1 \leq j \leq 4 \\ \sigma_1, \sigma_2, \sigma_X, \sigma_Y \geq 0 \end{array} \tag{8.3.7}$$

While (8.3.4) is equivalent to (8.3.1), the reader may question whether or not we have constructed a problem that is easier to solve. Certainly (8.3.4) has fewer constraints, but it has more variables, and its construction involved the determination of the vertices to solution spaces of two subproblems. However, the decomposition algorithm, which is based on (8.3.4) and (8.3.6), does not require that all the X_i^*'s and Y_j^*'s be determined initially. In fact, the algorithm proceeds by generating these vertices only as needed, as we now demonstrate.

Decomposition Algorithm Applied to (8.3.1)

STEP L_0 (L for Local)
Manager X ignores the global restrictions on raw materials and solves the problem of (8.3.2). He reports to the company director:

- Maximum profit attained at $X_1^* = (0, 80)$
- Profit earned by X_1^* will be $c_X X_1^* = \$320$
- Raw materials consumed $= M_X X_1^* = [240, 160]^t$

Similarly, Manager Y solves the problem of (8.3.3) and reports:

- Maximum profit attained at $Y_1^* = (75, 0)$
- Profit earned by Y_1^* will be $c_Y Y_1^* = \$225$
- Raw materials consumed $= M_Y Y_1^* = [300, 75]^t$

(The reader should verify the optimality of these solution points.)

STEP G_1 (G for Global)
Director Z determines the optimal combination of these two policies, that is, the director considers the linear programming problem of

$$\begin{aligned}
&\text{Maximizing } z_1 = 320\lambda_1 + 225\mu_1 \\
&\text{subject to} \\
&240\lambda_1 + 300\mu_1 \leq 456 \\
&160\lambda_1 + 75\mu_1 \leq 260 \\
&0 \leq \lambda_1 \leq 1 \\
&0 \leq \mu_1 \leq 1
\end{aligned} \tag{8.3.8}$$

and its dual of

$$\begin{aligned}
&\text{Minimizing } v_1 = 456\sigma_1 + 260\sigma_2 + \sigma_X + \sigma_Y \\
&\text{subject to} \\
&240\sigma_1 + 160\sigma_2 + \sigma_X \qquad\quad \geq 320 \\
&300\sigma_1 + 75\sigma_2 \qquad\quad + \sigma_Y \geq 225 \\
&\sigma_1, \sigma_2, \sigma_X, \sigma_Y \geq 0
\end{aligned} \tag{8.3.9}$$

The maximum of z_1 is 482 attained at $\lambda_1 = 1$, $\mu_1 = \frac{18}{25}$, with $\sigma_1 = \frac{3}{4}$, $\sigma_2 = 0$, $\sigma_X = 140$, $\sigma_Y = 0$, an optimal solution point to the dual. Since

$$\text{Max}\, z_1 = \text{Min}\, v_1 = 456(\tfrac{3}{4}) + 260(0) + 140 + 0 = 482$$

it follows that with this optimal solution, each unit of M_1 has value $\$(\frac{3}{4})$ and each unit of M_2 \$0, that is, the σ_1 and σ_2 provide marginal prices for these resources.

STEP L_1

The director sends this pricing information to the two managers. Manager X now reconsiders the operation of Plant X, subtracting from profits $\$(\frac{3}{4})$ for every unit of M_1 used and \$0 for every unit of M_2 used. His problem:

$$\begin{aligned} \text{Maximize} \quad & 2x_1 + 4x_2 - \tfrac{3}{4}(x_1 + 3x_2) - (x_1 + 2x_2) \\ & = \tfrac{1}{4}(5x_1 + 7x_2) \\ \text{subject to} \quad & \\ & 3x_1 + 5x_2 \leq 400 \\ & 5x_1 + x_2 \leq 300 \\ & x_1, x_2 \geq 0 \end{aligned} \tag{8.3.10}$$

Notice that the only change here from the problem of (8.3.2) is in the objective function. Manager X solves (8.3.10) and reports:

- Maximum profit attained at $X_2^* = (50, 50)$
- Profit earned by X_2^* will be $c_X X_2^* = \$300$
- Raw materials consumed $= M_X X_2^* = [200, 150]^t$

Similarly, Manager Y reconsiders the operation of Plant Y, the problem of (8.3.3), but with objective function now

$$3y_1 + y_2 - \tfrac{3}{4}(4y_1) - 0(y_1 + y_2) = y_2$$

She reports:

- Maximum profit attained at $Y_2^* = (0, 65)$
- Profit earned by Y_2^* will be $c_Y Y_2^* = \$65$
- Raw materials consumed $= M_Y Y_2^* = [0, 65]^t$

(Observe that the profits reported for the X_2^* and Y_2^* are the profits earned on the market, without the deductions for the present marginal costs of the raw materials. Thus, for example, in (8.3.10), the optimal value of the objective function is actually 150, $(\frac{3}{4})(200) + 0(150)$ less than $c_X X_2^* = 300$.)

STEP G_2

Director Z adds these two new proposals to the set of active proposals and considers the problem of

$$\begin{aligned} & \text{Maximizing } 320\lambda_1 + 300\lambda_2 + 225\mu_1 + 65\mu_2 \\ & \text{subject to} \\ & 240\lambda_1 + 200\lambda_2 + 300\mu_1 \leq 456 \\ & 160\lambda_1 + 150\lambda_2 + 75\mu_1 + 65\mu_2 \leq 260 \\ & \lambda_1 + \lambda_2 \leq 1,\ \lambda_i \geq 0 \\ & \mu_1 + \mu_2 \leq 1,\ \mu_j \geq 0 \end{aligned} \tag{8.3.11}$$

and its dual, the problem of

$$\begin{aligned}&\text{Minimizing } 456\sigma_1 + 260\sigma_2 + \sigma_X + \sigma_Y \\ &\text{subject to} \\ &240\sigma_1 + 160\sigma_2 + \sigma_X \geq 320 \\ &200\sigma_1 + 150\sigma_2 + \sigma_X \geq 300 \\ &300\sigma_1 + 75\sigma_2 + \sigma_Y \geq 225 \\ &65\sigma_2 + \sigma_Y \geq 65 \\ &\sigma_1, \sigma_2, \sigma_X, \sigma_Y \geq 0\end{aligned} \tag{8.3.12}$$

The problem of (8.3.11) is the problem of (8.3.4) with $\lambda_3 = \mu_3 = \mu_4 = 0$, that is, the master problem without X_3^*, Y_3^*, and Y_4^*. An optimal solution to (8.3.11) is

$$\lambda_1 = 0, \quad \mu_1 = \tfrac{64}{75}$$
$$\lambda_2 = 1, \quad \mu_2 = \tfrac{11}{75}$$

and to (8.3.12),

$$\sigma_1 = \tfrac{8}{15}, \quad \sigma_X = 193\tfrac{1}{3}$$
$$\sigma_2 = 0, \quad \sigma_Y = 65$$

As before, the σ_1 and σ_2 provide the marginal prices for the next iteration.

Moreover, the values of σ_X and σ_Y also provide useful information. In the optimal solution to (8.3.11), $\lambda_1 = 1 > 0$, and so, from complementary slackness, it follows that the slack in the second constraint of (8.3.12), when evaluated at the $(\sigma_1, \sigma_2, \sigma_X, \sigma_Y)$ optimal solution point, is 0, that is,

$$\begin{aligned}193\tfrac{1}{3} = \sigma_X &= 300 - 200\sigma_1 - 150\sigma_2 \\ &= 300 - 200(\tfrac{8}{15}) - 150(0) \\ &= c_X X_2^* - (M_X X_2^*)[\tfrac{8}{15}, 0]^t\end{aligned}$$

Thus $\sigma_X = 193\frac{1}{3}$ is the value of Manager X's next objective function at X_2^*. And from the first inequality in (8.3.12), we have

$$\begin{aligned}193\tfrac{1}{3} = \sigma_X &\geq 320 - 240(\tfrac{8}{15}) - 160(0) \\ &= c_X X_1^* - (M_X X_1^*)[\tfrac{8}{15}, 0]^t\end{aligned}$$

Hence, with these new marginal prices for the raw materials, σ_X is the maximum that can be earned using the presently active policies from Plant X. Similarly, $\sigma_Y = 65$ is the maximum attainable from Plant Y using Y_1^* and Y_2^* with these new prices.

STEP L_2

Director Z sends Manager X the revised prices of 185 and 0 for the raw materials and the present optimal value of $193\frac{1}{3}$. The manager considers the problem of

$$\begin{aligned}\text{Maximizing} \quad & 2x_1 + 4x_2 - \tfrac{8}{15}(x_1 + 3x_2) - 0(x_1 + 2x_2) \\ &= \tfrac{1}{15}(22x_1 + 26x_2)\end{aligned}$$

subject to the constraints of (8.3.2). The maximum is $193\frac{1}{3}$, and so he reports back to the director that he has no new policies to add to the master list at this time.

Manager Y is given the revised prices and the present optimal value of 65 for her plant and considers the problem of

$$\begin{aligned} \text{Maximizing} \quad & 3y_1 + y_2 - \tfrac{8}{15}(4y_1) - 0(y_1 + y_2) \\ & = \tfrac{1}{15}(3y_1 + 15y_2) \end{aligned}$$

subject to the constraints of (8.3.3). She determines that the maximum of this objective function is greater than 65 and is attained at $Y_3^* = (60, 40)$. She reports:

- Maximum profit attained at $Y_3^* = (60, 40)$
- Profit earned by Y_3^* will be $c_Y Y_3^* = 220$
- Raw materials consumed $= M_Y Y_3^* = [240, 100]'$

STEP G_3

Director Z considers the problem of

$$\begin{aligned} & \text{Maximizing } 320\lambda_1 + 300\lambda_2 + 225\mu_1 + 65\mu_2 + 220\mu_3 \\ & \text{subject to} \\ & 240\lambda_1 + 200\lambda_2 + 300\mu_1 \qquad\qquad + 240\mu_3 \le 456 \\ & 160\lambda_1 + 150\lambda_2 + 75\mu_1 + 65\mu_2 + 100\mu_3 \le 260 \\ & \lambda_1 + \lambda_2 \le 1,\ \lambda_i \ge 0 \\ & \mu_1 + \mu_2 + \mu_3 \le 1,\ \mu_j \ge 0 \end{aligned} \tag{8.3.13}$$

and its dual,

$$\begin{aligned} & \text{Minimizing } 456\sigma_1 + 260\sigma_2 + \sigma_X + \sigma_Y \\ & \text{subject to} \\ & 240\sigma_1 + 160\sigma_2 + \sigma_X \ge 320 \\ & 200\sigma_1 + 150\sigma_2 + \sigma_X \ge 300 \\ & 300\sigma_1 + 75\sigma_2 + \sigma_Y \ge 225 \\ & 65\sigma_2 + \sigma_Y \ge 65 \\ & 240\sigma_1 + 100\sigma_2 + \sigma_Y \ge 220 \\ & \sigma_1, \sigma_2, \sigma_X, \sigma_Y \ge 0 \end{aligned} \tag{8.3.14}$$

An optimal solution to (8.3.13) is

$$\begin{aligned} \lambda_1 &= \tfrac{2}{5}, \quad \mu_1 = \mu_2 = 0 \\ \lambda_2 &= \tfrac{3}{5}, \quad \mu_3 = 1 \end{aligned}$$

and to (8.3.14),

$$\begin{aligned} \sigma_1 &= \tfrac{1}{2}, \quad \sigma_X = 200 \\ \sigma_2 &= 0, \quad \sigma_Y = 100 \end{aligned}$$

STEP L_3
Director Z relays to Manager X the prices of $\frac{1}{2}$ and 0 and the optimal value for his plant of 200. The manager determines that the maximum of

$$2x_1 + 4x_2 - \tfrac{1}{2}(x_1 + 3x_2) = \tfrac{1}{2}(3x_1 + 5x_2)$$

subject to the constraints of (8.3.2) does not exceed 200 and reports this information to the director. Similarly, Manager Y determines that the maximum of

$$3y_1 + y_2 - \tfrac{1}{2}(4y_1) = y_1 + y_2$$

subject to the constraints of (8.3.3) does not exceed 100 and reports this to the director.

STEP G_F
The director now concludes that the optimal operation of the company has been determined. The maximum profit ($528) is the optimal value of the latest version of the master problem, the problem of (8.3.13), and the λ's and μ's in the solution of that problem provide the optimal operating schedules. Manager X is directed to the policy

$$(\tfrac{2}{5})X_1^* + (\tfrac{3}{5})X_2^* = (30, 62)$$

and Manager Y to the policy

$$Y_3^* = (60, 40)$$

The fact that the original problem (8.3.4) has been completed follows from duality. To see this, extend the optimal solution point of the latest master problem to a solution point of (8.3.4) by defining $\lambda_i = 0$ and $\mu_j = 0$ for those X_i^*'s and Y_j^*'s not included in the present master list (here, set $\lambda_3 = \mu_4 = 0$), and compare the value of the objective function of (8.3.4) at this point to the value of the objective function of (8.3.6) at the optimal solution point to (8.3.14), the point $\sigma_1 = \frac{1}{2}$, $\sigma_2 = 0$, $\sigma_X = 200$, $\sigma_Y = 100$. (They must both equal 528, the optimal value of the objective functions of (8.3.13) and (8.3.14).) That $\sigma_1 = \frac{1}{2}$, $\sigma_2 = 0$, $\sigma_X = 200$, $\sigma_Y = 100$ provides a feasible solution to (8.3.6) follows from the results in Step L_3. Manager X's final report to the director implies that

$$200 = \sigma_X \geq c_X X_i^* - (M_X X_i^*)[\tfrac{1}{2}, 0]^t, \quad 1 \leq i \leq 3$$

and Manager Y's final report implies that

$$100 = \sigma_Y \geq c_Y Y_j^* - (M_Y Y_j^*)[\tfrac{1}{2}, 0]^t, \quad 1 \leq j \leq 4$$

Problem Set 8.3

1. (a) In the example of this section, verify that the stated optimal solution points to the following problems are in fact optimal:
   (i) (8.3.8) and (8.3.9) of Step G_1.

(ii) The two local problems of Step L_1.
(iii) (8.3.11) and (8.3.12) of Step G_2.
(iv) The two local problems of Step L_2.
(v) (8.3.13) and (8.3.14) of Step G_3.
(vi) The two local problems of Step L_3.

(b) Show that in Step L_1, the optimal value of the objective function of (8.3.10) exceeds the value of σ_X (140) in the optimal solution to (8.3.9); and similarly, show that the optimal value of Manager Y's objective function in Step L_1 exceeds the previous σ_Y value of 0.

2. Suppose in the example of this section only 250 units of M_2 are available weekly. Solve the problem with this change, using the decomposition algorithm. (*Hint.* The optimal solution points generated in the application of the algorithm in the original problem should work here through Step G_2. In the dual problem for Step G_3, the optimal solution has $\mu_2 = 0$, and the other four variables are positive. With this information, and complementary slackness, the optimal solution points for the Step G_3 linear programming problems can be readily determined by solving systems of equations.)

3. Solve the following using the decomposition algorithm.

(a) Maximize $3x_1 + 8x_2 + 20y_1 + 7y_2$
subject to

$$x_1 + 6x_2 + 8y_1 + 4y_2 \leq 22$$
$$x_1 + 3x_2 + 4y_1 + 3y_2 \leq 14$$
$$2x_1 + 3x_2 \leq 6 \qquad 4y_1 + y_2 \leq 4$$
$$x_1, x_2 \geq 0 \qquad y_1, y_2 \geq 0$$

(b) Maximize $4x_1 + 7x_2 + 2y_1 + 3y_2$
subject to

$$8x_1 + 9x_2 + 3y_2 \leq 97$$
$$x_1 + 2x_2 \leq 10 \qquad 3y_1 + 2y_2 \leq 18$$
$$x_1, x_2 \geq 0 \qquad 5y_1 + 4y_2 \leq 32$$
$$y_1, y_2 \geq 0$$

(c) Maximize $5x_1 + 3x_2 + 2x_3 + 8y_1 + 7y_2$
subject to

$$3x_1 + x_2 + 2x_3 + 3y_1 + y_2 \leq 40$$
$$x_1 + x_2 + x_3 + 2y_1 + y_2 \leq 20$$
$$3x_1 + 2x_2 + x_3 \leq 18 \qquad 4y_1 + 3y_2 \leq 24$$
$$x_1, x_2, x_3 \geq 0 \qquad y_1, y_2 \geq 0$$

(d) Maximize $3x_1 + x_2 + 2y_1 + 5y_2$
subject to

$$\begin{aligned} 2x_1 \qquad\qquad + \; y_2 &\le 42 \\ x_2 + y_1 + 2y_2 &\le 48 \end{aligned}$$

$$\begin{array}{ll} 2x_1 + x_2 \le 36 & y_1 + 2y_2 \le 40 \\ -x_1 + x_2 \le 18 & 3y_1 + 2y_2 \le 60 \\ x_1, x_2 \ge 0 & y_1, y_2 \ge 0 \end{array}$$

4. In spite of our efforts, the small boat division of Example 2.3.1 on page 21 (which we revisited in Examples 5.1.2, 8.1.1, and 8.2.1) has been sold and is now a division of Water Sports Inc. The small boat group still makes rowboats and canoes using aluminum and their own pressing machine and labor (data as in Example 2.3.1), except that now WSI supplies the aluminum. WSI's other division makes two types of motorboats, with input and profit data for the manufacture and sale of a boat of each type as follows:

	Aluminum (lb)	*Trim (units)*	*Labor (hr)*	*Profit ($)*
Turboboat	140	10	12	200
Outboard	60	4	8	80

For the next quarter, the motorboat division has available 240 units of trim and 320 hr of labor; and WSI has 2 tons of aluminum to be shared by the two divisions. Determine, using the decomposition algorithm, the distribution of the supply of aluminum and the operating schedules of the two divisions that maximize WSI's total profit.

8.4 AN EXAMPLE IN DATA ENVELOPMENT ANALYSIS

The techniques of Data Envelopment Analysis (DEA) are used to measure the relative efficiency of a group of similar operating units with comparable inputs and outputs. DEA has been utilized to assess efficiency among hospitals, branch offices of a bank, schools, seaports, building supply outlets, dairy farms, and so on. The seminal paper by Charnes, Cooper and Rhodes [27] was published in 1978. Since then, the theory has evolved and the methods have been refined and expanded to encompass a variety of applications. We present here, via an elementary example, one approach to measuring efficiencies using DEA (and linear programming).

Example 8.4.1. The state Board of Education has been requested to measure the efficiency of the four local community colleges. The schools, with input coming from faculty, staff, an operating budget, state support, and a physical plant, generate output involving full-time students, part-time students, associate degrees awarded, certification program graduates, and so on. After some deliberation, the Board establishes two input categories and three output categories to compare the schools:

Input A — total staff size, in units of 10
Input B — annual operating budget, in units of \$20,000
Output C — number of students enrolled per year, in units of 10
Output D — number of associates degrees awarded per year, in units of 15
Output E — number of certificates awarded per year, in units of 30

(The unit measures chosen keep the data tractable. Since, for each input and output category, the unit of measurement is uniform throughout the four-college system, relative efficiencies are unaffected.) Using data from the last 8 years, the following input/output table of average annual values for the each category has been generated for the four colleges:

		Colleges			
		1	*2*	*3*	*4*
Inputs	*A*	30	20	55	45
	B	50	15	75	35
Outputs	*C*	150	100	600	225
	D	40	65	275	125
	E	25	20	30	10

For example, from the table we see that College 3 is the largest, with an average annual staff of 550, an operating budget of \$1.5 million, 6000 students, and 4125 associate degrees and 900 certificates awarded.

With these data at hand, to determine if a given college is inefficient compared to the others, we could attempt to find another college with less input and greater output, and if it exists, conclude that the given college is inefficient. But here, for each of the four schools, no simple comparison exists, as is easily seen.

We refine our approach. For a given college, instead of restricting our search to a single more efficient college, we expand our set of competitors to include composite units consisting of weighted combinations of the other colleges. And it is linear programming that allows this search to be made effectively. We state the procedure for the general case.

Suppose we have a system of n distinct operating units each with p inputs and q outputs. For unit k, $k = 1, 2, \ldots, n$, denote by [In k] the p-dimensional column vector with components given by the p input quantities for unit k, and similarly by [Out k] the q-dimensional column vector of output quantities. For example, in the above situation, for $k = 4 \leftrightarrow$ College 4, we would have

$$[\text{In } 4] = \begin{bmatrix} 45 \\ 35 \end{bmatrix} \qquad \text{and} \qquad [\text{Out } 4] = \begin{bmatrix} 225 \\ 125 \\ 10 \end{bmatrix}$$

Then, to measure the relative efficiency of a given unit, say unit k, we seek to determine if there exists n nonnegative numbers $x_1, x_2, \ldots, x_n$ such that

$$x_1 [\text{In } 1] + x_2 [\text{In } 2] + \ldots + x_k [\text{In } k] + \ldots + x_n [\text{In } n] \;<\; [\text{In } k]$$

and

$$x_1\,[\text{Out } 1] + x_2\,[\text{Out } 2] + \ldots + x_k\,[\text{Out } k] + \ldots + x_n\,[\text{Out } n] \;\geq\; [\text{Out } k]$$

If such numbers exist, then the composite unit defined by the x's has input strictly less than unit k's input but output at least equal to if not greater than unit k's output. We can conclude that unit k is inefficient. And to determine if such numbers exist for unit k, we solve the following linear programming problem:

$$\begin{aligned}
&\text{Minimize } e \text{ subject to} \\
&x_1\,[\text{In } 1] + x_2\,[\text{In } 2] + \ldots + x_k\,[\text{In } k] + \ldots + x_n\,[\text{In } n] \;\leq\; e\,[\text{In } k] \\
&x_1\,[\text{Out } 1] + x_2\,[\text{Out } 2] + \ldots + x_k\,[\text{Out } k] + \ldots + x_n\,[\text{Out } n] \;\geq\; [\text{Out } k] \\
&x_1, x_2, \ldots, x_n, e \geq 0
\end{aligned} \tag{8.4.1}$$

The problem has $n+1$ variables, an x for each of the operating units plus the variable e, and $p+q$ constraints, p constraints for the p input categories and q constraints for the q outputs categories. Note that the minimum value of e is at most 1, as setting $e = x_k = 1$ and the remaining x's to 0 is a feasible solution. If the minimum of e in fact proves to be 1, we can conclude that under this scheme unit k is efficient; and if the minimum is less than 1, then unit k is inefficient, with the value of e providing a measure of the inefficiency.

To apply this scheme to the above four-college system, we need to consider four distinct but very similar linear programming problems, one for each college. The problems would have five variables and five constraints and would differ only in the constant-term column, with the numbers for that column determined by the data for the college being reviewed. We illustrate using College 4.

Example 8.4.2 (Continuation of Example 8.4.1)**.** To determine if College 4 is efficient, consider the linear programming problem of

$$\begin{aligned}
&\text{Minimizing } e \text{ subject to} \\
&x_1\,[\text{In } 1] + x_2\,[\text{In } 2] + x_3\,[\text{In } 3] + x_4\,[\text{In } 4] \;\leq\; e\,[\text{In } 4] \\
&x_1\,[\text{Out } 1] + x_2\,[\text{Out } 2] + x_3\,[\text{Out } 3] + x_4\,[\text{Out } 4] \;\geq\; [\text{Out } 4] \\
&x_1, x_2, x_3, x_4, e \geq 0
\end{aligned}$$

that is, the problem of

$$\begin{aligned}
&\text{Minimizing } e \text{ subject to} \\
&30x_1 + 20x_2 + 55x_3 + 45x_4 \leq 45e \\
&50x_1 + 15x_2 + 75x_3 + 35x_4 \leq 35e \\
&150x_1 + 100x_2 + 600x_3 + 225x_4 \geq 225 \\
&40x_1 + 65x_2 + 275x_3 + 125x_4 \geq 125 \\
&25x_1 + 20x_2 + 30x_3 + 10x_4 \geq 10 \\
&x_1, x_2, x_3, x_4, e \geq 0
\end{aligned} \tag{8.4.2}$$

The problem, small in size, can be solved using LP Assistant (use of the dual simplex algorithm is recommended), or Microsoft Excel and Solver. A major advantage of Solver is that it allows a very easy transition from one problem to the next because of the similarities among the problems.

The spreadsheet solution of (8.4.2) is in Figure 8.4, and the accompanying sensitivity report appears in Figure 8.5. (*Comment.* To facilitate the later computation

	A	B	C	D	E	F	G
1	**College 4**						
2				**College**			
3		**Avg. Annual Values**	**1**	**2**	**3**	**4**	
4		Input A	30	20	55	45	
5		Input B	50	15	75	35	
6		Output C	150	100	600	225	
7		Output D	40	65	275	125	
8		Output E	25	20	30	10	
9							
10					**Variables**		
11			**1**	**2**	**3**	**4**	**Efficiency**
12			0.0000	1.1413	0.1848	0.0000	0.8851
13							
14		Minimize Efficiency (*e*)	0.8851				
15							College 4
16		**Constraints**	LHS		RHS		Data
17		Input A	32.9891	≤	39.8292		45
18		Input B	30.9783	≤	30.9783		35
19		Output C	225.0000	≥	225.0000		225
20		Output D	125.0000	≥	125.0000		125
21		Output E	28.3696	≥	10.0000		10

Figure 8.4

Adjustable Cells

Cell	Name	Final Value	Reduced Cost	Objective Coefficient	Allowable Increase	Allowable Decrease
\$C\$12	x1	0.0000	1.0000	0	1E+30	1
\$D\$12	x2	1.1413	0.0000	0	0.0779	0.0714
\$E\$12	x3	0.1848	0.0000	0	0.4286	0.3297
\$F\$12	x4	0.0000	0.1149	0	1E+30	0.1149
\$G\$12	Efficiency	0.8851	0.0000	1	1E+30	1

Constraints

Cell	Name	Final Value	Shadow Price	Constraint R.H. Side	Allowable Increase	Allowable Decrease
\$C\$17	Input A	32.9891	0.00000	0	1E+30	6.8401
\$C\$18	Input B	30.9783	-0.02857	0	5.3200	1E+30
\$C\$19	Output C	225.000	0.00186	225	47.7273	27.2250
\$C\$20	Output D	125.000	0.00373	125	17.2070	21.875
\$C\$21	Output E	28.3696	0.00000	10	18.3696	1E+30

Figure 8.5

of efficiencies for the other three colleges, we duplicate the data for College 4 at the lower right of the spreadsheet in Figure 8.4 and then base the constraints on these copied values.) We see that the operation of College 4 is inefficient, with an efficiency rating of about 88.5% when compared to the composite college consisting of 1.1413 times the input and output from College 2 plus 0.1814 times the input and output from College 3. For input B, the composite college uses only 88.5% of the 35 units used by College 4 and even less for input A. Output values are the same for outputs C and D, but the composite college delivers almost three times the output of College 4 in category E.

To provide information to College 4 on how to improve efficiency, we use the dual to the problem of (8.4.2), which has five variables corresponding to the two input and three output constraints. From the Shadow Price column of the sensitivity report (Figure 8.5), we note that the marginal values associated with outputs C and D are nonzero. Using these data, we see, for example, that if College 4 could increase output C from 225 to 250 (i.e., increase enrollment by 250 students), with no other changes made, the college's efficiency rating of 88.51% would increase to $.8851 + 25(.0019) = .9326 = 93.26\%$; if, instead, input D were decreased from 125 to 115 (150 fewer associate degrees awarded), the efficiency rating would drop to $.8851 - 10(.0037) = .8481 = 84.81\%$. (Note that these changes of $+25$ and -10 are each within appropriate allowable ranges of validity for the associated marginal values.)

Concerning inputs, from the spreadsheet solution of the problem, we see that there is positive slack in the input A constraint and no slack in the input B constraint. Thus decreasing input A would have no immediate effect on efficiency, but any decrease in input B would allow the value of e to increase, since the left side of the corresponding constraint would be unaffected as $x_4 = 0$ (see Lemma 8.4.1). However, we cannot gauge the rate of increase with the dual variables. In (8.4.2), after the variable terms $45e$ and $35e$ are shifted to the main body of equations, the constant terms of zero remain on the right; 45 and 35 do not serve as constant terms and therefore are not coefficients in the associated dual. The shadow prices from Solver of 0 and -0.0286 are not immediately applicable here.

To complete the review of the community college operation, we proceed just as above for the other three community colleges. We find that Colleges 2 and 3 are efficient under this scheme. The spreadsheet for College 2 is in Figure 8.6; College 3's is equivalent. However, College 1 is only 87.96% efficient when compared to the composite consisting of $\frac{7}{6}$ of College 2 and $\frac{1}{18}$ of College 3. The corresponding spreadsheet and sensitivity report are in Figures 8.7 and 8.8. From these we see that College 1 should attempt to improve outputs in categories C (student enrollment) and especially E (certificates awarded) and decrease input in category A (staff size).

In our model for each unit k, the competing composite unit constructed consisted of a sum of multiples of the units. The coefficients used were restricted only by nonnegativity. One variation in this procedure is to restrict the composite unit to be a weighted average of the units, that is, to add to the linear programming problem of (8.4.1) the constraint $\sum_{i=1}^{n} x_i = 1$. We do not address here which of the two methods is preferred (how does one measure?), but we do note that the methods

	A	B	C	D	E	F	G
1	**College 2**						
2					**College**		
3		**Avg. Annual Values**	**1**	**2**	**3**	**4**	
4		Input A	30	20	55	45	
5		Input B	50	15	75	35	
6		Output C	150	100	600	225	
7		Output D	40	65	275	125	
8		Output E	25	20	30	10	
9							
10					**Variables**		
11			**1**	**2**	**3**	**4**	**Efficiency**
12			0.0000	1.0000	0.0000	0.0000	1.0000
13							
14		Minimize Efficiency (*e*)	1.0000				
15							College 2
16		**Constraints**	LHS		RHS		Data
17		Input A	20.0000	≤	20.0000		20
18		Input B	15.0000	≤	15.0000		15
19		Output C	100.0000	≥	100.0000		100
20		Output D	65.0000	≥	65.0000		65
21		Output E	20.0000	≥	20.0000		20

Figure 8.6

	A	B	C	D	E	F	G
1	**College 1**						
2					**College**		
3		**Avg. Annual Values**	**1**	**2**	**3**	**4**	
4		Input A	30	20	55	45	
5		Input B	50	15	75	35	
6		Output C	150	100	600	225	
7		Output D	40	65	275	125	
8		Output E	25	20	30	10	
9							
10					**Variables**		
11			**1**	**2**	**3**	**4**	**Efficiency**
12			0.0000	1.1667	0.0556	0.0000	0.8796
13							
14		Minimize Efficiency (*e*)	0.8796				
15							College 1
16		**Constraints**	LHS		RHS		Data
17		Input A	26.3889	≤	26.3889		30
18		Input B	21.6667	≤	43.9815		50
19		Output C	150.0000	≥	150.0000		150
20		Output D	91.1111	≥	40.0000		40
21		Output E	25.0000	≥	25.0000		25

Figure 8.7

are not equivalent. If the constraint $\sum_{i=1}^{n} x_i = 1$ is added to (8.4.2) to determine the efficiency of College 4, and similarly for the other three colleges, we find that Colleges 2 and 3 remain efficient and College 4 remains inefficient, now with an efficiency rating of 91.84% (the relevant spreadsheet is in Figure 8.9). However,

Adjustable Cells

Cell	Name	Final Value	Reduced Cost	Objective Coefficient	Allowable Increase	Allowable Decrease
C12	x1	0.0000	0.1204	0	1E+30	0.1204
D12	x2	1.1667	0.0000	0	0.1032	0.3611
E12	x3	0.0556	0.0000	0	2.1667	0.8333
F12	x4	0.0000	0.8426	0	1E+30	0.8426
G12	Efficiency	0.8796	0.0000	1	1E+30	1

Constraints

Cell	Name	Final Value	Shadow Price	Constraint R.H. Side	Allowable Increase	Allowable Decrease
C17	Input A	26.3889	-0.0333	0	13.3889	9.591E+15
C18	Input B	21.6667	0.0000	0	1E+30	22.3148
C19	Output C	150.0000	0.0019	150	350	25
C20	Output D	91.1111	0.0000	40	51.1111	1E+30
C21	Output E	25.0000	0.0241	25	5	17.5

Figure 8.8

	A	B	C	D	E	F	G
1	College 4 (sum = 1)						
2				College			
3		Avg. Annual Values	1	2	3	4	
4		Input A	30	20	55	45	
5		Input B	50	15	75	35	
6		Output C	150	100	600	225	
7		Output D	40	65	275	125	
8		Output E	25	20	30	10	
9							
10					Variables		
11			1	2	3	4	Efficiency
12			0.0000	0.7143	0.2857	0.0000	0.9184
13							
14	Minimize Efficiency (*e*)		0.9184				
15							College 4
16		Constraints	LHS		RHS		Data
17		Input A	30.0000	≤	41.3265		45
18		Input B	32.1429	≤	32.1429		35
19		Output C	242.8571	≥	225.0000		225
20		Output D	125.0000	≥	125.0000		125
21		Output E	22.8571	≥	10.0000		10
22		sum	1.0000	=	1.0000		

Figure 8.9

College 1 now joins the ranks of the efficient units, with an efficiency rating of 1. We ask the reader to verify this in Problem 2.

Note that in our community college model we did not address the question of weighing or comparing the relative significance of the contributions from the different output and input categories. For example, 1 unit of output C was treated the same way as 1 unit of output D when comparing outputs. As with many DEA procedures, discussion of the relative merits of the different categories is precluded. This is especially helpful in applications where the bottom lines are not simply in monetary units, such as with nonprofit institutions.

On the other hand, some methods do consider the use of relative weights in the determination of efficiency. One general approach is to allow each unit to use the most advantageous weights in determining that unit's efficiency. Thus if unit k's efficiency is being measured, a nonnegative weight for each input and each output category is computed (using linear programming) so that unit k's measure of efficiency, defined as

$$\text{measure of efficiency} = \frac{k\text{'s total output value}}{k\text{'s total input value}} \tag{8.4.3}$$

is maximized. Some restrictions apply on these weights: unit k's total input value must equal 1, and the measure of efficiency of each of the units, using unit k's weights, must be less than or equal to 1. Interestingly, this general procedure and the scheme defined in this section are related by duality. See Problem 7.

We close the section with a proof of a result alluded to in the discussion for College 4 relating the effects that changes in input values have on the measure of efficiency. It was noted there that $x_4 = 0$ in the solution of (8.4.2). Similarly, in the corresponding solution for the inefficient College 1 (Figure 8.7), we find $x_1 = 0$. This seems reasonable. A unit should not be used to help prove itself inefficient.

Lemma 8.4.1. *If the solution of (8.4.1) demonstrates that unit k is inefficient with* $\text{Min}\, e = e_0 < 1$, *then* $x_k = 0$.

Proof. From the input and output constraints of (8.4.1) we have

$$\sum\nolimits_{\text{all } i} x_i[\text{In } i] \le e_0[\text{In } k] \qquad \text{and} \qquad \sum\nolimits_{\text{all } i} x_i[\text{Out } i] \ge [\text{Out } k].$$

In particular, the first set of inequalities shows that $x_k[\text{In } k] \le e_0[\text{In } k]$, and thus $x_k \le e_0 < 1$.

Assume now that $x_k > 0$. Define $\bar{x}_i = \frac{x_i}{1-x_k}$ for $i \ne k$, $\bar{x}_k = 0$, and $e_1 = \frac{e_0 - x_k}{1-x_k}$. Then $e_1 < e_0$ as

$$e_0 < 1 \Rightarrow -x_k e_0 > -x_k \Rightarrow e_0 - x_k e_0 > e_0 - x_k \Rightarrow e_0 > \frac{e_0 - x_k}{1 - x_k} = e_1$$

We claim that setting $x_i = \bar{x}_i$ for $1 \le i \le n$ and $e = e_1$ defines a feasible solution to (8.4.1). If so, then the minimum value of the objective function e is less than or equal to $e_1 < e_0$, a contradiction. To verify the claim, consider first the outputs. We have

$$\begin{aligned}
\sum\nolimits_{\text{all } i} \bar{x}_i[\text{Out } i] &= \frac{1}{1-x_k} \sum\nolimits_{i \ne k} x_i[\text{Out } i] \\
&= \frac{1}{1-x_k} \left(\sum\nolimits_{\text{all } i} x_i[\text{Out } i] \right) - \frac{x_k}{1-x_k}[\text{Out } k] \\
&\ge \frac{1}{1-x_k}[\text{Out } k] - \frac{x_k}{1-x_k}[\text{Out } k] = [\text{Out } k].
\end{aligned}$$

Considering inputs, we have

$$\begin{aligned}\sum\nolimits_{\text{all } i} \bar{x}_i[\text{In } i] &= \frac{1}{1-x_k}\sum\nolimits_{i\neq k} x_i[\text{In } i] \\ &= \frac{1}{1-x_k}\left(\sum\nolimits_{\text{all } i} x_i[\text{In } i]\right) - \frac{x_k}{1-x_k}[\text{In } k] \\ &\leq \frac{1}{1-x_k}\, e_0\, [\text{In } k] - \frac{x_k}{1-x_k}[\text{In } k] \\ &= \frac{e_0 - x_k}{1-x_k}[\text{In } k].\end{aligned}$$

□

Problem Set 8.4

1. The efficiency rating for College 1 of the examples of this section is 87.96% from the spreadsheet in Figure 8.7. Using the data from the associated sensitivity report in Figure 8.8, for each part below (assume the parts are independent) determine the college's efficiency rating, when possible, if the described change in data is made.
   (a) Output C is increased from 150 to 180;
   (b) Output C is decreased from 150 to 120;
   (c) Output D is increased from 40 to 80;
   (d) 75 more certificates are awarded.

2. Show that College 1's efficiency becomes 1 if the competing composite college must be a weighted average of the colleges.

3. (a) Write out the dual to the problem of (8.4.2), the problem used to determine the efficiency of College 4. The dual has five variables, two associated with the input constraints (use u_1 and u_2) and three associated with the output constraints (use v_1, v_2, and v_3).
   (b) From Solver's sensitivity report (Figure 8.5), the solution to the dual is $u_1 = 0$, $u_2 = .02857$, $v_1 = .00186$, $v_2 = .00373$, $v_3 = 0$. Using these values:
      (i) Evaluate the dual objective function at this point. (You should get the college's efficiency rating.)
      (ii) Evaluate the constraint in the dual corresponding to the variable e. (You should find zero slack in the constraint at this point.)

4. A system has three operational units, each utilizing one input and delivering two outputs, with data as follows:

		Unit		
		1	2	3
Input	*A*	12	16	9
Outputs	*B*	20	28	18
	C	32	30	14

(a) Show that unit 1 is efficient. (The computations for this problem can be easily done using LP Assistant.)
(b) Show that unit 2 is inefficient, with an efficiency rating of 93.75%.
(c) Write out the linear programming problem used in part (b) and also its dual.
(d) From your solution in part (b), determine the solution to the dual.
(e) Verify that when evaluated at your solution in part (d), the dual objective function equals .9375, and that the dual constraint associated with the variable x_1 (the variable measuring unit 1's contribution to the competing composite unit) is an equality.
(f) Determine the minimum amount by which unit 2's output B of 28 needs to increase, with no other changes made, so that unit 2 is efficient.

5. Show that unit 2 of Problem 4 is efficient if the competing composite unit must be a weighted average of the units.

6. Six of the city of Buffalo's eight branch libraries are under an efficiency review. With input categories involving number of holdings and staff size and output categories related to number of patrons, number of checkouts, and computer usage, the following data have been compiled:

		Branch					
		1	*2*	*3*	*4*	*5*	*6*
Inputs	*A*	75	180	210	120	45	160
	B	60	120	75	145	40	130
Outputs	*C*	150	200	185	225	95	225
	D	80	155	275	165	60	105
	E	90	75	40	100	25	50

In fact, two of the branches are inefficient. Determine which two are inefficient, and with spreadsheets and sensitivity reports in hand, suggest how both branches can improve their efficiency.

7. In the paragraph containing equation (8.4.3), an alternate procedure is described for defining relative efficiency based on the definition of efficiency in (8.4.3) and the use of weights assigned to the various input and output categories by the unit under review.

(a) Using u_1 and u_2 for the relative weights of inputs A and B, and v_1, v_2, and v_3 for the relative weights of outputs C, D, and E, show that the following linear programming problem corresponds to the problem of College 4 when using this procedure to determine its relative efficiency.

$$\begin{aligned}
&\text{Maximize } 225v_1 + 125v_2 + 10v_3 \\
&\text{subject to} \\
&150v_1 + \ \ 40v_2 + 25v_3 \leq 30u_1 + 50u_2 \\
&100v_1 + \ \ 65v_2 + 20v_3 \leq 20u_1 + 15u_2 \\
&600v_1 + 275v_2 + 30v_3 \leq 55u_1 + 75u_2 \\
&225v_1 + 125v_2 + 10v_3 \leq 45u_1 + 35u_2 \\
&\ 45u_1 + \ \ 35u_2 = 1 \\
&u_1, u_2, v_1, v_2, v_3 \geq 0
\end{aligned}$$

(For general interest, note that if this procedure were used to determine the efficiency of each of the units, only the objective function and the terms in the one equality constraint would vary in the associated problems, similar to the commonality encountered in the procedure of (8.4.1).)

(b) Show that the dual to the above problem is the problem of (8.4.2), the problem we used to determine College 4's efficiency rating.

(c) The spreadsheet solution of the problem of part (a) appears in Figure 8.10, and the accompanying sensitivity report appears in Figure 8.11. Given this relationship,
   (i) find in the data in Figure 8.10 College 4's efficiency rating and marginal values (as listed in the sensitivity report in Figure 8.5), and
   (ii) find in the data of Figure 8.11 the solution point to (8.4.2).

	A	B	C	D	E	F	G
1	College 4 (sum = 1)						
2				College			
3		**Avg. Annual Values**	**1**	**2**	**3**	**4**	
4		Input A	30	20	55	45	
5		Input B	50	15	75	35	
6		Output C	150	100	600	225	
7		Output D	40	65	275	125	
8		Output E	25	20	30	10	
9							
10					**Variables**		
11			**1**	**2**	**3**	**4**	**Efficiency**
12			0.0000	0.7143	0.2857	0.0000	0.9184
13							
14		Minimize Efficiency (*e*)	0.9184				
15							College 4
16		**Constraints**	LHS		RHS		Data
17		Input A	30.0000	≤	41.3265		45
18		Input B	32.1429	≤	32.1429		35
19		Output C	242.8571	≥	225.0000		225
20		Output D	125.0000	≥	125.0000		125
21		Output E	22.8571	≥	10.0000		10
22		sum	1.0000	=	1.0000		

Figure 8.10

Adjustable Cells

Cell	Name	Final Value	Reduced Cost	Objective Coefficient	Allowable Increase	Allowable Decrease
\$H\$4	u1	0.0000	-6.8401	0	6.8401	1E+30
\$H\$5	u2	0.0286	0.0000	0	1E+30	5.3200
\$H\$6	v1	0.0019	0.0000	225	47.7273	27.2250
\$H\$7	v2	0.0037	0.0000	125	17.2070	21.8750
\$H\$8	v3	0.0000	-18.3696	10	18.3696	1E+30

Constraints

Cell	Name	Final Value	Shadow Price	Constraint R.H. Side	Allowable Increase	Allowable Decrease
\$C\$13	Eff. 1 ≤ 1	0.4286	0.0000	0	1E+30	1
\$C\$14	Eff. 2 ≤ 1	0.4286	1.1413	0	0.0779	0.0714
\$C\$15	Eff. 3 ≤ 1	2.1429	0.1848	0	0.4286	0.3297
\$C\$16	Eff. 4 ≤ 1	0.8851	0.0000	0	1E+30	0.1149
\$C\$17	Invalue = 1	1.0000	0.8851	1	1E+30	1

Figure 8.11

CHAPTER 9

TWO-PERSON, ZERO-SUM GAMES

9.1 INTRODUCTION TO GAME THEORY

In this chapter and the next, some topics from the theory of games are discussed. In this section, the subject is introduced primarily by means of examples. The remainder of the chapter is devoted to a special class of games: two-person, zero-sum games. For these games, a complete theory can be presented. Starting with the definition of a two-person, zero-sum game, we will formulate some general principles on which to base our notion of a *solution* to the game. Then we will show, using the Duality Theorem of Chapter 4, that such games always have solutions, and we will develop techniques for finding these solutions. Thus, for these games, a mathematical model can be formulated precisely and analyzed completely. Although the applicability of this theory is limited due to the restrictive assumptions that we make, the theory does serve as the springboard for the study of the more general classes of games, some of which are introduced in the next chapter.

By a *game* we mean roughly a situation of conflict between two or more people, in which each contestant, player, or participant has some, but not total, control over the outcome of the conflict. We assume that all players have complete knowledge of all actions, moves, or choices available to themselves and their opponents, and knowledge of the results of the conflict associated with any given selection of actions. Assuming that each player acts rationally to maximize his or her gain, our basic problem is to develop a theory that will help us to understand and predict human behavior or economic phenomena. As we will see, however, the translation of the statement "Each player acts rationally to maximize his or her gain" into mathematical terms is not always straightforward, but can lead to various interpretations of, approaches to, and solutions of a game. This is in contrast to the situation in linear programming, in which the optimization problems lead to a well-defined and generally accepted theory.

Parlor games such as poker, tic-tac-toe, and chess are of the type of game described above. On the other hand, games such as roulette or craps are not, since one simply plays against certain odds. And even though some of our examples and problems are of the parlor game variety, the broader scope of the theory must be recognized and appreciated. Indeed, the book that established game theory as a

An Introduction to Linear Programming and Game Theory, Third Edition. By P. R. Thie and G. E. Keough.

mathematical theory of its own, with abundant potential applications, is the work of von Neumann and Morgenstern entitled *Theory of Games and Economic Behavior*, written in 1944 [28]. Since then, much work on the theory has been done by both mathematicians and social scientists, but its full range of applications is still being explored.

We now give examples of some elementary games.

Example 9.1.1. Two players each have two cards, a 1 (or ace) and a 2. Each player selects one of her cards, with her choice unknown to her opponent. They then compare the two selected cards. If the sum of the face values of the two selected cards is even, Player 1, denoted by P_1, wins that sum from Player 2, denoted by P_2. If the sum is odd, P_2 wins that amount from P_1.

This is an example of a *two-person, zero-sum* game: two-person, because there are exactly two participants, and zero-sum, because the sum of the payments made to the players at the end of the game (with a negative payment indicating a loss) is equal to zero. In a two-person game, *zero-sum* means simply that what one player wins, the other player loses.

Each player has two possible courses of action or *strategies*, in this case, to play either a 1 or a 2. In general, by the term *strategy* we mean a rule of action or set of instructions that tells a player what to do under all possible circumstances.

The information critical to this game can be recorded very easily using matrix notation. If we let s_1 and s_2 denote the two strategies for P_1, with s_i meaning that P_1 plays card i and, similarly, let t_1 and t_2 denote P_2's two strategies, then all the possible outcomes for the game from P_1's point of view can be recorded by the following tableau:

	t_1	t_2
s_1	2	−3
s_2	−3	4

In this tableau, the entry in the ith row and jth column represents the amount that P_2 pays P_1 if P_1 plays s_i and P_2 plays t_j, with a negative sign indicating that P_2 wins the associated amount from P_1. Note that since the game is zero sum, we need only record the outcomes for one player, and we fix that player to be P_1. The 2×2 matrix

$$\begin{bmatrix} 2 & -3 \\ -3 & 4 \end{bmatrix}$$

is called the *payoff matrix* of the above game.

Example 9.1.2 (Extension of Example 9.1.1). Here the rules are just as in the game of Example 9.1.1, with the only difference being that P_1 is given initially one additional card, a 3. P_1 has now three strategies: to play either the 1, 2, or 3. Denoting these strategies by s_1, s_2, and s_3, respectively, the tableau associated with this game is the following:

	t_1	t_2
s_1	2	−3
s_2	−3	4
s_3	4	−5

Example 9.1.3 (Extension of Example 9.1.1)**.** The rules are as in Example 9.1.1, with each player receiving two cards — but after the players have selected their card and before the cards are compared, P_1 declares "even" or "odd." P_1 wins if and only if the sum of the face values of the selected cards is of the parity she has declared. P_1 has now four strategies: to declare even and play the 1 or play the 2; or to declare odd and play the 1 or play the 2. Denote these strategies by (even; 1), (even; 2), (odd; 1), and (odd; 2). P_2 still has only the two strategies t_1 and t_2.

	t_1	t_2
(even; 1)	2	−3
(even; 2)	−3	4
(odd; 1)	−2	3
(odd; 2)	3	−4

Example 9.1.4 (Extension of Example 9.1.3)**.** The rules are as in Example 9.1.3, except that now P_1 must declare "even" or "odd" before P_2 selects her card. P_1's set of strategies remain as in Example 9.1.3, but P_2 has now the opportunity to react to P_1's declaration of parity. Since a strategy for P_2 is a set of instructions telling P_2 what to do in all possible circumstances, here some possible strategies for P_2 are to "play the 1 if P_1 declares even and the 2 if P_1 declares odd" or to "always play the 1," etc. P_2 has, in fact, a total of four strategies, and they can be denoted by the set of ordered pairs (i, j), $1 \leq i, j \leq 2$, where strategy (i, j) is defined to mean to play card i if P_1 declares even and card j if P_1 declares odd. The payoff tableau is as follows:

	(1, 1)	*(1, 2)*	*(2, 1)*	*(2, 2)*
(even; 1)	2	2	−3	−3
(even; 2)	−3	−3	4	4
(odd; 1)	−2	3	−2	3
(odd; 2)	3	−4	3	−4

Example 9.1.5. Two players have two cards. P_1 has a red 5 and a black 4; P_2 has a red 7 and a black 8. Each player selects one of his cards, with his choice unknown to his opponent, and then the players compare the selected cards. If the selected cards are of the same color, P_1 wins the difference in face values from P_2; if the selected cards are of different colors, P_2 wins the difference in face values from P_1. Denoting P_1's two strategies by *R5* and *B4*, and P_2's by *R7* and *B8*, the game tableau is:

	R7	*B8*
R5	2	−3
B4	−3	4

This payoff matrix is identical to the payoff matrix for the game of Example 9.1.1, and so the two games are essentially equivalent. Thus, to study both games, we simply need to consider the two-person game having two strategies for each player and payoff matrix

$$\begin{bmatrix} 2 & -3 \\ -3 & 4 \end{bmatrix}$$

Example 9.1.6. P_1 has two cards, a 1 and a 2. P_1 selects a card, and P_2 attempts to guess the card selected. After hearing P_2's guess, P_1 can either "stop" or "double." If P_1 elects to stop, he reveals his selected card. P_1 wins from P_2 the face value of this selected card if P_2's guess was incorrect; and P_2 wins 2 from P_1 if P_2's guess was correct. If P_1 elects to double, then P_2 must, before the selected card is revealed, either "accept" or "reject" the double. If P_2 accepts the double, the winner is determined just as before, with the stakes doubled (P_1 wins 2 or 4 on an incorrect guess; P_2 wins 4 with a correct guess). If P_2 rejects the double, then P_1 wins the face value of the selected card (as before) if P_2's guess was incorrect, but now P_2 wins only 1 if her guess of P_1's card was correct.

If P_1 did not have the option of doubling the stakes, P_2 would have a definite advantage — winning 2 with a correct guess but losing only 1 or 2 with an incorrect guess. However, P_1 can increase the stakes, which seem especially propitious if P_2 has guessed incorrectly. To counter this, P_2 can reject the increase. In fact, if P_1 increases the stakes only when P_2 has guessed incorrectly and P_2 becomes aware of this strategy, she can always reject the increase and maintain her edge. Thus P_1 should opt for doubling occasionally after a correct guess by P_2 (a bluffing strategy), so that a constant rejection strategy by P_2 could hurt her. (If P_1 always doubles and P_2 always rejects, P_2's initial advantage has become P_1's — P_1 winning 1 or 2 on an incorrect guess, P_2 winning only 1 with a correct guess.)

To construct the payoff matrix for this two-person, zero-sum game, we first need to describe the available strategies for the players. The game is played in a sequence of four moves: P_1 selects a card; P_2 states her guess; P_1 opts to stop or double; and if P_1 doubles, P_2 accepts or rejects the increase. A strategy for a player must direct the player's action at each relevant point in this sequence, and must allow for reaction to information gained by the player up to each point. Thus a strategy for P_1 must first prescribe his card selection and then, given his choice of card and the accuracy of P_2's guess, the strategy must prescribe a betting action (either stopping or doubling). A strategy for P_2 must include P_2's guess and then, given her guess, the strategy must prescribe P_2's reaction to the doubling of the stakes by P_1.

Specifically, we can denote a strategy for P_1 as an ordered triple of the form $(n;X_c,X_i)$, where $n \in \{1,2\}$ directs P_1's card selection, $X_c \in \{S,D\}$ represents P_1's action (stop or double) to a correct guess by P_2, and $X_i \in \{S,D\}$ represents P_1's action to an incorrect guess by P_2. There are eight such distinct strategies. P_2's strategies can be represented by ordered pairs of the form $(n;Y)$, where $n \in \{1,2\}$ represents P_2's guess and Y equal to either A (accept) or R (reject) prescribes P_2's reaction to a double by P_1. There are four distinct strategies for P_2.

Actually, for this game, we can simplify our model. Note that if P_2's guess is incorrect, P_1 has nothing to lose by increasing the stakes: he wins double if P_2 accepts the increase and the same 1 or 2 if P_2 rejects the increase. Thus P_1 should always offer to double the stakes when P_2's guess is incorrect, and so we can reduce P_1's viable strategy set to pairs of the form $(n;X_c)$, where the second component $X_c \in \{S,D\}$ directs P_1's action to a correct guess by P_2. The payoff tableau is as follows:

	(1; R)	(1; A)	(2; R)	(2; A)
(1; S)	−2	−2	1	2
(1; D)	−1	−4	1	2
(2; S)	2	4	−2	−2
(2; D)	2	4	−1	−4

These entries are easily computed. For example, if P_1 uses strategy $(1;S)$ and P_2 strategy $(2;A)$, P_1 selects the 1, P_2 guesses 2, P_1 automatically doubles, P_2 accepts the increase, and P_1 wins 2; if P_1 uses $(2;D)$ and P_2 uses $(2;R)$, P_1 selects the 2, P_2 guesses 2, P_1 doubles, P_2 rejects the increase, and P_2 wins 1. The reader should verify the accuracy of the other 14 entries. (See Problem 6 of Section 9.5 for a further analysis of this game.)

Example 9.1.7. Two major automobile manufacturers compete for a fixed market of new-car buyers. The buyers in this group are attracted to a particular automobile for two reasons: the styling, features, and quality of the automobile and the intensity of the manufacturer's advertising campaign. Each of the manufacturers has fixed amounts of money, say M_1 and M_2 dollars, to divide between their Research and Development Division and their Product Promotion Division. Moreover, suppose the number of these new-car buyers attracted (or lost) to Manufacturer 1 because of product development is given the function $f(x,y)$, where x and y are the amounts of money spent by Manufacturers 1 and 2, respectively, on research and development. Similarly, suppose the function $g(x,y)$ measures the number of buyers attracted (or lost) to Manufacturer 1 because of advertising. Assuming that both manufacturers have the above information, how should they allocate their resources?

This is an example of a *two-person, infinite game*. Each player, (i.e., each manufacturer) must decide how to divide her resources between the two critical divisions, and each player has essentially an infinite number of choices available. To determine the payoff for this game, suppose Manufacturer 1 allots x dollars to her Research and Development Division, $0 \le x \le M_1$, and the remainder to her Product Promotion Division. Similarly, define y for Manufacturer 2, $0 \le y \le M_2$. Then the total number of car buyers attracted (or lost) to Manufacturer 1 is given by the function

$$A(x,y) = f(x,y) + g(M_1 - x, M_2 - y)$$

Games of this type have been studied, and a complete theory can be developed as long as the payoff function $A(x,y)$ is reasonably well behaved (e.g., when A is continuous). However, we will say no more about them in this text. For more information, see the books on game theory listed in the Bibliography, especially those by Dresher [29] and McKinsey [30].

Example 9.1.8. A town puts up for closed bidding its annual trash collection contract. Three firms compete for the job by submitting sealed bids, with the job awarded to that firm submitting the lowest bid less than or equal to \$50,000, since the town itself will manage the collection operation if all bids received exceed that figure. Because of the differences in costs of labor and efficiency of equipment, the actual cost of providing the service to the town varies from firm to firm, and these overhead costs are known to all the parties. Specifically, the cost to Firm 1 to provide the service would be \$38,000, to Firm 2 \$40,000, and to Firm 3 \$44,000. Economic necessity demands that the bids submitted by each firm exceed their individual overhead costs. How should each of the firms bid?

This is an example of a *three-person game*. Each of the firms must submit a bid determined by their overhead costs and the profit they wish to realize. The payoff (i.e., the profit actually realized by each of the firms as a result of the three bids) can then be easily determined. For example, if Firms 1, 2, and 3 bid \$44,000, \$43,000, and \$48,000, respectively, Firm 2 would be awarded the contract and realize a profit of \$3000, while the other two firms earn nothing.

The distinguishing factor between two-person game theory and n-person theory ($n \geq 3$) is the existence of the potential for the players to form coalitions in n-person games. In fact, n-person games are usually described in terms of their *characteristic function*, a function that measures the strength of each of the possible coalitions of players. Suppose, for example, that it is possible for any subgroup of the three firms in this situation to agree beforehand on the bids to be set by the members of that subgroup. We can then define a function v that measures the maximum profit that the subgroup can guarantee itself if all members in the group cooperate. For example, if Firms 1 and 2 decide to form a coalition, Firm 1 can bid \$44,000 and Firm 2 can bid anything greater. Since it is known that the bid of Firm 3 must exceed \$44,000, Firm 1 would be awarded the contract and realize a profit of \$6000. Thus the coalition of Firms 1 and 2 is worth \$6000, or $v(\{1,2\}) = \$6000$. The reader may verify that

$$\begin{aligned} &v(\{1\}) = v(\{1,3\}) = \$2000 \\ &v(\{1,2,3\}) = \$12{,}000 \\ &v(\{2\}) = v(\{3\}) = v(\{2,3\}) = \$0 \end{aligned}$$

We have now seen examples of the major types of games. Examples 9.1.1–9.1.6 are examples of finite, two-person, zero-sum games, the class of games that will be studied in this chapter. The games described in Section 1.3 are two-person, non-zero-sum games, the class of games that will be considered in the next chapter. Example 9.1.7 is an example of an infinite game, and Example 9.1.8 an example of an n-person game. The theory of n-person games is still very much in the development stage; for more information refer again to the game theory books in the Bibliography, especially those by Luce and Raiffa [31] and by Owen [32].

Problem Set 9.1

1. Determine the payoff matrices for the following two-person, zero-sum games.

(a) P_1 has two cards, a red 1 and a black 7. P_2 has three cards, a red 3, a red 8, and a black 9. Each selects a card, with his choice unknown to his opponent. The selected cards are then compared: if they are of the same color, P_1 wins the sum in face values of the selected cards from P_2; otherwise, P_2 wins that amount from P_1.

(b) As in part (a), except that if the colors of the selected cards are not the same, P_2 wins an amount x from P_1.

(c) P_1 selects a number n from $A \cup B$, where $A = \{1,2\}$ and $B = \{3,4\}$, and P_2 attempts to guess the selected number. If P_2's guess is correct, P_2 wins $2n$ from P_1; and if his guess is incorrect but in the same set A or B as n, the game is a draw. If P_2's guess is not in the same set as n, P_1 wins n from P_2.

(d) P_1 selects a number n from $\{1,2,3\}$, and P_2 is given two guesses. (P_2's guesses must be from $\{1,2,3\}$ but need not be distinct.) After P_2 makes her two guesses, P_1 reveals his selected number n. If P_2 did not guess n, P_1 wins $2n$ from P_2; if P_2 did guess n, P_2 wins from P_1 an amount equal to P_2's other guess.

(e) *Colonel Blotto.* The Colonel has three divisions available to attempt to capture a town accessible by two different roads. The defender has four divisions to divide between the two routes to defend the town. If the Colonel's forces outnumber the defender's along a route, the Colonel wins the number of enemy divisions on that road plus the town, equivalent to two enemy divisions. If the defender's forces outnumber the attacker's, the defender wins the attacker's divisions. If the numbers are equal, it is a draw. The total payoff is the sum of the results along the two routes.

(f) *Morra.* Two players simultaneously show one or two fingers and shout out a number. If the number announced by a player is the same as the total number of fingers shown by both players, then she wins that amount from her opponent. If both players guess correctly, the game is a draw.

2. For each of the following games, describe the strategies available to the players and state the size of the corresponding payoff matrix.

(a) P_1 and P_2 each have three cards: a king, a queen, and a jack. They play their cards, one at a time, with the high card winning the trick ($K > Q > J$) and the playing of equal cards scoring a draw for that trick). The first and second tricks are each worth 1 point. The last cards are played only if there is a draw among the first two plays, and if so, the third trick is worth 2 points. At the conclusion, the player with the greater number of points wins from the other player an amount equal to the difference in point totals.

(b) P_1 and P_2 each ante \$1 into the pot, and then P_1 is dealt one card from a deck containing two red cards and one black card. P_1 looks at her card, and then exercises the option of either "Increasing" the stakes or "Passing." If P_1 opts to increase, each player must add another \$3 to the pot; if P_1 opts to pass, nothing is added to the pot. Then P_2 attempts to guess the color of P_1's card. If P_2's guess is correct, P_2 wins the pot; if the guess is incorrect, P_1 wins the pot.

(c) P_1 and P_2 each ante an amount a into the pot. Then, from a single deck of four black cards and one red card, the players are each dealt a card, each seeing only his or her own card. P_1 then can either "Pass" or "Raise." If P_1 opts to pass, the players show their cards; P_1 wins the pot if the red card has been dealt to either player, and P_2 wins the pot if the red card has not been dealt out, that is, if both players have black cards. If P_1 opts to raise, he adds an amount b to the pot, and now P_2 can either "Fold" or "Call." If P_2 folds, P_1 wins the pot (and therefore the amount a from P_2), no matter what cards the players hold. If P_2 calls, she too adds an amount b to the pot. Then, as before, the players show their cards, with P_1 winning the pot if the red card has been dealt out and P_2 winning if not.

3. What would happen in the following situation? A stranger offers to possibly give either \$120 or \$150 to a group of three friends. If two members of the group can decide on how to divide the money between only themselves, the stranger will give \$150 to these two. If all three can agree on a way of sharing the money, the stranger will give the three \$120. If there is no agreement between any two or all three members of the group, the stranger keeps the money.

4. A submarine has one attempt to intercept and destroy an enemy ship transporting strategic cargo while the ship passes through a small, unprotected area. The submarine is sailing directly at the enemy, and initially the two are 2000 yd apart. The submarine can fire a round of torpedoes at any time, but the probability of a hit increases as the ships come closer because of the increase in the accuracy of the weapon and the decrease in the available time for the enemy to evade the torpedoes. Suppose, in fact, that the probability of a hit is given by the function $p(x) = (2000 - x)/2000$, $0 \le x \le 2000$, where x is the distance in yards between the ships. However, the cargo ship has one antisubmarine missile at its disposal, and the ship's captain is aware of the approaching submarine. If the missile can be landed within 50 yd of the submarine, its effect would so disrupt the operation of the submarine that its torpedoes would not be able to be launched and safe passage through the unprotected waters would be guaranteed the cargo ship. However, the accuracy of this missile also increases as the target comes closer, with the probability of a hit given by the function

$$q(y) = \begin{cases} \dfrac{50}{y}, & y \ge 50 \\ 1, & y \le 50 \end{cases}$$

where y is again the distance in yards between the ships.

Assume that the cargo ship's captain knows at all times the distance between the two ships and also whether or not the submarine has released its torpedoes. Similarly, assume that the submarine commander knows at all times whether or not the missile has been fired from the ship, even if the missile does not land near the submarine.

The captain and the commander must decide at what point to fire their weapons. The goal of the submarine commander is to prevent the passage of the cargo ship; the goal of the cargo ship's captain is to deliver the strategic cargo. Determine a payoff function, a function of the points at which the two weapons are fired, that measures the probability of the success of the submarine commander.

9.2 Some Principles of Decision Making in Game Theory

All the problems discussed in this book fall into the general category of problems in decision making. By the term *decision making* we mean in general a situation in which, from a given set of possible courses of action, a specific course of action must be selected that is in some way preferred over the alternatives. The field of decision making is extremely broad, as is apparent from the not too restrictive definition given above. However, the field can be narrowed somewhat by partitioning according to whether a decision is made under conditions of certainty, risk, or uncertainty.

A decision is made *under conditions of certainty* if all the available courses of action lead to specific, fixed outcomes. Linear programming is an example. For instance, in the standard diet problem, the cost and nutritional content of any given diet were known, and the decision problem was reduced to a problem of optimizing a linear function on a domain restricted by a system of linear constraints. The optimization problems of freshman calculus are other examples of problems in decision making under conditions of certainty.

A decision is made *under conditions of risk* if a given course of action can lead not to a unique outcome, but to a set of possible outcomes, each outcome in the set occurring with a specified probability known to the decision maker. For example, when playing roulette, one must decide on what color or number or combination of numbers to bet on. The possible outcomes for these actions, either winning an amount determined by the type and amount of the original bet or losing the original bet, occur with known, definite probabilities depending (presumably) only on the type of bet.

A decision is made *under conditions of uncertainty* if a given course of action can lead to a set of possible outcomes but the probability of these outcomes occurring is unknown. In a general sense, game theory falls into this category, because the players in a game are ignorant of their opponents' moves. However, as we will see shortly, we strive to reduce this element of uncertainty in game theory by attempting to predict our opponents' courses of action, based on some reasonable principles. In fact, the phrase *decision making under uncertainty* is usually reserved for a definite theory that excludes game theory but includes a study of the experimentation and statistical analysis that can be used to reduce the element of uncertainty.

Our primary concern in the remainder of this chapter will be with two-person, zero-sum games, in which each player has only a finite number of possible courses of action (i.e., strategies). Let us agree to call these games *matrix games* since, as we have seen in the previous section, such a game can be represented by its payoff matrix. Recall that if $\{s_1, s_2, \ldots, s_m\}$ is P_1's strategy set and $\{t_1, t_2, \ldots, t_n\}$ is P_2's strategy set, the payoff matrix is that $m \times n$ matrix $A = (a_{ij})$ in which, for $1 \leq i \leq m$ and $1 \leq j \leq n$, the entry a_{ij} is the outcome of the game for P_1 if P_1 uses strategy s_i and P_2 uses strategy t_j. In the examples we have seen so far, the numerical values of the a_{ij}'s translated immediately into dollars or some other monetary unit, and, in general, in game theory we assume that the desirability of an outcome for a player can be measured by a real number. That this is always possible is not immediately obvious. Consider, for example, a game-type situation between two conglomerates where the outcome is either the successful elimination of the competition or an antitrust suit. However, the problem of translating the desirability of various alternatives into numerical values has been studied, and the resulting theory is called *utility theory*. This topic is discussed in the next chapter, but for the time being, we will assume that the outcomes of all of our games can be expressed using numerical values.

To illustrate the conditions of uncertainty present in these simple games, consider the game represented by the following payoff matrix:

	t_1	t_2
s_1	0	2
s_2	8	−8
s_3	−5	5

Here P_1 has three strategies and P_2 has two. Suppose now that we were in the position of P_1 and about to play the game. What strategy should we choose to maximize our winnings? We have three courses of action, but clearly, the resulting outcome of these actions is uncertain, being contingent on the play of P_2. For example, playing strategy s_2 will either win or lose for us 8 units, depending on whether P_2 plays t_1 or t_2. If we expect P_2 to play t_1, then s_2 would be our obvious choice, but if we expect P_2 to play t_2, then playing s_3 would realize our maximum gain. On the other hand, by playing s_1, we are assured of not losing, and possibly of winning 2. But if we do play this strategy, P_2 can simply play t_1, with no one winning and our seemingly advantageous position nullified. But expecting P_2 to play t_1 suggests the response s_2, as mentioned above.

We could go on and on with these circular arguments, leading us to no definite position. What we need are some reasonable and acceptable principles of play that, if followed, will enable us to predict our opponent's play and suggest our own play. Only with some such precise statement of the goals of the players can we hope to develop a mathematical model and, depending on the applicability of the principles, a theory that can assist in understanding and predicting human behavior.

A very simple rule for play that we could consider is that each player choose a strategy that has as a possible outcome the most favorable outcome for the game for that player. Thus P_1 should choose strategy s_i if the ith row of A contains the largest entry for A. For example, for the game with payoff matrix

	t_1	t_2
s_1	100	−50
s_2	6	5

P_1 would play s_1 if he is using this rule to guide his choice, because the largest entry of the payoff matrix, 100, is in the first row. Is this in any way a reasonable choice? Certainly for this game the answer must be "no!" P_2 is going to play t_2 and, by playing s_1, P_1 would lose 50 units, whereas by playing s_2, P_1 would win 5 units. Thus, this proposed rule of play is unacceptable.

Another suggestion, somewhat similar to the first, is that each player choose that strategy for which the sum of the entries of the payoff matrix in the associated row or column is most favorable. However, in the above example, the sum of the entries in the first row of the payoff matrix is 50 and the sum in the second row is 11. Thus the use of this rule as a guide places P_1 in the same untenable position as before, so this rule too must be rejected.

The above suggested principles proved unacceptable because they represented a rather naive approach to the game. A player following one of these principles, while attempting to maximize his gain, would be completely ignoring the potential moves of his opponent, whose interests are strictly opposed to his own, as the game is zero-sum — what one player wins, the other one loses. A more reasonable approach, especially for this type of game in which interests are in direct conflict, would be to use the strategy that can guarantee the largest gain regardless of what the opponent does. In other words, a player determines the least amount that he can gain from playing each of his strategies and then chooses that strategy corresponding to the maximum of these least amounts.

For example, in the above 2×2 game, P_1 could lose 50 by playing s_1 but is certain of winning at least 5 by playing s_2 Thus, using this principle as a guide, P_1 should play s_2, because this play guarantees the larger gain independent of P_2's action. Similarly, P_2 could lose as many as 100 units by playing t_1, but no more than 5 by playing t_2, and so his choice would be t_2.

This principle represents a much more conservative approach to the game. By following it, each player is giving complete recognition to his opponent's capabilities and then is acting to maximize his guaranteed gain, or better, his security level. The word *gain* indicates winnings, whereas we mean either maximized winnings or minimized losses, whichever the case may be. In a game with high stakes representing some economic, military, or social situation, it seems realistic to expect the players to play conservatively, not take chances, and give full recognition to the capabilities of their opponent. Thus we establish this principle as a rule of action governing the play of the participants of a game, and we refer to it as Principle I of Game Theory.

Principle I *Each player acts to maximize his or her security level.*

Example 9.2.1. Consider the game given by the tableau

	t_1	t_2	t_3	t_4
s_1	-1	3	-1	-2
s_2	1	2	0	1
s_3	-1	-3	0	2

By playing s_1, P_1 could lose up to 2 units; by playing s_2, P_1 would at least break even; and by playing s_3, P_1 could lose up to 3 units. Guided by Principle I, then, P_1 would choose strategy s_2. Similarly, P_2 could lose up to 1, 3, or 2 units by playing t_1, t_2, and t_4, respectively, but by playing t_3, she would at least break even. Thus P_2 is led to strategy t_3.

There is one obvious complication in using only this principle as a guide to the play of a game. Since we are assuming that both players are intelligent, the acceptance of this principle implies that each player can anticipate the opponent's move. Thus each player would certainly consider the possibility of improving his or her outcome for the game by changing strategies in anticipation of the expected move of the opponent. Consider, for example, the game of Example 9.2.1. P_1 realizes that P_2 maximizes her security level by playing t_3, and he thus asks if playing s_2 is the best response to strategy t_3. In this case the answer is "yes," since he has nothing to gain by playing s_3 and could lose 1 by playing s_1. Similarly, P_2 would expect P_1 to play s_2, but by considering all the entries in the second row of the payoff matrix, P_2 would still choose strategy t_3. However, this resulting stability with the strategy choices of s_2 and t_3 need not always be present.

Example 9.2.2. Consider the game given by the tableau

	t_1	t_2
s_1	-2	4
s_2	1	-3

Principle I leads P_1 to choose s_1 since, by playing s_1, P_1 could lose 2, but by playing s_2, he could lose 3. Similarly, P_2 is led to strategy t_1, where her losses are at most 1. However, if P_1 suspects that P_2 will play t_1, he should alter his strategy and play s_2, since he would then win 1 instead of losing 2. But now we can go on. If P_2 expects P_1 to play s_2, she should change her strategy from t_1 to t_2. And so forth.

The above example shows that Principle I alone does not adequately reflect the behavior of rational players for all games. We must also seek in a prescription for playing a game strategies for the players that are stable or in equilibrium. A strategy pair (s,t) is in *equilibrium* if P_1, expecting P_2 to play t, has nothing to gain by deviating from playing s, and P_2, expecting P_1 to play s, has nothing to gain by deviating from playing t. In Example 9.2.1, the strategy pair (s_2,t_3) is in equilibrium; in Example 9.2.2, the pair (s_1,t_1) is not. Since any reasonable description of the play of rational players should include this element of potential reaction of a player to the anticipated move of his opponent, we consider the need for stability in establishing Principle II of Game Theory.

Principle II *The players tend to use strategy pairs that are in equilibrium.*

In the next few sections, Principles I and II will be translated into precise mathematical statements. Then, once we broaden our concept of strategy to include what will be called *mixed strategies*, we will show that for any matrix game there is a pair of strategies that satisfies these two principles, and the outcome of the game played with these strategies measures the value or worth of the game to the two players.

Since we can find such a strategy pair, we say that matrix games are *solved*. But it must be emphasized that these games are completely determined and the play of the participants is accurately predicted only upon the acceptance of the two principles. Our model will provide a norm for human behavior and a measure of the value of a game only if the players use the two principles as a guide to play. In any mathematical model, the applicability of the theory is limited to those examples for which the underlying axioms of the theory are valid.

Problem Set 9.2

1. Find strategy pairs that satisfy Principles I and II for the games with the following payoff matrices:

(a) $\begin{bmatrix} 3 & 1 & 2 \\ 1 & 0 & 5 \end{bmatrix}$

(b) $\begin{bmatrix} 7 & 1 & 5 & 9 \\ 1 & 0 & 3 & 2 \\ 6 & 3 & 6 & 4 \end{bmatrix}$

2. Consider the game with payoff matrix

$$\begin{bmatrix} 1 & -1 \\ -1 & 1 \end{bmatrix}$$

How would you play it? What do you think the term *mixed strategy* means? Does your suggested definition apply to possible strategies for the game

$$\begin{bmatrix} 3 & -1 \\ -1 & 1 \end{bmatrix}$$

3. We all have an intuitive idea of what it means to say that a game is *fair*. Would you accept the following definition for matrix games: A matrix game is fair if the sum of all the a_{ij}'s is zero?

4. On a cloudy morning, a baseball fan must decide whether to travel 50 miles that day to see her favorite team play. Let us say that the fan feels that witnessing her team play is worth 20 units (of satisfaction?), but that traveling to the stadium only to have the game postponed on account of rain is worth −10 units. On the other hand, by staying home and working on her book, she can accumulate up to 3 units. Is the fan making a decision under conditions of certainty, risk, or uncertainty? Is she playing a game with nature as the opponent? If so, should she accept our two principles as guides to her action?

9.3 SADDLE POINTS

In this section, we characterize games for which strategies of the form s_i and t_j can be found to satisfy the two principles set out in the previous section. (Strategies calling for the play of a single row or single column are called *pure strategies*, in contrast to a more general type of *mixed strategy*, which will be defined in Section 9.4.)

Principle I states that the players seek to maximize their security levels. Guided by this principle, Player 1, as we have seen, should determine the minimum entry of each row of the payoff matrix and then consider playing that strategy corresponding to a row at which the maximum of these row minimums is attained. We define these terms precisely for a game with an $m \times n$ payoff matrix $A = (a_{ij})$.

Definition 9.3.1. The *security level* for the (pure) strategy s_i is the minimum of the entries in the ith row of A, that is,

$$\operatorname*{Min}_{1 \leq j \leq n} a_{ij}$$

Suppose the maximum of these row minimums occurs in row h, and define u_1 to be this maximum. Thus

$$u_1 = \operatorname*{Max}_{1 \leq i \leq m} \operatorname*{Min}_{1 \leq j \leq n} a_{ij} = \operatorname*{Min}_{1 \leq j \leq n} a_{hj}$$

Similarly, Player 2 should determine the maximum entry in each column and then consider using that strategy corresponding to a column at which the minimum of these column maximums is attained.

Definition 9.3.2. The *security level* for the (pure) strategy t_j is the maximum of the entries in the jth column of A, that is,

$$\operatorname*{Max}_{1 \leq i \leq m} a_{ij}$$

Suppose the minimum of the column maximums is attained in column k, and define u_2 to be this minimum. Thus

$$u_2 = \operatorname*{Min}_{1 \leq j \leq n} \operatorname*{Max}_{1 \leq i \leq m} a_{ij} = \operatorname*{Max}_{1 \leq i \leq m} a_{ik}$$

Example 9.3.1. In the following, the row minimums are written to the right of the payoff matrix, and the circled numbers are the maximum of these minimums. Similarly, the column maximums are written below the matrix, and the smallest are circled.

$$\text{(a)} \quad \begin{bmatrix} 10 & 5 & 5 & 20 & 3 \\ 10 & 15 & 10 & 17 & 25 \\ 7 & 12 & 8 & 9 & 8 \\ 5 & 13 & 9 & 10 & 5 \end{bmatrix} \begin{matrix} 3 \\ \textcircled{10} \\ 7 \\ 5 \end{matrix} \qquad \begin{matrix} u_1 = 10, h = 2 \\ u_2 = 10, k = 1 \text{ or } 3 \end{matrix}$$
$$\begin{matrix} \textcircled{10} & 15 & \textcircled{10} & 20 & 25 \end{matrix}$$

$$\text{(b)} \quad \begin{bmatrix} 1 & 3 \\ 4 & 2 \end{bmatrix} \begin{matrix} 1 \\ \textcircled{2} \end{matrix} \qquad \begin{matrix} u_1 = 2, h = 2 \\ u_2 = 3, k = 2 \end{matrix}$$
$$\begin{matrix} 4 & \textcircled{3} \end{matrix}$$

The question we must next consider, given Principle II, is, "When is the strategy pair (s_h, t_k) in equilibrium?" In Example 9.3.1(a), it can be seen that each of the pairs (s_2, t_1) or (s_2, t_3) is in equilibrium; if P_2 expects P_1 to play s_2, her best response is either t_1 or t_3, and if P_1 anticipates P_2 to play either t_1 or t_3, he can gain nothing by deviating from strategy s_2. However, in Example 9.3.1(b), the pair (s_2, t_2) is not in equilibrium, since P_1 can benefit by deviating from s_2, provided that P_2 plays t_2. Thus, only in Example 9.3.1(a) is it reasonable to say that we have a *solution* to the game. Notice that in the first game $u_1 = 10 = u_2$, but that in the second, $u_1 = 2 \neq 3 = u_2$.

In fact, equality of the u_1 and u_2 is the simple condition we seek for determining the existence of stability, as the corollary to the following theorem will imply.

Theorem 9.3.1. $u_1 \leq u_2$.

Proof. By our definition of h and k, the maximum of the row minimums occurs in row h and the minimum of the column maximums occurs in column k. Suppose entry a_{hj} is the minimum of row h and a_{ik} is the maximum of column k. Since entry a_{hk} is in row h and column k, we have $a_{hj} \leq a_{hk}$ and $a_{hk} \leq a_{ik}$. Thus

$$u_1 = a_{hj} \leq a_{hk} \leq a_{ik} = u_2$$

Note that the possibility of $j = k$ or $i = h$ has not been excluded.

A simple way to visualize this proof is to consider the payoff matrix

$$\begin{bmatrix} u_1 = a_{hj} \leq & a_{hk} \\ & \text{\rotatebox{90}{$\geq$}} \\ & a_{ik} = u_2 \end{bmatrix}$$ □

Corollary 9.3.1. *If $u_1 = u_2$, then $a_{hj} = a_{hk} = a_{ik}$, and entry a_{hk} is both a minimum of row h and a maximum of column k.*

Thus $u_1 = u_2$ implies that entry a_{hk} is both a row minimum and a column maximum. It follows immediately that the strategy pair (s_h, t_k) is in equilibrium, and so these strategies satisfy both of our principles. Thus, in this case, we say that the strategy pair (s_h, t_k) is a *solution to the game* and that the *value of the game* is this common value $u_1 = u_2$.

We single out this property of the entry a_{hk} by a special definition. It turns out that the existence of such an entry is not only a necessary but also a sufficient condition for equality of u_1 and u_2.

Definition 9.3.3. An entry a_{hk} of a payoff matrix A is a *saddle point* if $\text{Min}_{1\leq j\leq n}\, a_{hj} = a_{hk} = \text{Max}_{1\leq i\leq m}\, a_{ik}$, that is, if a_{hk} is the minimum of row h and the maximum of column k.

Theorem 9.3.2. $u_1 = u_2$ *if and only if the payoff matrix A has a saddle point.*

Proof. The above corollary proves that $u_1 = u_2$ implies the existence of a saddle point. Suppose now that A has a saddle point, say the entry a_{hk}. Since a_{hk} is a column maximum, all other entries in the kth column must be less than or equal to a_{hk}. Thus the minimum values of all the rows other than row h must be less than or equal to a_{hk}, since the kth entry in each row has this property. But since a_{hk} is also the minimum of row h, a_{hk} equals the maximum of the row minimums; that is, $a_{hk} = u_1$. Similarly, $a_{hk} = u_2$. Hence $u_1 = u_2$. □

In summary, we have shown that the play of those matrix games for which $u_1 = u_2$ is completely determined by our two principles. Player 1 should play any row in which the maximum of the row minimums is attained, and Player 2 should play any column in which the minimum of the column maximums is attained. The value of such a game is $u_1 = u_2$, since this is the expected outcome. And a necessary and sufficient condition for u_1 to equal u_2 is that the payoff matrix A has a saddle point. In fact, if a_{hk} is a saddle point, P_1 should play row h, P_2 should play column k, and the value of the game is $a_{hk} = u_1 = u_2$.

Problem Set 9.3

1. Do the following payoff matrices have saddle points? If they do, what is a solution and value of the corresponding game?

(a) $\begin{bmatrix} 9 & 7 & 8 & 10 \\ 6 & 5 & 12 & 8 \\ 8 & 10 & 5 & 9 \end{bmatrix}$

(b) $\begin{bmatrix} 2 & 6 & 1 & 2 \\ 3 & 5 & 4 & 3 \\ 1 & 0 & 2 & 4 \end{bmatrix}$

(c) $\begin{bmatrix} 1 & 1 & 1 & 2 \\ 1 & 2 & 0 & 0 \\ 1 & 2 & 1 & 1 \end{bmatrix}$

(d) $\begin{bmatrix} 4 & -3 & 7 & -4 & 9 \\ 5 & -2 & -1 & -2 & 0 \\ 6 & -8 & 8 & -2 & -3 \end{bmatrix}$

2. For each of the following payoff matrices, determine the set of values of x for which the game has a saddle point, and for x in this set, determine the saddle point.

   (a) $\begin{bmatrix} 1 & 2 \\ 3 & x \end{bmatrix}$

   (b) $\begin{bmatrix} 1 & 3 \\ 2 & x \end{bmatrix}$

   (c) $\begin{bmatrix} x & 1 \\ 3 & x \end{bmatrix}$

   (d) $\begin{bmatrix} 2 & 1 \\ 3 & x \end{bmatrix}$

   (e) $\begin{bmatrix} 2 & 3 \\ 1 & x \end{bmatrix}$

   (f) $\begin{bmatrix} 3 & 1 \\ 2 & x \end{bmatrix}$

3. Suppose a matrix has two saddle points. Prove that they have the same numerical value. (*Hint*. For a very short proof, use the result in the proof of Theorem 9.3.2.) Thus the value of any game with a saddle point is unique.

4. Why is the word *saddle* used to describe a point of a matrix that is both a row minimum and a column maximum?

5. Suppose entries a_{ij} and a_{hk} are saddle points of a matrix. What can you say about the entries a_{ik} and a_{hj}?

6. Show that if the strategy pair (s_h, t_k) is in equilibrium, the entry a_{hk} of the payoff matrix is a saddle point.

9.4 MIXED STRATEGIES

As we have seen, games with saddle points are completely determined by the two principles set out in Section 9.2. For other games, however, we can use Principle I to lead us to suggested strategies for each player, but these strategies do not turn out to be in equilibrium. Thus, for these games, we are still faced with the problem of determining if strategies of some sort exist for the two players that satisfy Principles I and II and, if so, how to find them.

Although the reader may think that we have placed ourselves in an impossible situation, because certainly no strategies that we know could be stable for, say, the game with payoff matrix

$$\begin{bmatrix} 1 & -1 \\ -1 & 1 \end{bmatrix}$$

the solution to our dilemma is both realistic and elementary and consists of simply broadening our concept of possible strategies. Consider, for example, the above game. Even though each player has only two choices, it is intuitively clear that the players would choose either of their strategies with equal probability, and in a manner that would not give their opponent an opportunity to predict their move. Thus Player 1, each time the game is played, would randomly select either s_1 or s_2, choosing s_1 with probability $\frac{1}{2}$ and s_2 with probability $\frac{1}{2}$. This extension of the concept of strategy is called a *mixed strategy*.

Definition 9.4.1. A *mixed strategy* for P_1 is a vector $X = (x_1, x_2, \ldots, x_m)$ of nonnegative real numbers satisfying the condition $x_1 + x_2 + \cdots + x_m = 1$, with the interpretation that P_1 plays strategy s_i with probability x_i, $1 \leq i \leq m$.

Similarly, a mixed strategy $Y = (y_1, y_2, \ldots, y_n)$ for P_2 is defined. The set of all mixed strategies for P_1 will be denoted by S and those for P_2 by T. A strategy calling for the play of only one row or column can be considered a special case in the set of mixed strategies. For example, the m-tuple $(1, 0, \ldots, 0)$ is a mixed strategy corresponding to the pure strategy of P_1 always playing s_1.

The interpretation of a mixed strategy is important. For example, a mixed strategy of $(\frac{2}{3}, \frac{1}{3})$ for P_1 does not mean that P_1 should play s_1 twice and then s_2 once, but that at each play of the game P_1 should play s_1 with probability $\frac{2}{3}$ and s_2 with probability $\frac{1}{3}$. One way to implement this would be for P_1 to roll a die at each play and use s_1 if the 1, 2, 3, or 4 comes up and s_2 if the 5 or 6 comes up.

Since the notion of a mixed strategy does embody an idea that a player may realistically use in a game, we need a way of evaluating the outcome of games for which mixed strategies are employed. For this we use the concept of *expected value* from probability theory. The expected value of an event is defined simply to be the sum of the values of each possible outcome of the event times the probability that the outcome occurs. For example, if a fair die were to be rolled and you were to win an amount in dollars equal to the number rolled if that number were even, and otherwise you were to lose \$3, the expected value to you would be $\$2(\frac{1}{6}) + \$4(\frac{1}{6}) + \$6(\frac{1}{6}) - \$3(\frac{1}{2}) = \$0.50$. And for a game with payoff matrix $A = (a_{ij})$, if P_1 uses strategy $X = (x_1, x_2, \ldots, x_m)$ and P_2 uses strategy $Y = (y_1, y_2, \ldots, y_n)$, the outcome a_{ij} will occur with probability $x_i y_j$, since this is the probability of both P_1 playing s_i and P_2 playing t_j. Thus the *expected payoff* for the game is the sum of all the products $x_i a_{ij} y_j$ that is, the sum

$$\sum_i \sum_j x_i a_{ij} y_j$$

It can be easily seen that this sum is simply the product XAY^t by a direct computation of XAY^t. (Note that since mixed strategies are expressed as row vectors, the transpose of the vector Y gives us the appropriate column vector for multiplication on the right of A.)

Definition 9.4.2. The *expected payoff* for a game with payoff matrix $A = (a_{ij})$, $1 \leq i \leq m$, $1 \leq j \leq n$, in which P_1 uses strategy $X = (x_1, x_2, \ldots, x_m)$ and P_2 uses strategy $Y = (y_1, y_2, \ldots, y_n)$, is

$$XAY^t = \sum_{1\le i\le m}\sum_{1\le j\le n} x_i a_{ij} y_j$$

Example 9.4.1. Consider the game with payoff matrix

$$A = \begin{bmatrix} 1 & 3 \\ 4 & 0 \end{bmatrix}$$

We have that $u_1 = 1$, so P_1 can secure 1 unit by always playing s_1. Consider, however, the result for P_1 of using the mixed strategy $X = (\frac{1}{2}, \frac{1}{2})$. If P_2 responds with t_1, the expected payoff is $1(\frac{1}{2}) + 4(\frac{1}{2}) = \frac{5}{2}$, and if P_2 responds with t_2, the expected payoff is $\frac{3}{2}$. In fact, if P_2 uses strategy $Y = (y_1, y_2)$, the expected payoff is

$$\begin{aligned}(\tfrac{1}{2}, \tfrac{1}{2}) \begin{bmatrix} 1 & 3 \\ 4 & 0 \end{bmatrix} \begin{bmatrix} y_1 \\ y_2 \end{bmatrix} &= (\tfrac{5}{2}, \tfrac{3}{2}) \begin{bmatrix} y_1 \\ y_2 \end{bmatrix} \\ &= \tfrac{5}{2}y_1 + \tfrac{3}{2}y_2 \\ &\ge \tfrac{3}{2}y_1 + \tfrac{3}{2}y_2 \\ &= \tfrac{3}{2}(y_1 + y_2) \\ &= \tfrac{3}{2}\end{aligned}$$

Thus, by using the strategy $(\frac{1}{2}, \frac{1}{2})$, P_1 can secure an expected payoff of at least $\frac{3}{2}$ because no matter what P_2 plays, we have $(\frac{1}{2}, \frac{1}{2})AY^t \ge \frac{3}{2}$. This is not to say that on any one play P_1 would win no less than $\frac{3}{2}$, but it means that by adhering to this particular mixed strategy, P_1 would win on the average at least $\frac{3}{2}$ per game. Thus, by using this strategy, P_1's security level has increased from 1 to $\frac{3}{2}$. Can P_1 do better? What is the best that P_1 can do?

As suggested by this example, we must reconsider the application of the two basic principles in light of the fact that each player has at his or her disposal an infinite set of mixed strategies. Our development initially will follow the same steps as the development in Section 9.3, with the difference being that the sets of pure strategies are replaced with their generalization, the sets of mixed strategies.

Definition 9.4.3. For a game with payoff matrix A, define the *security level* of a (mixed) strategy X_1 for P_1 to be

$$\operatorname*{Min}_{Y\in T} X_1 AY^t$$

and the security level of a (mixed) strategy Y_1 for P_2 to be

$$\operatorname*{Max}_{X\in S} XAY_1^t$$

Thus the security level of a strategy X_1 for P_1 is the worst possible expected payoff if P_1 uses X_1, allowing that P_2 has available the full set of mixed strategies T. According to Principle I, P_1 should now seek a strategy that has the maximum security level, that is, the maximum of these guaranteed minimal gains. But before we consider these terms, we simplify the definition of the security level of a strategy.

First, we need some notation. For a matrix A, let $A_{(i)}$ denote the ith row of A and $A^{(j)}$ the jth column (as in Section 5.2). With this notation, for a game with payoff matrix A, the expected payoff XAY^t can be expressed in two ways, as

$$\begin{aligned} XAY^t = (XA)Y^t &= \sum_{1\leq j\leq n} (XA^{(j)})y_j \\ = X(AY^t) &= \sum_{1\leq i\leq m} x_i(A_{(i)}Y^t) \end{aligned}$$

where $X = (x_1, x_2, \ldots, x_m)$ and $Y = (y_1, y_2, \ldots, y_n)$. Note that the term $XA^{(j)}$ is simply the expected payoff if P_1 uses strategy X and P_2 uses the pure strategy t_j, and similarly, the term $A_{(i)}Y^t$ is the expected payoff if P_1 uses the pure strategy s_i and P_2 the strategy Y.

Theorem 9.4.1. *For a fixed strategy* X_1 *for* P_1,

$$\operatorname*{Min}_{Y\in T} X_1AY^t = \operatorname*{Min}_{1\leq j\leq n} X_1A^{(j)}$$

Similarly, for a fixed strategy Y_1 *for* P_2,

$$\operatorname*{Max}_{X\in S} XAY_1^t = \operatorname*{Max}_{1\leq i\leq m} A_{(i)}Y_1^t$$

Proof. Take a fixed strategy X_1 for P_1, Then, for each j, $X_1A^{(j)}$ is a real number. Let

$$w = \operatorname*{Min}_{1\leq j\leq n} X_1A^{(j)}$$

Then, for any $Y \in T$,

$$\begin{aligned} X_1AY^t &= \sum_{1\leq j\leq n} (X_1A^{(j)})y_j \\ &\geq \sum_{1\leq j\leq n} wy_j \\ &= w\sum_{1\leq j\leq n} y_j \\ &= w\cdot 1 = w \end{aligned}$$

But clearly, the outcomes $X_1A^{(j)}$ are contained in the set $\{X_1AY^t : Y \in T\}$, since they correspond to the outcomes when P_2 uses her pure strategies. Thus

$$\operatorname*{Min}_{Y\in T} X_1AY^t = \operatorname*{Min}_{1\leq j\leq n} X_1A^{(j)}$$

Similarly,

$$\operatorname*{Max}_{X\in S} XAY_1^t = \operatorname*{Max}_{1\leq i\leq m} A_{(i)}Y_1^t$$

for a fixed $Y_1 \in T$. □

Thus, if P_1 wishes to determine his security level for a strategy X_1, he need not consider the set of all possible outcomes X_1AY^t for $Y \in T$, but only those outcomes corresponding to the use of pure strategies by P_2. Similarly, the security level for a strategy Y_1 for P_2 is the worst (for P_2, and therefore the maximum) outcome of the game with P_2 using Y_1 and P_1 restricted to using only pure strategies.

Here we make two other observations. First, in our definitions of security level for a strategy X for P_1 and a strategy Y for P_2, we used the words *minimum* and *maximum* to be taken over the infinite sets T and S, respectively. Since these sets are infinite, to be precise we should have used the terms *greatest lower bound* and *least upper bound* and questioned the boundedness of the set of possible values of the XAY^t. However, from Theorem 9.4.1 it follows that the use of *minimum* and *maximum* is completely justified. Determining the greatest lower bound and the least upper bound over the infinite sets is equivalent to determining a minimum and a maximum over finite sets.

Second, in the previous section, when we defined the security level of a pure strategy, the response of the other player was limited to the use of pure strategies. In the above, more general definition of security level, we allow the use of the full set of mixed strategies. Theorem 9.4.1 shows that these two definitions agree on their common domain.

Principle I suggests that the players seek strategies that deliver their optimal security levels. Thus we define:

Definition 9.4.4. For a game with payoff matrix A, define P_1's and P_2's *optimal security levels*, denoted by v_1 and v_2, respectively, by

$$\begin{aligned} v_1 &= \operatorname*{Max}_{X \in S}\{\text{ security level of } X\ \} \\ &= \operatorname*{Max}_{X \in S} \operatorname*{Min}_{Y \in T} XAY^t \\ &= \operatorname*{Max}_{X \in S} \operatorname*{Min}_{1 \le j \le n} XA^{(j)} \end{aligned}$$

and

$$\begin{aligned} v_2 &= \operatorname*{Min}_{Y \in T}\{\text{ security level of } Y\ \} \\ &= \operatorname*{Min}_{Y \in T} \operatorname*{Max}_{X \in S} XAY^t \\ &= \operatorname*{Min}_{Y \in T} \operatorname*{Max}_{1 \le i \le m} A_{(i)}Y^t \end{aligned}$$

Note that these terms are simply a generalization of the terms in the previous section, with the v_1 and v_2 comparing with the u_1 and u_2, and any mixed strategy X_0 with security level v_1 corresponding to the s_h, and any mixed strategy Y_0 with security level v_2 corresponding to the t_k. However, as we will see in the next section, we will be able to develop, using mixed strategies, a complete theory for all games based on Principles I and II. (We will also show in the next section that the use of the words *minimum* and *maximum* over the infinite sets T and S, respectively, is justified here also.)

We conclude this section by calculating v_1 and v_2 for the game of Example 9.4.1.

Example 9.4.2 (Continuation of Example 9.4.1). Consider the game of Example 9.4.1 with payoff matrix

$$A = \begin{bmatrix} 1 & 3 \\ 4 & 0 \end{bmatrix}$$

We have

$$\begin{aligned} v_1 &= \operatorname*{Max}_{X=(x_1,x_2)\in S} \operatorname*{Min}_{1\leq j\leq 2} XA^{(j)} \\ &= \operatorname*{Max}_{(x_1,x_2)\in S} \operatorname{Min}\{x_1+4x_2, 3x_1\} \end{aligned}$$

Now, for any $(x_1,x_2) \in S$, $x_1+x_2 = 1$ and $0 \leq x_1,x_2 \leq 1$. Hence $S = \{(x_1,x_2) : 0 \leq x_1 \leq 1 \text{ and } x_2 = 1-x_1\}$. Therefore

$$\begin{aligned} \operatorname*{Max}_{(x_1,x_2)\in S} \operatorname{Min}\{x_1+4x_2, 3x_1\} &= \operatorname*{Max}_{0\leq x_1\leq 1} \operatorname{Min}\{x_1+4(1-x_1), 3x_1\} \\ &= \operatorname*{Max}_{0\leq x_1\leq 1} \operatorname{Min}\{4-3x_1, 3x_1\} \end{aligned}$$

Consider the graph in Figure 9.1. The heavy line segments represent $\operatorname{Min}\{4-3x_1, 3x_1\}$. The maximum of this function occurs at the point $x_1 = \frac{2}{3}$ where the two lines intersect, and the value of the function at this point is 2. Thus $v_1 = 2$, and a strategy for P_1 with security level $v_1 = 2$ is the mixed strategy $(\frac{2}{3}, \frac{1}{3})$. Note that for any $Y = (y_1,y_2) \in T$,

$$(\tfrac{2}{3}, \tfrac{1}{3})AY^t = 2y_1 + 2y_2 = 2(y_1+y_2) = 2$$

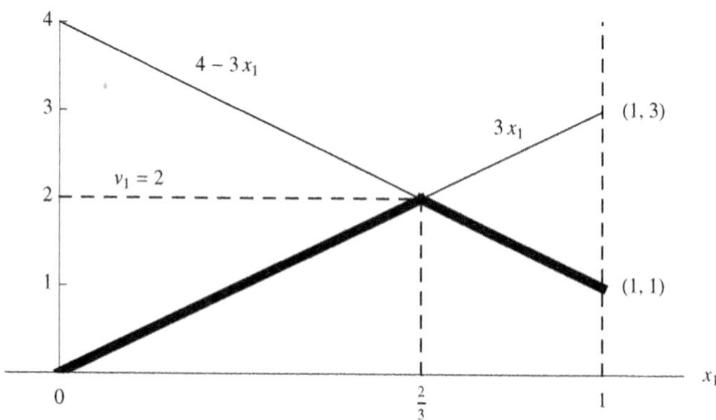

Figure 9.1

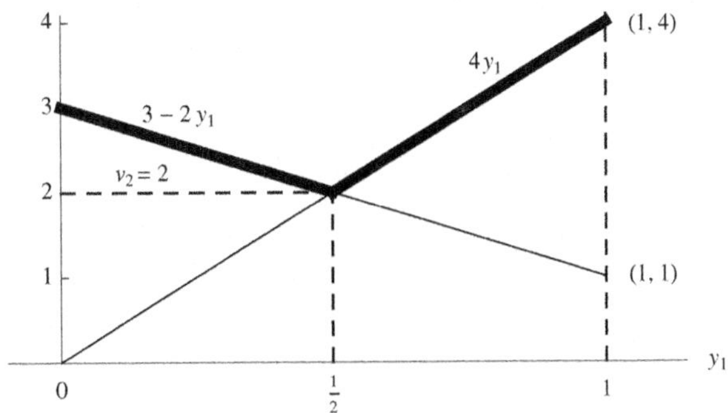

Figure 9.2

Similarly,

$$\begin{aligned} v_2 &= \underset{Y=(y_1,y_2)\in T}{\text{Min}} \ \underset{1\le i\le 2}{\text{Max}} A_{(i)}Y^t \\ &= \underset{(y_1,y_2)\in T}{\text{Min}} \text{Max}\{y_1+3y_2, 4y_1\} \\ &= \underset{0\le y_1\le 1}{\text{Min}} \text{Max}\{y_1+3(1-y_1), 4y_1\} \\ &= \underset{0\le y_1\le 1}{\text{Min}} \text{Max}\{3-2y_1, 4y_1\} \end{aligned}$$

From the graph in Figure 9.2, we see that v_2 also equals 2. A strategy for P_2 with security level $v_2 = 2$ is the mixed strategy (y_1, y_2) with $y_1 = \frac{1}{2}$, that is, the strategy $(\frac{1}{2}, \frac{1}{2})$.

Problem Set 9.4

1. For the matrix game A,

$$A = \begin{bmatrix} 1 & 2 & 3 & 4 \\ 6 & 5 & 2 & 1 \\ 7 & 0 & 1 & 8 \end{bmatrix}$$

(a) Compute P_1's security level for $X_1 = (\frac{1}{3}, \frac{1}{3}, \frac{1}{3})$ and for $X_2 = (\frac{2}{3}, \frac{1}{3}, 0)$. What can you now conclude about v_1?

(b) Compute P_2's security level for $Y_1 = (\frac{1}{6}, 0, \frac{5}{6}, 0)$ and for $Y_2 = (0, \frac{1}{3}, \frac{1}{2}, \frac{1}{6})$. What can you now conclude about v_2?

2. For the matrix game A,

$$A = \begin{bmatrix} -1 & 1 & 2 & 0 \\ 4 & -2 & -3 & 2 \\ 0 & 3 & 1 & -2 \end{bmatrix}$$

(a) Compute P_1's security level for $X_1 = (\frac{2}{3}, \frac{1}{3}, 0)$ and $X_2 = (\frac{1}{3}, \frac{1}{3}, \frac{1}{3})$.

(b) Compute P_2's security level for $Y_1 = (\frac{1}{4}, \frac{1}{4}, \frac{1}{4}, \frac{1}{4})$ and $Y_2 = (0, \frac{1}{2}, 0, \frac{1}{2})$
(c) What can you now conclude about v_1 and v_2?

3. For the game of Example 9.4.1, define $X_0 = (\frac{2}{3}, \frac{1}{3})$ and $Y_0 = (\frac{1}{2}, \frac{1}{2})$. Show that the strategy pair (X_0, Y_0) is in equilibrium, that is, if P_1 expects P_2 to play Y_0, he has no reason to deviate from X_0, and if P_2 expects P_1 to play X_0, she has no reason to deviate from Y_0.

4. Using the graphical technique introduced in the game of Example 9.4.2, determine v_1, v_2 and strategies that have these values as their security levels for the matrix game

$$\begin{bmatrix} -1 & 4 \\ 3 & -6 \end{bmatrix}$$

Is $v_1 = v_2$? Which player, if either, has the advantage?

5. Use the graphical approach of Example 9.4.2 to determine v_1, v_2, and strategies that have these values as their security levels for the matrix game

$$\begin{bmatrix} 1 & 6 \\ 3 & 4 \end{bmatrix}$$

Note that this game has a saddle point. Compare your answers to the above with the solution to the game as defined in Section 9.3.

6. Prove that for any game $u_1 \leq v_1$ and $v_2 \leq u_2$.

7. Suppose $A = (a_{ij})$ is an $m \times n$ game matrix such that the sum of the entries in each column is positive, that is,

$$\sum_{i=1}^{m} a_{ij} > 0 \text{ for each } j, 1 \leq j \leq n.$$

Show that $v_1 > 0$.

8. Suppose A is a game matrix such that the sum of the entries in each row is zero. What can you say about v_2?

9. Suppose $A = (a_{ij})$ is a 3×4 game matrix such that

$$6a_{1j} + 3a_{2j} + a_{3j} \geq 2 \text{ for each } j, 1 \leq j \leq 4$$

Show that $v_1 \geq \frac{1}{5}$.

9.5 The Fundamental Theorem

In this section we prove the Fundamental Theorem of two-person, zero-sum game theory. The theorem states that there exists a (mixed) strategy X_0 for P_1 with security level the optimal value v_1, a strategy Y_0 for P_2 with security level v_2, and that these optimal security levels are equal, that is, $v_1 = v_2$. Then, with the equality of v_1 and

v_2, we will be able to show that the strategy pair (X_0, Y_0) is in equilibrium, and so we will have proved that for any two-person, zero-sum game, there exist strategies for the players satisfying our two principles.

The proof of the Fundamental Theorem that we give is based on the Duality Theorem of linear programming and is due to G. Dantzig [33]. One advantage of this proof is that it is constructive; that is, the proof provides a practical technique for actually computing solution strategies X_0 and Y_0 and the value $v_1 = v_2$ for any game. The theorem was first proved by J. von Neumann [34] in 1928 using the fixed-point theorem of Brouwer. Since then, many other proofs, both topological and algebraic, have been developed.

Theorem 9.5.1 (The Fundamental Theorem of Game Theory). *For any matrix game A, there exist strategies X_0 for P_1 and Y_0 for P_2 such that*

$$v_1 = \underset{X \in S}{\text{Max}} \underset{1 \le j \le n}{\text{Min}} XA^{(j)} = \underset{1 \le j \le n}{\text{Min}} X_0 A^{(j)}$$

and

$$v_2 = \underset{Y \in T}{\text{Min}} \underset{1 \le i \le m}{\text{Max}} A_{(i)} Y^t = \underset{1 \le i \le m}{\text{Max}} A_{(i)} Y_0^t$$

and, moreover, $v_1 = v_2$.

Proof. Suppose the game has payoff matrix $A = (a_{ij})$, $1 \le i \le m$, $1 \le j \le n$. To apply linear programming techniques to the problem of determining v_1, v_2, and the strategies X_0 and Y_0, we need to assume that v_1 and v_2 are positive. This will certainly be the case if all the entries a_{ij} are positive. Thus we divide our proof into two cases, considering first the case in which all $a_{ij} > 0$ and then the general case. The heart of the proof is contained in the first case since, as we will see, in the second case a minor modification of an arbitrary matrix A will allow us to apply the results from the first part.

CASE 1. All $a_{ij} > 0$.

Consider the game from P_1's point of view. P_1's optimal security level is

$$v_1 = \underset{X \in S}{\text{Max}} \underset{1 \le j \le n}{\text{Min}} XA^{(j)}$$

To determine v_1, for any strategy $X \in S$ the minimum of the $XA^{(j)}$ must first be determined, that is, the minimum of the n quantities

$$\begin{aligned} &a_{11}x_1 + a_{21}x_2 + \ldots + a_{m1}x_m \\ &a_{12}x_1 + a_{22}x_2 + \ldots + a_{m2}x_m \\ &\vdots \\ &a_{1n}x_1 + a_{2n}x_2 + \ldots + a_{mn}x_m \end{aligned}$$

Notice that the minimum of these n quantities is less than or equal to each of the n quantities, and is in fact equal to at least one of them, the smallest one. Thus the minimum is the largest real number w satisfying the n inequalities

$$\begin{aligned} a_{11}x_1 + a_{21}x_2 + \ldots + a_{m1}x_m &\geq w \\ a_{12}x_1 + a_{22}x_2 + \ldots + a_{m2}x_m &\geq w \\ &\vdots \\ a_{1n}x_1 + a_{2n}x_2 + \ldots + a_{mn}x_m &\geq w \end{aligned} \tag{9.5.1}$$

Hence, for each $X \in S$, the maximum w satisfying (9.5.1) must first be determined. Next, consider v_1, which is the maximum over all the $X \in S$ of these w's. Since

$$S = \{(x_1, x_2, \ldots, x_m) : x_1 + x_2 + \cdots + x_m = 1, x_i \geq 0, 1 \leq i \leq m\}$$

it follows that v_1 is equal to the maximum w satisfying

$$\begin{aligned} a_{11}x_1 + a_{21}x_2 + \ldots + a_{m1}x_m &\geq w \\ &\vdots \\ a_{1n}x_1 + a_{2n}x_2 + \ldots + a_{mn}x_m &\geq w \\ x_1 + x_2 + \ldots + x_m &= 1 \\ x_i \geq 0,\ 1 \leq i \leq m & \end{aligned} \tag{9.5.2}$$

Moreover, a point $X = (x_1, x_2, \ldots, x_m)$ at which this maximum is attained would be a strategy for P_1 with a security level equal to this optimal value v_1.

Because we have assumed in this case that all $a_{ij} > 0$, we know that v_1 must be positive, and so we can restrict our attention to those $w > 0$ that satisfy (9.5.2). Dividing the equation and the inequalities in (9.5.2) by w, we have the problem of

$$\begin{aligned} &\text{Maximizing } w \\ &\text{subject to} \\ &a_{11}\left(\frac{x_1}{w}\right) + a_{21}\left(\frac{x_2}{w}\right) + \ldots + a_{m1}\left(\frac{x_m}{w}\right) \geq 1 \\ &\vdots \\ &a_{1n}\left(\frac{x_1}{w}\right) + a_{2n}\left(\frac{x_2}{w}\right) + \ldots + a_{mn}\left(\frac{x_m}{w}\right) \geq 1 \\ &\left(\frac{x_1}{w}\right) + \left(\frac{x_2}{w}\right) + \ldots + \left(\frac{x_m}{w}\right) = 1/w \\ &\frac{x_i}{w} \geq 0,\ 1 \leq i \leq m \end{aligned} \tag{9.5.3}$$

Let $x_i' = x_i/w$, $1 \leq i \leq m$. Since the problem of maximizing w is equivalent to the problem of minimizing $1/w$, the optimization problem of (9.5.3) is equivalent to the problem of

$$\begin{aligned} &\text{Minimizing } x_1' + x_2' + \cdots + x_m' \\ &\text{subject to} \\ &a_{11}x_1' + a_{21}x_2' + \ldots + a_{m1}x_m' \geq 1 \\ &\vdots \\ &a_{1n}x_1' + a_{2n}x_2' + \ldots + a_{mn}x_m' \geq 1 \\ &x_i' \geq 0,\ 1 \leq i \leq m \end{aligned} \tag{9.5.4}$$

To express this concisely in vector notation, we define $X' = (x'_1, x'_2, \ldots, x'_m)$, $b = (1,1,\ldots,1)^t$, and $c = (1,1,\ldots,1)^t$, where b is an m-tuple and c is an n-tuple. Then (9.5.4) is simply the problem of

$$\begin{aligned}&\text{Minimizing } b \cdot X' \\ &\text{subject to} \\ &A^t X' \geq c, X' \geq 0\end{aligned}$$

In summary, the reciprocal of the minimal value of the objective function $b \cdot X'$ of the linear programming problem of (9.5.4) is equal to v_1. Moreover, since $w(x'_1, x'_2, \ldots, x'_m) = (x_1, x_2, \ldots, x_m)$, multiplication of the coordinates of a point X' at which this minimum is attained by v_1 produces a strategy for P_1 with security level v_1.

We now show, by a completely parallel development, that the problem of determining v_2 and a strategy $Y_0 \in T$ with this value as its security level leads to the dual of the problem of (9.5.4). We have

$$v_2 = \operatorname*{Min}_{Y \in T} \operatorname*{Max}_{1 \leq i \leq m} A_{(i)} Y^t$$

For a fixed $Y = (y_1, y_2, \ldots, y_n) \in T$, the maximum of the m quantities $A_{(i)} Y^t$ is the smallest real number z satisfying

$$\begin{aligned}
a_{11}y_1 + a_{12}y_2 + \ldots + a_{1n}y_n &\leq z \\
a_{21}y_1 + a_{22}y_2 + \ldots + a_{2n}y_n &\leq z \\
&\vdots \\
a_{m1}y_1 + a_{m2}y_2 + \ldots + a_{mn}y_n &\leq z
\end{aligned}$$

Since v_2 is the minimum over all the $Y \in T$ of these z's, it follows that v_2 is equal to the minimum z satisfying

$$\begin{aligned}
a_{11}y_1 + a_{12}y_2 + \ldots + a_{1n}y_n &\leq z \\
&\vdots \\
a_{m1}y_1 + a_{m2}y_2 + \ldots + a_{mn}y_n &\leq z \\
y_1 + \quad y_2 + \ldots + \quad y_n &= 1 \\
y_j \geq 0, \ 1 \leq j \leq n &
\end{aligned} \tag{9.5.5}$$

and a point at which this minimal value is attained is a strategy for P_2 with security level v_2. Again, we know that $v_2 > 0$, and therefore all z's satisfying the constraints of (9.5.5) must be positive. Dividing the inequalities and equation of (9.5.5) by z gives the problem of

$$\begin{aligned}
&\text{Minimizing } z && (9.5.6)\\
&\text{subject to}\\
&a_{11}\left(\frac{y_1}{z}\right) + a_{12}\left(\frac{y_2}{z}\right) + \ldots + a_{1n}\left(\frac{y_n}{z}\right) \le 1\\
&\qquad\vdots\\
&a_{m1}\left(\frac{y_1}{z}\right) + a_{m2}\left(\frac{y_2}{z}\right) + \ldots + a_{mn}\left(\frac{y_n}{z}\right) \le 1\\
&\quad\left(\frac{y_1}{z}\right) + \quad\left(\frac{y_2}{z}\right) + \ldots + \quad\left(\frac{y_n}{z}\right) = \tfrac{1}{z}\\
&\frac{y_j}{z} \ge 0,\ 1 \le j \le n
\end{aligned}$$

Let $y'_j = y_j/z$, $1 \le j \le n$. Then (9.5.6) is equivalent to the problem of

$$\begin{aligned}
&\text{Maximizing } y'_1 + y'_2 + \cdots + y'_n && (9.5.7)\\
&\text{subject to}\\
&a_{11}y'_1 + a_{12}y'_2 + \ldots + a_{1n}y'_n \le 1\\
&\qquad\vdots\\
&a_{m1}y'_1 + a_{m2}y'_2 + \ldots + a_{mn}y'_n \le 1\\
&y'_j \ge 0,\ 1 \le j \le n
\end{aligned}$$

In vector notation, with $Y' = (y'_1 y'_2, \ldots, y'_n)^t$, this problem is to

$$\begin{aligned}
&\text{Maximize } c \cdot Y'\\
&\text{subject to}\\
&AY' \le b, Y' \ge 0
\end{aligned}$$

And, as before, the reciprocal of the maximum value of the objective function $c \cdot Y'$ of (9.5.7) is equal to v_2, and multiplication of the coordinates of a point Y' at which this maximum is attained by v_2 yields a strategy for P_2 with security level v_2.

But the problems of (9.5.4) and (9.5.7) are dual linear programming problems. Moreover, the problem of (9.5.4) must have a finite optimal solution, because the objective function $b \cdot X' = x'_1 + \cdots + x'_m$ is bounded below by zero and, since all the entries of A are positive, there exist feasible solutions to the system of constraints $A^t X' \ge c$. Thus it follows from the Duality Theorem of Section 4.4 that both problems have finite solutions attaining the same optimal value. Hence, the optimal security levels v_1 and v_2 are equal, and there are strategies X_0 for P_1 and Y_0 for P_2, each with a security level equal to this common value $v_1 = v_2$.

CASE 2. The General Case.

Suppose some entries a_{ij} are nonpositive. Choose any constant r with the property that $a_{ij} + r > 0$ for all i and j (e.g., r could equal $1 - \text{Min}_{i,j}\, a_{ij}$). Let E be the

$m \times n$ matrix with all entries equal to 1, and consider the game with payoff matrix $A + rE$. The expected payoff for any pair of strategies (X, Y) is

$$X(A+rE)Y^t = XAY^t + rXEY^t$$

However, X and Y are strategies, and thus XE will be the n-vector $(1, 1, \ldots, 1)$; further, $(1, 1, \ldots, 1)Y^t = 1$. The expected payoff is $XAY^t + r$, which is the expected payoff for the game with matrix A plus the constant r. Since these expected payoffs, XAY^t and $X(A+rE)Y^t$, differ only by the constant r, it follows that the games with payoff matrices A and $A + rE$ will have optimal security levels differing only by this constant, and that strategies delivering the optimal security level for one game will also deliver the optimal security level for the other. But all the entries of the matrix $A + rE$ are positive, and so the results of Case 1 can be applied to the corresponding game. Thus, for the game with payoff matrix A, the players' optimal security levels are equal, and strategies delivering this optimal value exist. □

Example 9.5.1. As an example of the technique developed in this proof, consider once again the game discussed in Example 9.4.1 (and then later in Example 9.4.2) with payoff matrix

$$\begin{bmatrix} 1 & 3 \\ 4 & 0 \end{bmatrix}$$

Since both entries in the first row are positive, v_1 is positive, and so no constant needs to be added to the entries of the matrix. The two linear programming problems associated with this game are simply to

$$\begin{array}{lcl} \text{Minimize } x_1' + x_2' & \text{and} & \text{Maximize } y_1' + y_2' \\ \text{subject to} & & \text{subject to} \\ \quad x_1' + 4x_2' \geq 1 & & \quad y_1' + 3y_2' \leq 1 \\ 3x_1' \qquad\quad \geq 1 & & 4y_1' \qquad\quad \leq 1 \\ x_1', x_2' \geq 0 & & y_1', y_2' \geq 0 \end{array}$$

Suppose we wish to solve these problems using the simplex method. The maximization problem associated with the determination of P_2's optimal security level and associated strategy can be handled without using artificial variables. Introducing two slack variables, y_3' and y_4', the simplex method leads to the tableaux of Table 9.1. The maximum of $y_1' + y_2'$ is $\frac{1}{2}$, and this value is attained at the point $(\frac{1}{4}, \frac{1}{4})$. Therefore $v_2 = 2$, and a strategy for P_2 with security level 2 is $2(\frac{1}{4}, \frac{1}{4}) = (\frac{1}{2}, \frac{1}{2})$. Furthermore, we know that $v_1 = 2$ and, using the entries in the bottom row of the slack variable columns, a strategy for P_1 with security level 2 is $2(\frac{1}{3}, \frac{1}{6}) = (\frac{2}{3}, \frac{1}{3})$.

While the game matrix in the above example is only 2×2, it should be clear that the simplex method can be applied to any matrix game. More will be said about this in the next section, in which computational techniques are discussed. In the remainder of this section, we consider some of the theoretical implications of the Fundamental Theorem.

Table 9.1

	y_1'	y_2'	y_3'	y_4'	
y_3'	1	(3)	1	0	1
y_4'	4	0	0	1	1
	-1	-1	0	0	0
y_2'	$\frac{1}{3}$	1	$\frac{1}{3}$	0	$\frac{1}{3}$
y_4'	(4)	0	0	1	1
	$-\frac{2}{3}$	0	$\frac{1}{3}$	0	$\frac{1}{3}$
y_2'	0	1	$\frac{1}{3}$	$-\frac{1}{12}$	$\frac{1}{4}$
y_1'	1	0	0	$\frac{1}{4}$	$\frac{1}{4}$
	0	0	$\frac{1}{3}$	$\frac{1}{6}$	$\frac{1}{2}$

Principle I states that the players should act to maximize their security levels, and the Fundamental Theorem guarantees that this is always possible, that is, that there always exists a strategy X_0 for P_1 with security level v_1 and a strategy Y_0 for P_2 with security level v_2. Principle II states that the players use strategies in equilibrium, a concept not yet formally defined over the complete strategy sets S and T.

Definition 9.5.1. For a matrix game A, a strategy pair (X_1, Y_1), with $X_1 \in S$ and $Y_1 \in T$, is in *equilibrium* if

$$XAY_1^t \leq X_1AY_1^t \quad \text{for all } X \in S \tag{9.5.8}$$

and

$$X_1AY_1^t \leq X_1AY^t \quad \text{for all } Y \in T \tag{9.5.9}$$

This definition is a direct translation of the suggested definition of equilibrium given in Section 9.2: (9.5.8) implies that P_1, expecting P_2 to play Y_1, has nothing to gain by deviating from the play of X_1; and (9.5.9) implies that Y_1 is P_2's best response to P_1 playing X_1. As a corollary to the Fundamental Theorem, we show that the strategy pair (X_0, Y_0) has this property.

Corollary 9.5.1. *Suppose for a matrix game A that X_0 is a strategy for P_1 with security level v_1 and Y_0 is a strategy for P_2 with security level v_2. Then the pair (X_0, Y_0) is in equilibrium.*

Proof. We have by definition

$$\begin{aligned} v_1 &= \text{security level of } X_0 = \operatorname*{Min}_{Y \in T} X_0AY^t \\ &\leq X_0AY_0^t \\ &\leq \operatorname*{Max}_{X \in S} XAY_0^t = \text{security level of } Y_0 = v_2 \end{aligned}$$

But $v_1 = v_2$. Therefore

$$\operatorname*{Max}_{X \in S} XAY_0^t = X_0AY_0^t = \operatorname*{Min}_{Y \in T} X_0AY^t$$

that is,

$$XAY_0^t \le X_0AY_0^t \quad \text{for any } X \in S$$

and

$$X_0AY_0^t \le X_0AY^t \quad \text{for any } Y \in T$$ □

Thus this strategy pair (X_0, Y_0) satisfies the two principles set out in Section 9.2. We have the solution to a matrix game.

Definition 9.5.2. For a matrix game A, the common value $v = v_1 = v_2$ is called the *value* of the game. Any strategy X_0 for P_1 with security level v is an *optimal strategy* for P_1, and any strategy Y_0 for P_2 with security level v is an *optimal strategy* for P_2. Such a strategy pair (X_0, Y_0) along with the value of the game $v = X_0AY_0^t$ is called a *solution* to the game. The game is *fair* if $v = 0$.

A solution is therefore a suggested course of play for both players given that their play is to be determined by Principles I and II. The value of the game is the optimal security level for both players and the expected outcome of the game if the players use the suggested strategies. Thus, for matrix games we have been able to develop a complete mathematical model based on the two principles. However, this model can be applied to a game-theoretic situation only if these principles are representative of the approach of the players to the situation. An excellent discussion of some of the limitations of the applicability of the theory from the viewpoint of a social scientist is presented in *Games and Decisions* by Luce and Raiffa [31].

Actually, for two-person, zero-sum games, the concept of equilibrium alone is sufficient to lead us to this solution of a game. This follows from the following theorem, a converse to the above corollary.

Theorem 9.5.2. *For the matrix game A, suppose that the strategy pair (X_1, Y_1) is in equilibrium. Then X_1 and Y_1 are optimal strategies, and $X_1AY_1^t$ is the value of the game.*

Proof. By the definition of equilibrium, we have, for any $X \in S$ and $Y \in T$,

$$XAY_1^t \le X_1AY_1^t \le X_1AY^t$$

Now, let v denote the value of the game. Then

$$\begin{aligned} v = v_1 &= \operatorname*{Max}_{X \in S} \operatorname*{Min}_{Y \in T} XAY^t \\ &\ge \operatorname*{Min}_{Y \in T} X_1AY^t = X_1AY_1^t \end{aligned}$$

and

$$\begin{aligned} v = v_2 &= \underset{Y \in T}{\text{Min}}\, \underset{X \in S}{\text{Max}}\, XAY^t \\ &\leq \underset{X \in S}{\text{Max}}\, XAY_1^t = X_1AY_1^t \end{aligned}$$

Therefore $v = X_1AY_1^t$ and $\text{Min}_{Y \in T} X_1AY^t = v$, and so, by definition, X_1 is an optimal strategy for P_1. Similarly, $\text{Max}_{X \in S} XAY_1^t = v$ implies that Y_1 is an optimal strategy for P_2. □

Problem Set 9.5

1. The following refer to the proof of Theorem 9.5.1.
   (a) In Case 1, it is claimed that feasible solutions exist for the linear programming problem of minimizing $b \cdot X'$ subject to $A^tX' \geq c$. Prove that this is true.
   (b) In Case 2, it is claimed that the security levels of the players differ only by the constant r for the two games with payoff matrices A and $A + rE$. Prove this using the definition of a security level.
2. Suppose A is a matrix game with saddle point a_{hk}. Let $X_1 = s_h$ and $Y_1 = t_k$.
   (a) By considering the security levels of X_1 and Y_1, show that $v_1 \geq a_{hk}$ and $v_2 \leq a_{hk}$. It follows from Theorem 9.5.1 that $v = a_{hk}$ and X_1 and Y_1 are optimal strategies. Why? (Thus the solution of a matrix game with a saddle point defined in Section 9.3 agrees with the more general definition in this section.)
   (b) Using only the definition of a saddle point, show that the strategy pair (X_1, Y_1) is in equilibrium, that is, for any $X \in S$ and $Y \in T$,

   $$XAY_1^t \leq X_1AY_1^t \leq X_1AY^t$$

   Now invoke Theorem 9.5.2 to give another proof that $v = a_{hk}$ and X_1 and Y_1 are optimal.
3. For the matrix game

   $$A = \begin{bmatrix} 0 & 0 & 3 & 1 \\ 2 & 1 & 4 & 2 \\ 1 & 3 & 0 & 0 \end{bmatrix}$$

   (a) Compute the security level for P_1 of $X_1 = (0, \frac{3}{4}, \frac{1}{4})$.
   (b) Compute the security level for P_2 of $Y_1 = (0, \frac{1}{2}, 0, \frac{1}{2})$.
   (c) What can you conclude?
4. Intuitively, the game with payoff matrix

   $$\begin{bmatrix} 1 & -1 \\ -1 & 1 \end{bmatrix}$$

should be fair and have optimal strategies $X_0 = (\frac{1}{2}, \frac{1}{2})$ and $Y_0 = (\frac{1}{2}, \frac{1}{2})$. Prove that this is the case.

5. For each of the following matrix games, prove or disprove that the given strategy pair is a solution to the game.

(a) $A = \begin{bmatrix} -1 & 2 & -3 \\ 3 & -4 & 2 \\ -2 & 0 & 1 \end{bmatrix}$ $\quad X_1 = (\frac{2}{5}, \frac{1}{5}, \frac{2}{5})$, $Y_1 = (\frac{1}{4}, \frac{1}{2}, \frac{1}{4})$

(b) $A = \begin{bmatrix} -1 & 1 & 2 & 0 \\ 4 & -2 & -3 & 2 \\ 0 & 3 & 1 & -2 \end{bmatrix}$ $\quad X_1 = \frac{1}{27}(18, 7, 2)$, $Y_1 = \frac{1}{27}(2, 12, 0, 13)$

(c) $A = \begin{bmatrix} 2 & -4 & 2 & -1 \\ -9 & -1 & 1 & 1 \\ -3 & 5 & -2 & 0 \end{bmatrix}$ $\quad X_1 = (\frac{1}{2}, 0, \frac{1}{2})$, $Y_1 = (\frac{1}{6}, 0, 0, \frac{5}{6})$

(d) $A = \begin{bmatrix} 5 & -2 & 4 & 7 \\ -3 & 1 & -2 & -5 \\ 1 & -4 & 8 & 3 \end{bmatrix}$ $\quad X_1 = (\frac{2}{5}, \frac{3}{5}, 0)$, $Y_1 = (\frac{1}{9}, \frac{2}{3}, \frac{1}{9}, \frac{1}{9})$

6. (a) Show that for the payoff matrix of Example 9.1.6 on page 340, the strategies $X_0 = (\frac{1}{36})(11, 10, 8, 7)$ and $Y_0 = (\frac{1}{36})(10, 5, 14, 7)$ are optimal strategies for P_1 and P_2, respectively. What is the value of the game? (Does the game now favor P_1?)

(b) P_1's optimal strategy X_0 translates as follows. P_1 should select the 1 with probability $\frac{7}{12}$ and the 2 with probability $\frac{5}{12}$. If P_2's guess is correct, P_1 should offer to double the stakes slightly less than one-half the time, with the exact probability depending on the card P_1 holds. For example, if P_2 has correctly guessed that P_1 holds a 1, P_1 should double with probability $\frac{10}{21}$. Provide a similar translation of P_2's optimal strategy Y_0.

7. Suppose that X_0 and Y_0 are optimal strategies for a game with payoff matrix A and value v. Prove that for any i and j, $X_0A^{(j)} \geq v$ and $A_{(i)}Y_0^t \leq v$.

8. Suppose that $X_0 = (x_1, x_2, \ldots, x_m)$ and $Y_0 = (y_1, y_2, \ldots, y_n)$ are optimal strategies and v is the value of a matrix game A. Show that

(a) $X_0A^{(k)} > v$ implies that $y_k = 0$.

(b) $x_i > 0$ implies that $A_{(i)}Y_0 = v$.

9. For the game of Problem 5(d), the given strategy X_1 is in fact an optimal strategy for P_1. Use this fact and the results of Problem 8 to determine an optimal strategy for P_2 and then verify the optimality of X_1.

10. Using the approach outlined in Problem 9, prove or disprove:

(a) $Y_1 = (\frac{1}{2}, 0, \frac{1}{2})$ is an optimal strategy for P_2 for the game

$$A = \begin{bmatrix} 7 & 8 & 4 \\ 1 & 3 & 6 \\ 9 & 5 & 2 \end{bmatrix}$$

(b) $X_1 = (\frac{1}{2}, 0, \frac{1}{2})$ is an optimal strategy for P_1 for the game

$$A = \begin{bmatrix} 1 & -5 & 3 & -4 \\ -1 & 0 & -3 & 1 \\ -2 & 5 & -2 & 3 \end{bmatrix}$$

11. True or false: If v is the value of a game and Y_0 is an optimal strategy for P_2, then $X_1 A Y_0^t = v$ implies that X_1 is an optimal strategy for P_1.

$$\textit{Hint.}\ \text{Let } A = \begin{bmatrix} 1 & 0 \\ 1 & 2 \end{bmatrix}$$

12. True or false: Suppose X_0 is an optimal strategy for P_1 for a matrix game A. Then, for any $X \in S$ and $Y \in T$, $XAY^t \leq X_0 AY^t$. (In other words, an optimal strategy is the best response to any strategy of the opponent.)

13. True or false: A game has a pair of optimal strategies (X_0, Y_0) that are both pure strategies if and only if the game has a saddle point.

14. True or false: One player has an optimal pure strategy if and only if the game has a saddle point.

15. True or false: $u_1 = v$ if and only if P_1 has an optimal pure strategy.

16. A matrix game A is said to be *symmetric* if $A^t = -A$. Prove that the value of a symmetric game is 0.

9.6 Computational Techniques

The solution of games with saddle points is straightforward, and the existence of saddle points can be easily determined by computing the u_1 and u_2 defined in Section 9.3. and using Corollary 9.3.1. For other games we list the following techniques.

Linear Programming

As seen in the previous section, the problem of determining optimal strategies and the value of a matrix game is equivalent to two dual linear programming problems that can be solved using the simplex method. However, this method is not directly applicable if the value of the game is not positive. For an arbitrary game, a constant must first be chosen such that when this constant is added to each entry of the original payoff matrix, the game corresponding to this new matrix has a positive value. Then the simplex method can be applied to this new game, with the value of the original game equal to the value of the new game less the constant. Note that it may not be necessary to make all the entries in the modified payoff matrix positive; for example, if the matrix has at least one row with all positive entries, the value of the corresponding game is positive (the possibility of P_1 using the pure strategy of playing that particular row shows that P_1's security level is positive).

Example 9.6.1. P_1 and P_2 each extend either one, two, or three fingers, and the difference in the numbers put forth is computed. If this difference is 0, the payoff is 0; if the difference is 1, the player putting forth the smaller amount wins 1; and if the difference is 2, the player putting forth the larger amount wins 2.

Each player has three pure strategies. Let s_i denote P_1's pure strategy of extending i fingers, $1 \le i \le 3$, and similarly define t_j, $1 \le j \le 3$, for P_2. The payoff tableau is

	t_1	t_2	t_3
s_1	0	1	-2
s_2	-1	0	1
s_3	2	-1	0

By symmetry it is reasonable to expect the value of this game to be 0. To verify this and compute optimal strategies, we first add 2 to each entry of the above matrix, giving the following matrix, which corresponds to a game with value at least 1, as all the entries in the last two rows are greater than or equal to 1.

$$\begin{bmatrix} 2 & 3 & 0 \\ 1 & 2 & 3 \\ 4 & 1 & 2 \end{bmatrix}$$

The associated linear programming problem corresponding to P_2's determination of an optimal strategy and security level is to

$$\begin{aligned} &\text{Maximize } y_1' + y_2' + y_3' \\ &\text{subject to} \\ &2y_1' + 3y_2' \qquad\quad \le 1 \\ &\ \ y_1' + 2y_2' + 3y_3' \le 1 \\ &4y_1' + \ \ y_2' + 2y_3' \le 1 \\ &y_1', y_2', y_3' \ge 0 \end{aligned}$$

Adding three slack variables and solving leads to the tableaux of Table 9.2.

The value of the modified game is 2, and so the value of the original game is 0, as suggested. Since the optimal value of the above problem is attained at $(y_1', y_2', y_3') = (\frac{1}{8}, \frac{1}{4}, \frac{1}{8})$, an optimal strategy for P_2 is $2(\frac{1}{8}, \frac{1}{4}, \frac{1}{8}) = (\frac{1}{4}, \frac{1}{2}, \frac{1}{4})$. Similarly, the solution to the dual problem, found in the bottom row in the slack variable columns, is $(x_1', x_2', x_3') = (\frac{1}{8}, \frac{1}{4}, \frac{1}{8})$, and so an optimal strategy for P_1 is also $2(\frac{1}{8}, \frac{1}{4}, \frac{1}{8}) = (\frac{1}{4}, \frac{1}{2}, \frac{1}{4})$.

2×2 Games

The solution of games with a 2×2 payoff matrix can be given by simple formulas. Before we state these, we state a theorem that enables one to immediately determine if a 2×2 game has a saddle point. The proof of this first theorem is outlined in Problem 3.

Table 9.2

	y'_1	y'_2	y'_3	y'_4	y'_5	y'_6	
y'_4	2	3	0	1	0	0	1
y'_5	1	2	(3)	0	1	0	1
y'_6	4	1	2	0	0	1	1
	-1	-1	-1	0	0	0	0
y'_4	2	(3)	0	1	0	0	1
y'_3	$\frac{1}{3}$	$\frac{2}{3}$	1	0	$\frac{1}{3}$	0	$\frac{1}{3}$
y'_6	$\frac{10}{3}$	$-\frac{1}{3}$	0	0	$-\frac{2}{3}$	1	$\frac{1}{3}$
	$-\frac{2}{3}$	$-\frac{1}{3}$	0	0	$\frac{1}{3}$	0	$\frac{1}{3}$
y'_2	$\frac{2}{3}$	1	0	$\frac{1}{3}$	0	0	$\frac{1}{3}$
y'_3	$-\frac{1}{9}$	0	1	$-\frac{2}{9}$	$\frac{1}{3}$	0	$\frac{1}{9}$
y'_6	($\frac{32}{9}$)	0	0	$\frac{1}{9}$	$-\frac{2}{3}$	1	$\frac{4}{9}$
	$-\frac{4}{9}$	0	0	$\frac{1}{9}$	$\frac{1}{3}$	0	$\frac{4}{9}$
y'_2	0	1	0	$\frac{5}{16}$	$\frac{1}{8}$	$-\frac{3}{16}$	$\frac{1}{4}$
y'_3	0	0	1	$-\frac{7}{32}$	$\frac{5}{16}$	$\frac{1}{32}$	$\frac{1}{8}$
y'_1	1	0	0	$\frac{1}{32}$	$-\frac{3}{16}$	$\frac{9}{32}$	$\frac{1}{8}$
	0	0	0	$\frac{1}{8}$	$\frac{1}{4}$	$\frac{1}{8}$	$\frac{1}{2}$

Consider the game with payoff matrix

$$A = \begin{bmatrix} a & b \\ c & d \end{bmatrix}$$

Theorem 9.6.1. *A has no saddle points if and only if a and d are both greater than or are both less than b and c, that is, if and only if either $a > b$, $a > c$, $d > b$, $d > c$ or $a < b$, $a < c$, $d < b$, $d < c$.*

Theorem 9.6.2. *Suppose A has no saddle points. Let $r = a + d - b - c$. Then the value v of the game is*

$$v = \frac{ad - bc}{r}$$

and optimal strategies X_0 and Y_0 for P_1 and P_2 are

$$X_0 = \left(\frac{d-c}{r}, \frac{a-b}{r} \right)$$
$$Y_0 = \left(\frac{d-b}{r}, \frac{a-c}{r} \right)$$

Proof. Note that by Theorem 9.6.1, $r \neq 0$ and X_0 and Y_0 are strategies. By direct calculation, we have

$$X_0 A = \left(\frac{ad - bc}{r}, \frac{ad - bc}{r} \right)$$

and

$$AY_0^t = \left(\frac{ad-bc}{r}, \frac{ad-bc}{r} \right)^t$$

Thus for any strategies $X = (x_1, x_2) \in S$ and $Y = (y_1, y_2) \in T$,

$$X_0AY^t = \frac{ad-bc}{r}(y_1+y_2) = \frac{ad-bc}{r}$$

and

$$XAY_0^t = (x_1+x_2)\frac{ad-bc}{r} = \frac{ad-bc}{r}$$

From Theorem 9.5.2, it follows that X_0 and Y_0 are optimal strategies, and the value of the game is

$$X_0AY_0^t = \frac{ad-bc}{r}$$

□

These formulas are easy to remember. Once r is determined, the value of the game is the determinant of A divided by r. The numerators in the optimal strategy for P_1 are the differences between the entries in the rows of A, with the difference of the entries in the second row going into the first component. Moreover, since the components of a strategy are nonnegative, all that is critical is the magnitude of these quantities, and X_0 can be remembered as

$$\left(\frac{|d-c|}{|r|}, \frac{|a-b|}{|r|} \right)$$

In a similar manner, P_2's optimal strategy Y_0 can be interpreted, simply replacing the word *row* with the word *column*. The fact that the sum of the components of a strategy is 1 provides a partial check of one's calculations.

Example 9.6.2. For the game with matrix

$$A = \begin{bmatrix} 1 & 3 \\ 4 & 0 \end{bmatrix}$$

we see that it has no saddle point, because 1 and 0 are both less than 3 and 4. Thus we can apply Theorem 9.6.2. We have $r = -6$, $v = -12/(-6) = 2$, and

$$X_0 = \left(\frac{-4}{-6}, \frac{1-3}{-6} \right) = \left(\frac{2}{3}, \frac{1}{3} \right) \text{ and } Y_0 = \left(\frac{-3}{-6}, \frac{1-4}{-6} \right) = \left(\frac{1}{2}, \frac{1}{2} \right)$$

Example 9.6.3. For the game with matrix

$$A = \begin{bmatrix} 10 & -8 \\ -7 & 5 \end{bmatrix}$$

Theorem 9.6.2 once again applies directly, with $r = 30$, $v = -6/30 = -1/5$,

$$X_0 = \left(\frac{12}{30}, \frac{18}{30} \right) = \left(\frac{2}{5}, \frac{3}{5} \right) \text{ and } Y_0 = \left(\frac{13}{30}, \frac{17}{30} \right)$$

Dominance

Consider the game with payoff matrix

$$A = \begin{bmatrix} 0 & -2 & 0 & 3 \\ 6 & 3 & -1 & -4 \\ 8 & -1 & 7 & 3 \end{bmatrix}$$

Notice that the four entries in the third row are all greater than or equal to the corresponding entries in the first row, and that the three entries in the third column are less than or equal to the corresponding entries in the first column. It follows intuitively that P_1 would never use strategy s_1, since P_1 can do at least as well with s_3 and, similarly, that P_2 would never use strategy t_1. In fact, row 1, which is said to be *dominated* by row 3, and column 1, which is dominated by column 3, can be deleted from the payoff matrix and optimal strategies and a value for the original game computed by using the resulting 2×3 matrix. We make this precise with the following definition.

Definition 9.6.1. Consider a game with an $m \times n$ payoff matrix $A = (a_{ij})$. Then row h is *dominated* by row i if $a_{hj} \leq a_{ij}$ for all j, $1 \leq j \leq n$, and column k is *dominated* by column j if $a_{ik} \geq a_{ij}$ for all i, $1 \leq i \leq m$.

Theorem 9.6.3. *Suppose row h of a payoff matrix A is dominated. Then there is an optimal strategy $X_0 = (x_1, x_2, \ldots, x_m)$ for P_1 with $x_h = 0$. An optimal strategy for P_1 for the game with payoff matrix A but with the hth row removed is an optimal strategy for the original game (after the addition of an hth component equal to zero). Similarly for a dominated column.*

The proof of Theorem 9.6.3 is left to the reader (see Problem 4). Certainly if our definition of optimal strategy is to be at all reasonable, Theorem 9.6.3 must hold.

Example 9.6.4. Consider the game with payoff matrix

$$A = \begin{bmatrix} 0 & -2 & -1 & 0 \\ 3 & 5 & 6 & -1 \\ 5 & -1 & -3 & -2 \end{bmatrix}$$

Column 1 is dominated by column 4; thus, finding a solution reduces to consideration of the game

$$\begin{bmatrix} 0 & -2 & -1 & 0 \\ 3 & 5 & 6 & -1 \\ 5 & -1 & -3 & -2 \end{bmatrix}$$

In this 3×3 game, row 3 is dominated by row 2, and so we consider

$$\begin{bmatrix} 0 & -2 & -1 & 0 \\ 3 & 5 & 6 & -1 \\ 5 & -1 & -3 & -2 \end{bmatrix}$$

Now column 3 is dominated by column 2, leaving just a 2×2 matrix

$$\begin{bmatrix} 0 & -2 & -1 & 0 \\ 3 & 5 & 6 & -1 \\ 5 & -1 & -3 & -2 \end{bmatrix}$$

Using the formulas of Theorem 9.6.2, we find that the value of the game with this 2×2 payoff matrix is $-\frac{1}{4}$, with $(\frac{3}{4}, \frac{1}{4})$ and $(\frac{1}{8}, \frac{7}{8})$ optimal strategies for P_1 and P_2, respectively. Thus the original game has value $-\frac{1}{4}$ and optimal strategies

$$X_0 = (\tfrac{3}{4}, \tfrac{1}{4}, 0) \text{ and } Y_0 = (0, \tfrac{1}{8}, 0, \tfrac{7}{8})$$

$2 \times n$ and $m \times 2$ Games

Games in which one player has only two pure strategies have several methods of solution. One method that uses a graph to determine the optimum of a set of linear functions was described in Example 9.4.2. To solve a game using this method, follow that example, using the definition of

$$v = v_1 = \operatorname*{Max}_{X \in S} \operatorname*{Min}_{1 \le j \le n} XA^{(j)}$$

if it is P_1 that has only the two pure strategies and the definition of

$$v = v_2 = \operatorname*{Min}_{Y \in T} \operatorname*{Max}_{1 \le i \le m} A_{(i)}Y$$

if it is P_2.

Another method makes use of the ease of determining solutions of 2×2 games. From Corollary 3.8.1 on page 109, it follows that for any game with an $m \times n$ payoff matrix, there are optimal strategies for both players with at most the minimum of m and n nonzero components. Thus, for games in which one player has only two pure strategies, there is an optimal strategy for the other player that has at most two nonzero components, and so the original game reduces to a 2×2 game. Moreover, that 2×2 game would correspond to the 2×2 submatrix of the original $m \times 2$ or $2 \times n$ matrix that attains the most favorable value for the player who has more than two pure strategies. In other words, in an $m \times 2$ game, P_1 should compute the values of all the 2×2 subgames (there are $\binom{m}{2} = m(m-1)/2$ of them) and play the game that has the largest value. Similarly, in a $2 \times n$ game, P_2 would select the 2×2 subgame with the smallest value.

Example 9.6.5. Consider the game with payoff matrix

$$A = \begin{bmatrix} 1 & -5 \\ -4 & 4 \\ -2 & 3 \\ 0 & -5 \end{bmatrix}$$

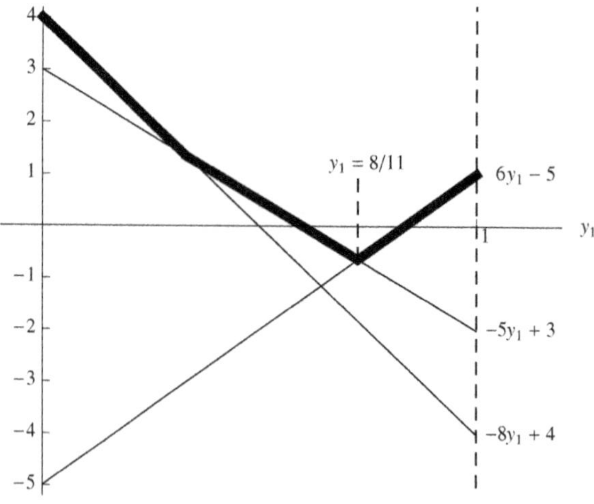

Figure 9.3

It can be easily seen that there is no saddle point but that the last row is dominated by the first row. Thus, to find the solution, we need only consider the first three rows. Using the first technique described for $m \times 2$ games, we have

$$\begin{aligned} v = v_2 &= \underset{Y \in T}{\text{Min}} \underset{1 \leq i \leq 3}{\text{Max}} A_{(i)} Y \\ &= \underset{Y \in T}{\text{Min}} \text{Max} \{y_1 - 5y_2, -4y_1 + 4y_2, -2y_1 + 3y_2\} \\ &= \underset{0 \leq y_1 \leq 1}{\text{Min}} \text{Max} \{6y_1 - 5, -8y_1 + 4, -5y_1 + 3\} \end{aligned}$$

Consider the graph in Figure 9.3. The heavy line represents the maximum of $\{6y_1 - 5, -8y_1 + 4, -5y_1 + 3\}$, and the minimum occurs when $Y_1 = \frac{8}{11}$ at the intersection of the lines determined by $6y_1 - 5$ and $-5y_1 + 3$. The common value of $6y_1 - 5$ and $-5y_1 + 3$ at $y_1 = \frac{8}{11}$ is $-\frac{7}{11}$. Thus the value of the game is $-\frac{7}{11}$, and an optimal strategy for P_2 is $(\frac{8}{11}, \frac{3}{11})$. The two intersecting lines determined by $6y_1 - 5$ and $-5y_1 + 3$ correspond to the first and third rows of the matrix. An optimal strategy for P_1 for the associated 2×2 subgame with matrix

$$\begin{bmatrix} 1 & -5 \\ -2 & 3 \end{bmatrix}$$

is $(\frac{5}{11}, \frac{6}{11})$, and so an optimal strategy for P_1 for the original game is $(\frac{5}{11}, 0, \frac{6}{11}, 0)$.

Using the second method described for $m \times 2$ games, we consider the three possible 2×2 subgame and compute their values.

$$\begin{bmatrix} 1 & -5 \\ -4 & 4 \end{bmatrix}, \quad \text{value} = \frac{4-20}{14} = -\frac{8}{7}$$
$$\begin{bmatrix} 1 & -5 \\ -2 & 3 \end{bmatrix}, \quad \text{value} = \frac{3-10}{11} = -\frac{7}{11}$$
$$\begin{bmatrix} -4 & 4 \\ -2 & 3 \end{bmatrix}, \quad \text{value} = -2$$

The second game corresponding to the first and third rows of A gives P_1 the largest value. It has value $-\frac{7}{11}$ and optimal strategies $(\frac{5}{11}, \frac{6}{11})$ and $(\frac{8}{11}, \frac{3}{11})$ for P_1 and P_2, respectively. Thus the value of the original game is $-\frac{7}{11}$ with optimal strategies

$$X_0 = (\tfrac{5}{11}, 0, \tfrac{6}{11}, 0) \text{ and } Y_0 = (\tfrac{8}{11}, \tfrac{3}{11})$$

Problem Set 9.6

1. The only technique discussed in this section that cannot be applied to matrix games with saddle points is the use of the formulas given in Theorem 9.6.2 for 2×2 games. Consider the game with payoff matrix

$$A = \begin{bmatrix} 2 & 3 \\ 1 & 4 \end{bmatrix}$$

   (a) Solve by using linear programming.
   (b) Solve by using only dominance.
   (c) Show that the formulas of Theorem 9.6.2 do not give the solution to this game. However, in this case the "r" does not equal 0. Where does the proof of Theorem 9.6.2 break down?

2. As suggested by the proof of the Fundamental Theorem in the previous section, the simplex method cannot be applied directly to a game with a value less than or equal to 0. Convince yourself of this by trying to use the simplex method without altering the original matrix to the games with payoff matrices

$$\begin{bmatrix} 2 & -1 \\ -3 & 1 \end{bmatrix} \text{ and } \begin{bmatrix} 2 & -1 \\ -4 & 2 \end{bmatrix}$$

3. *Proof of Theorem 9.6.1.* Consider the game with payoff matrix

$$A = \begin{bmatrix} a & b \\ c & d \end{bmatrix}$$

   (a) Prove that if $a < b, a < c, d < b, d < c$, then A has no saddle point.
   (b) Prove that if the two entries in any one row or column are equal, then A has a saddle point.
   (c) Prove that if A has no saddle point and $a < b$, then $a < c, d < c, d < b$.
   (d) Complete the proof of Theorem 9.6.1.

4. Prove Theorem 9.6.3. (*Hint.* We know that for any game, optimal strategies exist. Show that if one such optimal strategy for P_1 does not have $x_h = 0$, it can be modified to a strategy with the hth component equal to 0 that also satisfies the definition of optimality.)

5. True or false: A 2×2 game has a saddle point if and only if the game can be resolved using dominance.

6. True or false: Suppose A is an $m \times m$ matrix. Then the two games, one with payoff matrix A and the other with payoff matrix A^t, have the same value, and an optimal strategy for P_i for one game is an optimal strategy for P_{3-i} for the other game, $i = 1, 2$.

7. Determine the value, as a function of x, of each of the following games.

   (a) $\begin{bmatrix} 4 & x \\ 6 & 2 \end{bmatrix}$

   (b) $\begin{bmatrix} 1 & 2 \\ x & 3 \end{bmatrix}$

   (c) $\begin{bmatrix} 2x & 9 & 1 \\ 4 & 3 & 3 \\ 0 & 5x & 2 \end{bmatrix}$

   (d) $\begin{bmatrix} x & -3 \\ -1 & 2 \end{bmatrix}$

   (e) $\begin{bmatrix} 3 & 2 \\ x & 1 \end{bmatrix}$

   (f) $\begin{bmatrix} x+6 & x-2 \\ x-3 & x+1 \end{bmatrix}$

8. (a) Solve the games with the following payoff matrices using both methods of solution for $2 \times n$ and $m \times 2$ games.

   (i) $\begin{bmatrix} 2 & 0 & 1 & 4 \\ 1 & 3 & 2 & 0 \end{bmatrix}$

   (ii) $\begin{bmatrix} -1 & 7 \\ 4 & -1 \\ 2 & 1 \end{bmatrix}$

   (iii) $\begin{bmatrix} 5 & -1 & -2 \\ -2 & 0 & 1 \end{bmatrix}$

   (iv) $\begin{bmatrix} 1 & 6 \\ 5 & 2 \\ 3 & 4 \end{bmatrix}$

   (b) Show that for the game in (i), P_1's optimal strategy is not unique. In fact, when the game is solved graphically, it is found that the optimal security level for P_1 is attained at the intersection point of three of the relevant lines. Thus we have some freedom in selecting the associated 2×2 matrix to use to determine P_1's optimal strategy. However, we cannot arbitrarily use any two of the three columns corresponding to the intersecting lines. Explain why not. (This also occurs in the graphical analysis of the game in (iv).)

9. Solve the game of Problem 1(a) of Section 9.1.

10. For each of the following games, first determine x so that the game is fair, and then with x equal to this value, determine optimal strategies.

(a) Problem 1(b) of Section 9.1.

(b) P_1 selects a number from $\{1,2\}$ and P_2 a number from $\{1,2,3\}$. If the sum of the selected numbers is even, P_2 wins that amount from P_1. If the sum of the selected numbers is 3, P_1 wins 3 from P_2; if the sum is 5, P_1 wins the amount x ($x > 0$) from P_2.

(c) P_1 selects a number from $\{1,2\}$ and P_2 a number n from $\{1,2,3\}$. If the sum of the selected numbers is even, P_1 wins the amount nx ($x > 0$) from P_2. If the sum is odd, P_2 wins the amount $(2n-1)$ from P_1.

11. P_2 selects one of the following two games, A or B, to play. After making her selection, which is known to P_1, he sets the value of x, with the restriction that $0 \leq x \leq 50$. Which game should P_2 select, and why?

$$A = \begin{bmatrix} -4 & -2x \\ x & 8 \end{bmatrix} \quad \text{or} \quad B = \begin{bmatrix} x & -2x \\ -x & 3x \end{bmatrix}$$

12. Solve the games with the following payoff matrices.

(a) $$\begin{bmatrix} -12 & 9 & -5 & 0 \\ 4 & 7 & 2 & 1 \\ 0 & -5 & 10 & -3 \end{bmatrix}$$

(b) $$\begin{bmatrix} 8 & -1 & -3 & 6 \\ -3 & 10 & 8 & -4 \\ 3 & -4 & -5 & 4 \end{bmatrix}$$

(c) $$\begin{bmatrix} -1 & 0 & 1 & 1 \\ 2 & 1 & -2 & -1 \\ 1 & 3 & -1 & 0 \end{bmatrix}$$

(d) $$\begin{bmatrix} 4 & -2 & 5 & 3 \\ 2 & 3 & 0 & -2 \\ -1 & 4 & 2 & -1 \end{bmatrix}$$

(e) $$\begin{bmatrix} 4 & 1 & 2 & 8 \\ 6 & 4 & 4 & 5 \\ 0 & 7 & 3 & 1 \end{bmatrix}$$

(f) $$\begin{bmatrix} 5 & 8 & 6 & -2 \\ -5 & -3 & -2 & 1 \\ 3 & 1 & 8 & -4 \end{bmatrix}$$

13. For the game with payoff matrix

$$A = \begin{bmatrix} -2 & 1 & 0 \\ 1 & -2 & 1 \\ 0 & 1 & -2 \end{bmatrix}$$

it seems reasonable to assume that any optimal strategy for P_2 must use all three columns of A (and similarly for P_1 and the rows of A). If this is so, then any optimal strategy $X_0 = (x_1, x_2, x_3)$ for P_1 must satisfy $X_0 A^{(j)} = v$ for each j, $1 \leq j \leq 3$, where v is the value of the game (see Problem 8 of Section 9.5). While we do not yet know v, this does imply that $X_0 A^{(1)} = X_0 A^{(2)} = X_0 A^{(3)}$. Thus the (x_1, x_2, x_3) should satisfy the system of equations

$$\begin{aligned} -2x_1 + x_2 \quad &= x_1 - 2x_2 + x_3 \\ -2x_1 + x_2 \quad &= \quad x_2 - 2x_3 \\ x_1 + x_2 + x_3 &= 1 \end{aligned}$$

Solving, we find that $X_0 = (x_1, x_2, x_3) = (\frac{1}{10})(3, 4, 3)$. The security level of X_0 is $\text{Min}\{-\frac{1}{5}, -\frac{1}{5}, -\frac{1}{5}\} = -\frac{1}{5}$, and so we know for a fact that $v \geq -\frac{1}{5}$.

Now we could go through the same procedure to determine a strategy Y_0 for P_2. If the security level for this strategy were $-\frac{1}{5}$, we would have justified our original assumption and solved the game. Verify that this is the case. (Actually, here the symmetry of A also suggests the strategy Y_0 to use to verify the optimality of X_0.)

14. (a) Use the procedure outlined in Problem 13 to solve the following games.

(i) $\begin{bmatrix} -2 & 1 & 1 \\ 1 & -4 & 1 \\ 1 & 1 & -6 \end{bmatrix}$

(ii) $\begin{bmatrix} -1 & 1 & 1 \\ 2 & -2 & 2 \\ 3 & 3 & -3 \end{bmatrix}$

(iii) $\begin{bmatrix} 1 & -1 & -1 & 0 \\ -1 & 2 & 0 & -1 \\ 0 & -1 & -1 & 3 \end{bmatrix}$

(b) If the assumption that the optimal strategies use all the rows or columns is false, this procedure breaks down. Verify this for the game

$$\begin{bmatrix} 2 & 2 & -3 \\ 1 & 0 & -2 \\ -2 & -1 & 4 \end{bmatrix}$$

15. (a) Use the procedure outlined in Problem 13 to solve the following *diagonal game*, where the constants λ_1, λ_2, and λ_3 are all positive.

$$\begin{bmatrix} \lambda_1 & 0 & 0 \\ 0 & \lambda_2 & 0 \\ 0 & 0 & \lambda_3 \end{bmatrix}$$

(b) Generalize. Solve the game

$$\begin{bmatrix} \lambda_1 & 0 & \dots & 0 \\ 0 & \lambda_2 & \dots & 0 \\ \vdots & \vdots & \ddots & \vdots \\ 0 & 0 & \dots & \lambda_n \end{bmatrix}$$

where $\lambda_i > 0$, $1 \leq i \leq n$.

(c) Analyze the following game. P_2 has a \$1 bill, a \$5 bill, a \$10 bill, and a \$20 bill. P_2 selects one of the bills, and P_1 attempts to guess its denomination. If P_1 guesses correctly, P_1 wins the bill from P_2; if P_1 guesses incorrectly, no money is exchanged.

(d) Is P_1's position in the above game improved if P_2 also has a \$100 bill to use? A \$1000 bill?

Analyze the following games.

16. (a) P_1 selects a number from $\{1,2\}$ and P_2 a number from $\{1,2,3\}$. If the sum of the selected numbers is even, P_1 wins that amount from P_2; if the sum is odd, P_2 wins the sum from P_1.

(b) Example 9.1.2 of Section 9.1.

17. P_1 and P_2 each select a number from $\{1,2,3\}$. If they select the same number, P_2 wins an amount equal to the common selection from P_1. If not, P_1 wins an amount equal to the difference of the two selections.

18. The games of

(a) Problem 1(c) of Section 9.1.

(b) Problem 1(d) of Section 9.1.

(c) Problem 1(e) of Section 9.1.

(d) Problem 1(f) of Section 9.1.

19. (a) P_1 and P_2 each have two red cards and one black card. They play their cards, one by one, with P_1 winning the trick if the colors of the played cards match and P_2 winning if the colors do not match. Each trick is worth 1 point. At the completion of the play, the player with the greater number of points wins from the other an amount equal to the difference in point totals.

(b) As in (a), but now assume that the last trick is worth 2 points. (If, at the end of the play, the players have the same number of points, the game is a draw.)

20. P_1 and P_2 each have two cards, a red card and a black card. Each selects a card. If the colors of the selected cards do not match, P_2 wins 1 from P_1, and if they each selected their red card, P_1 wins 2. If they each selected their black card, they pick up their cards and play the game again. In the second play, all payoffs are as before, except that now, if each player selects their black card, P_1 wins 3 from P_2. (*Hint.* One way to analyze the game is to interpret the play as a sequence of two 2 x 2 games.)

21. P_1 has one red card and two black cards. P_1 selects a card, and P_2 attempts to guess its color. P_1 reveals his card and then selects a second card from his remaining two. P_2 attempts to guess the color of the second card. If both of P_2's guesses are correct, P_2 wins 4 from P_1; if only one is correct, P_1 wins 1 from P_2; and if both are wrong, P_1 wins 11. (*Hint.* This game also can be considered to be a sequence of two 2 x 2 games.)

9.7 Games People Play

In this section, we will consider several games modeled on the common card game of poker. Poker is an example of a zero-sum game in which the outcome of the game is determined by both chance and the skill of the players — the element of chance in determining the deal and the skill of the players in successfully balancing conservative play with bluffing. In fact, in the first book on game theory by von Neumann and Morgenstern [28], a section was devoted to a study of a form of poker; since then, various other articles on the game have appeared.

The standard game of poker is beyond our analysis, even after restricting the game to two players, because of the many different hands a player can be dealt. In the examples we develop, the range of possible draws will be much more limited — in fact, one player will receive only one card, and that will be either a "high" or a "low" card. However, our examples will preserve the opportunity for the players to bluff or to play conservatively, and so will preserve the element of poker that makes the game interesting.

Note, too, that these games are the first that we have encountered in which outcomes are partially determined by chance. For such games, the entries in the payoff matrix are the expected value or weighted average for Player 1 of all possible outcomes.

Game 1

Rules. Players 1 and 2 ante an amount $a > 0$ into the pot. One card is dealt to P_1 from a deck containing an equal number of high and low cards. After looking at his card, P_1 can either "pass" or "raise." If P_1 passes, P_1 wins the pot if he had been dealt a high card and loses the pot if he had been dealt a low card. If P_1 raises, he adds an amount $b > 0$ to the pot, and then Player 2 has two options. P_2 can either "fold" or "call." If P_2 folds, P_1 wins the pot (without revealing his hand). If P_2 calls, she also adds the amount b to the pot, and then P_1 wins or loses the pot if he has been dealt a high or low card, respectively.

Thus, if P_1 is dealt a high card, he is guaranteed to win at least the amount a from P_2, and he can win $a+b$ if he raises and P_2 elects to call. On the other hand, if P_1 is dealt a low card, he can pass and lose the amount a or he can raise (i.e., bluff), with the hope that P_2 will assume that P_1 has a high card and elect to fold. If P_2 folds, P_1 wins a even though he has a low card. However, if P_2 suspects that P_1 does not have a high card but is bluffing, P_2 can call, and now P_1 loses not a but $a+b$.

To apply our theory to this game, we must first list all the possible strategies for the two players. Recall that a strategy is a rule that tells a player what to do in any possible situation the player may encounter during the game. Consider P_1's situation. He is dealt either a high or low card, and so a strategy for him must tell him what to do in either case. Thus a possible strategy for P_1 would be to raise if he has a high card and pass if he has a low card. We will denote this strategy by (R,P), where the first component of the ordered pair directs P_1 if he has a high card and the second component if he has a low card, with R corresponding to raising and P to passing. It can be seen that P_1 has four possible strategies, denoted by (R,R), (R,P), (P,R), and (P,P). Thus (P,P) directs P_1 to pass no matter what he has been dealt, and (P,R) directs him to pass on a high card and raise on a low card. The strategy set for P_2 is simpler because she is unaware of the card dealt P_1. If P_1 elects to pass, P_2 has no options and the game is terminated. If P_1 elects to raise, P_2 can either call or fold, and so a strategy for P_2 is a rule that directs P_2's response to a raise by P_1. Denote these two possible strategies by *call* and *fold*.

Thus the payoff matrix associated with this game is 4×2. Since the outcome of the game is contingent not only on the strategies employed by the players but also on the card dealt P_1, the entries of the matrix are found by computing the expected value of the outcome to P_1 for the eight possible strategy pairs. For example, suppose P_1 uses strategy (P,R) and P_2 uses strategy call. Then if P_1 is dealt a high card, he passes and wins a, and if he is dealt a low card, he raises and P_2 calls, and so P_1 loses $a+b$. Since we are assuming that the deck contains an equal number of high and low cards and that the card dealt P_1 is randomly chosen, the probabilities of both a high-card hand and a low-card hand are $\frac{1}{2}$. Thus the expected value of the outcome corresponding to this strategy pair is $a/2+(-a-b)/2=-b/2$. Similarly, if P_1 uses strategy (R,P) and P_2 uses strategy fold, on a high-card deal P_1 raises, P_2 folds, and P_1 wins a, and on a low-card deal, P_1 passes and loses a. The expected outcome is therefore $a/2+(-a)/2=0$.

The other six entries in the payoff matrix are similarly computed. The result is

	Call	*Fold*
(R, R)	0	a
(R, P)	$b/2$	0
(P, R)	$-b/2$	a
(P, P)	0	0

With this the translation of the game into a matrix game is complete, and we can apply the theory as developed in this chapter. Of course, we must assume that the decisions of the players are governed by the two principles set out in Section 9.2 in order for our notion of a solution to the game to be meaningful and applicable. Is this reasonable?

An initial inspection of the above game matrix indicates that the last two rows are dominated by the first row and so can be deleted without affecting the solution. Notice that the two associated strategies correspond to P_1 passing if dealt a high card. Actually, we could have reasoned to the ineffectiveness of such an action by P_1 by just considering the rules of the game. If P_1 is dealt a high card, he is guaranteed to win at least the amount a regardless of P_2's play, and he could win $a+b$ if he raises and P_2 elects to call. Thus P_1, if dealt a high card, has nothing to lose and something to gain by raising, and so should always raise in this case.

The game matrix is thus reduced to the 2 x 2 matrix

	Call	*Fold*
(R, R)	0	a
(R, P)	$b/2$	0

The first row corresponds to P_1 raising with a low card (i.e., bluffing) and the second row to passing with a low card (i.e., conservative play). Similarly, the first column corresponds to P_2 challenging a raise and the second to P_2 playing conservatively. Since a and b are both positive, the solution to this game can be computed using the formulas of Theorem 9.6.2. The value v of the game and optimal strategies X_0 and Y_0 are given by

$$v = \frac{ab}{2a+b}$$

$$X_0 = \frac{1}{2a+b}(b, 2a)$$

$$Y_0 = \frac{1}{2a+b}(2a, b)$$

Hence no matter what a and b are, the game favors P_1. Both players should mix bluffing with conservative play, with the amounts of each in reverse order. For example, if $a = 1$ and $b = 1$, the value of the game is $\frac{1}{3}$, and to realize this value, P_1 must bluff $\frac{1}{3}$ of the time and P_2 must call $\frac{2}{3}$ of the time. If $a = 1$ and $b = 2$, the value of the game is $\frac{1}{2}$ and both players should play conservatively $\frac{1}{3}$ of the time.

Game 2

Rules. The ante, the deal, and the options for P_1 are just as in Game 1, and if P_1 elects to raise, P_2's options are also as in Game 1. However, if P_1 elects to pass, the play is not terminated, but P_2 can either pass or raise. If P_2 elects to pass, play is terminated and the pot distributed as before. If P_2 elects to raise, both she and P_1 add an amount b to the pot, and then the pot is distributed as before.

Thus the only difference between this game and Game 1 is that if P_1 chooses to pass, P_2 now has the opportunity to increase the pot to $a+b$, and P_1 cannot withdraw if P_2 wishes to increase the pot. P_1's strategy set remains the same, but now a strategy for P_2 must direct her response to both possible plays of P_1. Thus a possible strategy for P_2 would be to fold if P_1 raises and raise if P_1 passes. We will denote this strategy by (F,R), where the first component of the ordered pair is P_2's response to a raise by P_1 and the second component is her response to a pass, with the obvious abbreviations for fold, call, pass, and raise. P_2 has four pure strategies, denoted by (C,R), (C,P), (F,R), and (F,P). For example, the strategy (C,P) directs P_2 to call if P_1 raises and pass if P_1 passes.

For this game the payoff matrix is 4×4. The entries again are expected values and are computed just as before. The following tableau results.

	(C, R)	(C, P)	(F, R)	(F, P)
(R, R)	0	0	a	a
(R, P)	0	$b/2$	$-b/2$	0
(P, R)	0	$-b/2$	$(2a+b)/2$	a
(P, P)	0	0	0	0

While seeming to offer more variety, this game is not as interesting as the first game; the payoff matrix has saddle points, the first and fourth entries of the first column. The game has value 0, and optimal strategies for P_1 are either to raise all the time or pass all the time. The optimal strategy for P_2 is to call if P_1 raises and raise if P_1 passes. Notice that this strategy for P_2 ensures that the game is always played for the amount $a+b$ and that the winner is determined by the deal, and so P_1's threat of bluffing is effectively nullified.

Game 3

Rules. This game is played just like Game 2, with one minor difference. If P_1 elects to pass and P_2 elects to raise, both players must add the amount $2b$ to the pot before it is distributed in the usual manner.

Thus, if P_1 raises and P_2 calls, the stakes are $a+b$; but if P_1 passes and P_2 raises, the stakes are $a+2b$. It is not at all obvious how P_1 should proceed if he is dealt a high card. If he raises, he wins either a or $a+b$, depending on P_2's actions; if he passes, he wins at least a; and if P_2 raises, he wins $a+2b$. But if P_2 expects P_1 to always pass if he has a high card, then a raise by P_1 indicates that he has a low card, and in such circumstances a raise by P_2 will net P_2 the amount $a+2b$. Thus, to still bluff successfully with a low card by raising on a low-card hand, it seems that P_1 must occasionally raise on a high-card hand.

To develop a more precise analysis of this game, we need to consider the payoff matrix. The strategy sets for both players are identical to those in Game 2, and the resulting 4×4 payoff matrix this time is given by

	(C, R)	(C, P)	(F, R)	(F, P)
(R, R)	0	0	a	a
(R, P)	$-b/2$	$b/2$	$-b$	0
(P, R)	$b/2$	$-b/2$	$a+b$	a
(P, P)	0	0	0	0

As can be seen, there are no saddle points, but the fourth row is dominated by the first. However, no other simplifications are possible unless the ratio of a to b is known (and then, the only case of domination occurs between the first and last columns if $b/2 \leq a$). Thus we have a game with essentially a 3×4 payoff matrix to evaluate. However, an analysis is not difficult in this case. Since all the entries in the first row are either 0 or a, P_1's security level v_1 and therefore the value of the game is at least 0. Suppose now that P_2 restricts her choice of mixed strategy to those strategies involving the pure strategies (C,R) and (C,P), that is, the two strategies corresponding to the first two columns of the payoff matrix. P_2 would then be forcing P_1 to play the game with the 3×2 payoff matrix,

$$\begin{bmatrix} 0 & 0 \\ -b/2 & b/2 \\ b/2 & -b/2 \end{bmatrix}$$

But as can be easily seen, the value of this game is 0, and so P_2's security level v_2 and therefore the value of the original game is at most 0. Hence the game has in fact value 0, and optimal strategies can be determined by considering the above 3×2 game. For example, an optimal strategy for P_2 would be $(\frac{1}{2}, \frac{1}{2}, 0, 0)$, and optimal strategies for P_1 would be either $(1,0,0,0)$ or $(0, \frac{1}{2}, \frac{1}{2}, 0)$ regardless of the value of b.

Game 4

Rules. This game is played just like Game 2, the only difference being the amounts bet. If P_1 elects to raise, he must add $2b$ to the pot, and if P_2 elects to call, she must add only b to the pot. Similarly, if P_1 elects to pass and P_2 elects to raise, P_2 adds $2b$ to the pot and P_1 only b.

Thus, for this game, the player who wishes to increase the stakes must risk $2b$ and his or her opponent only b. Otherwise, Game 4 is identical to Game 2, with P_2 having the option of folding if P_1 raises; but if P_1 passes and P_2 raises, P_1 cannot withdraw (but only adds b to the pot, whereas P_2 adds $2b$).

The strategy sets for the two players are the same as those in Games 2 and 3. The 4×4 payoff matrix, which can be easily computed, is

	(C, R)	(C, P)	(F, R)	(F, P)
(R, R)	$-b/2$	$-b/2$	a	a
(R, P)	0	$b/2$	$-b/2$	0
(P, R)	0	$-b$	$a+b$	a
(P, P)	$b/2$	0	$b/2$	0

This matrix has no saddle points, and there is no domination regardless of the ratio of a to b. The last row of the matrix indicates that the value of the game is at least 0, but further elementary analysis does not seem possible in this case.

However, the simplex method can be used to solve this game, even in this general form with arbitrary positive values for a and b. See Problem 1. The value and optimal strategies for the game are given by

$$v = \frac{ab}{c}$$
$$X_0 = \frac{1}{c}(b, 2a+b, 0, 2a+b)$$
$$Y_0 = \frac{1}{c}(2a, 2a, 0, 3b)$$

where $c = 4a + 3b$.

Thus the value of the game is always positive, and therefore the game favors P_1. For example, if $a = 2$ and $b = 1$, $v = \frac{2}{11}$, $X_0 = (\frac{1}{11}, \frac{5}{11}, 0, \frac{5}{11})$, and $Y_0 = (\frac{4}{11}, \frac{4}{11}, 0, \frac{3}{11})$. In this case, then, in order to attain his maximum security level of $\frac{2}{11}$, P_1 should use strategy (R,R) with probability $\frac{1}{11}$ and strategies (R,P) and (P,P), both with probability $\frac{5}{11}$. Hence P_1 should raise with a high-card deal $\frac{6}{11}$ of the time and should raise with a low-card deal only $\frac{1}{11}$ of the time. Similarly, to keep her security level down to $\frac{2}{11}$, P_2 should respond to a raise by P_1 by calling with probability $\frac{8}{11}$ and folding with probability $\frac{3}{11}$, and should respond to a pass by P_1 by raising with probability $\frac{4}{11}$ and passing with probability $\frac{7}{11}$.

Problem Set 9.7

1. The following refers to Game 4.
   (a) Use the simplex method to solve the game when $a = 1$ and $b = 2$.
   (b) As in part (a), with $a = 2$ and $b = 1$.
   (c) Do these two operations suggest a sequence of pivot operations that could be used in the general case (arbitrary a and b)? If they do, try them. (If not, try pivoting for the first step in the first row, fourth column; second step, second column; third step, first column.)

2. (a) By computing security levels, verify that the strategies X_0 and Y_0 of Game 4 are optimal.
   (b) Suppose $a = 1$ and $b = 2$ in Game 4. Following his optimal strategy, with what probability should P_1 raise with a high card? With a low card? Similarly, interpret P_2's optimal strategy.

3. In the standard game of poker, the probability of getting a good hand is low. Analyze Game 1 under the assumption that the probability of P_1 being dealt a high card is only $\frac{1}{3}$. (The value of the game is $(a/3)[(b-2a)/(2a+b)]$.) Intuitively, why does this game become more favorable to P_2 if a is increased and b fixed?

4. In Game 1, suppose that $a = b = 1$ and that the probability of P_1 drawing a high card is p. Determine p so that the game is fair.

5. (a) Analyze the following variation of Game 1. Initially, one card is dealt to each player. Assume that each of the four possible deals (two high cards dealt; two low cards dealt; high card to P_1 and low card to P_2; low card to P_1

and high card to P_2) are equiprobable. The options for the players and the betting are as in Game 1. However, the pot is won by the player with the higher card and, in the event of a tie, the pot is distributed equally. (Note that now a strategy for P_2 must also include consideration of the card in P_2's hand.)

(b) You are P_1, playing the above game for big stakes in Las Vegas. The ante is one red chip ($1000), and the only permissible raise is one blue chip. It has been a while since you purchased your chips, and all you can remember is that the blue chips are worth either $999 or $1001. You are dealt a low card. What do you do?

6. Consider the following card game. Initially, both players ante an amount a into the pot.

From a three-card deck consisting of a king, queen, and jack, each player is dealt one card. After looking at his card, P_1 can either raise or pass. If he passes, the player with the higher card wins the pot. If P_1 raises, he adds an amount b to the pot, and then P_2, after looking at her card, can either call or fold. If she folds, P_1 wins the pot. If P_2 calls, she adds b to the pot, and then the player with the higher card wins the pot.

(a) Show that there are six possible deals.
(b) A strategy for P_1 must instruct him to either raise or pass in the event that he has either a king, queen, or jack. Show that P_1 has eight pure strategies.
(c) Similarly, show that P_2 has eight pure strategies.
(d) Before constructing the game matrix, consider domination. Convince yourself that P_1 has nothing to gain by passing when he has a king, and P_2 has nothing to gain by folding when she has a king. Thus the payoff matrix is essentially reduced to a 4×4 matrix.
(e) Convince yourself that P_2 has nothing to gain by calling when she has a jack, and that then, after these considerations, P_1 has nothing to gain by raising when he has a queen.
(f) The viable pure strategies for each player have been reduced to two: P_1 raises on a king, passes on a queen, and either raises or passes on a jack; P_2, in response to a raise by P_1, calls on a king, folds on a jack, and either calls or folds on a queen. Compute the associated 2×2 payoff matrix.
(g) Solve the game.
(h) Just as in 5(b), but you have moved to the table playing the three-card game of this problem. (The ante here is also one red chip, but the raise must be two blue chips.) You, as P_1, have made your ante and have been dealt a jack. What do you do?

7. Analyze the game of Problem 2(a) of Section 9.1. (*Hint.* While each player has 24 pure strategies, the value of the game is suggested by symmetry, and there are various strategies with this value as security level. In fact, each player's first play is irrelevant!)

8. Analyze the following. The only two clothing stores in a shopping center compete for the weekend trade. On a clear day, the larger store gets 60% of the business; on a rainy day, the larger store, being closer to the parking lot, gets 80% of the business. However, either or both retailers may hold a "sidewalk sale" on any given weekend, but the decision to hold such a sale must be made a week in advance and in ignorance of the competitor's plans. If both retailers conduct sidewalk sales, the breakdown in business is just as above. If, however, one holds the sale and the other does not, the one conducting the sale gets 90% of the business on a clear day and 10% on a rainy day. During the present season, it rains 40% of the time. How frequently should each retailer conduct sales?

9. Consider the game in Problem 2(b) of Section 9.1. Without P_1's option of raising, the game favors P_2. (P_2 could always guess red, with an expected gain of $\frac{1}{3}$.) To offset this, P_1 is given the option of increasing the stakes significantly. This option would seem especially useful when he has been dealt the "less likely to be guessed" black card. (Of course, he must bluff occasionally. Otherwise, P_2 automatically guesses black after a raise by P_1.) Analyze the game precisely, and in particular, determine by how much the option to raise has helped P_1.

10. Consider the game in Problem 2(c) of Section 9.1. P_1 wins if either player holds the red card, but the probability of the red card being dealt is only $\frac{2}{5}$. To compensate for this disadvantage, P_1 has the option of increasing the stakes, an obvious move when he has the red card and a (viable?) bluffing option when he has the black card. Analyze the game precisely. In particular, determine for what values of a and b, if any, the game is fair.

11. Reconsider the game of Problem 2(c) in Section 9.1 (the game of Problem 10 above), but with the following two modifications. First, suppose that the deck consists of three black cards and one red card; second, suppose that P_2 wins if the red card is dealt.

(*Observation*. The analysis here is a bit more difficult. The probability of the red card being out is $\frac{1}{2}$, but P_1 has the first move and so may have an advantage. Of course, if he never raises when he holds the losing red card, P_2 gains the advantage by accepting P_1's raise if she has the red card and rejecting the raise if she has a black card. Does this game in fact favor P_1?)

12. Analyze the following two-stage games.

(a) Two players play Game 1, with the exception that if P_1 elects to pass, he returns his card to the deck, no payments or additional antes are made, and Game 1 as originally described is played, but with the roles of P_1 and P_2 interchanged. Thus, if P_1 passes, the initial ante remains at a for each player, but now P_2 draws the card and makes the first move.

(b) As above, but suppose that now the deck contains only four cards, two high cards and two low cards, and that if P_1 elects to pass, he shows the card he has drawn but does not return it to the deck before P_2 draws her card.

13. Can the following variations of Game 1 be analyzed using the theory of this chapter?

(a) Game 1 played with a deck consisting of $p\%$ high cards, but with p unknown to both players.

(b) As in part (a), but with p known by only one player.

(c) P_2 is given a deck of four cards, two high cards and two low cards. She extracts a card of her choice without revealing her selection to P_1. Then Game 1 is played with this modified deck of the remaining three cards.

Chapter 10

Other Topics in Game Theory

10.1 Utility Theory

One of the basic assumptions that we have made in the study of game theory is that each of the possible outcomes of a game can be assigned a numerical value that represents the value or worth for a particular player of that outcome over the other possible outcomes. However, it is not immediately obvious that this is always possible, as was pointed out in Section 9.2. Even for parlor games in which the payoffs are in terms of money, it may be that these monetary payoffs cannot be used directly to measure a player's preferences. For example, one may derive much more satisfaction in a game of poker from winning \$2 by bluffing an opponent as opposed to winning \$5 with a hand of four aces. Or, would not the significance of a \$5 loss be different for a player already up by \$15 as opposed to a player already down by \$15? These difficulties are not insurmountable, at least theoretically. The body of knowledge developed to deal with this problem is called *utility theory*. In this section, we provide a brief introduction to the theory. What we propose to do is to indicate how one might go about assigning appropriate values to three different outcomes, regardless of their nature. The intuitive ideas that we develop form the foundation of utility theory.

Let us denote the three possible outcomes or events by the letters A, B, and C. We want to assign numerical values, or utilities, to each of these events that will in some way represent their relative desirability for an individual. Denote these numbers to be assigned by $u(A)$, $u(B)$, and $u(C)$. Out first step is to order the events linearly, that is, to determine the order of preference between the events. It could be that we are indifferent to two of the events. For example, we may not be able to make any distinction between A and B, believing that we would derive the same amount of satisfaction from either. In that case $u(A)$ should equal $u(B)$, and our problem would reduce to the problem of assigning values to only two distinct events. Thus we assume that the events can be strictly ordered, and that A is preferred over B and B is preferred over C. (We therefore assume that A is preferred over C because, intuitively, an ordering of events by preference must be transitive.) This ordering demands that $u(A) > u(B) > u(C)$. Our next task is to determine how our preference for A over B compares with our preference for B over C.

The key idea used to make this comparison is a *lottery*. Consider another event, a lottery, in which there are two possible outcomes, A and C. Suppose the lottery will result in A with probability r and C with probability $1-r$. For example, suppose the circumference of a wheel is divided into two arcs, one arc of length the fraction r of the entire circumference, and a pointer located at the center of the wheel spun. If the pointer comes to rest in the arc of length r of the whole, A occurs; otherwise, C occurs. This lottery is an event, and we can also assign to it a numerical value that measures its desirability. This value would depend on r. If r is near 0, the outcome of the lottery would more likely be C, and so the value of the lottery would be closer to $u(C)$ than to $u(A)$. And as r increases to 1, this value would approach $u(A)$.

The desirability of B must lie somewhere between the desirability of C and of A. It seems reasonable to assume that there exists a particular r, $0<r<1$, such that we are indifferent to the events B and the lottery with outcomes A and C, A occurring with probability r and C with probability $1-r$. Let us denote this particular lottery by the symbol $rA+(1-r)C$.

Since we are indifferent to these two events, the utility of B, $u(B)$, should equal the utility of the lottery, denoted by $u(rA+(1-r)C)$. We now make another basic assumption concerning the assignment of utilities, in this case, concerning the assignment of utilities to lotteries. Analogous to the definition of expected value in probability theory, as discussed in Section 9.4, it is reasonable to assume that the desirability of the lottery $rA+(1-r)C$ should be given by $ru(A)+(1-r)u(C)$, since this can be considered to be the *expected utility value* of the lottery. Thus, to determine $u(B)$, all we need determine is $u(A)$, $u(C)$, and the above r, and then set

$$u(B)=ru(A)+(1-r)u(C)$$

Next, consider assignment of the values $u(A)$ and $u(C)$. We have seen that the crucial quantity to be measured is how our preference for A over B compares with our preference for B over C. These preferences can be measured by the differences $u(A)-u(B)$ and $u(B)-u(C)$, suggesting that, unless otherwise restricted, arbitrary values can be assigned to $u(A)$ and $u(C)$ as long as $u(A)>u(C)$. Once such values are chosen, the lottery system described above can be used to determine $u(B)$.

The phrase *unless otherwise restricted* deserves some elaboration. By this we mean that our three events A, B, and C exist by themselves and cannot be compared with any other events or standards, and so we can freely assign the numbers $u(A)$ and $u(C)$. However, in many situations, the values of the outcomes are compared, consciously or subconsciously, to some external standards. For example, if we are involved in a game in which we will either win \$5 (event A) or lose \$5 (event C), the demands of the obvious preferential ordering are satisfied by any assignment of $u(A)$ and $u(C)$ such that $u(A)>u(C)$. In this case, though, we naturally compare A and C with the standard of not winning or losing anything (event B), intuitively setting $u(B)=0$. Everything else being equal, we would want $u(A)>0$ and $u(C)=-u(A)$. Moreover, if the value of the game is to be directly translated into dollars, we would want $u(A)=5$.

In summary, if we have three events, with no external standards to be imposed, we can assign utilities to these events using the described method. If there are more than three, it should be clear how this method can be extended. Once two events are assigned utilities, the lottery method can be applied to each of the remaining events separately until appropriate utilities have been assigned to all events. (It is not necessary initially to single out the two events at the extremes of the linear preferential ordering. See Problem 1.) In case there are also external standards to be imposed on our system, these events and their associated utilities would simply be added to the set of possible outcomes, and those events with preassigned utilities would provide the starting point.

Although the above discussion has been more suggestive than axiomatic, utility theory can be developed rigorously from a system of axioms. Refer to the books by von Neumann and Morgenstern [28], Luce and Raiffa [31], or Owen [32].

Example 10.1.1. You are involved in a game of chess with a grand master. There are three possible outcomes: you win, event A; you draw, event B; or you lose, event C. Setting $u(A) = 1$ and $u(C) = -1$, what should $u(B)$ be? Because of the abilities of your opponent, much satisfaction would be gained from a draw, and so clearly, $u(B)$ should be positive. More precisely, suppose you feel equally disposed to a draw and a lottery in which you have a probability of $\frac{19}{20}$ of winning over the grand master and a probability of $\frac{1}{20}$ of losing. Then

$$\begin{aligned} u(B) &= u(\tfrac{19}{20}A + \tfrac{1}{20}C) \\ &= \tfrac{19}{20}u(A) + \tfrac{1}{20}u(C) \\ &= \tfrac{19}{20} - \tfrac{1}{20} \\ &= \tfrac{9}{10} \end{aligned}$$

Problem Set 10.1

1. Utilities are to be assigned to four events, A, B, C, and D. Event A is preferred over B, B over C, and C over D. $u(B)$ is to be set equal to 1 and $u(C)$ 0. It is determined that B is indifferent to the lottery $rA + (1 - r)C$, and C is indifferent to the lottery $sB + (1 - s)D$. Compute $u(A)$ and $u(D)$.

2. Let A be the event that you are given \$100, B the event that your status quo is maintained, and C the event that you are elected president of the student body. Set $u(A) = 1$, $u(B) = 0$, and determine your personal utility for C.

10.2 Two-Person, Non-Zero-Sum Games

In the next four sections we will discuss two-person, non-zero-sum games, that is, two-person games for which the sum of the payoffs to the two players for each of the various possible outcomes of the game is not necessarily always zero. We have seen examples of such games in Section 1.3. Our primary purpose in studying these games is to demonstrate some of the difficulties that arise when attempting to develop mathematical models of more complex situations. Indeed, although non-zero-sum games reflect the types of situations encountered much more frequently in real-world applications, the formulation of a widely-accepted, all-encompassing mathematical model is, as we will see, not at all straightforward.

For non-zero-sum games, the payoff for one player is not necessarily the negative of the payoff for the other player. Thus, to express the payoffs for such games, ordered pairs will be used, where the first component represents the payoff to Player 1 and the second the payoff to Player 2.

Game 1

In the first game described in Section 1.3, both players had two pure strategies and the payoff tableau was given by

	t_1	t_2
s_1	$(0,0)$	$(12,-12)$
s_2	$(-12,12)$	$(6,6)$

Thus, if P_1 uses s_2 and P_2 uses t_1, P_1 would lose 12 units and P_2 would gain 12 units. Note that the (6,6) payoff is the only non-zero-sum payoff.

Zero-sum games are strictly competitive — what one gains, the other loses. However, this is not the case for non-zero-sum games. For these games, both players may be able to ensure for themselves an advantage by cooperating with the other player. Thus the possibility of preplay communication and cooperation adds a new dimension to the study of non-zero-sum games. For example, consider Game 2.

Game 2

	t_1	t_2
s_1	$(0,0)$	$(1,1)$
s_2	$(1,1)$	$(0,0)$

In Game 2, there is no reason for either player to choose one strategy over the other unless, of course, they can communicate beforehand. Permitted such communication, the players would coordinate their strategies to ensure a $(1,1)$ payoff for themselves.

Repeated playing of the same game may achieve the same effect as pregame communication. Certainly, if two people played Game 2 100 times without communication, after several turns a pattern would be established providing a constant $(1,1)$ payoff.

However, cooperation, achieved either through preplay communication or repeated play, does not begin to resolve the difficulties inherent in non-zero-sum games. Consider Game 3.

Game 3

	t_1	t_2
s_1	$(10,1)$	$(2,2)$
s_2	$(2,2)$	$(1,10)$

In Game 3, P_1 would prefer the $(10,1)$ payoff and P_2 the $(1,10)$ payoff. Following these preferences, P_1 would play s_1 and P_2 would play t_2, resulting in a $(2,2)$ payoff. It is not clear how preplay communications could resolve these conflicting preferences. However, if the game were played twice, the players might agree on using strategy pair (s_1,t_1) for the first game and (s_2,t_2) for the second. In this way, both players would gain 11 units. But, after the game is played once with outcome $(10,1)$, what is to prevent P_1 from disregarding the agreement and playing s_1 again, gaining at least a total of 12 units?

Game 1 provides another example of the problems arising from incoercible agreements. In that game, as a result of preplay discussion, both players may agree to the strategy pair (s_2,t_2) with payoff $(6,6)$. But if either player expects the other to abide by this agreement, that player can gain 12 units by breaking his part of the bargain and using his first strategy. In fact, notice that strategy s_1 dominates s_2 and t_1 dominates t_2 (i.e., $0 > -12$ and $12 > 6$). Thus, no matter what the opponent does, each player has more to gain by using his or her first strategy. The concept of dominance, so reasonable for zero-sum games, leads to the somewhat unreasonable outcome of $(0,0)$ in this case. In general, dominance plays a minor role in nonzero theory. (If you are not yet convinced, consider the game in Problem 1.)

The above examples show that preplay communication can lead to other problems, but in some cases it may not even be desirable. Consider Game 4.

Game 4

	t_1	t_2
s_1	$(1,10)$	$(10,1)$
s_2	$(0,-10)$	$(0,-9)$

In this game, P_2 would prefer the $(1,10)$ outcome and P_1 the $(10,1)$ outcome. However, without preplay discussion, P_1 has no reason to use strategy s_2, and so the game would probably result in the $(1,10)$ payoff following the use of the strategy pair (s_1,t_1). However, if the players can communicate, P_1 could demand that P_2 use strategy t_2, threatening the use of s_2 if P_2 does not agree to play t_2. Thus, with preplay discussion, P_1 can attempt to force P_2 to the $(10,1)$ outcome, and so the notion of a threat becomes a component in the theory of cooperative games.

Another factor for consideration in non-zero-sum games is whether or not utility can be transferred; that is, can one player make a side payment to his opponent after the game is played? Consider Game 5.

Game 5

	t_1	t_2
s_1	(50,0)	(1,5)
s_2	(1,0)	(1,5)

Here, if side payments are not possible, P_2 has no reason to do anything but play t_2, leading to the $(1,5)$ outcome. However, if the players can cooperate and transfer utility, P_1 could offer to share her 50 units in some way with P_2 in return for P_2 using t_1 (and P_1 using s_1). Of course, this assumes that the utility is divisible and transferable and that the utility scales are comparable.

In fact, the comparison of utilities between the players raises various questions that can be critical in the solution of a game. Does one player derive as much satisfaction from a gain of, say, 25 units as the other player does? Is one player concerned not only with his or her own payoff but also with that of the other player, attempting perhaps to make it as small as possible or, on the other hand, to ensure that it remain above a certain level? Are negative payoffs always undesirable and positive payoffs always desirable? In any application of the theory, these questions would have to be considered. In the remainder of our discussion of non-zero-sum games, we will assume that the payoffs represent monetary units, say dollars, or \$10, and so on, to players in equivalent financial positions.

Given the many difficulties with non-zero-sum games, how can we attempt to develop a mathematical model of these games that in some way reflects rational behavior (whatever that is)? To begin, any model needs a precise starting point. Thus, the conditions under which a game is to be played must be made precise. This suggests that for non-zero-sum games, various types, such as cooperative games versus noncooperative games, should be considered separately. Moreover, mathematicians are inclined to study the extreme cases first since, on the one hand, these cases may be the most susceptible to a reasonable theory and, on the other hand, descriptions of the extreme cases may provide insights applicable to the intermediate cases. The study of non-zero-sum games thus contains two major divisions: the study of *noncooperative games*, in which no preplay communication is permitted, and the study of *cooperative games*, in which preplay communication and binding agreements are permitted. In the next section noncooperative games will be considered briefly, and in Sections 10.4 and 10.5 an introduction to cooperative games will be presented.

Problem Set 10.2

1. In the following, row and column dominance leads to what outcome? Is this outcome reasonable?

	t_1	t_2
s_1	$(-99,-99)$	$(100,-100)$
s_2	$(-100,100)$	$(99,99)$

2. The following game appears at first glance to be symmetric. Show that this is not the case and that one of the players has a more favorable position.

	t_1	t_2
s_1	$(0,5)$	$(-1,1)$
s_2	$(1,-1)$	$(5,0)$

10.3 Noncooperative Two-Person Games

In this section, we assume that no form of communication or cooperation is permitted between the players of the non-zero-sum game. Situations such as business competition between bitter rivals or market competition between large companies restricted by antitrust legislation may fall into this category.

In zero-sum games, the elements of communication and cooperation are not present. In this respect zero-sum and noncooperative non-zero-sum games are similar, and so it would seem that a reasonable starting point for the study of non-zero-sum games would be the two principles set forth in Section 9.2. These two principles, concerning the maximization of security levels and the tendency to equilibrium strategies, provided the foundation for a complete theory of zero-sum games. In this section, we will consider the role of security levels and equilibrium strategies in noncooperative non-zero-sum games. However, we will see that in this instance, these two principles alone are incapable of leading to a complete theory.

To define equilibrium strategies and security levels for non-zero-sum games, it is easiest first to single out the two payoff matrices. For a non-zero-sum game with m pure strategies for P_1 and n pure strategies for P_2, let A be the $m \times n$ matrix with entries equal to the payoffs to P_1, and let B be the corresponding $m \times n$ matrix of payoffs for P_2. As before, by a (mixed) strategy for P_1 (or P_2), we mean an m-component (or n-component) vector with nonnegative coordinates, the sum of which is 1. Again, we will use S and T to denote the sets of strategies for P_1 and P_2, respectively.

A strategy pair is said to be in equilibrium if neither player can gain by deviating from his or her prescribed strategy as long as the opponent's strategy remains fixed. Formally, for non-zero-sum games, we have the following definition.

Definition 10.3.1. The strategy pair (X_0, Y_0) is in *equilibrium* if, for all $X \in S$ and $Y \in T$,

$$XAY_0^t \leq X_0AY_0^t \quad \text{and} \quad X_0BY^t \leq X_0BY_0^t$$

To determine the players' security levels, we can use the theory of zero-sum games. Consider the zero-sum game with payoff matrix A. The value of this game is the maximum amount P_1 can be guaranteed regardless of the play of P_2 since, as we saw in the previous chapter, the value of the game is equal to P_1's optimal security level v_1, as defined in Section 9.4. And an optimal strategy for the first player of the zero-sum matrix game A would provide P_1 with a *security level strategy*, that is, a strategy that will enable P_1 to realize this security level.

To define P_2's security level, a minor adjustment in the above process must first be made. The matrix B represents the payoffs to P_2, the column player in the non-zero-sum game. Thus, in general, P_2 seeks the larger entries and not the smaller entries, as the column player does in a zero-sum game. In fact, P_2's position is identical to that of the row player of the zero-sum game with matrix B^t, since the transpose operation simply interchanges the rows and columns. Hence P_2's security level is defined in terms of the matrix game B^t.

Definition 10.3.2. P_1's *security level* is the value of the zero-sum matrix game A, and an optimal strategy for the row player of that game provides P_1 with a *security level strategy*.

Similarly, P_2's *security level* is the value of the zero-sum matrix game B^t, and an optimal strategy for the row player of that game provides P_2 with a *security level strategy*.

Example 10.3.1. Consider Game 4 of the previous section. The payoff tableau was

	t_1	t_2
s_1	$(1,10)$	$(10,1)$
s_2	$(0,-10)$	$(0,-9)$

For this game

$$A = \begin{bmatrix} 1 & 10 \\ 0 & 0 \end{bmatrix}, B = \begin{bmatrix} 10 & 1 \\ -10 & -9 \end{bmatrix}, \text{ and } B^t = \begin{bmatrix} 10 & -10 \\ 1 & -9 \end{bmatrix}$$

Matrix A has a saddle point at the 1 entry. Therefore P_1's security level is 1, and the associated security level strategy is $X_0 = (1,0)$. Matrix B^t has a saddle point at -9, and so P_2's security level is -9 with the associated security level strategy $Y_0 = (0,1)$, the optimal strategy for the row player of the matrix game B^t. However, the pair (X_0, Y_0) is not in equilibrium; if P_1 uses s_1, P_2's gain is maximized by using only t_1. That is, if $Y_1 = (1,0)$,

$$10 = X_0 B Y_1^t \not\leq X_0 B Y_0^t = 1$$

Moreover, the first row of A dominates the second, and so it follows that any equilibrium strategy pair must have P_1's strategy equal to X_0. But P_2's best response to this is Y_1. Thus (X_0, Y_1) is the only strategy pair in equilibrium. Notice that the payoff $(1,10)$ associated with this pair maintains P_1's security level, provides P_2 with 19 units more than P_2's security level, and seems to be a reasonable solution of the game if played without communication and cooperation.

In this game, the pair of security level strategies is not in equilibrium, as contrasted with the situation in zero-sum games (see Corollary 9.5.1 in Section 9.5). So already differences in the results of the application of the two basic principles begin to appear. In the remainder of this section, we will consider games in which these differences inhibit the solution of the games.

Game 6

	t_1	t_2
s_1	$(10,1)$	$(0,0)$
s_2	$(0,0)$	$(1,10)$

For this game, both strategy pairs $((1,0),(1,0))$ and $((0,1),(0,1))$ are in equilibrium, as can be easily seen. However, the payoffs associated with these two equilibrium pairs are quite distinct, with P_1 benefiting more from the $(10,1)$ payoff and P_2 from the $(1,10)$ payoff. Thus we come to the major problem of equilibrium pairs: different pairs can provide different payoffs. (Again, this is in contrast to the zero-sum case. See Theorem 9.5.2 on page 367.)

This problem cannot be solved in general. Certainly in Game 6, if we know nothing about the nature of the players themselves, there is no reason to choose one of these equilibrium pairs over the other. The security level strategies for P_1 and P_2 are $(\frac{1}{11},\frac{10}{11})$ and $(\frac{10}{11},\frac{1}{11})$, respectively, with both players having as security level $\frac{10}{11}$. However, this pair of strategies is not in equilibrium; if P_1 uses s_2 with frequency $\frac{10}{11}$, P_2 would be better off using t_2 more frequently than $\frac{1}{11}$ of the time, and conversely. On the other hand, the strategy pair $((\frac{10}{11},\frac{1}{11}),(\frac{1}{11},\frac{10}{11}))$ is in equilibrium, with the expected payoff of $(\frac{10}{11},\frac{10}{11})$. Would the players tend to use this mixed strategy pair?

Notice that if the players in Game 6 could communicate and cooperate, they might agree to coordinate their strategies so that half the time the payoff is $(10,1)$ and the other half it is $(1,10)$. For example, they could agree to flip a coin, and if it turns up heads they both play their first strategy, and if it is tails they play their second strategy. In this way the $(0,0)$ payoffs would be completely avoided, and the expected outcome for each player then would be $5\frac{1}{2}$, a considerable improvement over $\frac{10}{11}$.

Game 7

	t_1	t_2
s_1	$(0,0)$	$(10,-1)$
s_2	$(-1,10)$	$(9,9)$

This game is of the prisoner's dilemma type, as in Game 1 of the previous section. In Game 6, there were three different pairs of equilibrium strategies, and none of these pairs corresponded to the security level strategies. However, in this game there are no such complications. Strategy s_1 dominates s_2 ($0 > -1$ and $10 > 9$) and t_1 dominates t_2. Thus $((1,0),(1,0))$ is the only equilibrium pair. The matrices A and B^t associated with this game have saddle points at the 0 entries, so the strategies $(1,0)$ and $(1,0)$ are also the security level strategies, providing the security level of 0 for both players. Hence the two basic principles lead in this game to one unique strategy pair. However, even in this case, is this strategy pair a reasonable solution of the game, since the $(9,9)$ payoff offers much more to the players? We emphasize this dilemma in the next game.

Game 8

	t_1	t_2	t_3
s_1	$(0,0)$	$(0,-1)$	$(9,-10)$
s_2	$(-1,1)$	$(9,9)$	$(8,10)$

This game is quite similar to Game 7. As can be easily shown, the security level of both players is 0 with security level strategies of $X_0 = (1,0)$ and $Y_0 = (1,0,0)$. Moreover, the strategy pair (X_0, Y_0) is in equilibrium (see Problem 6). However, is not the $(9,9)$ outcome much more likely, especially in this game?

Even if Game 8 is played without cooperation repeatedly between the same two players, it would seem that the (s_1, t_1) strategy pair would occur most infrequently, if ever. Suppose the players start by using the pair (s_2, t_2) several times. If P_2 attempts to increase her payoff by moving to t_3, after several plays of (s_2, t_3), would not P_1 move to penalize P_2 by playing s_1? The pair (s_1, t_3) would result in a -10 payoff to P_2, and so would she not immediately move to t_2 in order to reestablish as quickly as possible the mutually beneficial outcome of $(9,9)$?

To summarize, these examples show that different equilibrium strategy pairs may provide different payoffs, that a pair of security level strategies may not be in equilibrium, and that in some cases, a solution of the game in terms of equilibrium pairs may not seem reasonable even when an equilibrium pair is unique and corresponds to the security level strategies. Thus, as adequate and reasonable as they were for zero-sum games, the two basic principles of Section 9.2 are incapable of providing a complete theory for noncooperative non-zero-sum games.

Even with the above limitations, much of the existing analysis of noncooperative games employs the concept of equilibrium pairs. The role of the players security levels, however, is not critical in noncooperative theory. One reason for this is that the payoffs to the two players associated with any equilibrium pair of strategies are at least as great as their security levels, as proved by the following theorem.

Theorem 10.3.1. *Given a two-person, non-zero-sum game with payoff matrices A and B, let u and v denote the security levels of P_1 and P_2, respectively. Let (X_0, Y_0) be any strategy pair in equilibrium. Then*

$$u \leq X_0 A Y_0^t \quad \text{and} \quad v \leq X_0 B Y_0^t$$

Proof. Let S and T denote the strategy sets of P_1 and P_2, respectively. Then, for any $X \in S$,

$$\operatorname*{Min}_{Y \in T} XAY^t \leq XAY_0^t \quad \text{and} \quad XAY_0^t \leq X_0 A Y_0^t$$

Thus

$$\begin{aligned} u &= \operatorname*{Max}_{X \in S} \operatorname*{Min}_{Y \in T} XAY^t \\ &\leq \operatorname*{Max}_{X \in S} XAY_0^t \\ &= X_0 A Y_0^t \end{aligned}$$

Similarly,

$$\begin{aligned} v &= \operatorname*{Max}_{Y \in T} \operatorname*{Min}_{X \in S} Y B^t X^t \\ &\leq \operatorname*{Max}_{Y \in T} Y B^t X_0^t \\ &= \operatorname*{Max}_{Y \in T} X_0 B Y^t \\ &= X_0 B Y_0^t \end{aligned}$$

□

While not significant, we know that security levels and security level strategies always exist in noncooperative two-person, non-zero-sum games. This suggests the more important question — do equilibrium strategy pairs always exist? The answer is "yes." J. Nash, in his 1950 dissertation at Princeton, proved the existence equilibrium strategy pairs using Brouwer's Fixed-Point Theorem ([35]; see also [36]). We outline a proof, stating first a version of Brouwer's Fixed Point Theorem.

Theorem 10.3.2 (Brouwer's Fixed Point Theorem). *Any continuous map F from the unit cube $K^n = \{(x_1, \ldots, x_n) : 0 \leq x_i \leq 1$, for all $i\}$ into K^n has at least one fixed point, that is, a point $X \in K^n$ such that $F(X) = X$.*

The proof of this often-used theorem is nontrivial. The theorem does extend to any set equivalent to (homeomorphic to) a unit cube. In particular, if S and T are the strategy sets for P_1 and P_2, respectively, in a noncooperative non-zero-sum game, then any continuous map from $S \times T$ to $S \times T$ has at least one fixed point. It is in this context that we use the theorem. Also, one other result will be used, which we state in a preliminary lemma.

Lemma 10.3.1. *For an $m \times n$ non-zero-sum game with P_1's game matrix A and strategy set S and P_2's game matrix B and strategy set T, the strategy pair (X_1, Y_1) is in equilibrium if and only if $A_{(i)} Y_1^t \leq X_1 A Y_1^t$ for all $1 \leq i \leq m$ and $X_1 B^{(j)} \leq X_1 B Y_1^t$ for all $1 \leq j \leq n$.*

Proof. By definition, (X_1, Y_1) is in equilibrium if

$$XAY_1^t \leq X_1 A Y_1^t \text{ for all } X \in S \quad \text{and} \quad X_1 B Y^t \leq X_1 B Y_1^t \text{ for all } Y \in T$$

Applying the first inequality to P_1's pure strategies yields $A_{(i)} Y_1^t \leq X_1 A Y_1^t$ for all $1 \leq i \leq m$; and applying the second inequality to P_2's pure strategies gives $X_1 B^{(j)} \leq X_1 B Y_1^t$ for all $1 \leq j \leq n$.

Conversely, if $A_{(i)} Y_1^t \leq X_1 A Y_1^t$ for all i, then for any $X \in S$,

$$XAY_1^t = \sum_i x_i (A_{(i)} Y_1^t),$$

being a weighted average of the $\{A_{(i)} Y_1^t\}$, is less than or equal to $X_1 A Y_1^t$. Similarly for $Y \in T$. □

Theorem 10.3.3. *For any noncooperative two-person, non-zero-sum matrix game there exists at least one strategy pair in equilibrium.*

Proof. Let A, S, B, and T be defined as in the above lemma. To prove the theorem, we will first define a continuous map $Q: S \times T \to S \times T$, apply Brouwer's Fixed Point Theorem to Q, and then show that any fixed point of Q is an equilibrium strategy pair. The essence of the proof lies in the definition of Q.

To define Q for a point $(X,Y) \in S \times T$, we must first introduce some auxiliary terms. For a strategy pair $X \in S$ and $Y \in T$, and for a fixed i and j, we define

$$c_i = \text{Max}\{A_{(i)}Y^t - XAY^t, 0\} \text{ and } d_j = \text{Max}\{XB^{(j)} - XBY^t, 0\}$$

Then

- all $c_i \geq 0$ and all $d_j \geq 0$;
- $c_i = 0 \Leftrightarrow A_{(i)}Y^t - XAY^t \leq 0 \Leftrightarrow A_{(i)}Y^t \leq XAY^t$;
- $d_j = 0 \Leftrightarrow XB^{(j)} - XBY^t \leq 0 \Leftrightarrow XB^{(j)} \leq XBY^t$;
- (X,Y) is in equilibrium if and only if all c_i and d_j are 0.

Now define $Q: S \times T \to S \times T$ by

$$Q(X,Y) = (X',Y') = ((x'_1, x'_2, \ldots), (y'_1, y'_2, \ldots))$$

where

$$x'_i = \frac{x_i + c_i}{1 + \sum c_k} \text{ and } y'_j = \frac{y_j + d_j}{1 + \sum d_k}$$

Then

- X' is a strategy as each $x'_i \geq 0$ and

$$\sum_i x'_i = \sum_i \frac{x_i + c_i}{1 + \sum c_k} = \frac{1}{1 + \sum c_k}\left(\sum_i x_i + \sum_i c_i\right) = 1$$

- Y' is a strategy;
- Q is continuous.

From Brouwer's Fixed Point Theorem, Q has at least one fixed point, that is, there is a strategy pair (X,Y) such that $Q(X,Y) = (X,Y)$. We claim that the fixed point (X,Y) of the map Q is in equilibrium.

Since $Q(X,Y) = (X,Y)$, for any i we have

$$x'_i = \frac{x_i + c_i}{1 + \sum c_k} = x_i$$

and so $x_i + c_i = x_i + x_i \sum c_k$, which implies that $c_i = x_i \sum c_k$; and similarly, for any j, we have $d_j = y_j \sum d_k$.

Suppose that (X,Y) is not in equilibrium. Then there exists $\hat{i}$ such that $A_{(\hat{i})}Y^t > XAY^t$ or $\hat{j}$ such that $X_1 B^{(\hat{j})} > X_1 BY_1^t$. Assume that $A_{(\hat{i})}Y^t > XAY^t$, and so $c_{\hat{i}} > 0$. Then

$$0 < c_{\hat{i}} = x_{\hat{i}} \sum c_k \Rightarrow x_{\hat{i}} > 0 \text{ and } \sum c_k > 0$$

However, $XAY^t = \sum_i x_i (A_{(i)}Y^t)$ is a weighted average of the $\{A_{(i)}Y^t\}$; and since $A_{(\hat{i})}Y^t > XAY^t$ and $x_{\hat{i}} > 0$, there is an $\bar{i}$ such that $x_{\bar{i}} > 0$ and $A_{(\bar{i})}Y^t < XAY^t$. Therefore $c_{\bar{i}} = 0$. But

$$0 = c_{\bar{i}} = \underbrace{x_{\bar{i}}}_{>0} \underbrace{\sum c_k}_{>0}$$

which is a contradiction. Conclusion: (X,Y) is in equilibrium. □

Problem Set 10.3

1. Why do we not consider the possibility of cooperation in zero-sum games?

2. Show that the following game has a unique strategy pair in equilibrium.

	t_1	t_2
s_1	$(2,1)$	$(6,7)$
s_2	$(8,3)$	$(4,5)$

3. Consider the game with the payoff tableau

	t_1	t_2
s_1	$(0,5)$	$(1,1)$
s_2	$(-1,-1)$	$(5,0)$

(a) Show that both players have a security level of 0.
(b) Show that $X=(1,0)$ and $Y=(0,1)$ are security level strategies. Is this pair (X,Y) in equilibrium?
(c) Find two obvious pairs of equilibrium strategies that employ pure strategies. Can you find the one other equilibrium pair that uses mixed strategies?
(d) Compare the payoffs associated with the equilibrium pairs to the players' security levels.

4. Consider the game of Problem 2 of Section 10.2. (It is identical to the above game if the 1 and -1 payoffs for P_1 are interchanged.) Show that P_1's security level is greater than P_2's security level and that the unique equilibrium strategy pair leads to the $(5,0)$ payoff. Thus, for this game, these concepts demonstrate P_1's more favorable position.

5. Consider the game with the payoff tableau

	t_1	t_2
s_1	$(2,7)$	$(-5,-1)$
s_2	$(0,-2)$	$(7,2)$

(a) Show that P_1's position is more favorable.
(b) Compute the security levels.
(c) Show that each of the three strategy pairs $((1,0),(1,0))$, $((0,1),(0,1))$ and $((\frac{1}{3},\frac{2}{3}),(\frac{6}{7},\frac{1}{7}))$ is in equilibrium. Find the associated expected payoffs.
(d) Show that there are no other equilibrium strategy pairs.
(e) Do either of these concepts distinguish P_1's more favorable position?

6. Prove that Game 8 has three strategy pairs in equilibrium. (*Hint*. Lemma 10.3.1.)

7. Do the concepts of security levels and equilibrium strategy pairs assist in the solution of the following game? (There are three equilibrium strategy pairs.)

	t_1	t_2
s_1	$(1,2)$	$(0,0)$
s_2	$(-2,-1)$	$(4,1)$

8. In the proof of Theorem 10.3.3, it was shown that if (X,Y) is a fixed point of Q, then (X,Y) is in equilibrium. Establish the converse of this conclusion, that is, that if (X,Y) is in equilibrium, then (X,Y) is a fixed point of Q.

10.4 Cooperative Two-Person Games

In this and the next section, we assume that the two players of the non-zero-sum game can enter into preplay discussion and binding agreements. Conflict situations of this type could be encountered, for example, in the negotiations between labor and management over a labor contract or between two countries over a trade agreement.

As pointed out in the discussion of Game 6 in the previous section, the players can usually expand the set of possible payoffs of the game by cooperating. In Game 6, if the players coordinate the use of the strategy pairs (s_1,t_1) and (s_2,t_2), using each with probability $\frac{1}{2}$, the expected payoff is $(5\frac{1}{2},5\frac{1}{2})$. However, if each player independently uses the mixed strategy $X = Y = (\frac{1}{2},\frac{1}{2})$, the expected payoff is

$$(XAY^t, XBY^t) = (\tfrac{11}{4}, \tfrac{11}{4})$$

where A and B are defined in the usual manner. Note that the independent use of these strategies would result in a $(0,0)$ payoff with frequency $\frac{1}{2}$.

In fact, since the players are now permitted to discuss the game beforehand and correlate their play, the question of what strategy to use is subordinate to the question of what mutually beneficial payoff the players can agree to. Thus we determine first the set of all payoffs possible with the use of cooperation. If the game tableau is $m \times n$, then a payoff, a point in $\mathbb{R}^2$, is produced by the coordinated use of the pure strategy pairs (s_i,t_j), $1 \le i \le m$, $1 \le j \le n$. With each pair (s_i,t_j) is associated the outcome (a_{ij},b_{ij}), where the a_{ij}'s and b_{ij}'s are the entries of the payoff matrices A and B, respectively. If the players agree to use each pair (s_i,t_j) with probability or frequency r_{ij}, where $0 \le r_{ij} \le 1$, then the expected payoff is simply $\sum r_{ij}(a_{ij},b_{ij})$. Thus the set of all possible payoffs, denoted by M, is given by

$$M = \left\{ \sum_{\substack{1 \le i \le m \\ 1 \le j \le n}} r_{ij}(a_{ij},b_{ij}) : 0 \le r_{ij} \le 1, \sum_{\substack{1 \le i \le m \\ 1 \le j \le n}} r_{ij} = 1 \right\}$$

This set M, which we will call the *cooperative payoff set*, is the smallest convex region containing all the points (a_{ij},b_{ij}). It can be easily determined from the set of points (a_{ij}, b_{ij}), and its boundary consists of line segments with terminal points from this set.

Example 10.4.1. For Game 6, the cooperative payoff set M is the shaded region:

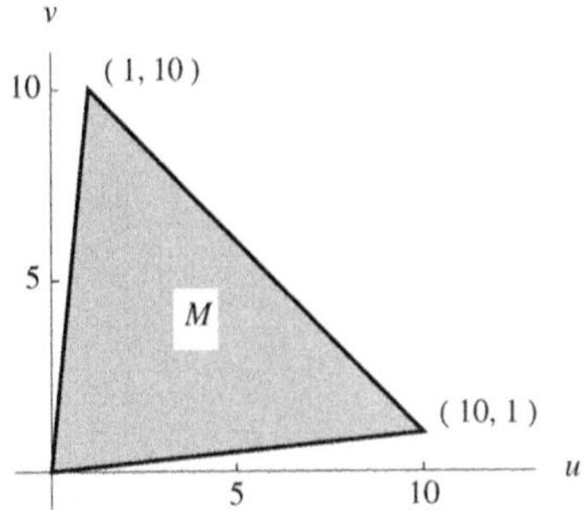

Example 10.4.2. For the game in Problem 7 of Section 10.3, the set M is sketched as follows:

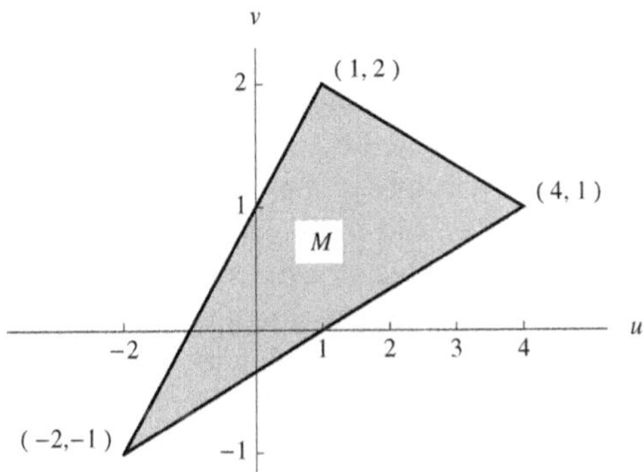

If side payments are permitted, the cooperative payoff set can be easily altered to reflect this fact. For example, suppose the utility units are infinitely divisible and comparable between the players, and the payoff set without side payments contains points in the first quadrant. If (u', v') is such a point, then the total utility $u' + v'$ can be divided in any way between the two players, and so the set of possible payoffs would contain all the points in the first quadrant of the form (u, v), where $u + v = u' + v'$.

Example 10.4.3. For the game of Example 10.4.1, if side payments are permitted, the payoff set would be the shaded region of Figure 10.1. For the game of Example 10.4.2, if side payments are permitted, the first quadrant of the payoff region would be the set illustrated in Figure 10.2.

In either case, any cooperative two-person game has a corresponding cooperative payoff set M. The question we are faced with now is whether or not it is possible to develop a theory based on acceptable (to whom?) axioms that will lead to a point, or at least a subset of M, that represents a reasonable (to whom?) solution of the game. As we have seen in Section 10.2, the concepts of security levels and equilibrium

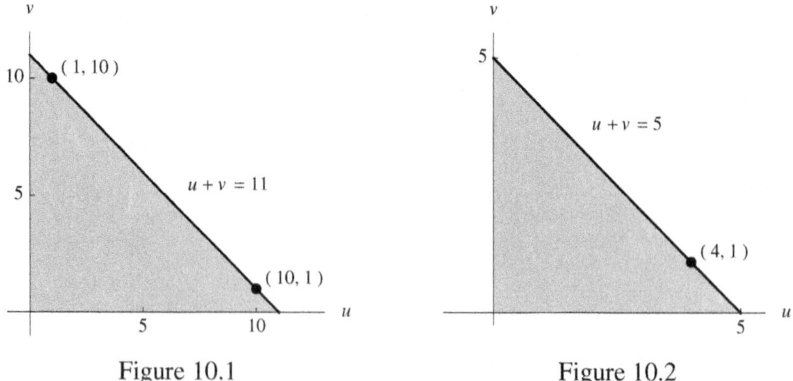

Figure 10.1

Figure 10.2

strategies, which worked so well in the zero-sum case, are incapable of producing a conclusive theory for noncooperative non-zero-sum games, and it is reasonable to assume that they again are inadequate by themselves for cooperative games, which resemble the zero-sum games even less. Thus other principles must be introduced, and associated with them are always the problems of reasonableness and acceptability. In remainder of this section we develop one possible solution of cooperative games, called the *negotiation set*. It is based on two very plausible principles (one concerning security levels) formulated by von Neumann and Morgenstern, but the solution leads in general only to a subset of M and not to a unique point.

First, the point (u,v) of M is said to be *dominated by* the point (u',v') of M if both $u' \geq u$ and $v' \geq v$. Since the players, by acting jointly, can attain as a payoff any point in M, it is reasonable to assume that they would not find acceptable any dominated payoff. Thus we restrict our attention to the undominated points of M.

Second, although the game is cooperative, each player, by using a security level strategy, can attain at least his or her security level payoff regardless of the play of the other player. From this it is reasonable to conclude that P_1 would find acceptable only those payoffs (u,v) with the property that u is at least as large as P_1's security level, and similarly for P_2 and P_2's security level.

These two principles together lead to that subset of M consisting of all undominated payoffs (u,v) of M such that $u \geq P_1$'s security level and $v \geq P_2$'s security level. This subset is called the *negotiation set*.

Example 10.4.4. Consider the game of Problem 2 of Section 10.2 (see also Problem 4 in Section 10.3.) The security levels of P_1 and P_2 are 1 and 0, respectively. The graph, assuming that side payments are not permitted, is sketched in Figure 10.3. The set of undominated payoffs consists of the entire line segment between the points $(0,5)$ and $(5,0)$. The negotiation set, however, is the subset of that line between the points $(1,4)$ and $(5,0)$.

A possible solution, then, of cooperative two-person games is to say that the solution is simply this negotiation set. It seems that any just or fair solution would certainly be contained in this set, but it seems also that we may be leaving too much to be decided by the bargaining powers of the two players. In the next section, we

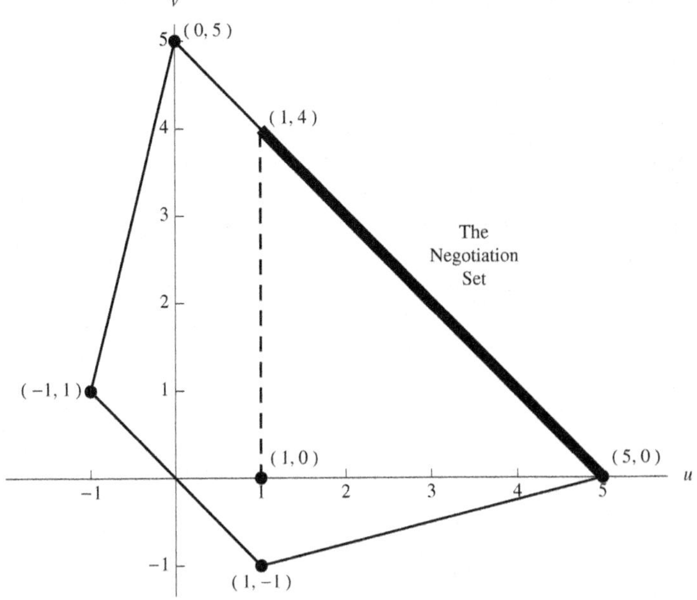

Figure 10.3

outline procedures based on a family of axioms by J. Nash that lead to a unique point in the negotiation set for the solution of the cooperative game.

Problem Set 10.4

(Assume in the following problems that side payments are not permitted.)

1. Determine the negotiation set of Game 4 of Section 10.2. (This game was also discussed in the example of Section 10.3.) Does the negotiation set in any way reflect P_1's stronger bargaining position?
2. Determine the negotiation set of Game 8 of Section 10.3.
3. Suppose preplay discussion between the players is permitted, but agreements reached are not enforceable. Does the negotiation set still provide a reasonable solution of the game? Consider this question for, say, Game 1 of Section 10.2.
4. The negotiation set may be small. Determine this set for Game 2 of Section 10.2.
5. True or false: If (X,Y) is a strategy pair in equilibrium, then the associated expected payoff (XAY^t, XBY^t) is a point of the negotiation set.
6. Show that if a two-person game is, in fact, zero-sum, the negotiation set consists of 1 point, the solution of the zero-sum game.

10.5 THE AXIOMS OF NASH

The axioms of Nash provide a procedure for determining a unique solution to every cooperative two-person game. Mathematically, this solution of such games is quite attractive. The axioms seem reasonable enough, and a simple theorem shows that a unique point, easily determined by considering the maximal value on the set M of an elementary function, satisfies the axioms. A modification of the procedure leads to a scheme that incorporates the familiar notions of security levels and equilibrium strategies and reflects the threat potential of the players. In this section, we provide only an outline of the techniques. More details can be found in the papers of Nash [37, 38] and the books by Luce and Raiffa [31] and Owen [32].

As we have seen, with every cooperative two-person game we can associate its cooperative payoff set M. We can also associate a point (u^*, v^*) of M, a *status quo point*, consisting of the minimally acceptable payoffs for the two players. For example, the security levels of the players could serve as the components of such a point. Now we propose to construct a function, denoted by $F[M,(u^*,v^*)]$, that assigns to the set M and the point (u^*,v^*) a "solution" of the game, a payoff point that is in some way a solution of the cooperative game. We ask initially, what sorts of properties should this function satisfy? The following axioms provide an answer to this question.

Axioms of Nash ([37])

Denote the point $F[M,(u^*,v^*)]$ by (u',v').

1. (u',v') is an undominated point of M such that $u' \geq u^*$ and $v' \geq v^*$.
2. If L is a linear transformation from $\mathbb{R}^2$ to $\mathbb{R}^2$ of the form $L(u,v) = (c_1 u + d_1, c_2 v + d_2)$, where c_1 and c_2 are positive, then
$$F[L(M), L(u^*,v^*)] = L(u',v')$$
3. If $N \subset M$, $(u^*,v^*) \in N$, and $(u',v') \in N$, then
$$F[N,(u^*,v^*)] = F[M,(u^*,v^*)]$$
4. If $(u,v) \in M$ implies that $(v,u) \in M$, and if $u^* = v^*$, then $u' = v'$.

The first axiom states that this bargaining solution F is a feasible payoff, dominated by no other payoff, and at least as acceptable as the status quo point. The second axiom is concerned with changes in the utility functions. Recall that in Section 10.1, it was pointed out that a utility function is determined once its value for two distinct outcomes is defined, but that the values for these two outcomes can be arbitrarily set as long as their linear preferential ordering is maintained. If different values are used, the resulting equivalent utility function will differ by a transformation of the form $cu + d$, where $c > 0$.

The second axiom states that changes such as these should be reflected in the obvious manner in the payoff function F, so that F is essentially invariant under utility transformations.

The third axiom is called the *independence of irrelevant alternatives* axiom. It states that if the bargaining solution of a game $[M,(u^*,v^*)]$ is also a point in the subset N of M, then this point is also the solution of the game $[N,(u^*,v^*)]$. Another way of saying this is that if (u',v') is the solution of the game $[N,(u^*,v^*)]$ and if the set N is enlarged to the set M containing other possible payoffs, the solution of $[M,(u^*,v^*)]$ must be either the former solution (u',v') or a new point in M, not in N.

The fourth axiom is straightforward. If M is symmetric about the line $u = v$, and if $u^* = v^*$, then the positions of the players are equivalent and the payoff to P_1 should equal the payoff to P_2.

The remarkable thing about these four axioms is that for any given set M and point $(u^*,v^*) \in M$, there exists a unique point of M satisfying the axioms. (Thus they work admirably!) This point can be easily determined. As long as M contains points (u,v) such that $u > u^*$ and $v > v^*$, $F[M,(u^*,v^*)]$ is the unique point of that subset of M with $u > u^*$ and $v > v^*$ at which the function $(u-u^*)(v-v^*)$ attains its maximal value. (Notice that the level curves of the function $(u-u^*)(v-v^*)$ are hyperbolic, and so uniqueness follows from the convexity of M.) A proof that this prescription provides the unique point satisfying the axioms is contained in [37] (see also [31] or [32]).

We now give some examples illustrating this procedure. We again assume that side payments are not permitted and, for the status quo point (u^*,v^*), we use the security levels of the two players.

Example 10.5.1. Consider Game 6 of Section 10.3. The security level for both players is $\frac{10}{11}$, and the graph of set M, as described in Example 10.4.1, is sketched in Figure 10.4. On M, the maximum of the function $(u-10/11)(v-10/11)$ is attained at the point $(\frac{11}{2},\frac{11}{2})$, and so we have $F[M,(10/11,10/11)] = (\frac{11}{2},\frac{11}{2})$. This seems to be a reasonable solution of the game. Note that the first and fourth axioms also imply in this case that $F[M,(10/11,10/11)]$ is $(\frac{11}{2},\frac{11}{2})$.

Example 10.5.2. Consider the game in Problem 2 of Section 10.2. As determined in Problem 4 of Section 10.3, the security levels for P_1 and P_2 are 1 and 0, respectively. The set M, described in Example 10.4.4, is illustrated in Figure 10.5.

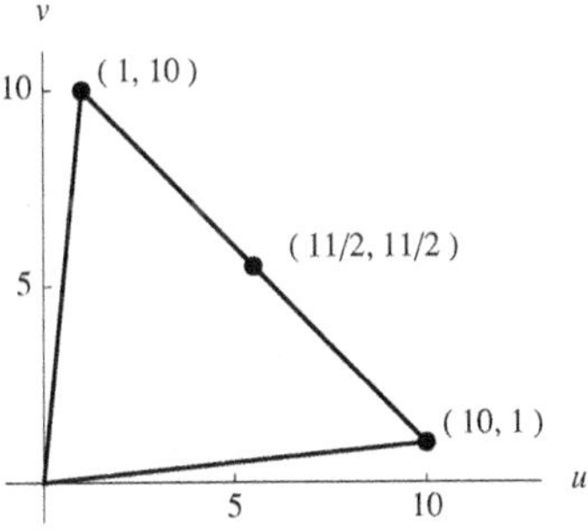

Figure 10.4

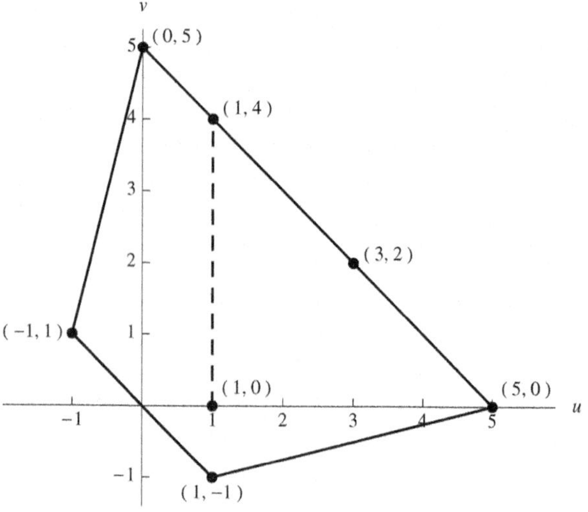

Figure 10.5

The maximal value of the function $(u-1)v$ on M is attained at the point $(3,2)$. (See Problem 1.) Thus this solution also reflects P_1's stronger bargaining position.

Example 10.5.3. Consider Game 4 of Section 10.2. The security levels are 1 and -9 (see Example 10.3.1), and the negotiation set M is depicted in Figure 10.6 (see also Problem 1 of Section 10.4). On M, the maximum of $(u-1)(v+9)$ is attained at

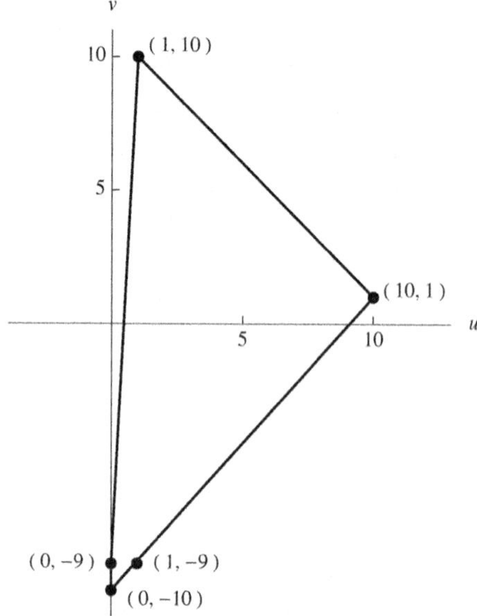

Figure 10.6

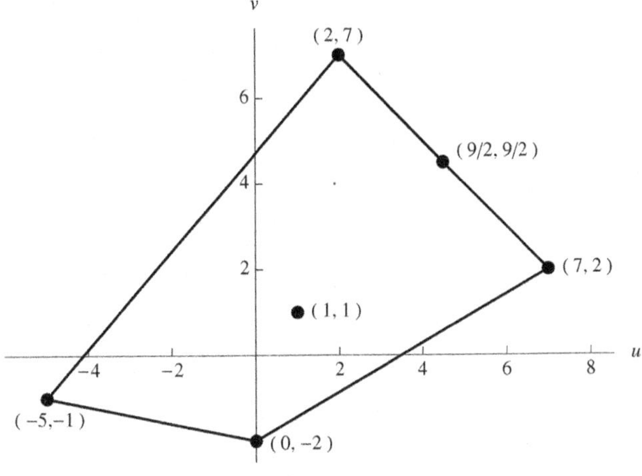

Figure 10.7

$(10,1)$ (see Problem 2), and so this resolution of the game, in contrast to the concept of the negotiation set, clearly recognizes the strong bargaining position of P_1.

Example 10.5.4. Consider the game of Problem 5 of Section 10.3. The security level of both players is 1, and the set M is given in Figure 10.7. For this game, $F[M,(1,1)] = (\frac{9}{2}, \frac{9}{2})$. Note that the negotiation set is the line segment between the points $(2,7)$ and $(7,2)$. Moreover, as determined in Problem 5, the payoffs associated with the equilibrium strategy pairs do not distinguish the two players. Thus, for this game, all the concepts so far developed in no way distinguish P_1 from P_2. Do you still consider P_1's position to be stronger? If so, read on.

John Nash [38] has extended this bargaining procedure to a scheme that considers the full threat potential for the players. Suppose two players are about to play a fixed, non-zero-sum game in which discussion, cooperation, and binding agreements are permitted. The payoff set M is fixed, determined by the payoff tableau. Assume further that the bargaining function F described above accurately represents the payoff mutually agreed on by the two players once a status quo point has been determined. But P_1 questions the use of his security level for determining the status quo point. P_1 wonders, is there a strategy X, a threat strategy, say, such that no matter what strategy Y used by P_2, if the expected payoff (XAY^t, XBY^t) corresponding to the strategy pair (X,Y) is used as the status quo point in the function F, the payoff to P_1 will be larger? Extending this, P_1 asks, what strategy X will guarantee the largest first component for the function $F[M,(XAY^t, XBY^t)]$ regardless of the strategy Y employed by P_2?

Similarly, P_2 considers her potential threat strategies. Hence each player desires to choose a strategy that he or she is willing to play and that will provide a status quo point (XAY^t, XBY^t) for the bargaining function $F[M,(XAY^t, XBY^t)]$ that will guarantee for the player the most profitable outcome independent of the opponent's strategy.

We can indicate this process as follows. Since M is fixed, $F[M,(XAY^t,XBY^t)]$ can be considered a two-step function depending only on $X \in S$ and $Y \in T$. Given a strategy pair (X,Y), first, the expected payoff (XAY^t,XBY^t), a point in M, is determined; and second, the value of F is determined using this point as the status quo point. The first component of $F[M,(XAY^t,XBY^t)]$ is the associated payoff to P_1, and the second is the payoff to P_2. Denote these two components by $F_1(X,Y)$ and $F_2(X,Y)$. Thus

$$F[M,(XAY^t,XBY^t)] = (F_1(X,Y),F_2(X,Y))$$

The players certainly wish to maximize their security levels. (Principle I of Section 9.2.) Paralleling the development for zero-sum games, the security levels of P_1 and P_2 are given by

$$\operatorname*{Max}_{X\in S}\operatorname*{Min}_{Y\in T} F_1(X,Y) \quad \text{and} \quad \operatorname*{Max}_{Y\in T}\operatorname*{Min}_{X\in S} F_2(X,Y)$$

Suppose $X_0 \in S$ and $Y_0 \in T$ are strategies that realize these security levels; that is,

$$\operatorname*{Max}_{X\in S}\operatorname*{Min}_{Y\in T} F_1(X,Y) = \operatorname*{Min}_{Y\in T} F_1(X_0,Y)$$

and

$$\operatorname*{Max}_{Y\in T}\operatorname*{Min}_{X\in S} F_2(X,Y) = \operatorname*{Min}_{X\in S} F_2(X,Y_0)$$

But now P_1 expects P_2 to use Y_0, and P_2 anticipates that P_1 will use X_0. Thus each asks, is my choice of strategy the best against the strategy I anticipate my opponent will use? That is, is the pair (X_0,Y_0) stable or in equilibrium, as defined using this scheme? (Principle II of Section 9.2.) In this context, the strategy pair (X_0,Y_0) is in equilibrium if, for any $X \in S$ and $Y \in T$,

$$F_1(X,Y_0) \le F_1(X_0,Y_0) \quad \text{and} \quad F_2(X_0,Y) \le F_2(X_0,Y_0)$$

It can be shown (see [38]) that there always exist strategies X_0 and Y_0 that realize these security levels (and the use of Max and Min is justified), and that any such strategy pair (X_0,Y_0) is in equilibrium, as defined above. (Actually, the proof is somewhat reversed. It can be shown first that there are equilibrium strategy pairs and then that they deliver the maximal security levels.) Thus these security levels, the payoff point of $F[M,(X_0AY_0^t,X_0BY_0^t)]$, provide another solution of the cooperative game, a solution that incorporates the full threat potential of the players. Moreover, in some cases, such as games permitting side payments, the zero-sum theory can be used to compute easily these strategies and the corresponding solution (see, for example, [32]).

Example 10.5.5 (Continuation of Example 10.5.4). For the game of Example 10.5.4, the strategies corresponding to the above can be determined to be $X_0 = (0,1)$ and $Y_0 = (1,0)$. The associated status quo point $(X_0AY_0^t,X_0BY_0^t)$ is $(0,-2)$. Using the graph in Figure 10.7 (and Problem 1), we have $F[M,(0,-2)] = (\frac{11}{2},\frac{7}{2})$. Thus this solution of the game of Problem 5 of Section 10.3 distinguishes P_1's stronger threat potential.

In fact, these results can be reasoned to by simply considering the game tableau and the corresponding set M. Note that if P_1 threatens to use strategy s_2, no matter what P_2 does, the corresponding expected payoff would lie on the line segment between $(0,-2)$ and $(7,2)$. The bargaining function F maps these points onto the segment between $(\frac{11}{2},\frac{7}{2})$ and $(7,2)$. Thus P_1's security level is at least $\frac{11}{2}$, and P_2's can be at most $\frac{7}{2}$. If P_2 threatens to use strategy t_1, no matter what P_1 does, the corresponding expected payoff would lie on the line segment between $(0,-2)$ and $(2,7)$. F maps these points onto the segment between $(\frac{11}{2},\frac{7}{2})$ and $(2,7)$. Hence P_2's security level is at least $\frac{7}{2}$, and P_1's is at most $\frac{11}{2}$. It follows that the payoff $(\frac{11}{2},\frac{7}{2})$ is the solution of the game under this scheme. Note also that this strategy pair (s_2,t_1) is in equilibrium under this scheme.

In this section, we have outlined briefly two related methods for solving cooperative two-person games. There are other techniques. Some of these are described in the books by Luce and Raiffa [31] and Rapoport [39]. Although these two methods based on the seemingly plausible axioms of Nash lead to unique solutions, the reasonableness and relevance of these axioms must be studied closely. For a critical evaluation of the applicability of these schemes, refer to the excellent book by Luce and Raiffa [31].

Problem Set 10.5

1. (a) Show that if the set M is of the form

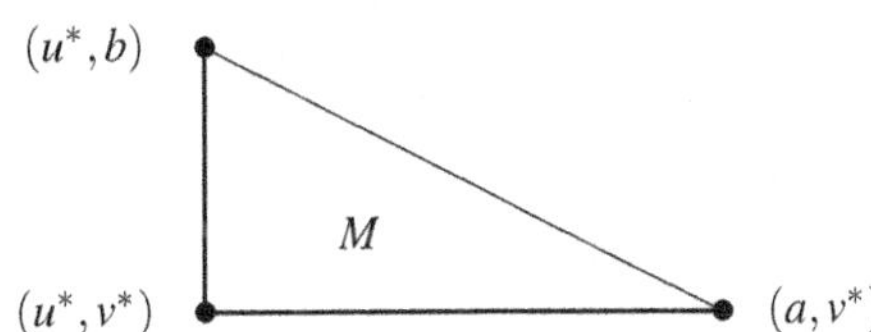

   the maximum of the function $(u-u^*)(v-v^*)$ is attained at the point $P = ((u^*+a)/2, (v^*+b)/2)$.

   (b) Show thai if N is a subset of M containing the point P, the maximum of that same function on N is also attained at P.

2. Show that in Example 10.5.3, the maximum of $(u-1)(v+9)$ is attained at the point $(10,1)$.

3. Show that for the game of Example 10.5.2, the pair of threat strategies (s_2,t_1) is in equilibrium according to the second scheme. What is the solution of this game using this scheme?

4. Determine the solutions of Problem 7 of Section 10.3 (and Example 10.4.2 on page 405) under both schemes of this section. For the second, consider the results of P_1 using s_2 as the threat strategy.

10.6 An Example

In this section, we will develop an example of a two-person game that will illustrate some of the concepts discussed in the previous two chapters. We hope that this example will help the reader understand how these concepts might be applied and realize the significance of the questions that have been raised. The model should also point out how tenuous the status of a solution to a game may be.

Consider the position of two food wholesalers competing for the public dollar through their respective supermarket chains. Each fall the wholesalers must decide on whether or not they, through their supermarkets, will conduct a promotion campaign the following winter. For example, their stores could offer their customers items such as dinnerware, silverware, computer games, DVDs, and so on, either free of charge or at reduced costs, or they could offer their customers participation in a game or contest with small cash prizes. In general, if only one chain of supermarkets has such a campaign, its business will be increased significantly. However, if both wholesalers decide on promotion campaigns, the effects are nullified. (Nevertheless, such campaigns would still provide an abundance of material for their advertising agencies.) Thus the decision to be made each fall by each wholesaler is to have or not have a promotion campaign that winter.

One wholesaler, the larger of the two, attempts to formulate this decision problem in terms of a two-person game. From past records, the wholesaler knows that her chain handles approximately 60% of what she at first considers a fixed segment of the business and her competitor 40%. If she conducts a promotion campaign and the competitor does not, her business increases to 90%. If the situation is reversed, the competitor's volume is increased to 70%. If both offer sales gimmicks, the business breakdown ratio is again 60:40. The wholesaler initially considers this as a zero-sum game, with 30% of the food-purchasing market to be gained or lost on account of promotion campaigns. Thus the wholesaler sets up the following tableau for the corresponding zero-sum game, with her firm represented by Player 1:

	No Promotion	*Promotion*
No Promotion	0	−30
Promotion	30	0

Upon further reflection, the wholesaler realizes that this model is totally inadequate. There are two obvious reasons why the game is not zero-sum. On the one hand, she knows that it cannot be assumed that between these two major wholesalers there is one fixed market. It is true that they control the majority of the market, but there are other small, independent food stores from which business can be attracted by an effective promotion campaign. On the other hand, the cost of conducting a promotion campaign — the costs of the items involved, the associated advertising, and the labor administrating the campaign — must also be considered.

Thus the wholesaler decides to formulate the problem as a non-zero-sum game. The wholesaler uses as utility units the volume of business, measured in thousands of dollars, less the cost of a promotion campaign if one is conducted. The result is the following tableau, where again her firm is represented by Player 1:

	No Promotion	*Promotion*
No Promotion	$(60,40)$	$(40,70)$
Promotion	$(90,20)$	$(50,30)$

The wholesaler believes that this table more accurately reflects the situation. If she conducts a promotion campaign and the competitor does not, she attracts business not only from the competitor but also from the other independent food stores, and the combined volume of the two major wholesalers is increased by 10 units, with her volume increased 30 units and the competitor's volume decreased 20 units. A similar relationship holds if the competitor is the only one to conduct a campaign. Moreover, the futility of both wholesalers conducting campaigns is now measured. The increases in business, in this case coming only from the independent stores, is not adequate to compensate for the expense of such campaigns, and both wholesalers lose 10 units from their income under routine operations.

The result is a two-person, non-zero-sum game of the prisoner's dilemma type. The "promotion" strategy of each player dominates the other strategy, and so the game has only one equilibrium strategy pair, and the outcome corresponding to this pair is the $(50,30)$ payoff. This payoff also represents the security levels of the two players. Thus, if the game is played noncooperatively, the seemingly expected result favors the advertising agencies.

The wholesaler notes, however, that the outcome $(60,40)$ is more beneficial to both parties than the outcome $(50,30)$. Can she bring about this outcome by attempting to cooperate with her competitor? Since they are not yet bitter rivals, this possibility exists; if they could mutually agree not to hold promotion campaigns in the winter, they could realize this advantageous payoff. Note, however, the importance of entering into a binding agreement. If either wholesaler can convince the competitor not to conduct a promotion campaign, that wholesaler has much to gain, at least that winter, by being the only one to conduct a promotion campaign. Actually, legislation restricting the use of gimmicks by food retailers could have the same effect as the possibility of entering into binding agreements.

Consider now the solution concepts applied to this game for cooperative two-person games. Remember that this theory assumes the possibility of the players entering into binding agreements. The negotiation set for this game is the line segment in the plane connecting the points $(80,30)$ and $(50,60)$. The outcome $(60,40)$ is not on this segment. Both solution concepts developed in the previous section lead to the outcome $(65,45)$, which is also the midpoint of the negotiation set. To realize this outcome, the wholesalers must enter into an even more involved collusion, each agreeing to conduct promotion campaigns in alternate years and not in competition with each other. In this case such campaigns maintain their novelty potential, and the independent store owners are hurt the most.

This suggests the possibility of the owners of the independent stores uniting and attempting to counteract their weaker position, either by promotion campaigns of their own or by adjustments in the selling prices of their products. But this leads to n-person game theory (in this case $n = 3$) and so is beyond the scope of this text. As is frequently the case, models involving human behavior quickly become complicated. We refer the food wholesaler to the references.

APPENDIX A

VECTORS AND MATRICES

This appendix provides a brief listing of the linear algebra topics used in this book. These topics include the basic concepts of vectors and matrices and the transpose of a matrix.

A point $X = (x_1, x_2, \ldots, x_n)$ of $\mathbb{R}^n$ is said to be an *n-dimensional vector* or simply a *vector*. Multiplication of vectors by real numbers is defined in the obvious manner; that is, for $r \in \mathbb{R}$, rX is defined to be the vector $(rx_1, rx_2, \ldots, rx_n)$ in $\mathbb{R}^n$. If X and $Y = (y_1, y_2, \ldots, y_n)$ have the same dimension, then we define their *sum* $X + Y = (x_1 + y_1, x_2 + y_2, \ldots, x_n + y_n)$ and *dot product* $X \cdot Y = x_1y_1 + x_2y_2 + \ldots + x_ny_n$. Note that the dot product of two vectors is simply a real number and that this operation is commutative, that is, $X \cdot Y = Y \cdot X$.

Example A.1. For $X = (3, 0, -1, 5)$ and $Y = (-2, 6, 7, 0)$, we have

$$\begin{aligned} 3X &= 3(3,0,-1,5) = (9,0,-3,15) \\ X + Y &= (3-2, 0+6, -1+7, 5+0) = (1,6,6,5) \\ X \cdot Y &= 3(-2) + 0(6) + (-1)(7) + 5(0) = -6 - 7 = -13 \end{aligned}$$

An $m \times n$ *matrix* is simply a rectangular array of real numbers, with the array having m rows and n columns. The matrix

$$A = \begin{bmatrix} a_{11} & a_{12} & \cdots & a_{1n} \\ a_{21} & a_{22} & \cdots & a_{2n} \\ \vdots & \vdots & \ddots & \vdots \\ a_{m1} & a_{m2} & \cdots & a_{mn} \end{bmatrix}$$

can be denoted by (a_{ij}), where a_{ij} is the element of matrix A appearing in the ith row and jth column of the array.

If $A = (a_{ij})$ and $B = (b_{ij})$ are matrices of the same dimensions, we define their sum $A + B = (a_{ij} + b_{ij})$, a matrix with the same dimension. Multiplication of a matrix by a real number is defined as follows: For $r \in \mathbb{R}$ and $A = (a_{ij})$, define $rA = (ra_{ij})$.

An Introduction to Linear Programming and Game Theory, Third Edition. By P. R. Thie and G. E. Keough.

Example A.2. For $A = \begin{bmatrix} 3 & 0 & 1 \\ -1 & 2 & 5 \end{bmatrix}$ and $B = \begin{bmatrix} 0 & -1 & -2 \\ 4 & 1 & -3 \end{bmatrix}$, we have

$$A+B = \begin{bmatrix} 3+0 & 0-1 & 1-2 \\ -1+4 & 2+1 & 5-3 \end{bmatrix} = \begin{bmatrix} 3 & -1 & -1 \\ 3 & 3 & 2 \end{bmatrix}$$

and

$$3A = \begin{bmatrix} 3(3) & 3(0) & 3(1) \\ 3(-1) & 3(2) & 3(5) \end{bmatrix} = \begin{bmatrix} 9 & 0 & 3 \\ -3 & 6 & 15 \end{bmatrix}$$

To define multiplication between matrices, suppose $A = (a_{ij})$ is an $m \times n$ matrix and $B = (b_{ij})$ is an $n \times p$ matrix. Then the product $AB = C = (c_{ij})$ is defined to be that $m \times p$ matrix with the ijth element $c_{ij} = \sum_{k=1}^{n} a_{ik} b_{kj}$. Thus the ijth element of the product is the dot product of the ith row of A with the jth column of B.

By convention, we will allow ourselves to treat a single real number as a 1×1 matrix, a matrix having just one row and one column — and vice versa. Thus an expression of the form $\begin{bmatrix} r \end{bmatrix}$, where r is a real number, will be written simply as r. (Such an extension of the matrix notation for real numbers is easily seen to be consistent with all of the definitions given so far.)

Example A.3. We have

$$\underbrace{\begin{bmatrix} 6 & -8 \end{bmatrix}}_{1\times 2} \underbrace{\begin{bmatrix} 1 & -2 \\ 4 & 5 \end{bmatrix}}_{2\times 2} = \underbrace{\begin{bmatrix} 6\cdot 1 - 8\cdot 4 & 6\cdot(-2) - 8\cdot 5 \end{bmatrix}}_{1\times 2}$$
$$= \begin{bmatrix} -26 & -52 \end{bmatrix}$$

$$\begin{bmatrix} 6 & -8 \end{bmatrix} \begin{bmatrix} 1 & -2 \\ 4 & 5 \end{bmatrix} \begin{bmatrix} -3 \\ 1 \end{bmatrix} = \underbrace{\begin{bmatrix} -26 & -52 \end{bmatrix}}_{1\times 2} \underbrace{\begin{bmatrix} -3 \\ 1 \end{bmatrix}}_{2\times 1}$$
$$= \underbrace{\begin{bmatrix} (-26)\cdot(-3) + (-52)\cdot(1) \end{bmatrix}}_{1\times 1}$$
$$= 26$$

$$\underbrace{\begin{bmatrix} 3 & 0 \\ -1 & 2 \\ 6 & -3 \end{bmatrix}}_{3\times 2} \underbrace{\begin{bmatrix} 1 & -2 & 3 & 7 \\ 4 & 5 & 0 & -5 \end{bmatrix}}_{2\times 4} = \underbrace{\begin{bmatrix} 3 & -6 & 9 & 21 \\ 7 & 12 & -3 & -17 \\ -6 & -27 & 18 & 57 \end{bmatrix}}_{3\times 4}$$

However, the matrix product

$$\underbrace{\begin{bmatrix} 1 & -2 & 3 & 7 \\ 4 & 5 & 0 & -5 \end{bmatrix}}_{2\times 4} \underbrace{\begin{bmatrix} 3 & 0 \\ -1 & 2 \\ 6 & -3 \end{bmatrix}}_{3\times 2}$$

is undefined, since the number of columns in the first matrix (four) is not the same as the number of rows in the second matrix (three).

Let I be the $n \times n$ matrix with 1s on the main diagonal and 0s elsewhere; that is,

$$I = \begin{bmatrix} 1 & 0 & \dots & 0 \\ 0 & 1 & \dots & 0 \\ \vdots & \vdots & \ddots & \vdots \\ 0 & 0 & \dots & 1 \end{bmatrix}$$

The matrix I is called the *identity matrix* (of order n). Notice that for any $n \times n$ matrix A, we have $AI = IA = A$, and thus multiplication of an $n \times n$ matrix A by the matrix I is the matrix version of multiplication of a real number by 1.

An $n \times n$ matrix A is said to be *invertible* or *nonsingular* if there exists an $n \times n$ matrix B such that $AB = BA = I$. Such a matrix B is called the *inverse of* A and is denoted by A^{-1}. Notice that a symmetry holds in this definition: if B is the inverse of A, then A must necessarily be the inverse of B.

It can be shown that for *square* ($n \times n$) matrices A and B, if one of the equations $AB = I$ or $BA = I$ holds, then so does the other. Thus an $n \times n$ matrix A will be invertible if there exists B such that either $AB = I$ or $BA = I$, in which case $B = A^{-1}$ (and $A = B^{-1}$).

Example A.4. If $A = \begin{bmatrix} 6 & -2 \\ 4 & -1 \end{bmatrix}$ and $B = \begin{bmatrix} -\frac{1}{2} & 1 \\ -2 & 3 \end{bmatrix}$, then

$$AB = BA = \begin{bmatrix} 1 & 0 \\ 0 & 1 \end{bmatrix}$$

and so $A^{-1} = B$, $B^{-1} = A$.

Let $A = (a_{ij})$ be an $m \times n$ matrix. Then the *transpose* of A, denoted by A^t, is that $n \times m$ matrix with its ijth element equal to the jith element of A; that is, $A^t = (a_{ji})$. Thus the rows of A^t are simply the columns of A and the columns of A^t are the rows of A. Notice that if this operation is performed twice, the resulting matrix will be the original matrix; that is, $(A^t)^t = A$.

Example A.5.

$$\begin{bmatrix} 1 & -2 & 3 \\ -4 & 5 & -6 \end{bmatrix}^t = \begin{bmatrix} 1 & -4 \\ -2 & 5 \\ 3 & -6 \end{bmatrix} \text{ and } \begin{bmatrix} 6 \\ 7 \\ 8 \end{bmatrix}^t = \begin{bmatrix} 6 & 7 & 8 \end{bmatrix}$$

Suppose $X = (x_1, x_2, \dots, x_n)$ and $Y = (y_1, y_2, \dots, y_n)$ are n-dimensional vectors. Expressed this way, they can also be considered $1 \times n$ matrices, and their transposes would thus be $n \times 1$ matrices. In fact, we have

$$X \cdot Y = Y \cdot X = XY^t = YX^t,$$

where the last two products are ordinary matrix multiplications.

Example A.6.

$$
\begin{aligned}
(3,1,-2)\cdot(0,4,8) = (0,4,8)\cdot(3,1,-2) \\
&= \begin{bmatrix} 3 & 1 & -2 \end{bmatrix} \begin{bmatrix} 0 \\ 4 \\ 8 \end{bmatrix} \\
&= \begin{bmatrix} 0 & 4 & 8 \end{bmatrix} \begin{bmatrix} 3 \\ 1 \\ -2 \end{bmatrix} \\
&= -12
\end{aligned}
$$

The following result concerning the transpose operator is used in Section 4.4.

Theorem A.1. *Suppose A is an $m \times n$ matrix and B is an $n \times p$ matrix. Then the two products AB and $B^t A^t$ are defined, and $(AB)^t = B^t A^t$.*

Proof. Note first that AB is an $m \times p$ matrix, so its transpose is a $p \times m$ matrix. Further, B^t is a $p \times n$ matrix while A^t is an $n \times m$ matrix; thus the product $B^t A^t$ is defined and is also a $p \times m$ matrix.

To show that these two $p \times m$ matrices are equal, consider the ijth entry of each ($1 \leq i \leq p$, $1 \leq j \leq m$). Using only definitions and the fact that dot product multiplication is commutative, we have

$$
\begin{aligned}
ij\text{th element of } (AB)^t &= ji\text{th element of } AB \\
&= (j\text{th row of } A)\cdot(i\text{th column of } B) \\
&= (j\text{th column of } A^t)\cdot(i\text{th row of } B^t) \\
&= (i\text{th row of } B^t)\cdot(j\text{th column of } A^t) \\
&= ij\text{th element of } B^t A^t
\end{aligned}
$$

□

APPENDIX B

AN EXAMPLE OF CYCLING

The following linear programming problem, from Beale [40], has seven variables, three constraints, and two zero constant terms.

$$
\begin{aligned}
&\text{Minimize } -\tfrac{3}{4}x_4 + 20x_5 - \tfrac{1}{2}x_6 + 6x_7 \\
&\text{subject to} \\
&x_1 \qquad\quad + \tfrac{1}{4}x_4 - 8x_5 - x_6 + 9x_7 = 0 \\
&\quad x_2 \qquad + \tfrac{1}{2}x_4 - 12x_5 - \tfrac{1}{2}x_6 + 3x_7 = 0 \\
&\qquad x_3 \qquad\qquad\qquad + x_6 \qquad\quad = 1 \\
&x_1, x_2, x_3, x_4, x_5, x_6, x_7 \geq 0
\end{aligned}
$$

If we apply the simplex algorithm to the problem, agreeing to always pivot in the column with the smallest c_j term and, when $\text{Min}\{b_i/a_{is} : a_{is} > 0\}$ is attained in more than one row, to pivot in the eligible row higher up in the tableau (or in the row with the basic variable of smallest index), we cycle after six iterations. See Table B.1. We leave to the reader the determination of a sequence of pivot steps that completes the problem.

An Introduction to Linear Programming and Game Theory, Third Edition. By P. R. Thie and G. E. Keough.

Table B.1

	x_1	x_2	x_3	x_4	x_5	x_6	x_7	
x_1	1	0	0	$\textcircled{\frac{1}{4}}$	−8	−1	9	0
x_2	0	1	0	$\frac{1}{2}$	−12	$-\frac{1}{2}$	3	0
x_3	0	0	1	0	0	1	0	1
	0	0	0	$-\frac{3}{4}$	20	$-\frac{1}{2}$	6	0
x_4	4	0	0	1	−32	−4	36	0
x_2	−2	1	0	0	$\textcircled{4}$	$\frac{3}{2}$	−15	0
x_3	0	0	1	0	0	1	0	1
	3	0	0	0	−4	$-\frac{7}{2}$	33	0
x_4	−12	8	0	1	0	$\textcircled{8}$	−84	0
x_5	$-\frac{1}{2}$	$\frac{1}{4}$	0	0	1	$\frac{3}{8}$	$-\frac{15}{4}$	0
x_3	0	0	1	0	0	1	0	1
	1	1	0	0	0	−2	18	0
x_6	$-\frac{3}{2}$	1	0	$\frac{1}{8}$	0	1	$-\frac{21}{2}$	0
x_5	$\frac{1}{16}$	$-\frac{1}{8}$	0	$-\frac{3}{64}$	1	0	$\textcircled{\frac{3}{16}}$	0
x_3	$\frac{3}{2}$	−1	1	$-\frac{1}{8}$	0	0	$\frac{21}{2}$	1
	−2	3	0	$\frac{1}{4}$	0	0	−3	0
x_6	$\textcircled{2}$	−6	0	$-\frac{5}{2}$	56	1	0	0
x_7	$\frac{1}{3}$	$-\frac{2}{3}$	0	$-\frac{1}{4}$	$\frac{16}{3}$	0	1	0
x_3	−2	6	1	$\frac{5}{2}$	−56	0	0	1
	−1	1	0	$-\frac{1}{2}$	16	0	0	0
x_1	1	−3	0	$-\frac{5}{4}$	28	$\frac{1}{2}$	0	0
x_7	0	$\textcircled{\frac{1}{3}}$	0	$\frac{1}{6}$	−4	$-\frac{1}{6}$	1	0
x_3	0	0	1	0	0	1	0	1
	0	−2	0	$-\frac{7}{4}$	44	$\frac{1}{2}$	0	0
x_1	1	0	0	$\frac{1}{4}$	−8	−1	9	0
x_2	0	1	0	$\frac{1}{2}$	−12	$-\frac{1}{2}$	3	0
x_3	0	0	1	0	0	1	0	1
	0	0	0	$-\frac{3}{4}$	20	$-\frac{1}{2}$	6	0

APPENDIX C

EFFICIENCY OF THE SIMPLEX METHOD

In Section 3.4 it was noted that a linear programming problem with m constraints and n variables has at most $\binom{n}{m} = \frac{n!}{m!(n-m)!}$ basic solutions. Thus the simplex algorithm applied to such a problem must terminate in at most $\binom{n}{m}$ iterations, as long as cycling is avoided and a basis is never repeated. However, the binomial coefficient grows very rapidly. For example, for a linear programming problem with 100 variables and 50 constraints we have the binomial coefficient $\binom{100}{50}$, which equals approximately 10^{29}. (To help put this number in perspective, note that the earth is only about 4×10^9 years old.) Thus the question: can we improve upon the rate of convergence of the simplex algorithm?

In considering this question, first we must make precise our meaning of the simplex algorithm, that is, we must state precisely the rules used to select at each iteration the pivoting term. For example, in selecting the column in which to pivot, we could choose the column with the smallest c_j^* term, or the column that would yield the greatest reduction in the value of the objective function, or enter the eligible variable with the smallest index. Whatever set of rules we select must be unambiguous, directing our selection of pivot term when ties occur among the choices; and the rules must be such that cycling is impossible. (As mentioned in Section 3.8, pivoting rules exist that guarantee that cycling will not occur. One possible set of rules would be to combine procedures, using a simple and efficient rule for our primary prescription but invoking a cycling-prevention procedure when a basis is repeated.)

Now the question in the first paragraph can be more precisely stated. Does there exist a well-defined, noncycling prescription for the simplex algorithm and a function $f(m,n)$ such that, when used, the number of iterations to solve any linear programming problem with n variables and m constraints is no more than $f(m,n)$, where the growth rate of f is slower that that of $\binom{n}{m}$? For example, can we characterize the "worst-case" behavior of the simplex method in terms of, say, a polynomial function in m and n? Since the simplex method works so well in practice, it would seem that there should exist a version of the method that has a polynomial bound on the maximum number of iterations necessary to solve any problem.

But in 1972, V. Klee and G. Minty [41] published an example of a linear programming problem with n nonnegative variables and $2n$ inequality constraints that took $2^n - 1$ iterations along the vertices of the feasible set to solve, each improving the

An Introduction to Linear Programming and Game Theory, Third Edition. By P. R. Thie and G. E. Keough.

value of the objective function. A modification showed that the worst-case behavior of the simplex algorithm using the simple rule "pivot in the column with the smallest c_j^*" is also exponential. Since then, comparable examples have been constructed for other variants of the simplex algorithm (see, for example, [42], [43], or [44]). Of course, these examples do not answer definitely the question raised above. There could still be a pivoting rule for the simplex algorithm that has a polynomial bound in its worst-case behavior, but the question of whether one exists has remained unanswered over the last five decades. (Progress, however, has been reported – see [45].) It is fitting that the simplex method, one of the top 10 algorithms "with the greatest influence on the development and practice of science and engineering in the 20th century" [46], would have an associated problem listed among the great problems for the 21st century [47].

Given these theoretical results, it is even more intriguing that the simplex method works so well in practice. In his 1963 book, G. Dantzig states ([7], p. 160): "For an m-equation problem with m different variables in the final basic set, the number of iterations may run anywhere from m as a minimum, to $2m$ and rarely to $3m$. The number is usually less than $3m/2$ when there are less than 50 equations and 200 variables (to judge from informal empirical observations)." Experimental results support these observations. For example, D. Avis and V. Chvatal [48] record the average number of iterations (and mean time) necessary to complete 100 linear programming problems (coefficients randomly selected) of various fixed sizes (m and n equal to 10, 20, 30, 40, and 50, with $m \leq n$) using three different pivoting rules (and the revised simplex algorithm). For the "pivot in the column with the smallest c_j^*" rule with $n = 50$, the mean number of iterations is very close to $2m$ for each of the five values of m.

Work has been done on developing a theoretical explanation for this propitious average case behavior of the simplex method. By placing a probability measure on the space of coefficients for a linear programming problem, statements on the expected number of steps necessary to complete a problem using a well-defined version of the simplex method can be made precise. See, for example, the articles by S. Smale [49], I. Adler and N. Megiddo [50], and M. Todd [51] and the monograph by H. K. Borgwardt [52].

Our story cannot end here. While we have not yet established a variant of the simplex method that can be proved to be polynomial time in its worst-case behavior, there are other algorithms that can solve a linear programming problem in polynomial time. In a 1979 article, the Russian mathematician Leonid Khachian [53] announced a polynomial time algorithm for the solution of the linear programming problem. The algorithm is quite distinct from the simplex method. It uses a sequence of ellipsoids to drive to a solution point of a linear programming problem. The announcement of this polynomial time algorithm generated considerable excitement. (See, for example, the *New York Times* article by M. Browne [54] or the article by E. Lawler [55] which gives an account of some of the events and misconceptions associated with Khachian's announcement.) The ellipsoid algorithm was a major theoretical development; however, this algorithm has not replaced the simplex method as the practical tool for solving linear programming problems. It did not match the simplex method's success in efficiently and effectively computing solutions to linear programming problems.

In 1984, the mathematician Narendra Karmarkar [56], while working at AT&T Bell Laboratories, published another polynomial time algorithm for the linear programming problem. (This announcement too received considerable attention; see, for example, [57] or [58].) Karmarkar's interior point algorithm uses spheres and projective geometry to construct a sequence of points converging to a solution of a linear programming problem. While this algorithm is not *the* successor to the simplex method, work continues utilizing theses ideas and others to develop practical and efficient algorithms to solve the general linear programming problem and specific types of the problem.

APPENDIX D

LP ASSISTANT

Overview

LP Assistant is software designed for use with this text that allows a student to create an electronic representation of a simplex tableau, and then to easily manage the arithmetic associated with both pivoting and the introduction of artificial variables. The screen and the print format are essentially identical to those shown for tableaux throughout the text. Included among its features are:

1. no fixed limits on the number of constraints or the number of variables;
2. spreadsheet-like entry of coefficients and easy designation of basic variables;
3. automatic management of artificial variables and the associated w-function;
4. execution of pivot steps by clicking the mouse;
5. display of the ratio necessary to help determine pivot terms, shown in real time as the mouse is moved over a tableau; and
6. the ability to initiate a new problem from another, which allows not only the w-function coefficients and artificial variables to be removed (when deemed no longer necessary), but also easy implementation of the Gomory Plane Cutting and Branch and Bound techniques developed in Chapter 6 (where new constraints must be added or problems subdivided).

Complete documentation for using LP Assistant is available on the Internet and is accessible from within the program. Thus we will show only a worked example to demonstrate the basic capabilities and ease of use of LP Assistant and conclude with a few comments about using LP Assistant.

Example D.1. We will work through Problem 3 of Problem Set 3.5 on page 92. However, LP Assistant should be used only for problems written in canonical form. Thus we will instead consider the problem for this exercise to be

$$\begin{array}{l} \text{Minimize } -x_4+x_5 \\ \text{subject to} \\ \begin{array}{rrrrrcl} x_1 & & & + \; x_4 & - \; 2x_5 & = & 1 \\ & x_2 & & + \; x_4 & & = & 6 \\ & & x_3 & + \; 2x_4 & - \; 3x_5 & = & 4 \end{array} \\ x_1,x_2,x_3,x_4,x_5 \geq 0 \end{array}$$

Upon opening LP Assistant, a Workspace Window appears on the user's desktop, and within this, a Problem Window is opened for a new, untitled problem having the

An Introduction to Linear Programming and Game Theory, Third Edition. By P. R. Thie and G. E. Keough.

default size of two constraints and three variables. Our problem has one additional constraint and two additional variables, so the user should once select *Tableau* → *Add Constraint* and twice select *Tableau* → *Add Regular Variable* from the menu bar displayed in the Problem Window. The LP Assistant Workspace Window now appears, as in Figure D.1.

The user is currently in **Edit Mode**, and can thus enter the problem coefficients at the desired locations using ordinary data entry techniques associated with spreadsheets. The variables x_1, x_2, and x_3 can serve as basic variables, and the user indicates this by clicking the mouse on the appropriate X in the Basis column and selecting the intended variable from the pop-up menu, as shown in Figure D.2.

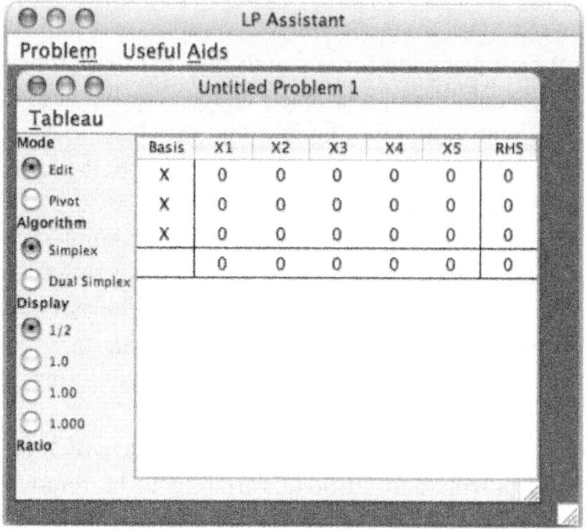

Figure D.1

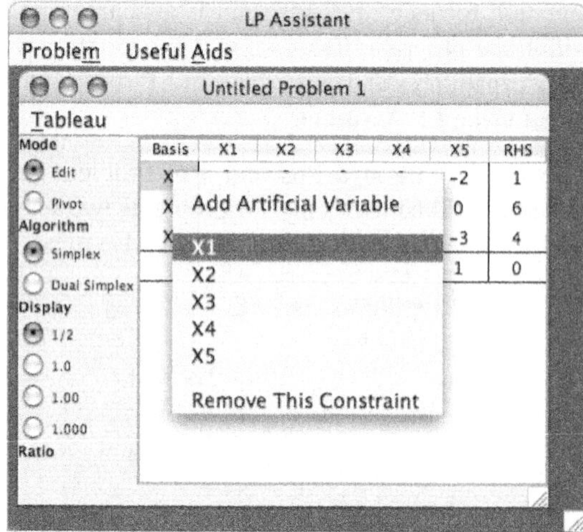

Figure D.2

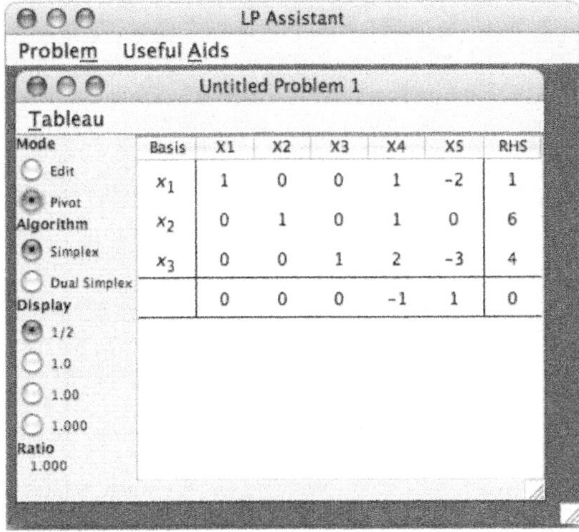

Figure D.3

After all coefficients are entered and all basic variables designated, the user should then change to **Pivot Mode** (in the left panel of the Problem Window). The display will change slightly; but the user will now notice that as the mouse moves over certain coefficient locations in the tableau, numbers will appear below the word **Ratio** at the bottom of the left pane in the Problem Window. Here, the appropriate pivot term to select to apply the simplex algorithm is in the x_4 column, in the first row. The ratio shown when the mouse is positioned there is 1.000, which is $b_1/a_{14} = 1/1 = 1$ (other ratios in the x_4 column are 6.000 and 2.000, for the second and third rows, respectively). See Figure D.3.

Clicking at the 1 in the first row, x_4 column will execute a pivot at that point, extending the display to show the resulting tableau. After scrolling and resizing the Workspace Window and the Problem Window within it as needed, and executing subsequent pivots at the third row, x_5 column, and then the second row, x_1 column, the problem will reach its solution of $\text{Min}\, z = -\frac{10}{3}$, at $x_1 = \frac{1}{3}$, $x_4 = 6$, and $x_5 = \frac{8}{3}$. Figure D.4 shows the completed LP Assistant Problem Window.

Notice, by the way, that the solution to Problem 3 of Problem Set 3.5 is that $\text{Max}\, z = \frac{10}{3}$ occurs at at $x_1 = \frac{1}{3}$, $x_2 = 0$, $x_3 = 0$, $x_4 = 6$, and $x_5 = \frac{8}{3}$.

The example above demonstrates the ease of use of LP Assistant yet only touches upon the full capabilities of the program, which are completely described in the documentation included with the program. We end this Appendix by mentioning several points about the design of LP Assistant.

1. LP Assistant is free (but copyrighted) software written for the Java 2 Platform Standard Edition version 1.4.2, using Swing and AWT, and thus will run on most platforms and under most operating systems.
2. Source code can be made available upon request. In fact, we look forward to working with users to improve both the design and educational usefulness of this application.

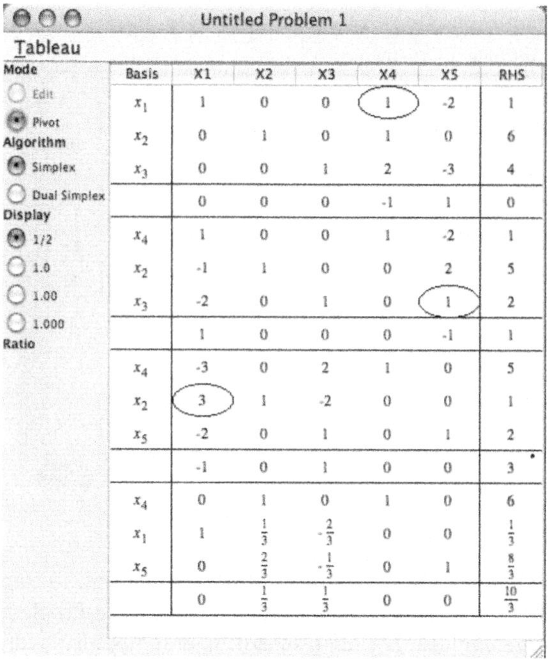

Untitled Problem 1

Tableau

Mode: Edit, Pivot
Algorithm: Simplex, Dual Simplex
Display: 1/2, 1.0, 1.00, 1.000
Ratio

Basis	X1	X2	X3	X4	X5	RHS
x_1	1	0	0	1	-2	1
x_2	0	1	0	1	0	6
x_3	0	0	1	2	-3	4
	0	0	0	-1	1	0
x_4	1	0	0	1	-2	1
x_2	-1	1	0	0	2	5
x_3	-2	0	1	0	1	2
	1	0	0	0	-1	1
x_4	-3	0	2	1	0	5
x_2	3	1	-2	0	0	1
x_5	-2	0	1	0	1	2
	-1	0	1	0	0	3
x_4	0	1	0	1	0	6
x_1	1	$\frac{1}{3}$	$-\frac{2}{3}$	0	0	$\frac{1}{3}$
x_5	0	$\frac{2}{3}$	$-\frac{1}{3}$	0	1	$\frac{8}{3}$
	0	$\frac{1}{3}$	$\frac{1}{3}$	0	0	$\frac{10}{3}$

Figure D.4

3. Numerical values are maintained internally as rational numbers having numerators and denominators of arbitrary precision. This allows for *exact* pivot computations, and numerical roundoff errors are *never* introduced. Coefficients in the initial tableau are required to be rational (in practice, this restriction is not a limitation) and can be entered in any exact, rational format (e.g., $\frac{1}{2}$ is "1/2" or "0.5;" $\frac{1}{3}$ is "1/3," which is not the same as either "0.3" or "0.33").
4. LP Assistant is intended to be used by students to help them learn the mechanics of the simplex algorithm and is designed to *not* be yet one more solver of linear programming problems. Many fine applications, both commercial and in the public domain, already exist for this task.
   Indeed, LP Assistant was created to provide only an interface and a reliable bookkeeping mechanism for executing the steps of the simplex algorithm. Consequently, the user has complete responsibility to:
   (a) formulate a given problem in canonical form, supplying slack variables as required to do so;
   (b) determine which variables are basic and whether artificial variables are needed (LP Assistant will manage the arithmetic associated with artificial variables);
   (c) indicate where to pivot according to the algorithm in use;
   (d) remove artificial variables when no longer needed; and
   (e) recognize when a problem has been completed.

APPENDIX E

MICROSOFT EXCEL AND SOLVER

In this appendix, we demonstrate the procedure used to solve a linear programming problem with Microsoft Excel and Solver. We assume that the reader is familiar with the standard spreadsheet techniques and formulas.

Implementing Microsoft Excel and Solver to solve a linear programming problem is accomplished in four basic steps:

1. The data for the problem are entered on the spreadsheet.
2. A representation of the mathematical model for the problem is constructed on the spreadsheet, usually below the data section.
3. The representation of the problem is transferred to Solver.
4. Using Solver, the problem is solved.

Note that the problem is defined on the spreadsheet in the first two steps and that Solver is brought into the solution process only in the last two steps. We illustrate these steps in detail with the following example.

Example E.1. Division P is responsible for the manufacture of two components of the parent company's final product. The division manager has available four different processes to produce the two parts. Each process uses varying amounts of labor and two raw materials, with inputs, outputs, and cost of 1 hr operation of each process given in the following table.

		Process 1	*Process 2*	*Process 3*	*Process 4*
	Labor (worker-hrs)	8	10	6	12
Input	*Material A (lb)*	160	100	200	75
	Material B (lb)	30	35	60	80
Output	*Units of Part 1*	35	45	70	0
	Units of Part 2	55	42	0	90
Cost	*($/hr)*	400	575	620	590

Each week the division is responsible for producing at least 1300 units of Part 1 and 2600 units of Part 2. The division manager has at her disposal weekly up to 2.1 tons of Raw Material A, 1 ton of Raw Material B, and 450 hr of labor. The manager

An Introduction to Linear Programming and Game Theory, Third Edition. By P. R. Thie and G. E. Keough.

can also purchase any number of units of Part 2 from an independent supplier at \$18/unit. To determine the minimum cost of the weekly operation, the manager defines variables x_i = number of hours that Process i is used, $i = 1,2,3,4$, and x_5 = number of units of Part 2 purchased from the outside vendor, and formulates the following linear programming problem:

$$\begin{array}{lr} \text{Minimize } 400x_1 + 575x_2 + 620x_3 + 590x_4 + 18x_5 & \text{(E.1)} \\ \text{subject to} & \\ \begin{array}{rrrrrll} 8x_1 + & 10x_2 + & 6x_3 + & 12x_4 & & \leq 450 & \textit{Labor (hr)} \\ 160x_1 + & 100x_2 + & 200x_3 + & 75x_4 & & \leq 4200 & \textit{Material A (lb)} \\ 30x_1 + & 35x_2 + & 60x_3 + & 80x_4 & & \leq 2000 & \textit{Material B (lb)} \\ 35x_1 + & 45x_2 + & 70x_3 + & 0x_4 & & \geq 1300 & \textit{Units of Part 1} \\ 55x_1 + & 42x_2 + & 0x_3 + & 90x_4 & + x_5 & \geq 2600 & \textit{Units of Part 2} \end{array} & \\ x_1, x_2, x_3, x_4, x_5 \geq 0 & \end{array}$$

Now, with the data and the linear programming problem at hand, we turn to Microsoft Excel. The initial spreadsheet representation for the problem, with steps 1 and 2 already completed, is in Figure E.1. The data are entered in the upper half of the spreadsheet, as the reader can see. The values of all the coefficients and constant terms of (E.1) are contained in the tables, and the rows, columns, and cells are labeled for easy identification.

	A	B	C	D	E	F	G
1	**Division P**						
2				**Process**			
3		**Input**	1	2	3	4	**Limit**
4		Labor (hr)	8	10	6	12	450
5		Material A (lb)	160	100	200	75	4200
6		Material B (lb)	30	35	60	80	2000
7		**Output**					**# Required**
8		# units Part 1	35	45	70	0	1300
9		# units Part 2	55	42	0	90	2600
10		**Cost (\$/hr)**	\$400	\$575	\$620	\$590	
11		Part 2 vendor cost/unit -->		\$18			
12							
13				**Variables**			
14		Process #	1	2	3	4	
15		Hours used					
16		# Units Part 2 purchased -->					
17							
18		**Minimize cost**					
19							
20		**Constraints**	LHS		RHS		
21		Labor	0	≤	450		
22		Material A	0	≤	4200		
23		Material B	0	≤	2000		
24		Part 1	0	≥	1300		
25		Part 2	0	≥	2600		

Figure E.1

	A B	C	D	E	F	G
1	**Division P**					
2			**Process**			
3	**Input**	1	2	3	4	**Limit**
4	Labor (hr)	8	10	6	12	450
5	Material A (lb)	160	100	200	75	4200
6	Material B (lb)	30	35	60	80	2000
7	**Output**					**# Required**
8	# units Part 1	35	45	70	0	1300
9	# units Part 2	55	42	0	90	2600
10	**Cost ($/hr)**	400	575	620	590	
11		Part 2 vendor cost/unit -->	18			
12						
13			**Variables**			
14	Process #	1	2	3	4	
15	Hours used					
16		# Units Part 2 purchased -->				
17						
18	**Minimize cost**	=SUMPRODUCT(C10:F10,C15:F15)+D11*D16				
19						
20	**Constraints**	LHS		RHS		
21	Labor	=SUMPRODUCT(C4:F4,C$15:F$15)	≤	=G4		
22	Material A	=SUMPRODUCT(C5:F5,C$15:F$15)	≤	=G5		
23	Material B	=SUMPRODUCT(C6:F6,C$15:F$15)	≤	=G6		
24	Part 1	=SUMPRODUCT(C8:F8,C$15:F$15)	≥	=G8		
25	Part 2	=SUMPRODUCT(C9:F9,C$15:F$15)+D16	≥	=G9		

Figure E.2

The representation of the actual programming problem of (E.1) is contained in the lower half of the spreadsheet. The construction of this representation consists of three parts.

(p1) The designation of the cells to be used as placeholders for the variables (here cells C15:F15 and D16), the objective function (cell C18), the left-hand sides of the constraints (cells C21:C25), and the right-hand sides of the constraints (cells E21:E25).

(p2) The entering of the appropriate formulas in the objective function and constraints cells, usually through the use of Microsoft Excel's Formula Bar. The region of cells containing formulas for this example (columns C through F, rows 18 through 25) are shown in Figure E.2. Microsoft Excel's SUMPRODUCT function (read "dot product of row vectors" if you wish) is especially helpful in expressing the linear forms of mathematical programming problems, and frequently the formulas can be effectively drag-copied.

(p3) The completion of the listing of the constraints, designating for each constraint the relationship between the left-hand and right-hand sides (cells D21:D25).

The last two steps in solving the problem involve Solver. Clicking on Solver in the Tools pull-down menu superimposes the Solver Parameters window (shown in Figure E.3) on the initial spreadsheet. In this window we enter the spreadsheet locations of the components of the problem to be solved. To be designated in the window are the locations of the cells in the spreadsheet containing the following:

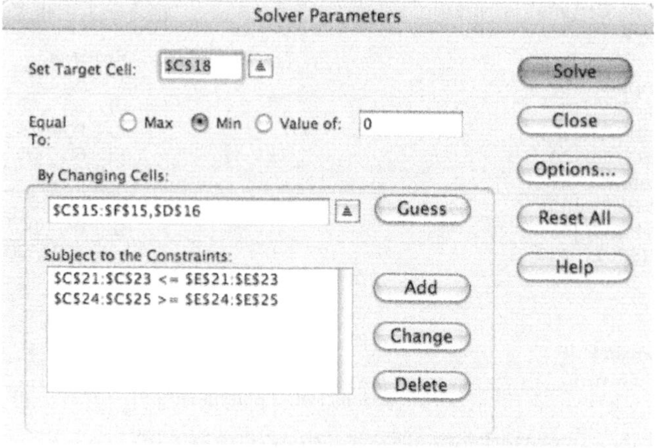

Figure E.3

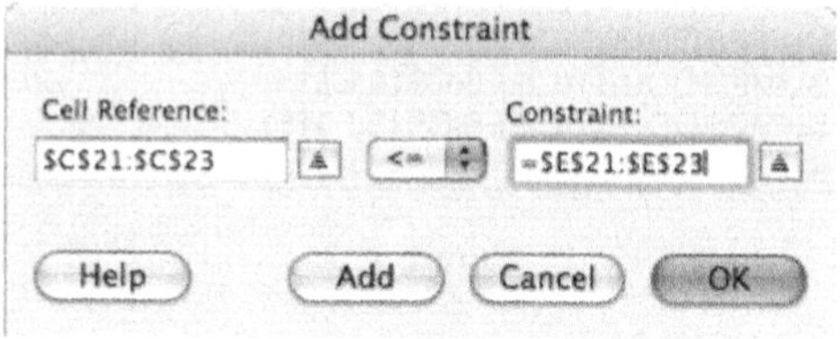

Figure E.4

(s1) The Target Cell, that is, the cell containing the objective function formula (with auxiliary buttons for designating the goal: to maximize or to minimize).

(s2) The Changing Cells, that is, the cells designated for the decision variables.

(s3) The Constraints Cells, both left- and right-hand sides and the type of the constraint. These are added, adjusted, or deleted in the "Subject to the Constraints" area in the lower, left of the Solver Parameters window, utilizing the corresponding pop-up subwindow (the Add Constraint subwindow is shown in Figure E.4). As the reader will see, all the appropriate assignments are in place in the Solver Parameters window of Figure E.3.)

After these steps are completed, a click on the Options button in the Solver Parameters window brings the Solver Options window to the screen, as displayed in Figure E.5. Here, for a linear programming problem we check the "Assume Linear Model" box; and checking the "Assume Non-Negative" box eliminates the need to enter in the constraints set window the nonnegativity restrictions on the variables (if called for in the problem).

That completes the entering of the specifics of the problem into Solver. Clicking the Solve button in the Solver Parameters window will now generate the "Solver Results" window displayed in Figure E.6. Since a solution exists for this problem, the Solver Results window shows the message "Solver found a solution. All constraints and optimality conditions are satisfied." The solution values for the variables, objec-

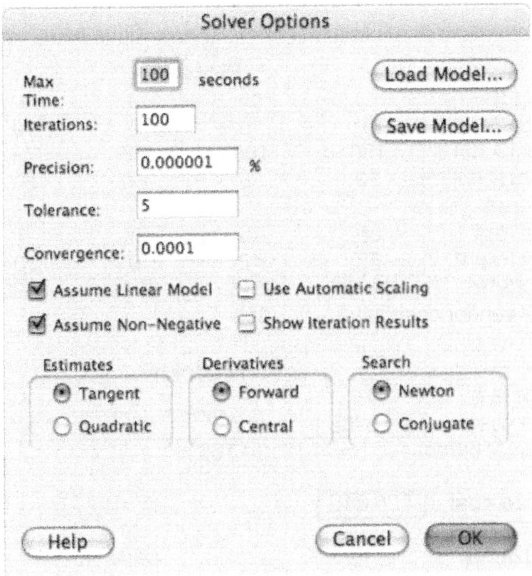

Figure E.5

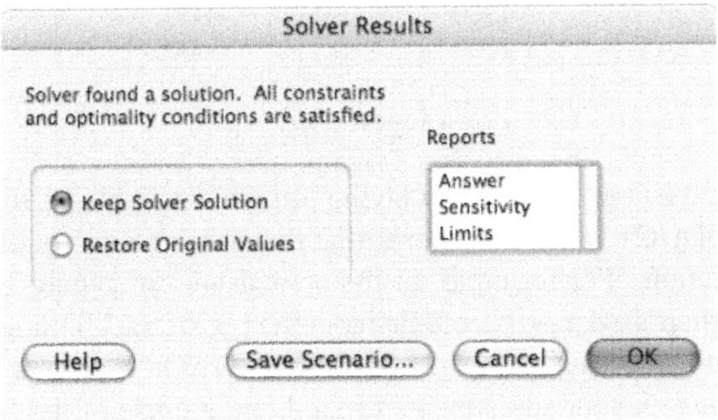

Figure E.6

tive function, and constraints will be displayed on the original spreadsheet, as seen in Figure E.7. The user here also has the option of generating the associated Sensitivity Report by clicking the corresponding word in the Reports window. The nature of this report is discussed at some length in Sections 5.1 and 5.3.

Two other messages can be displayed when the Solver Results window appears, indicating either that the objective function is unbounded ("The Set Cell values do not converge") or that the problem has no feasible solution ("Solver could not find a feasible solution"). *One must carefully read the message in the Solver Results Window before clicking OK to dismiss it*, since each of these outcomes may modify the data on the original spreadsheet; the hurried user might then unwittingly believe that a solution has been found upon returning to the spreadsheet.

We close with some helpful comments on using Solver and Microsoft Excel:

	A	B	C	D	E	F	G
1	**Division P**						
2				**Process**			
3		**Input**	1	2	3	4	**Limit**
4		Labor (hr)	8	10	6	12	450
5		Material A (lb)	160	100	200	75	4200
6		Material B (lb)	30	35	60	80	2000
7		**Output**					**# Required**
8		# units Part 1	35	45	70	0	1300
9		# units Part 2	55	42	0	90	2600
10		**Cost ($/hr)**	$400	$575	$620	$590	
11		Part 2 vendor cost/unit -->		$18			
12							
13				**Variables**			
14		Process #	1	2	3	4	
15		Hours used	4.82338	25.13737	0	12.19363	
16		# Units Part 2 purchased -->		181.51766			
17							
18		**Minimize cost**	$26,845				
19							
20		**Constraints**	LHS		RHS		
21		Labor	436	≤	450		
22		Material A	4200	≤	4200		
23		Material B	2000	≤	2000		
24		Part 1	1300	≥	1300		
25		Part 2	2600	≥	2600		

Figure E.7

1. A factor to be considered when laying out the data tables is that the use of the SUMPRODUCT function requires that the arrays being combined flow in the same direction. For example, in the spreadsheet of Figure E.1, the variable cells and their associated coefficients in the constraints both read horizontally, allowing for the easy use of SUMPRODUCT. On the other hand, you may want to make layout adjustments to facilitate the use of the SUMPRODUCT (see, for example, Figure 8.10 of Section 8.4 on page 335, where the variable cells are placed vertically to accommodate the data table structure).
2. Placing the characters ($\leq$) and ($\geq$) in Column D of the initial spreadsheet to indicate the direction of the inequality in each of the five constraints provides only a (very helpful) visual aid. (*The entry of these characters is system-dependent; you may instead prefer to write simply the two-character sequences* $<=$ *or* $>=$.) Solver makes no use of these entries, however; the appropriate inequality relations must still be entered directly in the Add Constraints window in step (s3) above.
3. The solution to (E.1) on the spreadsheet in Figure E.7 calls for nonintegral values for the variables. If integral values are required, one could, on the spreadsheet, round off the value of each of the variables to the nearest integer and then note the feasibility or nonfeasibility of this set of integers using the spreadsheet's adjusted values for the left-hand sides of the constraints. Here, in fact, the results would show that feasibility is maintained for the first four

constraints and that the output of Part 2 is only 13 units short of the required 2600. These can be purchased from the outside vendor, yielding an integral solution that costs only $120 more than the original minimum cost, as can be easily determined with the spreadsheet. Of course, this procedure in no way guarantees that the optimal integral solution has been found here or that, for a general problem, the procedures even leads to a feasible integral solution. Integer programming techniques may be required. Integer programming, along with applications using Solver, is discussed in Chapter 6.

Bibliography

[1] G. J. Stigler. The cost of subsistence. *J. of Farm Econ.*, 27:303–314, 1945.

[2] L. V. Kantorovich. *Mathematical Methods in the Organization and Planning of Production*. Publication House of the Leningrad State University, 1939. Translated in *Management Sci.*, 6:366–422, 1960.

[3] F. L. Hitchcock. The distribution of a product from several sources to numerous localities. *J. Math. Phys. Mass. Inst. Tech.*, 20:224–230, 1941.

[4] G. B. Dantzig. Maximization of a linear function of variables subject to linear inequalities. In T. C. Koopmans, ed., *Activity Analysis of Production and Allocation*, pp. 339–347. John Wiley & Sons, Inc., New York, 1951.

[5] D. Gale, H. W. Kuhn, and A. W. Tucker. Linear programming and the theory of games. In T. C. Koopmans, ed., *Activity Analysis of Production and Allocation*, pp. 317–329. John Wiley & Sons, Inc., New York, 1951.

[6] G. B. Dantzig. Inductive proof of the simplex method. Technical Report P-1851, The RAND Corporation, December 28 1959. Published in *IBM J. Res. Dev.*, 4:505–506, 1960.

[7] G. B. Dantzig. *Linear Programming and Extensions*. Princeton Univ. Press, Princeton, NJ, 1963.

[8] P. Wolfe. A technique for resolving degeneracy in linear programming. *J. Soc. Indust. Appl. Math.*, 11:205–211, 1963.

[9] R. G. Bland. New finite pivoting rules for the simplex method. *Math. Operations Res.*, 2(2):103–107, 1977.

[10] T. C. T. Kotiah and D. I. Steinberg. Occurrences of cycling and other phenomena arising in a class of linear programming models. *Commun. A.C.M.*, 20(2):107–112, 1977.

[11] V. Chvatal. *Linear Programming*. W. H. Freeman and Company, New York, 1983.

[12] G. Hadley. *Linear Programming*. Addison-Wesley Series in Industrial Management. Addison-Wesley, Reading, MA, 1962.

[13] C. E. Lemke. The dual method of solving the linear programming problem. *Naval Res. Logist. Quart.*, 1:36–47, 1954.

[14] R. E. Gomory. Outline of an algorithm for integer solutions to linear programs. *Bull. Am. Math. Soc.*, 64:275–278, 1958.

[15] R. E. Gomory. An algorithm for integer solutions to linear programs. In R. Graves and P. Wolfe, eds., *Recent Advances in Mathematical Programming*, pp. 269–302. McGraw–Hill Book Company, New York, 1963.

[16] G. Hadley. *Non-linear and Dynamic Programming*. Addison-Wesley, Reading, MA, 1964.

[17] A. H. Land and A. G. Doig. An automatic method of solving discrete programming problems. *Econometrica*, 28:497–520, 1960.

[18] R. J. Dakin. A tree-search algorithm for mixed integer programming problems. *Comput. J.*, 8:250–255, 1965.

[19] L. R. Ford and D. R. Fulkerson. *Flows in Networks*. Princeton Univ. Press, Princeton, NJ, 1962.

[20] L. R. Ford and D. R. Fulkerson. A simple algorithm for finding maximal network flows and an application to the Hitchcock problem. *Can. J. Math.*, 9:210–218, 1957.

[21] W. W. Jacobs. The caterer problem. *Naval Res. Logist. Quart.*, 1:154–165, 1954.

[22] G. B. Dantzig and D. R. Fulkerson. Minimizing the number of tankers to meet a fixed schedule. *Naval Res. Logist. Quart.*, 1:217–222, 1954.

[23] A. Ferguson and G. Dantzig. The allocation of aircraft to routes—an example of linear programming under uncertain demand. Technical Report P-727, The RAND Corporation, December 7 1956. Published in *Management Sci.*, 3:45–73, 1956.

[24] J. P. Ignizio. *Goal Programming and Extensions*. D. C. Heath, Lexington, MA, 1976.

[25] S. M. Lee. *Goal Programming for Decision Analysis*. Auerbach, Philadelphia, 1972.

[26] G. B. Dantzig and P. Wolfe. Decomposition principle for linear programs. *Operations Res.*, 8:101–111, 1960.

[27] A. Charnes, W. W. Cooper, and E. Rhodes. Measuring the efficiency of decision-making units. *Eur. J. of Operational Res.*, 2(6):429–444, 1978.

[28] J. Von Neumann and O. Morgenstern. *Theory of Games and Economic Behavior*. Princeton Univ. Press, Princeton, NJ, 1944.

[29] M. Dresher. *Games of Strategy, Theory and Applications*. Prentice–Hall, Englewood Cliffs, NJ, 1961.

[30] J. C. McKinsey. *Introduction to the Theory of Games.* McGraw-Hill Book Company, New York, 1952.

[31] R. D. Luce and Howard Raiffa. *Games and Decisions: Introduction and Critical Survey.* John Wiley & Sons, Inc., New York, 1957.

[32] G. Owen. *Game Theory, 2nd ed.* Academic Press, Inc., New York, 1982.

[33] G. B. Dantzig. A proof of the equivalence of the programming problem and the game problem. In T. C. Koopmans, ed., *Activity Analysis of Production and Allocation*, pp. 330–335. John Wiley & Sons, Inc., New York, 1951.

[34] J. Von Neumann. Zur theorie der gesellschaftsspiele. *Math. Annal.*, 100(1):295–320, 1928.

[35] J. F. Nash. Equilibrium points in n-person games. *Proc. Natl. Acad. Sci. USA*, 36:48–49, 1950.

[36] J. F. Nash. Non-cooperative games. *Ann. Math. (2)*, 54:286–295, 1951.

[37] J. F. Nash. The bargaining problem. *Econometrica*, 18:155–162, 1950.

[38] J. F. Nash. Two-person cooperative games. *Econometrica*, 21:128–140, 1953.

[39] A. Rapoport. *Two-Person Game Theory: The Essential Ideas.* University of Michigan Press, Ann Arbor, 1966.

[40] E. M. Beale. Cycling in the dual simplex algorithm. *Naval Res. Logist. Quart.*, 2:269–275, 1955.

[41] V. Klee and G. J. Minty. How good is the simplex algorithm? In O. Shisha, ed., *Inequalities III*, pp. 159–175. Academic Press Inc., New York, 1972.

[42] R. J. Jeroslow. The simplex algorithm with the pivot rule of maximizing criterion improvement. *Discrete Math.*, 4:367–377, 1973.

[43] K. G. Murty. Computational complexity of complementary pivot methods. In M. L. Balinski and R. W. Cottle, eds., *Mathematical Programming Study*, no. 7, pp. 61–73. North-Holland, Amsterdam, 1978.

[44] D. Goldfarb and W. Sit. Worst case behavior of the steepest edge simplex method. *Discrete Appl. Math.*, 1(4):277–285, 1979.

[45] J. A. Kelner and D. A. Spielman. A randomized polynomial-time simplex algorithm for linear programming. In *Proceedings of the 38th Annual ACM Symposium on Theory of Computing*, pp. 51–60, New York, 2006.

[46] J. Dongarra and F. Sullivan. The top ten algorithms. *Comput. Sci. Eng.*, 2(1):22–23, 2000.

[47] S. Smale. Mathematical problems for the next century. *Math. Intelligencer*, 20(2):7–15, 1998.

[48] D. Avis and V. Chvatal. Notes on Bland's pivoting rule. In M. L. Balinski and A. J. Hoffman, eds., *Mathematical Programming Study 8*, pp. 24–34. North-Holland, Amsterdam, 1978.

[49] S. Smale. On the average number of steps of the simplex method of linear programming. *Math. Programming*, 27(3):241–262, 1983.

[50] I. Adler and N. Megiddo. A simplex algorithm whose average number of steps is bounded between two quadratic functions of the smaller dimension. *J. ACM*, 32:871–895, 1985.

[51] M. Todd. Polynomial expected behavior of a pivoting algorithm for linear complementary and linear programming. *Math. Programming*, 35(2):173–192, June 1986.

[52] K. H. Borgwardt. *The Simplex Method: A Probabilistic Analysis*. Springer-Verlag, New York, Berlin, Heidelberg, 1987.

[53] L. G. Khachian. A polynomial algorithm in linear programming. *Dokl. Akad. Nauk SSSR*, 244(5):1093–1096, 1979. Translated in *Sov. Math. Dokl.*, 20:191–194 (1979).

[54] M. W. Browne. Mathematical problem-solving discovery reported. *New York Times*, 7 November 1979.

[55] E. L. Lawler. The great mathematical sputnik. *Math. Intelligencer*, 2:191–198, 1980.

[56] N. Karmarkar. A new polynomial-time algorithm for linear programming. *Combinatorica*, 4(4):373–395, 1984.

[57] J. Gleick. Breakthrough in problem solving. *New York Times*, 19 November 1984.

[58] G. Kolata. A fast way to solve hard problems. *Science*, 225(4668):1379–1380, 1984.

Solutions to Selected Problems

Problem Set 2.2

5. There is no change in the optimal solution; all the points of the shaded region in Figure 2.3 satisfy the inequality $4x + 2y \geq 40$.

7. (a) See Example 5.1.1 on page 161.
   (b) There is no change in the optimal diet if $\frac{3}{5} \leq$ the ratio of the cost of Feed 1 to Feed 2 $\leq \frac{5}{3}$.

11. Let x_i denote the amount in pounds of Mineral i used in the production of 100 lb of paint. The problem:

$$\begin{aligned}
&\text{Minimize } 4x_1 + 7.5x_2 + 3x_3 \\
&\text{subject to} \\
&\qquad\qquad\quad 0.06x_2 + 0.07x_3 \geq 5 \\
&0.05x_1 + 0.08x_2 \qquad\qquad\quad \geq 3 \\
&0.30x_1 + 0.30x_2 + 0.25x_3 \geq 26 \\
&0.20x_1 + 0.10x_2 + 0.16x_3 \leq 15 \\
&\qquad x_1 + \qquad x_2 + \qquad x_3 = 100 \\
&x_1, x_2, x_3 \geq 0
\end{aligned}$$

Problem Set 2.3

1. See Example 8.1.1 on page 299.

3. (a) The function to be maximized does not accurately measure profit when less than 2000 lb of aluminum is used.
   (b) The function to be maximized does not accurately measure profit when less than 1500 lb of aluminum is used.
   (c) The first constraint forces the use of at least 1500 lb of aluminum.

5. Replace the function f in (2.3.1) with

$$\begin{aligned}
f(x_1, x_2, x_3, x_4) = {} & 690x_1 + 545x_2 + 1020x_3 + 785x_4 \\
& - 3(35x_1 + 45x_2 + 70x_3 - 2100) \\
& - 2(55x_1 + 42x_2 + 90x_4 - 1800)
\end{aligned}$$

6. Let $x_6 \geq 0$ denote the amount in pounds of Raw Material A purchased and modify the problem of (2.3.2) as follows. Replace the function g with

$$g(x_1, x_2, x_3, x_4, x_5, x_6) = 30x_5 + 690x_1 + 545x_2 + 1020x_3 + 785x_4 + 4x_6$$

and the second constraint with

$$160x_1 + 100x_2 + 200x_3 + 75x_4 \le 8000 + x_6$$

9. The maximum profit is \$54, attained by making 108 dozen muffins and no brownies.

12. See Example 4.3.2 on page 134.

13. See Problem 4 of Section 4.3 on page 137.

17. Let C, T, B, P, and K denote the number of acres planted of corn, tomatoes, beans, peas, and carrots, respectively; U the number of acres of unused land; L the hours of labor employed; and M the amount of money borrowed. The problem:

$$\begin{array}{l}
\text{Maximize } (60-20)C + 800T + 145B + 185P + 250K - 7.25L - 9U - 0.03M \\
\text{subject to} \\
C + T + B + P + K + U = 100 \\
5C + 120T + 25B + 35P + 40K + 2U = L \\
20C + 200T + 55B + 40P + 75K + 9U + 3.25L \le 3000 + M \\
0 \le L \le 3600,\ 0 \le M \le 12000 \\
C, T, B, P, K, U \ge 0
\end{array}$$

Problem Set 2.4

2. (a) See Example 4.3.3 on page 135.

3. Let x_{ij} denote the number of cases shipped from Plant i to Outlet j and x_{i6} the number of surplus cases at Plant i, $1 \le i \le 3$, $1 \le j \le 5$. The problem:

$$\begin{array}{ll}
\text{Minimize} & 6.2x_{11} + 5.1x_{13} + 10.1x_{14} + 8x_{15} \\
& +6.5x_{21} + 10.5x_{22} + 4.3x_{23} + 11.3x_{24} + 6.5x_{25} \\
& +6.3x_{31} + 9x_{32} + 10.8x_{34} \\
& -120x_{16} - 110x_{26} - 114x_{36}
\end{array}$$

subject to

$$\begin{array}{l}
\sum_{j=1}^{6} x_{1j} = 4000 \quad (x_{12} = 0) \\
\sum_{j=1}^{6} x_{2j} = 2000 \\
\sum_{j=1}^{6} x_{3j} = 3000 \quad (x_{33} = x_{35} = 0) \\
\sum_{i=1}^{3} x_{ij} = 1000, 1200, 3000, 400, 2200 \quad (j = 1, 2, 3, 4, 5, \text{respectively}) \\
x_{i,j} \ge 0
\end{array}$$

Problem Set 2.5

1. Equalities would force each D_i to be at least 1000.

4. For month i ($i = 1$, Aug.; $i = 2$, Sept.; $i = 3$, Oct.), let

$$R_i(V_i) = \text{number of refrigerators (ovens) bought}$$
$$S_i(W_i) = \text{number of refrigerators (ovens) sold}$$
$$T_i(X_i) = \text{number of refrigerators (ovens) stored}$$

The problem:

$$\begin{aligned}\text{Minimize}\quad & 90S_1 + 110S_2 + 105S_3 + 200W_1 + 250W_2 + 240W_3 \\ & -(60R_1 + 65R_2 + 68R_3 + 150V_1 + 175V_2 + 200V_3) \\ & -7(T_1 + T_2 + X_1 + X_2)\end{aligned}$$

subject to

$$\begin{array}{ll} 25 + R_1 = S_1 + T_1 & V_1 = W_1 + X_1 \\ T_1 + R_2 = S_2 + T_2 & X_1 + V_2 = W_2 + X_2 \\ T_2 + R_3 = S_3 & X_2 + V_3 = W_3 \\ T_1 + X_1 \leq 45, T_2 + X_2 \leq 45 & \\ 0 \leq R_i \leq 65 & 0 \leq V_i \leq 35 \\ 0 \leq S_i \leq 100 & 0 \leq W_i \leq 55 \\ R_i, S_i, T_i, V_i, W_i, X_i \geq 0 & \end{array}$$

Problem Set 3.1

1. (a) $x_1 = 4, x_2 = 12, x_3 = 0, x_4 = -1$
(b) Any point $(x_1, x_2, x_3, x_4', x_4'', x_5, x_6)$ of the form $(1, 3, 5, 2+\lambda, \lambda, 3, 15)$ where $\lambda \geq 0$

3. (a) Minimize $-3x_1 + 2x_2$
subject to

$$\begin{array}{rcl} 5x_1 + 2x_2 - 3x_3 + x_4 + x_5 & = & 7 \\ 3x_2 - 4x_3 + x_6 & = & 6 \\ x_1 + x_3 - x_4 - x_7 & = & 11 \end{array}$$

$x_1, \ldots, x_7 \geq 0$

(b) Minimize $-x_2' + x_3' - x_3'' + x_4' - x_4''$
subject to

$$\begin{array}{rcl} x_1 + x_2' - x_5 & = & 6 \\ -x_2' + x_3' - x_3'' - x_4' + x_4'' + x_6 & = & 1 \\ 5x_1 + 6x_2' + 7x_3' - 7x_3'' - 8x_4' + 8x_4'' - x_7 & = & 2 \end{array}$$

$x_1, x_2', x_3', x_3'', x_4', x_4'', x_5, x_6, x_7 \geq 0$

(d) Minimize $-6x_1 + 2x_2' - 2x_2'' - 9x_3 - 300$
subject to

$$\begin{array}{rcl} 2x_1 - 6x_2' + 6x_2'' - x_3 + x_4 & = & 100 \\ x_1 + x_2' - x_2'' + 9x_3 + x_5 & = & 200 \\ x_1 + x_6 & = & 50 \\ x_2' - x_2'' - x_7 & = & -60 \\ x_3 - x_8 & = & 5 \end{array}$$

$x_1, x_2', x_2'', x_3, x_4, x_5, x_6, x_7, x_8 \geq 0$

4. (a) $\{(0,0,\lambda,0) : \lambda \geq 11\}$
(b) $\{(5,0,6,0)\}$
(c) $\emptyset$

Problem Set 3.2

1. (a) $(1,2,-3)$
(b) Arbitrarily selecting x_1 and x_2 to use as basic variables, two pivot steps yield the following equivalent system:

$$\begin{array}{rcl} x_2 + \frac{11}{17}x_3 & = & \frac{13}{17} \\ x_1 \quad - \frac{47}{17}x_3 & = & -\frac{3}{17} \end{array}$$

Thus the solution set is

$$\{(-\tfrac{3}{17} + \tfrac{47}{17}\lambda, \tfrac{13}{17} - \tfrac{11}{17}\lambda, \lambda) : \lambda \in \mathbb{R}\}$$

2. The system is equivalent to various systems of equations in canonical form. For example, an equivalent system with basic variables x_1 and x_3 is the system

$$\begin{array}{rcl} x_1 - 8x_2 \quad & = & -41 \\ -3x_2 + x_3 & = & -16 \end{array}$$

4. (a)

$$\begin{array}{rcl} x_2 & = & 9 \\ x_1 - x_3 & = & 4 \end{array}$$

(b) No
(c) $b = (17,4)^t$ can be expressed as a linear combination of $A^{(1)} = (2,1)^t$ and $A^{(2)} = (1,0)^t$, but not as a linear combination of $A^{(1)}$ and $A^{(3)} = (-2,-1)^t$

6. (b) $(0,6,2,0)$ and $(0,0,2,2)$
(d) The minimum value of the objective function is 8, attained at $(0,0,2,2)$

7. Min $f = \frac{15}{4}$ attained at$(\frac{45}{8}, 0, 0, \frac{3}{8})$

Problem Set 3.3

1. (a) $x_1 = 8 - 2x_4, x_2 = 6 - 3x_4, x_3 = 18 - 6x_4$
   (b) $0 \leq x_4 \leq 2$
   (c) x_2
   (d) We should extract x_2 from the basis; therefore, pivot at the $3x_4$ term of the second equation. Pivoting here yields:
   (e)
   $$\begin{aligned} x_1 - \tfrac{2}{3}x_2 \qquad\qquad &= 4 \\ \tfrac{1}{3}x_2 \qquad + x_4 &= 2 \\ -\ 2x_2 + x_3 \qquad\quad &= 6 \end{aligned}$$
   The associated basic solution, $(4,0,6,2)$, is feasible.
   (f) The minimum of $\frac{8}{2}$, $\frac{6}{3}$, and $\frac{18}{6}$ is $\frac{6}{3}$, attained with the data from the second equation.

4. Pivoting at the $2x_4$ term of the first constraint yields the equivalent problem of minimizing z with

$$\begin{aligned} \tfrac{1}{2}x_2 - 3x_3 + x_4 &= 3 \\ x_1 + \tfrac{1}{2}x_2 - x_3 &= 8 \\ 3x_2 - 14x_3 &= 18 + z \\ x_1, x_2, x_3, x_4 &\geq 0 \end{aligned}$$

The expression for z suggests putting x_3 into the basis, but there is no positive x_3 coefficient in the constraints. In fact, from this representation of the constraints, we see that the set of feasible solutions contains the set

$$\{(8 + x_3, 0, x_3, 3 + 3x_3) : x_3 \geq 0\}$$

What happens to z on this set?

Problem Set 3.4

1. $\text{Min}\, z = -67\frac{1}{3}$ attained at $(0, \frac{97}{3}, 0, \frac{17}{3}, \frac{4}{3})$
2. (a) $\text{Min}\, z = 0$ attained at $(5,10,0,0)$. No pivots necessary.
   (b) $\text{Min}\, z = 0$ attained at $(5,10,0,0)$. No pivots necessary.
   (c) Unbounded objective function.
   (d) Unbounded objective function.
   (e) $\text{Min}\, z = -5$ attained at $(5,0,5,0)$
   (f) $\text{Min}\, z = 0$ attained at $(0,10,0,0)$. One pivot necessary.
   (g) Unbounded objective function. No pivots necessary.
5. When the $\text{Min}\{b_i/a_{is} : a_{is} > 0\}$ is attained in more than one row.

Problem Set 3.5

2. (a) $\text{Min}\, z = -200$ attained at $(0,0,50,0)$

(c) Unbounded objective function
(d) $\text{Max}\, z = 90$ attained at $(250, 10, 0, 40, 0, 0)$

3. See Example D.1 on page 427.

5. (a) In the final tableau, $c_2^* = 0$ and at least one $a_{i2}^* > 0$. Thus x_2 can be inserted into the basis. Similarly for x_7.
(b) $(0, 0, 0, 25, 0, 15, 15)$
(c) $(10, 30, 0, 20, 0, 0, 0)$

8. Maximum income is \$7020, attained by producing 240 radios, 85 televisions, and 0 stereos.

Problem Set 3.6

1. (a) Applying the simplex algorithm to the problem of

$$\begin{aligned} &\text{Minimizing } w = x_4 + x_5 \\ &\text{subject to} \\ &x_1 - x_2 \qquad\quad + x_4 \qquad = 1 \\ &2x_1 + x_2 - x_3 \qquad + x_5 = 3 \\ &x_1, x_2, x_3, x_4, x_5 \geq 0 \end{aligned}$$

generates the solution point $(\frac{4}{3}, \frac{1}{3}, 0)$ to the original system.

2. (a) $\text{Min}\, z = \frac{3}{2}$ attained at $(0, \frac{29}{2}, \frac{11}{2})$
(b) $\text{Min}\, z = -\frac{36}{5}$ attained at $(0, \frac{21}{5}, \frac{12}{5})$. (Only one artificial variable required.)
(c) No feasible solutions.

4. The row corresponds to the expression for the function $w = x_5 + x_6$ in terms of the nonbasic variables for that tableau, namely, x_2, x_4, x_5, and x_6.

6. Follows from the definition of w and from Problem 9 of Section 3.4 on page 84.

8. Minimal cost is \$1950 attained by using Process 2 for $\frac{3}{2}$ hr and Process 3 for $\frac{9}{2}$ hr.

Problem Set 3.7

3. (a) $\text{Min}\, z = 50$ attained at $(50, 0, 0, 0)$. No redundant equations.
(c) $\text{Min}\, z = -\frac{5}{3}$ attained at $(0, 0, \frac{1}{3}, \frac{5}{3})$. One redundant equation.
(d) $\text{Max}\, z = -6$ attained at $(0, 1, 2, 0)$. No redundant equations.

4. True. If any artificial variables remained in the basis, they would be at zero level. The elimination of these variables from the basis would lead to a degenerate solution to the original system.

Problem Set 3.8

6. (a) Changing the constant-term column entries to 0 in the tableaux of Table 3.4, we have $\text{Max}\, z = 0$ attained at $(0, 0, 0)$.

(b) From the modified tableaux of Table 3.5, the objective function is unbounded.

Problem Set 4.1

1. Maximum gain is \$475, attained at $(25,100)$.
2. (a) Minimum cost is \$475, attained at $(0,\frac{3}{2},\frac{5}{12})$.

Problem Set 4.2

1. (a) Minimize $100y_1+90y_2+500y_3$
   subject to
   $$\begin{aligned} 5y_1 - \quad y_2 \qquad &\geq 20 \\ -4y_1 + 12y_2 + y_3 &\geq 30 \\ y_1,y_2,y_3 &\geq 0 \end{aligned}$$

   (b) Maximize $-30y_1-50y_2-80y_3$
   subject to
   $$\begin{aligned} 6y_1 - 2y_2 \qquad &\leq \ \ 4 \\ 11y_1 + 7y_2 - y_3 &\leq -3 \\ y_1,y_2,y_3 &\geq 0 \end{aligned}$$

   (c) Minimize $60y_1-10y_2+20y_3$
   subject to
   $$\begin{aligned} 5y_1 - 3y_2 + \ y_3 &\geq -1 \\ y_1 + 8y_2 + 7y_3 &\geq \ \ 2 \\ y_1,y_2 \geq 0,\ y_3 &\text{ unrestricted} \end{aligned}$$

   (f) Maximize $50x_1-70x_2-15x_3$
   subject to
   $$\begin{aligned} 4x_1 \qquad\qquad\qquad &\leq 1 \\ 2x_2 \qquad &\geq 1 \\ -x_1 - \ x_2 + x_3 &\geq 4 \\ x_1 \text{ unrestricted},\ x_2,x_3 &\geq 0, \end{aligned}$$

3. (b) Min $b\cdot Y$ is $41\frac{1}{4}$, attained at $(\frac{7}{4},\frac{3}{4})$.
   (c) Max $c\cdot X$ is $41\frac{1}{4}$, attained at $(\frac{25}{8},0,\frac{45}{8})$.

Problem Set 4.5

3. (a) $(1,1,0,0)$ optimal; complementary slackness generates $(2,2,0)$, a feasible solution to the dual.
   (b) $(0,4,0,2)$ optimal; complementary slackness generates $(3,2,0)$, a feasible solution to the dual.

(c) $(3,0,1,0,5)$ not optimal; complementary slackness generates $(0,1,3)$, but this point is not a feasible solution to the dual.

Problem Set 5.1

2. (b) Upper limit equals 14.25.

4. No. Since $\frac{4}{5} \leq \frac{124}{120} \leq \frac{10}{3}$, the daily minimum cost of an adequate diet would be 124 + 2(120), that is, \$3.64. This is an increase of \$1.20 over the original daily minimum cost of \$2.44 and so, over 2 weeks, would cost \$16.80.

10. An increase of 1% in the bluegrass requirement should increase the cost of producing 100 lb of the composition by \$0.50; and an increase of 1% in the fescue requirement should have no effect on costs.

Problem Set 5.2

3. (a) $B = \begin{bmatrix} 3 & 2 \\ -6 & 4 \end{bmatrix}, \quad B^{-1} = \frac{1}{24}\begin{bmatrix} 4 & -2 \\ 6 & 3 \end{bmatrix}, \quad c_B = [5,-4]$

(b) $b^* = B^{-1}b = [5,8]^t$. Therefore $b = 5A^{(3)} + 8A^{(4)}$

(c) $c^* = c - c_B B^{-1}A = [\frac{4}{3}, \frac{55}{12}, 0, 0, \frac{1}{4}] \geq 0$

Problem Set 5.3

2. The modified $c^* = [-\frac{1}{2}, 0, \frac{23}{4}, 0, \frac{7}{2}, \frac{1}{4}]$. Starting from the adjusted second tableau of Table 5.5, one iteration of the simplex algorithm yields the new optimal solution point of $(\frac{4}{11}, 0, 0, \frac{74}{11})$ and $\text{Max}\, z = 110\frac{15}{22}$.

3. (a) $\text{Max}\, z = 107\frac{1}{5}$ attained at $(0, 25\frac{3}{5}, \frac{6}{5}, 0)$

(b) $\text{Max}\, z = 115\frac{1}{2}$ attained at $(0,0,0,7)$

(c) $\text{Max}\, z = 100$ attained at $(0,4,0,6)$

6. $44 \leq$ number of required units of $A \leq 84$

$60 \leq$ number of required units of $B \leq 132$

number of required units of $C \leq 144$

12. (a)
$$\begin{aligned} c_1^* &= c_1 - c_B B^{-1} A^{(1)} \\ &= -11 - [-15,4]\begin{bmatrix} 2 & -1 \\ -7 & 4 \end{bmatrix}\begin{bmatrix} 3 \\ \alpha \end{bmatrix} \\ &= -5 + \alpha \end{aligned}$$

Problem Set 5.4

1. (a) $\text{Max}\, z = 106$ attained at $(x_1, x_2, x_3, x_4; x_7) = (0,4,0,6;0)$

(b) $\text{Max}\, z = 110$ attained at $(x_1, x_2, x_3, x_4; x_7) = (0,0,\frac{10}{3},0;\frac{32}{3})$

(c) z unbounded

(d) $\text{Max}\, z = 109$ attained at $(x_1, x_2, x_3, x_4; x_7) = (0,7,0,0;3)$

4. (b) Minimum monthly cost is \$40,875, attained by using Process 3 for 187.5 hr and the new process for 37.5 hr.

Problem Set 5.5

1. (a) Min $z = 15$ attained at $(9,1,0,0)$
(b) Min $z = 16$ attained at $(8,0,0,0)$

3. For $-3 \le \lambda \le \frac{4}{7}$, the optimal solution point is $(0, 4-7\lambda, 0, 6+2\lambda)$ and Max $z = 106 + 2\lambda$

Problem Set 5.6

3. (a) Max $= 105$ attained at $(0,0,0,7)$
(b) Max $= 107\frac{2}{3}$ attained at $(0,0,\frac{1}{6},\frac{43}{6})$
(c) No feasible solutions

6. Element A: $\frac{13}{2}$ cents; element B: 0 cents; element C: $\frac{1}{6}$ cents.

Problem Set 5.7

1. Check our arithmetic for the error. (The constant term -2 measures the value of the slack variable x_7 when the new constraint is evaluated at $(0,4,0,6,0,0)$ (Why?). If this slack were nonnegative, $(0,4,0,6,0,0)$ would have satisfied the new constraint.)

2. (a) The point $(19,8,0,0)$ remains an optimal solution point
(b) Min $z = 16$ attained at $(12,3,1,0)$
(c) No feasible solutions
(d) Min $z = 16$ attained at $(12,3,1,0)$

Problem Set 6.2

3. The optimal production schedule for Example 6.2.6 is 0 *A*'s and 27.5 *B*'s, with profit \$727.50. For Example 6.2.7, producing 30 *A*'s, and 12.5 *B*'s is optimal, resulting in a profit of \$862.50. Optimal production for Example 6.2.8 will be 55 *A*'s and 0 *B*'s, from which the company earns a \$900 profit.

5. For Figure 6.3:

$$\begin{aligned} &5x_1 + 4x_2 - 20 \le 40(1-y_1) \\ &3x_1 + 8x_2 - 24 \le 40(1-y_2) \\ &y_1 + y_2 \ge 1 \\ &0 \le x_1, x_2 \\ &0 \le y_1, y_2 \le 1 \text{ and integral} \end{aligned}$$

Problem Set 6.3

1. (a) Min $z = -1$ attained at $(0,1)$

(b) Max $z = 10$ attained at $(4,3)$

2. (a) Max $z = 13$ attained at $(4,3)$

Problem Set 6.5

2. (b) (i) Minimum cost is \$152, attained by using four bags of Feed 1, one bag of Feed 2, and four bags of Feed 3.
(ii) Minimum cost is \$125.75, attained by using 195 lb of Feed 1 and 40 lb of Feed 2.

Problem Set 7.1

1. (a) Feasible flows exist.
(b) Not feasible. $R' = \{1\}, C' = \{1,2,5\}$
(c) Not feasible. $R' = \{2,3\}, C' = \{2,4\}$
(d) Feasible
(e) Not feasible. $R' = \{2,3\}, C' = \{3,5\}$
(f) Feasible
(g) Feasible

Problem Set 7.2

1. (a) Min cost = 205
(b) Min cost = 146
(c) Min cost = 199

3. (a) Min cost = 173
(b) Min cost = 322

4. (a) $\lambda v_1 + (-\lambda)v_5 = 7\lambda$
(b) Cost increase = 14
(c) Cost increase = 22

6. (a) $3\lambda + (-3)\lambda = 0$
(b) $\lambda = 5$

8. Min cost = 183

20. (a) Min cost = \$522.50
(b) $\lambda > c_{13} - (u_1 + v_3) = 4$
(c) Min cost = \$511.50

Problem Set 7.3

1. (a) Min cost = 502
(b) Min cost = 508

2. (a) Min cost = 2455
(b) (i) Estimate = $(-\lambda)u_3 + (-\lambda)v_1 = 0$

(ii) Estimate = λ

3. (a) Min cost = \$14,345
(b) Month 2; estimated savings of \$32/unit produced.

4. (a) Min cost = 320
(b) Min cost = 338
(c) Min cost = 334
(d) Min cost = 358
(e) Min cost = 342

21. Min cost = \$22.40

23. (a) Min cost = 54
(b) Min cost = 12
(c) Min cost = 60

24. (a) Yes
(b) Min cost = 13

32. The minimal number of tankers necessary is 295. One possible assignment is to assign the two Route 1 tankers to Route 4, one Route 2 tanker to Route 1 and two to Route 4, one Route 3 tanker to Route 3, and one Route 4 tanker to Route 1 and three to Route 2

Problem Set 8.1

2. (a) Max expected profit = \$2293.33 at $R = 28$, $C = 20$
(b) Max expected profit = \$2341.67 at $R = 31$, $C = 15$

5. (a) Max $z = 20\frac{1}{4}$ attained at $(\frac{15}{4}, 4, \frac{5}{4}, 0)$

Problem Set 8.2

2. (a) Goals 1 and 2 achieved ($v_1 = u_2 = 0$) and Min $u_3 = 3$ at $x = 6$, $y = 8$.
(b) Goals 1–3 achieved ($v_1 = v_2 = u_3 = 0$) and Min $v_4 = 1$ at $x = 10$, $y = 5$.

3. (a) Goals 1–3 achieved and within 38 units of Goal 4 using 5 lb of Feed 1 and 9 lb of Feed 2.

Problem Set 8.3

2. Max profit is \$525.50 attained at $(x_1, x_2) = (37\frac{1}{2}, 57\frac{1}{2})$, $(y_1, y_2) = (61\frac{1}{2}, 36)$

3. (a) Max $z = 38\frac{2}{3}$ attained at $(x_1, x_2) = (1, \frac{4}{3})$, $(y_1, y_2) = (\frac{3}{8}, \frac{5}{2})$
(b) Max $z = 63$ attained at $(x_1, x_2) = (8, 1)$, $(y_1, y_2) = (0, 8)$

Problem Set 8.4

1. (a) $.8796 + 30(.0019) = .9366$; efficiency increases to 93.66%.

(b) $.8796 - 25(.0019) = .8321$; efficiency now is at most 83.21%, but no more information can be derived from the data at hand. The decrease of 30 units is outside of the range of validity of the marginal value by 5 units.
(c) No change in efficiency.
(d) Efficiency is now 93.99%.

Problem Set 9.1

1. (a)

	R3	*R8*	*B9*
R1	4	9	−10
B7	−10	−15	16

(c)

	1	*2*	*3*	*4*
1	−2	0	1	1
2	0	−4	2	2
3	3	3	−6	0
4	4	4	0	−8

2. (a) The strategies for each player can be denoted using ordered four-tuples $(X;Y_K,Y_Q,Y_J)$, where $X \in \{K,Q,J\}$ indicates the first card to be played, and the Y_K, Y_Q, and Y_J denote the play on the second trick, given that the opponent played a K, Q, or J, respectively, on the first trick. The payoff matrix would be 24×24.

Problem Set 9.3

1. (a) No
(b) Yes. Solution: (s_2, t_1); value: 3

2. (a) For $2 \leq x \leq 3$, saddle point x; for $x \geq 3$, saddle point 3
(b) For $x \geq 2$, saddle point 2
(c) For $1 \leq x \leq 3$, saddle point at the x in row 2, column 2

Problem Set 9.4

1. (a) The security level of X_1 is 2; the security level of X_2 is $\frac{8}{3}$. Therefore $v_1 \geq \frac{8}{3}$.
(b) Security level of Y_1 is $\frac{8}{3}$; security level of Y_2 is $\frac{17}{6}$; $v_2 \leq \frac{8}{3}$.

4. $v_1 = v_2 = \frac{3}{7}$; a strategy for P_1 with security level $\frac{3}{7}$ is $(\frac{9}{14}, \frac{5}{14})$; for P_2, $(\frac{5}{7}, \frac{2}{7})$

Problem Set 9.5

3. (a) $\frac{3}{2}$
(b) $\frac{3}{2}$
(c) Value of the game is $\frac{3}{2}$; and X_1 and Y_1 are optimal strategies for P_1 and P_2, respectively.

5. (a) Security level of X_1 is $-\frac{3}{5}$; of Y_1, 0. (X_1, Y_1) is not a solution.
(b) Security level of X_1 = security level of $Y_1 = \frac{10}{27}$. (X_1, Y_1) is a solution to the game, and the value of the game is $\frac{10}{27}$.

Problem Set 9.6

7. (a)

$$v = \begin{cases} 2, & x \leq 2 \\ x, & 2 \leq x \leq 4 \\ 6 - 8/x, & x \geq 4 \end{cases}$$

(c) $v = 3$

8. (a) (i) $v = \frac{3}{2}$; $X_0 = (\frac{1}{2}, \frac{1}{2})$; $Y_0 = \lambda Y_1 + (1-\lambda) Y_2$ for any $0 \leq \lambda \leq 1$, where $Y_1 = (\frac{3}{4}, \frac{1}{4}, 0, 0)$ and $Y_2 = (\frac{1}{2}, 0, \frac{1}{2}, 0)$
(ii) $v = \frac{27}{13}$; $X_0 = (\frac{5}{13}, \frac{8}{13}, 0)$; $Y_0 = (\frac{8}{13}, \frac{5}{13})$

9. $v = -\frac{9}{10}$; P_1 should select the red 1 with probability $\frac{13}{20}$ and the black 7 with probability $\frac{7}{20}$; P_2 should select the red 3 with probability $\frac{13}{20}$ and the black 9 with probability $\frac{7}{20}$.

12. (a) $v = 1$; $X_0 = (0, 1, 0)$; $Y_0 = (0, 0, 0, 1)$
(b) $v = \frac{12}{7}$; $X_0 = (\frac{4}{7}, \frac{3}{7}, 0)$; $Y_0 = (0, 0, \frac{10}{21}, \frac{11}{21})$
(c) $v = 0$; $Y_0 = (\frac{1}{2}, 0, \frac{1}{2}, 0)$; $X_0 = \lambda X_1 + (1-\lambda) X_2$ for $0 \leq \lambda \leq 1$, where $X_1 = (\frac{1}{2}, 0, \frac{1}{2})$ and $X_2 = (\frac{2}{3}, \frac{1}{3}, 0)$

14. (a) (i) $v = -\frac{34}{71}$; $X_0 = Y_0 = (\frac{1}{71})(35, 21, 15)$

16. (a) $v = -\frac{1}{12}$; P_1 should select the 1 with probability $\frac{7}{12}$ and the 2 with probability $\frac{5}{12}$; ditto for P_2.
(b) $v = \frac{1}{16}$; P_1 should select the 2 with probability $\frac{9}{16}$ and the 3 with probability $\frac{7}{16}$; P_2 should select the 1 with probability $\frac{9}{16}$ and the 2 with probability $\frac{7}{16}$.

Problem Set 9.7

5. (a)

$$v = \begin{cases} 0, & a \leq b \\ \dfrac{b(a-b)}{4}, & a \geq b \end{cases}$$

(b) Ascertain the value of the blue chips. (If they are worth \$1001, you should pass; if they are worth \$999, you should raise with probability 0.999 and pass with probability 0.001.)

Problem Set 10.1

1. $u(A) = l/r$; $u(D) = s/(s-1)$

Problem Set 10.3

3. (c) (s_1,t_1), (s_2,t_2), and (X_1,Y_1) are in equilibrium, where $X_1 = (\frac{1}{5}, \frac{4}{5})$ and $Y_1 = (\frac{4}{5}, \frac{1}{5})$.

5. See Example 10.5.4 on page 411

Problem Set 10.4

1. The line segment between the points $(1,10)$ and $(10,1)$
4. The point $(1,1)$

Index

CPSIA information can be obtained at www.ICGtesting.com
Printed in the USA
BVOW06*0636210915

418364BV00006B/86/P